T0180852

Advances in Intelligent Systems and Computing

Volume 1352

The series "Advances in Intelligent Systems and Computing" contains publications on theory, applications, and design methods of Intelligent Systems and Intelligent Computing. Virtually all disciplines such as engineering, natural sciences, computer and information science, ICT, economics, business, e-commerce, environment, healthcare, life science are covered. The list of topics spans all the areas of modern intelligent systems and computing such as: computational intelligence, soft computing including neural networks, fuzzy systems, evolutionary computing and the fusion of these paradigms, social intelligence, ambient intelligence, computational neuroscience, artificial life, virtual worlds and society, cognitive science and systems, Perception and Vision, DNA and immune based systems, self-organizing and adaptive systems, e-Learning and teaching, human-centered and human-centric computing, recommender systems, intelligent control, robotics and mechatronics including human-machine teaming, knowledge-based paradigms, learning paradigms, machine ethics, intelligent data analysis, knowledge management, intelligent agents, intelligent decision making and support, intelligent network security, trust management, interactive entertainment, Web intelligence and multimedia.

The publications within "Advances in Intelligent Systems and Computing" are primarily proceedings of important conferences, symposia and congresses. They cover significant recent developments in the field, both of a foundational and applicable character. An important characteristic feature of the series is the short publication time and world-wide distribution. This permits a rapid and broad dissemination of research results.

Indexed by DBLP, EI Compendex, INSPEC, WTI Frankfurt eG, zbMATH, Japanese Science and Technology Agency (JST), SCImago.

All books published in the series are submitted for consideration in Web of Science.

More information about this series at http://www.springer.com/series/11156

Tatiana Antipova
Editor

Advances in Digital Science

ICADS 2021

 Springer

Editor
Tatiana Antipova
Institute of Certified Specialists (ICS)
Perm, Russia

ISSN 2194-5357 ISSN 2194-5365 (electronic)
Advances in Intelligent Systems and Computing
ISBN 978-3-030-71781-0 ISBN 978-3-030-71782-7 (eBook)
https://doi.org/10.1007/978-3-030-71782-7

This Springer imprint is published by the registered company Springer Nature Switzerland AG
The registered company address is: Gewerbestrasse 11, 6330 Cham, Switzerland

Preface

This book is based on the selected papers accepted for presentation and discussion at the 2021 International Conference on Advances in Digital Science (ICADS 2021) that was held on February 19–21, 2021. This conference had the support of the Institute of Certified Specialists (ICS), Russia, and Springer.

The 2021 International Conference on Advances in Digital Science (ICADS 2021) is a network of scholars interested in natural and social sciences original research results. The network was established to promote cooperation between scholars of different countries. An important characteristic feature of the conference is the short publication time and worldwide distribution. This conference enables fast dissemination so the conference participants can publish their papers in print and electronic format, which is then made available worldwide and accessible by numerous researchers.

The Scientific Committee of ICADS 2021 was composed of a multidisciplinary group of 110 experts from 38 countries around the world. Fifty-one invited reviewers, united by science, have had the responsibility for papers' evaluating in a 'double-blind review' process. All the selected papers went through a round of strictly peer review process. The main topics of the included papers are the following: Advances in Digital Agriculture and Food Technology, Advances in Digital Economics, Advances in Digital Education, Advances in Public Health Care, Hospitals and Rehabilitation, Advances in Digital Social Media, Advances in Digital Technology and Applied Sciences, Advances in E-Information Systems, and Advances in Public Administration.

The papers accepted for presentation and discussion at the conference are published by Springer (this book) and will be submitted for indexing by ISI, SCOPUS, among others. We acknowledge all of those that contributed to the staging of ICADS 2021 (authors, committees, reviewers, organizers, and sponsors). We deeply appreciate their involvement and support that were crucial for the success of ICADS 2021 and hope they continue to be safe as we all work through these extraordinary times.

January 2021 Tatiana Antipova

Contents

Advances in Digital Social Media

Advances in Digital Technology and Applied Sciences

Advances in E-Information Systems

Advances in Public Administration

Advances in Digital Agriculture
and Food Technology

Means of Estimation the Anthropogenic Actions with Negative Effects on Hydrobionts

Denis Anatolevich Yurin[1]([✉]) [ID], Natalya Alexandrovna Yurina[1,2] [ID],
Boris Vladimirovich Khorin[1] [ID], Denis Vasilievich Osepchuk[1] [ID],
Marina Petrovna Semenenko[1] [ID], and Elena Vasilievna Kuzminova[1] [ID]

[1] Krasnodar Research Centre for Animal Husbandry and Veterinary Medicine,
4 Pervomaiskaya St., Town. Znamensky, 350055 Krasnodar, Russian Federation
4806144@mail.ru
[2] Federal State-Funded Educational Institution of Higher Professional Education,
Kuban State Agrarian University named after I. T. Trubilin, 13 Kalinina Str.,
350044 Krasnodar, Russian Federation

Abstract. The Federal Fisheries Agency of the Russian Federation has adopted a methodology for calculating anthropogenic actions with negative effects on aquatic bioresources and hydrobionts. This methodology is adopted in order to calculate the amount of harmful effects due to which losses of bioresources were incurred, taking into account the minimization of damage and elimination of possible consequences. The harm calculation caused to bioresources in a consequence of violation of legislative documents in fish breeding and fishery and ways of their completion and also as a result of natural cataclysms, technogenic and other catastrophes is separately carried out. In this methodology there are sections on the basis of which data on the amount of harm are calculated depending on the causes that caused it. When determining the harm caused to aquatic bioresources, special formulas are used to make calculations based on the specificity of possible damage. To minimize the process of calculating the amount of damage to the studied bioresources from conducting predicted, industrial or other activities that affect the state of hydrobionts and their habitat according to the methodology approved by the Federal Fisheries Agency, employees of the Federal State Budgetary Scientific Institution "Krasnodar Research Centre for Animal Husbandry and Veterinary Medicine" created an electronic template. The template for calculating the amount of damage to bioresources can be used in fish breeding enterprises of various forms of ownership, as well as in secondary and higher educational institutions as a training manual.

Keywords: Bioresources · Environmental impact assessment · Fish farming · Habitat pollution · Water reservoirs

1 Introduction

Anthropogenic exposure is a significant problem for the quality of aquatic bioresources. Access to clean water is becoming increasingly important. UN resolution 64/292,

T. Antipova (Ed.): ICADS 2021, AISC 1352, pp. 3–10, 2021.
https://doi.org/10.1007/978-3-030-71782-7_1

adopted in 2010, called the water "the new gold of the 21st century". A stable prolonged supply of the necessary quantity and a certain quality of water is associated with the state of aquatic and adjacent ecosystems, which the decomposition of waste, the binding of nutrients, soil and water treatment depend on. Active use of water resources and continued pollution of water bodies are currently reducing the quantity and quality of water in important water sources.

The increased capacity on aquatic bioresources has led to numerous research and legislative activities, both based on ecotoxicological evidence and the precautionary principle. A wide view of natural and technical water systems is required, including different scales and involving fundamental and applied science, which contributes to the development of an innovative, effective, convenient and unified research programme to calculate the amount of damage caused to hydrobiontes and their habitats [3, 4, 7, 14].

Problems in the field of conservation and management of bioresources require the use of research methods to quantify harmful effects on populations, as well as make political decisions about the permissible limits of the caused damage. Few studies quantify the population effects of individual harmful factors, and only some are synergistic effects [6, 12, 13].

Modern technological and methodological technologies have increased the diversity of available hydrological, biogeochemical and ecological indicators of water biopotential of water reservoirs [1, 5, 8].

The methodology for calculating the negative impact on aquatic bioresources has been approved. It is intended to determine the amount of harm caused to aquatic bioresources, taking into account the separation of the causes of possible harm. Firstly: damage is separately determined, which contributed to a decrease in the number of bioresources of the reservoir due to violations of legislative documents when fishing, the possibility of their preservation and minimization of losses during this activity, as well as after natural disasters, technogenic disasters and other accidents. Secondly: a separate calculation is made of the harm from conducting production activities (maintaining the fish farming industry), which can negatively affect the number of bioresources and the quality of their habitat. Also in this methodology there are sections according to which the rules for calculating the amount of damage depending on its causes are determined. The determination of the amount of damage provides for its calculation both in kind (for example, the number of individuals, grams, kilograms, tons) taking into account the consequences of negative factors on the state of aquatic bioresources, and in value terms (in rubles) based on the costs of restoring the disturbed state of bioresources taking into account the losses received, including lost profits [11].

2 Methods

The purpose of this work is to create a template for calculating the harm caused to aquatic bioresources as a result of human activity.

To fulfill the goal, employees of the Krasnodar Research Centre for Animal Husbandry and Veterinary Medicine created and tested a template in the Visual Basic for Applications (VBA) programming language.

Calculations are carried out in the listed cases: when planning construction work, capital repairs of facilities, planning the construction of industrial or other objects, the introduction of new technological methods for production that affect the state of biore-sources and their area of residence, in order to analyze the possible consequences of damage from this activity when assessing environmental risks. When determining the harm caused to aquatic bioresources, special formulas are used to make calculations based on the specificity of possible damage [2, 4, 10].

Calculation of amount of damage caused to bioresources of reservoirs as a result of violation of the law when conducting activity of fishery and also as a result of emergence of natural disasters, the abnormal phenomena, emergencies of natural and technogenic character, is applied in cases of death of water bioresources, reduction of a fish pro-ductivity of a water object and also pollution of the habitat of water bioresources by hazardous substances (acids, alkalis, pesticides, agrochemicals and other chemicals), production and consumption wastes, hydrocarbon raw materials and their derivatives, discharges into water bodies of fisheries significance and conservation zones of harmful substances whose maximum permissible concentrations in the waters of water reservoir of fisheries significance are not established.

The following indicators are used as input data for the calculation of damage to aquatic bioresources:

- changes due to negative effects of quantitative and qualitative composition of bioresources;
- feed coefficients of plankton and benthos;
- the number of dead adults, as well as caviar, fry and so on;
- area of the caused damage (places for spawning, wintering, chest ponds, passage to other reservoirs);
- weight and length of the adult fish body;
- changes due to negative effects in hydrochemical and hydrological indicators of the reservoir and water object;
- fishery return rates from dead bioresources;
- fish productivity of the reservoir for each type of bioresource;
- the percentage of females in the population to males, their reproductive qualities, fertility and other indicators;
- the cost of products obtained from 1 kg or one individual of bioresources;
- financial costs of repairing damage to the reservoir, bioresources and their habitat.

Sources of extraction of initial data used in damage calculations are data of scientific research, reports, laboratory analyses performed in the form of administrative investiga-tions of the facts of bioresource death and contamination of their habitat, environmental control, scientific data of subordinate research organizations and federal state budgetary institutions (basin departments) of the Federal Fisheries Agency, as well as tables of the appendix to the methodology [11].

The composition of the bioresource population on the studied territory is obtained according to the analysis of the modern state of the environment.

The section on the protection of bioresources should complement the previous sec-tions of the document and should not contradict them. It is necessary to assess the impact on soil and vegetation, as well as trees, on the atmospheric air, environmental and economic assessment, and other sections.

Compensation payments for animal objects are not provided for by the current legislation of the Russian Federation, but this is necessary in order to assess the damage and calculate the possibility and deadlines of its restoration.

Damage to aquatic bioresources is mainly caused during the construction of various facilities, when land is seized, which leads to the complete disappearance of biocenoses.

In addition, in the lands adjacent to the drainage lanes, that is, in the zone of influence (from 1.5 to 3.0 km each way from the objects under construction), the number of most living species of animals and birds during construction work decreases due to their increased anxiety. However, it can be predicted that at the end of construction work, animals and birds can adapt to changed conditions.

Modern publications on the calculation of the damage in the Russian Federation can be conditionally divided into several types: on the issues of applied taxa, analysis of mechanisms for calculating damage to bioresources and issues of application in judicial practice. There is a lack of publications on the practice of calculating damage, which can be used in the training of specialists [9, 15].

Environmental impact assessment is an integral part of the design documentation. This procedure should be carried out before the start of the project, and if the results of the environmental impact have a significant negative effect, then the project should be corrected or canceled.

At the beginning of work at any existing enterprise or designing a new one, the first question is: what is the flow rate of water, its hydrochemical parameters. Depending on the results, and only when the surveys are completed, developments, technological instructions are given and, sometimes, the possibility of carrying out fish-breeding activities is generally confirmed.

3 Results and Discussions

In FSBSU "Krasnodar Research Centre for Animal Husbandry and Veterinary Medicine" state topics are being studied to research the state of the steppe rivers of the Kuban, Lake Khansky. A working group of scientists has been created. Each reservoir is unique, in each case an individual approach is required. A separate issue has always been the problem of coastal strengthening of the Yeisk, Albash, Beisug estuaries. Moreover, this is an important economic and social component. There are several ways to develop and return the normal functioning of estuaries and steppe rivers, such as Yaseni, Eya, Albashi, Chelbas, Beisug.

To facilitate the calculation of the amount of damage to aquatic bioresources from the implementation of the planned economic and other activities, affecting the state of aquatic bioresources and their habitat, as well as calculating the amount of damage to aquatic bioresources from the implementation of the planned economic and other activities, influencing the state of aquatic bioresources and their habitat according to the methodology, approved by the Federal Fisheries Agency, employees of the Federal State Budgetary Scientific Institution "Krasnodar Research Centre for Animal Husbandry and Veterinary Medicine" created an electronic template.

You can use Visual Basic for Applications (VBA) to run the template. It is a visual object-oriented programming system designed to create programs - objects embedded in various other software systems and interfaces that support object binding and implementation technology (OLE), whose objects are executed inside Microsoft Office applications and software products of other manufacturers that support OLE technology.

For carrying out calculation of amount of damage to water bioresources during violation of the law in the field of fishery and preservation of water biological resources and also as a result of natural and technogenic catastrophes (and other disasters) 20 formulas (see Fig. 1) are used.

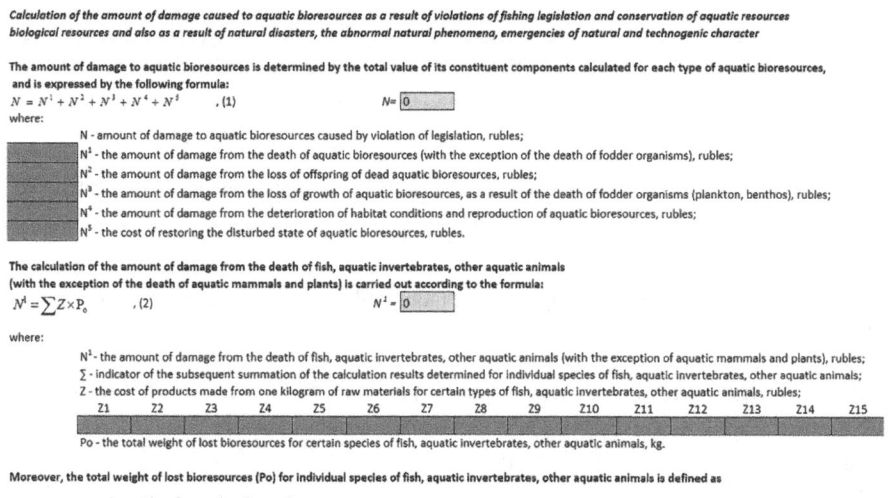

Fig. 1. Calculation of the amount of damage caused to aquatic bioresources as a result of violations of fishing legislation and conservation of aquatic resources, biological resources and also as a result of natural disasters, the abnormal natural phenomena, emergencies of natural and technogenic character.

To calculate the amount of damage to aquatic bioresources from the introduction of planned production activities or other activities affecting bioresources and their habitat was used 28 formulae (see Fig. 2).

For ease of use, the input field template is colored green and the results are displayed on a yellow background. An instruction has been made for users to quickly figure out the calculation method and get started.

The template for calculating the amount of damage to aquatic bioresources can be used in fish breeding enterprises, as well as in secondary and higher educational institutions as a tutorial.

Using the example of Albashi LLC of the Leningrad region of the Krasnodar Territory, the amount of damage to aquatic biological resources was calculated, which is determined by the total value of its constituent components, calculated for each type of aquatic biological resources,

The calculation of the amount of damage from the death of fish, for example, when cleaning a reservoir from silt, is presented in Table 1.

Calculation of the amount of damage to aquatic bioresources from the implementation of the planned economic and other activities affecting the state of aquatic bioresources and their habitat

Determination of annual losses of aquatic bioresources due to the negative impact of the proposed activity in case of irreversible complete or partial loss of fishery value of the water reservoir or its part:

$$N = P_o \times S \times d \times 10^{-3} \quad , (1) \qquad N = \boxed{0,95063}$$

where:

N - loss (amount of harm) of aquatic bioresources, kg or t;

Po is a fish productivity (annual) of a water object, g/m2, kg/sq.km, kg/hectare;

S - area of the water reservoir of fishery value (or its part), which loses the fishery value, m2, km2, ha;

d - the degree of exposure, or the proportion of the amount (biomass) of dying aquatic bioresources of their total amount, in fractions of one;

10^{-3} - multiplier for converting grams to kilograms or kilograms to tons.

	N1	N2	N3	N4	N5	N6	N7	N8	N9	N10	N11	N12	N13	N14	N15
	0,950625	0	0	0	0	0	0	0	0	0	0	0	0	0	0
Po	50														
S	63,375														
d	0,3														

The amount of harm is calculated for each type (or group of ecologically close species) of aquatic bioresources separately according to the formula:

$$N = P_o \times S \times \frac{F_1}{F_o} \times q \times \Theta \times 10^{-3} \quad , (2) \qquad N = \boxed{7,08296}$$

Fig. 2. Calculation of the amount of damage to aquatic bioresources from the implementation of the planned economic.

Table 1. Calculation of the total weight of lost biological resources for certain species of fish, aquatic invertebrates, and other aquatic animals.

Fish species	n, pcs.	n1, pcs.	n2, pcs.	n3, pcs.	p, kg	k1, %	k2, %	k3, %	Po, kg
Grass carp	0	0	0	0	5	0,01	0,01	5	0
Silver carp	0	0	0	0	4	0,01	0,01	5	0
Common carp	0	0	0	0	1	0,01	0,01	0,5	0
Pike perch	0	0	0	0	1,5	0,0012	0,002	0,23	0
Pike	0	0	0	0	1,3	0,014	0,025	1	0
Crucian	0	0	0	0	0,2	0,00004	0,0004	0,0004	0

The total weight of lost biological resources (Po) for individual fish species is determined by the formula:

$$P_0 = \sum (n \times p) + \frac{n^1 \times p \times k^1}{100} + \frac{n^2 \times p \times k^2}{100} + \frac{n^3 \times p \times k^3}{100} \tag{1}$$

Where:

$\sum$ – indicator of the subsequent summation of the calculation results determined for individual fish species;

n – number of dead adults of aquatic biological resources, pcs.;

n^1 – the number of dead caviars, pcs.;

n^2 – number of dead larvae, pcs.;
n^3 – number of dead juveniles, pcs;
p – average weight of adult fish, kg;
k^1 – coefficient of replenishment of the fishing stock (fishing return) from caviar, %;
k^2 – coefficient of replenishment of the fishing stock (fishing return) from larvae, %;
k^3 – coefficient of replenishment of the fishing stock (fishing return) from juveniles, %.

According to calculations, the enterprise plans to stock the ponds with commercial fish species: carp, silver carp, grass carp in the amount of 2–4 tons annually.

4 Conclusion

Proper monitoring of damage to bioresources should help reduce pollution of the territory, reduce environmental damage, preserve bioresources, it is necessary that enterprises planning their activities or construction in the immediate vicinity of water reservoirs pay attention to the conservation of biological diversity and strive to preserve it. The calculation of damage may be subject to environmental audits of firms along with other issues in calculating the financial costs of pollution. To simplify the calculation of the amount of damage to aquatic bioresources, as well as the calculation of the amount of damage to aquatic bioresources and their habitat according to the methodology approved by the Federal Fisheries Agency, employees of the Federal State Budgetary Scientific Institution "Krasnodar Research Centre for Animal Husbandry and Veterinary Medicine" created an electronic template. The template for calculating the amount of damage to aquatic bioresources can be used in fish breeding enterprises, as well as in secondary and higher educational institutions as a training manual.

References

1. Abbott, B.W., Baranov, V., Mendoza-Lera, C., Nikolakopoulou, M.: Using multi-tracer inference to move beyond single-catchment ecohydrology. Earth Sci. Rev. **160**, 19–42 (2016). https://doi.org/10.1016/j.earscirev.2016.06.014
2. Fletcher, W.J., Wise, B.S., Joll, L.M., Hall, N.G., et al.: Refinements to harvest strategies to enable effective implementation of Ecosystem Based Fisheries Management for the multisector, multi-species fisheries of Western Australia. Fish. Res. **183**, 594–608 (2016). https://doi.org/10.1016/j.fishres.2016.04.014
3. Gerbersdorf, S.U., Cimatoribus, C., Class, H., Engesser, K., et al.: Anthropogenic Trace Compounds (ATCs) in aquatic habitats—research needs on sources, fate, detection and toxicity to ensure timely elimination strategies and risk management. Environ. Int. **79**, 85–105 (2015). https://doi.org/10.1016/j.envint.2015.03.011
4. Gosset, A., Ferro, Y., Durrieu, C.: Methods for evaluating the pollution impact of urban wet weather discharges on biocenosis: a review. Water Res. **89**, 330–354 (2016). https://doi.org/10.1016/j.watres.2015.11.020
5. Jia, Z., Cai, Y., Chen, Y., Zeng, W.: Regionalization of water environmental carrying capacity for supporting the sustainable water resources management and development in China. Resour. Conserv. Recycl. **134**, 282–293 (2018). https://doi.org/10.1016/j.resconrec.2018.03.030
6. Jones, C.A., DiPinto, L.: The role of ecosystem services in USA natural resource liability litigation. Ecosyst. Serv. **29**, 333–351 (2018). https://doi.org/10.1016/j.ecoser.2017.03.015

7. Kasimov, D.V., Pinaev, V.E.: Assessment of the impact on the land cover - the practice of assessing the impact on the environment. Internet J. Naukovedenie **6** (2014). https://doi.org/10.15862/121EVN614. https://naukovedenie.ru/PDF/121EVN614.pdf

8. Kennen, J.G., Sullivan, D.J., May, J.T., Bell, A.H.: Temporal changes in aquatic-invertebrate and fish assemblages in streams of the north-central and northeastern US. Ecol. Ind. **18**, 312–329 (2012). https://doi.org/10.1016/j.ecolind.2011.11.022

9. Krymov, V.G., Galicheva, M.S., Semenenko, M.P., et al.: Possibilities of implementation of polyculture for optimization of industrial sturgeon aquaculture on the basis of closed water supply facilities. Res. J. Pharm. Biol. Chem. Sci. **9**(6), 540–545 (2018)

10. Majdi, N., Hette-Tronquart, N., Auclair, E., Bec, A., et al.: There's no harm in having too much: a comprehensive toolbox of methods in trophic ecology. Food Webs. **17**, e00100 (2018). https://doi.org/10.1016/j.fooweb.2018.e00100

11. Pinaev, V.E., Jakunin, S.A.: Review of modern methodologies of harm calculation, caused for living species, in Russian Federation. Russian J. Resour. conservation and recycling **2**(4) (2017). (in Russian). https://doi.org/10.15862/02RRO217. https://resources.today/PDF/02RRO217.pdf

12. Posthuma, L., Dyer, S.D., de Zwart, D., Kapo, K., et al.: Eco-epidemiology of aquatic ecosystems: separating chemicals from multiple stressors. Sci. Total Environ. **573**, 1303–1319 (2016). https://doi.org/10.1016/j.scitotenv.2016.06.242

13. Williams, R., Thomas, L., Ashe, E., Clark, C.W., Hammond, P.S.: Gauging allowable harm limits to cumulative, sub-lethal effects of human activities on wildlife: a case-study approach using two whale populations. Marine Policy **70**, 58–64 (2016). https://doi.org/10.1016/j.marpol.2016.04.023

14. Wu, Z., Di, D., Wang, H., Wu, M., He, C.: Analysis and emergy assessment of the eco-environmental benefits of rivers. Ecol. Ind. **106**, 105472 (2019). https://doi.org/10.1016/j.ecolind.2019.105472

15. Yang, Q., Liu, G., Casazza, M., Hao, Y., Giannetti, B.F.: Emergy-based accounting method for aquatic ecosystem services valuation: a case of China. J. Clean. Prod. **230**, 55–68 (2019). https://doi.org/10.1016/j.jclepro.2019.05.080

Rural Territories of Russia: Realities and Prospects

Anatoly E. Shamin⬡, Dmitry V. Proskura⬡, Nadezhda V. Denisova⬡,
Natalia V. Proskura⬡, and Olga A. Frolova^(✉) ⬡

Nizhny Novgorod State University of Engineering and Economics, Knyaginino, Russia
ekfakngiei@yandex.ru

Abstract. Rural development is also a social task for any region of Russia and the country as a whole. Assessment of the degree of rural development and development of measures to address the identified problems are the key points in this article. Based on the analysis of the problems of rural territories of one of the regions of Russia-the Nizhny Novgorod region, the crisis characteristics of the current situation and the causes of the systemic crisis of socio – economic development of these territories are revealed. The refined method of integrated assessment makes it possible to determine key parameters, specify the value of factors that take into account the peculiarities of rural development and the specifics of the market under study, make a forecast for key management parameters and, as a result, determine the main directions of sustainable development.

Keywords: Rural areas · Indicators · Methodology · Indicators · Digitalization · Quality of life

1 Introduction

Close attention to the problems of development of rural territories at the present time in Russia is connected with catastrophic reduction in the number of rural population, a decline in the level and quality of life in rural areas, increasing income gap between rural and urban households, the declining share of the total area of the arranged well dwellings in rural settlements. In many regions of Central Russia and the North, almost every fifth village has been depopulated. According to Rosstat's forecast, the number of rural residents will decrease by 2.8 million people, or 7.3%, by 2030. As the results of the all-Russian agricultural census of 2016 showed, the share of young workers in the structure of those employed in agriculture has decreased and the share of pensioners has increased. Rural unemployment and the lack of developed social infrastructure continue to be acute problems. Thus, almost 90% of the country's rural areas need improvement. According to the monitoring of rural development conducted in 2018, the availability of places in pre-school organizations for rural children is extremely low: only 48.2% of rural children aged one to six years are covered, which is 31.4% lower than the urban level. The development of non-agricultural activities in rural areas remains weak, which leads to a narrow scope of employment in rural areas. In General, the level and quality of life in rural areas remain extremely low [1–5].

Development of a comprehensive system of indicators for assessing rural development and their typology is one of the activities of the state program of the Russian Federation 2020–2025.

Rural territories include:

– rural settlements or rural settlements and inter-settlement territories United by a common territory within the borders of a municipal district;
– rural localities that are part of urban settlements, municipal districts, and urban districts (with the exception of urban districts where the administrative centers of the constituent entities of the Russian Federation are located);
– rural localities that are part of inner-city municipalities of Sevastopol;
– settlements with the status of urban settlements [6].

The strategic direction of the program for sustainable development of rural areas is to achieve continuous improvement of life support, living conditions of people and transition to a qualitatively new standard of living in the era of the digital economy, in which the ecosystem is not destroyed and the natural basis is preserved.

Socio-economic development of rural areas in Russia will be considered in one of the regions of Russia - the Nizhny Novgorod region.

Nizhny Novgorod region is one of the twenty largest regions of Russia (14th place in the Russian Federation) in terms of GRP. The largest sector of the region's economy is the manufacturing industry - 31% of GRP. Due to the absence of fuel and energy and ore minerals in the region, the share of the mining sector in industry is insignificant. By the volume of products shipped in the manufacturing sector, the Nizhny Novgorod region ranks 6–7 in the Russian Federation.

The economy has a fairly high share of high-tech and knowledge-intensive industries in GRP (31.3% of GRP - 4th place among the regions of the Russian Federation; in Russia-20.7% of GDP).

In terms of exports, the Nizhny Novgorod region ranks 20th in Russia, and last among regions with comparable economies (the top fifteen in terms of GRP). Currently, the total volume of exports is only 13% of the volume of products shipped by processing enterprises.

The region's economy remains dependent on equipment imports (as in Russia as a whole). Machine-building products are the main component of imports to the Nizhny Novgorod region - their share is 37.1% (in the Russian Federation-45.6%). This is the 35th place in the Russian Federation and the 7th place among regions of comparable economic scale.

In General, the region ranks on average 32nd in the ratings under review, which may indicate that it is a middle-class region that tends to be more developed regions due to a relatively developed sphere of innovation, the labor market and a fairly good quality of life.

2 Materials and Methods

Assessment of the level of socio-economic development of municipal areas and city environ-GOV in the Nizhny Novgorod region is carried out in accordance with the methodology approved by the decree of the government of Nizhny Novgorod region dated March 1, 2006 № 60 "Methodology of assessment of socio-economic development of municipal areas and city districts of Nizhny Novgorod region" [7].

This methodology includes:

1. Indicators that characterize tax capacity building, including economic and financial indicators (weight coefficient −0.7).
2. Indicators of the population's standard of living (weight coefficient −0.3).

To make an assessment for each municipal district and urban district of the Nizhny Novgorod region, the value of the integral indicator is calculated using the method.

Based on the integral indicator, the overall rating of the territory is determined (based on all indicators), as well as ratings for blocks of indicators that characterize the increase in tax potential and the standard of living of the population. The methodology provides for determining the rating by groups of territories, based on the number of people living in the territories.

Based on this methodology, the following data were obtained for the 1st half of 2020 in the Nizhny Novgorod region:

- with an above-average level of development − 11 municipal districts (municipal districts and urban districts);
- with an average level of development − 32;
- with a level of development below average − 9.
- According to the results of the 1st half of 2020 compared to the results for the 1st quarter of 2020:
- 5 territories moved to the group of territories with a higher level of development;
- Koverninsky, Lyskovsky municipal districts and the city district of Kulebaki moved from the category of territories with an average level of development to the category with an above-average level of development.

Changes in indicators that have affected the improvement of the situation in these territories are associated with an increase in:

- the growth rate of investment in fixed assets (in the real sector of the economy) in the Kovensky and Lyskovsky municipal districts;
- growth rates of shipped products (across the entire range of organizations) and profits of large and medium-sized organizations of the koverninsky municipal district;
- profit size of profitable organizations in Kovernino, Lyskovsky municipal district, Kulebaki city district;
- growth rates of tax payments to the budget system of the Russian Federation in the city of Kulebaki;

– the share of receipts of the unified imputed income tax and the unified tax paid in connection with the application of the simplified tax system, and receipts from the application of the patent tax system in tax revenues collected in the consolidated budget of the koverninsky and Lyskovsky municipal districts.

At the same time, the territories of koverninsky, Lyskovsky municipal districts and Kulebaki city district saw an increase in housing starts per capita during the reporting period.

Pilninsky and Tonkinsky municipal districts moved from the category with a level of development below average to the category of territories with an average level of development (Fig. 1).

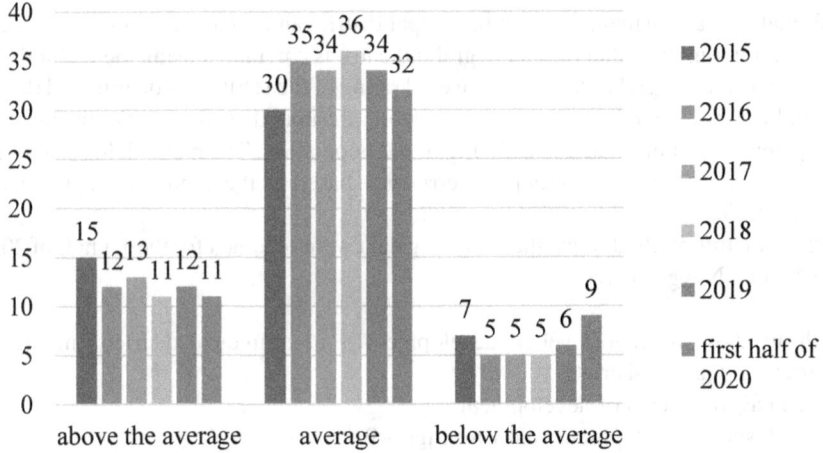

Fig. 1. Ranking of municipal districts and urban districts by indicator blocks and territory groups for the 1st half of 2020

These changes in the territories are associated with an increase in:

– the size and growth rate of profit of profitable organizations per employee, as well as the ratio of the average monthly salary of one employee to the subsistence minimum of both territories;
– growth rates of tax payments to the budget system of the Russian Federation, investment in fixed assets and average monthly wages in the Pilninsky municipal district;
– the volume of housing commissioning per capita in the Pilninsky and tonkinsky municipal districts.
– 3 territories moved to the group of territories with a lower level of development;
– Voznesensky, lukoyanovsky, and Krasnooktyabrsky municipal districts moved from the category of territories with an average level of development to the category of territories with a lower average level of development.

The main changes in indicators that affected the deterioration of the situation in these territories:

- decline decrease in the volume of tax payments to the budget system of the Russian Federation and slowdown in the growth rate of average monthly wages per employee in Voznesensky, lukoyanovsky and Krasnooktyabrsky municipal districts;
- slowing growth in fixed capital investment in these territories;
- decrease in the rate of profit growth of profitable organizations and a significant increase in the registered unemployment rate in the Voznesensky municipal district.

In our opinion, the decline in the main socio-economic indicators in the territories was also affected by the spread of the CoViD-19 coronavirus infection and the restrictive measures introduced in this regard.

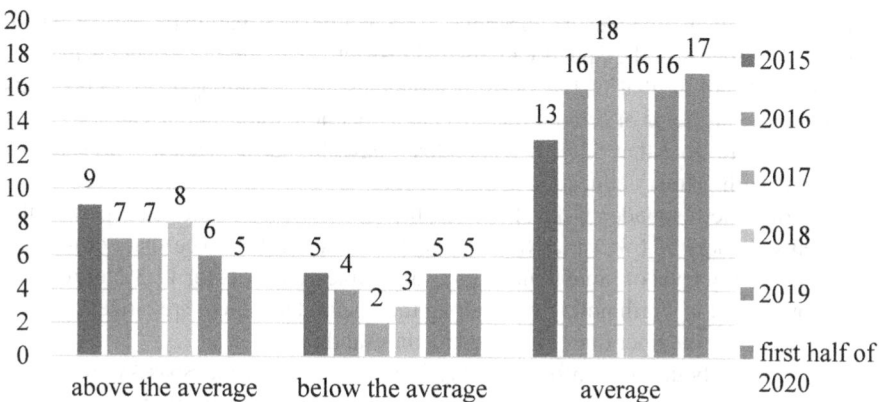

Fig. 2. Trends in the socio-economic development of municipal districts

Trends in the socio-economic development of territories over a number of years are shown in Fig. 2.

During the period under review (from 2015 to the 1st half of 2020), the following trends in the development of territories were noted:

- the socio-economic situation has improved in Bolshemurashkinsky, Gaginsky, Sechenovsky, tonkinsky municipal districts, Koverninsky municipal district, Kulebak city district;
- the socio-economic situation has worsened in Voznesensky, Volodarsky, Gorodetsky, Krasnooktyabrsky, lukoyanovsky, Shatkovsky municipal districts, Pochinkovsky municipal district and Chkalovsky city district.

The level of socio-economic development above average is consistently observed in the Kstovsky municipal district, Pavlovsk municipal district, as well as in the city districts of Arzamas, Bor, Vyksa, Dzerzhinsk, Nizhny Novgorod, and Sarov.

At the end of 2019, a regional Center for supporting sustainable rural development was established at the regional University of the Nizhny Novgorod state University of engineering and Economics in order to implement measures to implement the strategy for the development Of the North-urban region.

The peculiarity of this center is that it was created on the basis of our University and the University is located in one of the municipal districts of the region, where there are 2,000 students per 8,000 thousand inhabitants. Research and teaching staff and students are residents of rural areas of the region and are daily confronted with the factors that hinder the successful development of rural areas.

The center together with the Ministry of economic development and investment of the Nizhny Novgorod region conducted a survey of one of the municipal districts of the region "the Degree of satisfaction of the population with the level and quality of life".

The purpose of the survey is to determine the level of satisfaction of residents of the Spassky district with the standard of living and identify the most important problems for developing a strategy for the development of the municipal district.

The questionnaire included 42 questions, 1,500 people took part in the survey, which is 18.2% of residents of the Spassky municipal district. Analyzing respondents by gender, it is worth noting that 66.0% of all respondents are women aged 21 to 66 years working in various fields of activity (health care, education, trade, public catering, etc.). it is Important to note that 52% of respondents have higher education. The share of respondents with primary education is only 4.5%.

Analyzing the respondents' level of well-being, the salary of the main part of the respondents, namely 60.6%, varies from 11 to 20 thousand rubles. The share of respondents with a salary level of more than 20 thousand rubles accounts for 11.3% of respondents. However, it is worth noting that among the respondents there are citizens whose income per family member does not exceed 5 thousand rubles.

When asked about the comfort of living in the territory of the Spassky municipal district, 14.6% of respondents said that they are satisfied with the level of development of the territory where they live. Rather, they are satisfied with the level of comfort - 69.3% of respondents and 16.1% say that living is not comfortable and note a number of problems that would increase the level of attractiveness of rural areas. Among the problems are the underdeveloped transport infrastructure, problems related to youth employment, the underdeveloped system of additional education, the lack of a sufficient number of cultural and leisure institutions, issues in the housing and utilities sector, health care, etc.

Despite the fact that the survey was attended by residents of the Spassky municipal district who are at an economically active age, only 44.6% of respondents have information about the development Strategy of the Spassky district until 2035. It is worth noting that 74.4% of the respondents who do not have information would like to participate in its development in order to improve the quality and standard of living in rural areas, which indicates a high degree of social responsibility of those living in the territory and a low level of awareness of citizens on the part of the authorities.

3 Discussion

Many agricultural scientists note an insufficiently complete and comprehensive list of indicators and indicators in the existing methods of assessing socio-economic development, taking into account the current situation. As a result of the analysis of economic, environmental, social and digital problems of rural territories, it is proposed to improve the methodology by adding indicators (digitalization, municipal management, ecology) and corresponding indicators of sustainable development of rural territories (Table 1, 2). It is based on the methodology for assessing the level of socio-economic development of municipal districts and urban districts of the Nizhny Novgorod region [8], the development of scientists [8–14] and regulatory documents [15–22].

4 Results

The indicators presented in Table 1 characterize the degree of digitalization of the territory, the innovation climate of the region and the environmental environment, and contribute to increasing the tax potential. Indicators that characterize indicators and their normative values can be improved as the economic situation changes in a given territory, thereby increasing or decreasing its competitiveness.

Table 1. Updated indicators and indicators that characterize the increase in tax potential

Indicator	Indicators	Unit	Calculation
Digitalization	Share of organizations that used information and communication technologies	%	Ratio of the number of organizations that used information and communication technologies in their activities to the total number of organizations surveyed
Innovative activity of organizations	Share of organizations that implemented technological, organizational, and marketing innovations in the reporting year in the total number of organizations surveyed)	%	Ratio of the number of organizations that implemented technological, organizational, and marketing innovations in the reporting year to the total number of organizations surveyed
	Share of innovative goods, works, and services in the total volume of goods shipped, works performed, and services rendered	%	The ratio of the cost of innovative goods, works, services to the cost of the total volume of goods shipped, works performed, services
Ecology	Share of protected areas from the entire territory	%	Ratio of the area of protected areas to the total area of the territory
	Air pollution index (ISA)		The ratio of the average annual concentration of the pollutant to its average daily maximum permissible concentration (taking into account the coefficient that depends on the degree of harmfulness of the pollutant)

Of the 18 indicators proposed in the Strategy, only one – item 16 "the share of rural households that have access to the information and telecommunications network "Internet" from a home computer"-indicates the need for the presence of telecommunications services (hereinafter referred to as TLC services).

In the remaining seventeen, there is no direct or indirect indication of the need for quantitative and qualitative TLC services. In fact, almost no one indicator can be executed without availability of the telecommunications wrappers.

In addition, the following services should be provided as priority: broadband access (hereinafter referred to as broadband) to the Internet at a minimum speed of 100 Mbit/s, IP telephony, geoservices, video surveillance, video conferencing systems and unified electronic document management, smart electricity and water meters, smart street lighting, interactive digital TV vision, etc. All these TLC services are designed not only to assist in the production of agricultural products, but also to significantly save resources through systematic monitoring.

Table 2. Augmented indicators and quality of life indicators

Indicator	Indicators	Unit	Calculation
Digitalization	Percentage of the population that used the Internet to receive state and municipal services in the total population that received state and municipal services	%	The ratio of the number of people who used the Internet to receive state and municipal services to the total number of people who received state and municipal services
	Percentage of households with broadband Internet access in the total number of households	%	Ratio of the number of households with broadband Internet access to the total number of households in the analyzed territory
	Percentage of the population who used the Internet to order goods and/or services in the total population	%	The ratio of the population that used the Internet to order goods and/or services during the analyzed period to the total population at the reporting date
Municipal management	Number of programs implemented to improve the quality of life of the population per 1000 people	Units	The ratio is the number of programs implemented to improve the quality of life of the population per 1000 people
	The level of public satisfaction with the activities of OmSU (survey)	%	The ratio of the number of respondents who are satisfied with the activities of the OmSU to the total number of respondents
	Percentage of requests on the official websites of the city/district on the Internet in the total number of requests for the year	%	The ratio of the number of requests (requests) of citizens to official sites on the Internet to the total number of requests (requests) of citizens

Indicators of digitalization, as well as indicators describing them, characterize trends in the development of the digital economy, namely the degree of use by business, government and social sphere of cloud services, the Internet at a speed of at least 2 Mbit/s, CRM and ERP systems, which allows you to effectively manage sales, resources, finances, logistics processes, etc.

Today, an organization's innovation activity should be viewed from several perspectives, since some economic entities conduct research and development, while others actively introduce elements of technological, organizational and marketing innovations into production activities. Consequently, innovation activity is characterized not only by the frequency of use of innovative technologies, but also by the degree of involvement of business units in their development.

An important indicator in the analysis of socio-economic development of the region is the degree of formation of the ecological environment. Namely, the implementation of measures to reduce the impact of negative factors on the environment will help to improve the quality and standard of living of the population, will allow you to correlate the level of pollution with the number of diseases of the population within the analyzed territory.

Indicators and indicators of quality of life characterize the degree of penetration of digital technologies into the daily life of each person in terms of solving everyday issues (obtaining state and municipal services using the Internet, purchasing goods and services, etc.), which in turn increases the level of comfort of living conditions in a particular territory, expands the possibility of purchasing goods and services regardless of territorial remoteness.

5 Conclusions

Based on the above, it can be concluded that rural areas play an important role in the activities of any State. The state should pay considerable attention to the development of villages and villages, because the "extinction" of these territories will lead to irreparable consequences. Overpopulation of cities, environmental problems, social problems due to the fatigue of citizens from the rapid pace of urban life are only a small part of what the state can face. However, rural development currently still faces a number of challenges, the main ones being:

– mass Exodus of the population;
– reducing the attractiveness of small businesses in the agro-industrial complex;
– low social and transport infrastructure of settlements. All these problems should be analyzed and identified on the basis of modern methods of socio-economic development of rural areas.

References

1. Glebov, O.V., Chudakova, K.A.: Problems and prospects of development of rural territories of Russia. In: Economics and Management: Current Trends. Collected Papers. Cheboksary, pp. 8–12 (2019)
2. Adukova, R.H., Adukova, A.N.: Public-private partnership in rural development: expediency and risks. Econ. Labor Manag. Agric. 4(13), 29–32 (2012)
3. Alekseeva, O.L.: Management of social infrastructure development in the agro-industrial complex: Diss. for the degree of Cand. Econ. Sciences 08.00.05, p. 184 (2007)
4. Zabelina, N.V.: Assessment of the level of development of the social infrastructure of the Ivanovo region. Agrarian Bull. Upper Volga Region 3, 43–49 (2014)
5. Stoyanova, T.A., Zabelina, N.V.: Methodology for determining the level of development of social infrastructure in rural municipalities. Econ. Agric. Russia 5, 89–95 (2015)
6. Avtushkova, E.P., Avtushko, A.M.: Organization, assessment and management of the rural territories (based on materials Yalutorovski District). In the collection: Modern Scientific and Practical Solutions in Agriculture. Collection of Articles of the All-Russian Scientific and Practical Conference, pp. 823–838 (2017)
7. Shamin, A.E., Proskura, D.V., Denisova, N.V., Proskura, N.V.: Digital transformation of economy: the need for integrated intro-duction of availability rate of telecommunication services within the sustainable development of rural territories. In: Advances in Economics, Business and Management Research. Proceedings of the International Conference on Policies and Economics Measures for Agricultural Development, vol. 147, 302–307 (2020)

8. Akifieva, L.V., Bolshakova, Y.A., Ilicheva, O.V., Kuchin, S.V., Belousova, O.A.: Problems of use of information technologies for management of social and economic development of the municipal unit. In: Proceedings of the International Conference «The 2018 International Conference on Digital Science», DSIC 2019: Digital Science 2019, pp. 470–479 (2019)

9. Frolova, O.A., Shamin, A.E., Shavandina, I.V., Kutaeva, T.N.: SmartVillage. Problems and prospects in Russia. In: Proceedings of the international conference «The 2018 International Conference on Digital Science», DSIC 2019: Digital Science 2019, pp. 400–486 (2019)

10. Krasovskaya, Yu.V., Lesnyak, A.Yu., Podlevskaya, O.M.: Sustainable ecological and economic development of agriculture in cross-border territories on the example of Ukraine and Belarus. Zbilansowana prirodokoristuvannya, pp. 24–29 (2017)

11. Minenko, A.V.: The Territory of advanced socio-economic development as a model for improving the infrastructure for the development of the agro-industrial complex of a rural municipality. Vector Econ. **12**(30), 66 (2018)

12. Blokhin, V.N.: Strategic management of territories: the possibility of transition to sustainable development. In the collection: Actual Problems of Human Potential Development in Modern Society. Materials of the V International Scientific and Practical Conference, Perm, pp. 183–187 (2018)

13. Surikova, N.V.: The Role of K(f) X in the development of rural territories. In the collection: Innovative Trends in the Development of Russian Science Materials of the IX International Scientific and Practical Conference of Young Scientists, pp. 305–307 (2016)

14. Gazizov, R.M.: Sustainable development of rural territories: a method of assessment and typology (on the example of the Krasnoyarsk territory). Actual Probl. Econ. Law **3**, 34–42 (2014)

15. Resolution of the Government of the Russian Federation from 31.05.2019 No. 696 "On approval of the state program of the Russian Federation "integrated development of rural areas" and on amendments to some AK-you're the government of the Russian Federation"

16. Resolution of the government of Nizhny Novgorod region dated 1 March 2006, no. 60 "On approval of the research Institute of methodology of assessment of socio-economic development of municipal areas and city districts of Nizhny Novgorod oblast" (as amended on 19 June 2019) [Electronic resource]. https://minec.government-nnov.ru/?id=145/

17. Decree of the government of the Nizhny Novgorod region No. 546-R dated 20 March 2013 "on evaluating the effectiveness of local government bodies in urban districts and municipal districts of the Nizhny Novgorod region". [Electronic resource]. https://government-nnov.ru/?id=124111

18. Decree of the President of the Russian Federation No. 203 of 09 May 2017 on the development strategy of the information society in the Russian Federation for 2017–2030

19. Decree of the President of the Russian Federation No. 204 of 07 May 2018 on national goals and strategic objectives of the development of the Russian Federation for the period up to 2024

20. Forecast of long-term socio-economic development of the Russian Federation for the period up to 2030 (developed by the Ministry of economic development of the Russian Federation) [Electronic resource]. https://base.garant.ru/70309010/#ixzz2xtlAS9En

21. Order of the Government of the Russian Federation of November 30, 2010 N 2136-R "on approval of the Concept of sustainable development of rural territories of the Russian Federation for the period up to 2020" [Electronic resource]. https://www.garant.ru/products/ipo/prime/doc/2073544/

22. Federal target program "Social development of the village until 2013", approved by the decree of the Government of the Russian Federation of December 3, 2002 No. 858 (as amended by the resolutions of the Government of the Russian Federation of April 29, 2005 No. 271, of April 3, 2006 No. 190, of September 17, 2007 No. 596, of March 5, 2008 No. 143, of January 31, 2009 No. 83) [Electronic resource]. https://base.garant.ru/194365/#ixzz2xtj

Determination of Territory Specialization as a Factor in Formation of Agricultural Cooperatives System

Dmitri Ganin⬤, Natalia Sidorova⬤, Vladimir Makarychev⬤, Tatiana Kutaeva⬤, and Darya Gorshkova$^{(\boxtimes)}$ ⬤

Nizhny Novgorod State Engineering and Economic University, Knyaginino 606340, Russia
Fleur-me@yandex.ru

Abstract. Agriculture is an important part in agricultural and industrial complex of Russia. Its effective development influences on ensuring food security in the country, as well as solving social and economic problems of territorial and sectoral levels. It is generally accepted that the main factor of sustainable functioning of agricultural complex is the formation of agricultural cooperation. In economic literature it is established that cooperation contributes significantly to the production of consumer goods, organization of processing of agricultural raw materials, promotion to the market and sale of finished products. Cooperation is devoted a lot to solving social issues in the municipalities of the country, primarily maintaining the employment of citizens and ensuring the reduction of social and economic tensions in rural areas.

The research shows a slow development of agricultural cooperation in the country and the region. In this paper the authors analyzed the current state of agricultural cooperation in Russia and noted the main trends in the functioning of cooperatives in Volga Federal district (in particular in Nizhny Novgorod region). This made it possible to determine that the process of formation and development of cooperation is constrained by many factors such as legal, organizational, economic and others.

During research authors revealed the chaotic formation of cooperatives in the country without detailed study and adaptation of the system to the realities of the modern economy and specialization of agricultural territories. The article presents an algorithm that includes three stages of determining the specialization of agricultural territories with preferential conditions for the formation and development of agricultural cooperative system, as well as detailed measures for each stage.

Keywords: Agriculture · System · Consumer cooperation

1 Introduction

Nowadays special attention is paid to the development of agriculture. The research shows that all sectors of the Russian economy interact directly or indirectly with agricultural and industrial complex.

© The Author(s), under exclusive license to Springer Nature Switzerland AG 2021
T. Antipova (Ed.): ICADS 2021, AISC 1352, pp. 21–30, 2021.
https://doi.org/10.1007/978-3-030-71782-7_3

The development of many sectors of the economy depends on agricultural and industrial complex; the role of agricultural sector in social and economic development of the country is constantly increasing.

It is necessary to consider the issues that exist in creating of organizational and economic conditions, which are important for the sustainable functioning and effective development of agricultural consumer cooperatives system [1–3]. The relevance of this question determined the choice of the research topic.

Studying the main trends in the development of agricultural production, many economists note the need to take into account "the formation and development of effectively functioning food market and the sustainable development of rural areas" [4].

In modern economy the development of cooperative relations and, first of all, effective rural consumer cooperation is the priority direction in reforming of agricultural and industrial complex. These are cooperatives based on the voluntary association of small business entities into cooperatives for processing, storing, marketing products, and logistics. This is due to the fact that the development of agribusiness requires the creation of an adequate economic system aimed at reducing production costs, increasing profits and increasing its efficiency.

This process is closely related to the rational use of land and other productive resources of agriculture, that is, to obtain maximum yield of quality products at the lowest cost. In this regard, the intensification of agricultural production requires solving problems in the management of this type of business, organizing promising forms of financial support and clear regulatory guidance of activities.

Foreign experience of research on the development of agriculture and other natural resource industries is considered taking into account three models - "sectoral (development of agriculture itself), territorial (formation of relationships within the local economy) and redistributive (assistance to rural development as a way of reducing the gap between regions and sectors of economy)" [5]. According to scientists all these aspects can be considered as the necessary conditions for the sustainable development of rural areas.

2 Materials and Methods

2.1 Researches

In modern period S. N. Poblitsyn, V. N. Lazhentsev, A. V. Makarov made a significant contribution to the study of agricultural cooperation.

Certain aspects of the development of agricultural consumer cooperation system can be traced in the works of A. V. Chayanov, A. Petrikov, F. Mantino, L. Bondarenko.

2.2 Research of Information Base

The theoretical and methodological basis of the research is based on the results of fundamental and applied researches of domestic and foreign scholars-economists in the field of agricultural and industrial complex, including the field of cooperation. The empirical base of research is based on official statistics, reference and reporting data on the state of agricultural consumer cooperation system.

2.3 Methods

The following methods of research were used: systemic approach, generalization, economic and statistical analysis, method of comparisons.

2.4 Purpose

The purpose of research is to make an algorithm for determining the specialization of agricultural territories with preferential conditions, which will identify municipal entities that have favorable conditions for the formation and development of consumer cooperatives system using modern methods of analysis and mathematical modeling.

The implementation of this purpose required solving the following tasks that determine the logic of scientific research:

1) to analyse the current state of agricultural consumer cooperatives and identify trends and prospects for development;
2) to make an algorithm for determining the specialization of the territory, based on a balanced system of indicators reflecting the main aspects of agricultural consumer cooperatives activities;
3) to evaluate its practical significance and the possibilities for differentiated application.

3 Results and Discussion

The experience of agricultural cooperation in developed countries shows the diversity of its institutional forms and structures. However, the organizational structure of agricultural cooperative movement (under all conditions) is based on individual membership of primary cooperative organizations. They are the main element of the cooperative business system by connecting individual farm production with related sectors of economy within agricultural and industrial complex [6].

In order to improve efficiency and protect interests, primary cooperatives form alliances and associations, creating cooperatives of cooperatives. In the practice of most developed countries, this association is carried out according to sectoral, territorial or territorial and sectoral principle.

Agricultural cooperatives (in the interest of their members) help to accelerate an agricultural industrialization. In addition, by developing links between agriculture and related industries, cooperatives make the most cost-effective use of their accumulated financial resources, which helps to strengthen their position in the fight against usurious and intermediary capital. The creation of their own cooperative agricultural system (in a number of countries) has to some extent reduced their dependence on commercial banks. Organizing farmers service system of agricultural production, distribution and social activities cooperatives operate at the same time in the interests of the whole society, the normal functioning of which is impossible without a high level of agricultural and rural development [7].

At the present stage one of the important purposes of the development of agricultural consumer cooperation system is strengthening positions in the competitive space,

increasing the share of presence in the market segment (Table 1) [8–11]. During the analyzed period the turnover of organizations by economic activities increased 5.5 times,
which determines the development of the country's economy as a whole.

Table 1. The turnover of organizations by economic activities, million rubles

Indicators	2005	2015	2016	2017	2018
Total for organizations (legal entities) - million rubles According to main types of activities	28287	111795	120158	133989	154625
Agriculture, forestry, hunting, fishing and fish farming	539, 9	1918	2057	2283	2629
Mining of minerals	3145, 2	11365	11851	13995	17637
Manufacturing activity	9977,7	31573	33534	36504	43450
Provision of electric power, gas and steam; air conditioning	1969,4	8262	8961	9227	9643
Water supply; water disposal, organization of waste collection and disposal, pollution disposal activities	187	3129	1570	860	990
Construction	963	3438	3626	4031	4390
Wholesale and retail trade; repair of motor vehicles and motorcycles	7077,3	37135,7	40,7137	46023	51611,8
Shipping and storage	908	541	549	9613,7	11225,8
Activities of hotels and public catering	132	445	496	590	652
Information and communications activities	2986	1023,6	1098,3	3146,0	3576,39
Real estate operations	1137	628,5	727,1	171,6	2136,9
Professional, scientific and technical activities	...	...	...	446,9	4967,9
Administrative activities and related additional services	...	...	...	7642,27	1004,6
Public administration and military security; social security	324,8	1604,1	1951,2	5570,2	4603,3
Education	3285,5	8946,8	8730,1	9114,4	466,9
Health and social services activities	5660,9	2732,66	3067,79	3319,63	3128,98
Cultural, sports, leisure and entertainment activities	...	...	...	1598,92	2283,25
Provision of other types of services	2135,85	5348,30	5793,85	1237,68	1155,07

The turnover of agricultural organizations in the total turnover of organizations was
1.7%. This situation is due to the seasonality of agricultural production, heavy losses
due to adverse weather conditions and presence of large proportion of regions with

unfavourable social conditions for the development of rural areas and extensive zones of social and economic depression.

To achieve the purpose of increasing the turnover of agricultural organizations, there is an objective need to use, first of all, the creation of cooperative system in order to obtain mutually beneficial results. Nowadays agricultural cooperation of the Russian Federation is a multisectoral branch of economy (Table 2).

Table 2. Structure of agricultural products by farm category, %

Indicators	2000	2015	2016	2017	2018
Farms of all categories, including:	100	100	100	100	100
Agricultural organizations, including:	45,2	54,0	55,1	55,2	56,5
Agricultural consumer cooperation	21,3	24,9	26,1	26,5	27,3
Households	51,6	34,5	32,5	32,4	31,0
Farming enterprises	3,2	11,5	12,4	12,4	12,5

The share of products produced by agricultural consumer cooperatives in the structure of agricultural products in 2018 was 27.3%. At the same time, there is an annual dynamics of share growth, which for the analyzed period amounted to 6.0%.

Farming enterprises are the main competitors of the system, which are potential participants in cooperation.

In 2018 the number of farming enterprises in Russia was 210.3 thousand, with the total number of organizations of small forms equal to 2754.6 thousand. Due to the difficult economic situation in the world, the number of organizations is declining. Most experts emphasize that the main problem with the functioning of farms is the complexity of marketing products. This leads to financial difficulties, inability of farmers to meet their obligations.

According to statistics in 2018 small forms of farming (which are potential participants in cooperation) are represented by 31.0% of the population and 12.5% - by farming enterprises. Despite the growth in 2018 in share of large and medium-sized organizations in the structure of agricultural products, in personal households, individual farms and farming enterprises were produced 43.5%

Despite the growth in share of large and medium-sized organizations in gross agricultural production, the share of small forms of farming in total agricultural production in the Russian Federation amounted to 47.2%. Most of the regions with the highest concentration of small forms of farming (up to 90%) are located in favorable regions for farming. This also confirms the importance of territorial specialization in the formation of agricultural consumer cooperatives system.

According to the state registration on January 1, 2020 5.742 agricultural consumer cooperatives were registered in the Russian Federation, including 1.199 supply and household, 1.501 processing, 849 credit and 2.193 cooperatives engaged in other activities (servicing, gardening and livestock cooperatives) (Table 3).

The largest number of agricultural cooperatives is registered in the Volga Federal District – 1.644 units.

This concentration is due to the predominance of agricultural regions in the district, as well as a high degree of cooperation, the development of personal households, following traditions (the district mainly consists of national republics). The Volga Federal District includes one of the largest agricultural regions of Russia - the Republic of Tatarstan.

Table 3. Number of agricultural consumer cooperatives in the constituent entities of the Russian Federation on January 1, 2020

Constituent entity of the Russian Federation	Consumer cooperatives						
	Processing	Servicing	Marketing	Supply	Credit	Others	Total
Russian Federation	1501	860	868	331	849	1333	5742
Central Federal District	234	187	208	131	365	350	1475
Northwestern Federal District	50	29	16	9	38	40	182
Southern Federal District	103	95	105	22	102	106	533
North Caucasian Federal District	107	33	34	8	21	164	367
Volga Federal District	535	306	223	98	132	350	1644
Ural Federal District	55	62	82	19	38	55	311
Siberian Federal District	243	75	152	26	34	104	634
Far Eastern Federal District	174	73	48	18	119	164	596

This concentration is due to the predominance of agricultural regions in the district, as well as a high degree of cooperation, the development of personal households, following traditions (the district mainly consists of national republics). The Volga Federal District includes one of the largest agricultural regions of Russia - the Republic of Tatarstan.

The greatest concentration of agricultural cooperatives largely depends on the geographical location, proximity of rivers and land fertility. One of the largest rivers (the Volga) flows through the territory of the district, while the Volga Federal District also includes regions geographically located in the Urals (the Republic of Bashkortostan, Udmurt Republic, Perm Krai, Orenburg Oblast and Kirov Oblast). The share of agricultural consumer cooperatives in Nizhny Novgorod region (in the total number of organizations of the Volga Federal District) is 6.0% (Fig. 1).

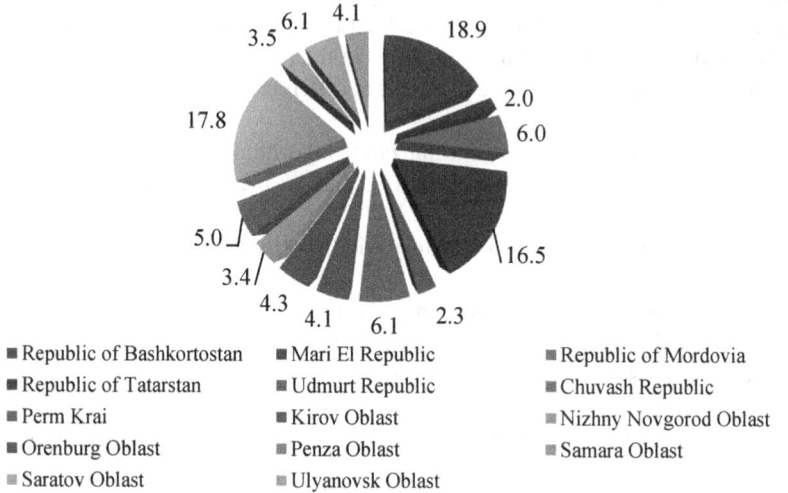

Fig. 1. The structure of the quantitative composition of agricultural consumer cooperatives of the Volga Federal District in 2020, %

It can be concluded (based on the analysis) that nowadays there is a chaotic formation of agricultural consumer cooperatives. Activity arrangement of cooperatives should depend on the type of territories in the Russian Federation.

Many scientists addressed the issue of specialization (specialization of territories). But it should be noted that the vast majority of them considered the legal content of zoning institution, the peculiarities of establishing a legal regime in certain territories, and often did not take into account that the content of relations in the field of zoning territories is broader than the subject of legal regulation of land relations.

It is necessary to determine the specialization of the territory for the purpose of determining agricultural areas favourable to the formation and development of agricultural cooperative system (including the characteristics of the region). This specialization could be based on factors influencing the formation and development of the cooperative system, and the criterion or criteria of specialization (such as efficiency of operation or indicators of social and economic development) may be a resulting indicator, or a set of criteria [12].

4 Results

Methods of specialization of agricultural territory of the region should be considered to establish the possibility of forming certain forms of economic activity, to develop a sustainable form of management and control over the use of the territory, aimed at dynamic development of the territories, attracting investments, stimulating business activity, using public-private partnership tools, improving the living conditions of the rural population, as a result of improving the quality of agricultural products [13–17].

Special agricultural zoning of the territory of the constituent entity of the Russian Federation for the purpose of forming of agricultural consumer cooperatives system is a scientifically based grouping of municipalities in the region of the Russian Federation, carried out by clustering methods and aimed at differentiating the region's production policy.

The main stages of the territory specialization algorithm in order to create the determination of groups of territories with preferential business conditions are presented below (Fig. 2).

The most significant features of territory with priority features are:

– presence of geographical concentration (location);
– commonality of products (resources, technologies, services);
– close relationships between firms within the entity.

Therefore, the following indicators were selected for the specialization of agricultural territories:

1. Indicators of geographical proximity of economic entities: geographical coordinates of entities (latitude, longitude).
2. Organizational and economic indicators: profitability of agricultural production; turnover of agricultural production; number of potential participants; number of agricultural consumer cooperatives.

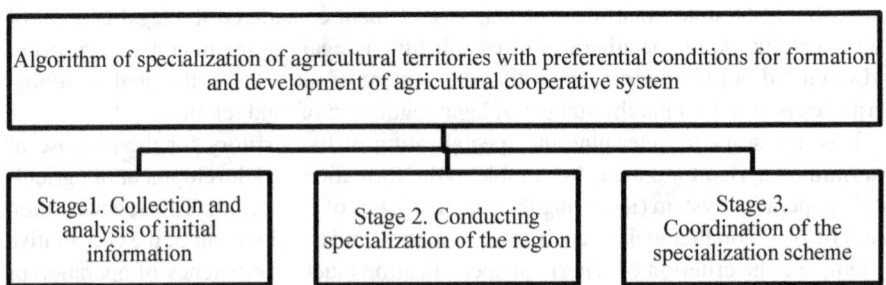

Fig. 2. Algorithm of specialization of agricultural territories with preferential conditions for formation and development of agricultural cooperative system

To implement the algorithm, it is necessary to consider in detail the activities of each stage, including determining the most convenient and qualitative methods of analysis (Table 4):

Table 4. Detailing the stages of specialization algorithm

Stage	Process description
Stage 1. Collection and analysis of initial information	1.1 Analysis of the state of agriculture in the region and agricultural consumer cooperatives
	1.2 Determination of indicators required for specialization
	1.3 Data collection for each indicator within municipalities in the region
Stage 2. Conducting specialization of the region	2.1 Determining the specialization of region's territory using analysis
	2.2 Construction of preliminary territory specialization dendrogram
Stage 3. Coordination of the specialization scheme	3.1 Preliminary dendrogram analysis, identification of data inconsistencies, scheme correction
	3.2 Approval of regional specialization scheme

5   Conclusion

The problem of developing of agricultural consumer cooperatives system is the most important task of regions and state. The solution to this problem can be achieved by determining the specialization of region's territory.

To summarize the above, we can conclude that the specialization of agricultural territories should also be considered in differentiated way in order to highlight the specific features of each group of territories.

Nowadays there are no comprehensive studies in Russia considering the specialization of the constituent entities of the Russian Federation in order to create groups of agricultural territories with preferential conditions for the formation and development of agricultural consumer cooperatives.

Identifying the territories of the region by social and economic development, level of concentration of agricultural cooperatives and presence of potential participants in the system, it is possible to significantly differentiate the development strategy of agricultural consumer cooperatives, directing financing precisely to improve the strengths and correct the weaknesses of specific municipalities in group.

The presented methodological approach (algorithm) of specialization of rural areas is universal and can be used to determine zones for creating the system of consumer cooperatives in various entities.

References

1. Shamin, A., Frolova, O., Klychova, G., Nigmatullina, N., Iskhakov, A.: Formation and development of clusters in the Russian regional agro-industrial complex. In: E3S Web of Conference, p. 06005 (2019)
2. Shamin, A., Frolova, O., Makarychev, V., Yashkova, N., Kornilova, L., Akimov, A.: Digital transformation of agricultural industry. In: IOP Conference Series on Earth and Environmental Science, p. 012029 (2019)
3. Shamin, E., Frolova, O., Zimina, E., Kholkin, A., Zonova, A., Valerianov, A., Levanova, T.: Conditions and prerequisites of the activity based costing method application. In: Prospects for the development of agricultural Sciences, Materials of the Interface science and practice conference, pp. 172–175 (2019)
4. Polbitsyn, S.N.: Russia's rural entrepreneurial ecosystems. Econ. Region **15**(1), 298–308 (2019)
5. Lazhentsev, V.N.: Rural development strategy of the northern region. Econ. Region **16**(3), 696–711 (2020). https://doi.org/10.17059/ekon.reg.2020
6. Makarov, A.V., Trapeznikov, V.A.: Forming the development program of industrial cooperation in the region. Econ. Region **3**(27), 175–183 (2011)
7. Chapter 14: Promotion Sustainable Agriculture and Rural Development. Agenda 21. FAO, Rome (1996)
8. Agriculture in Russia. 2019: Statistical collection. Rosstat – S 29 Moscow, 91 p. (2019)
9. Agriculture in Russia. 2016 [Electronic resource]: Statistical collection. Rosstat – S 29 Moscow (2016). https://www.gks.ru/storage/mediabank/sh_2016.pdf
10. Agriculture in Russia. 2017 [Electronic resource]: Statistical collection. Rosstat – S 29 Moscow (2017) https://www.gks.ru/storage/mediabank/sh_2017.pdf. Accessed 12 Apr 2020
11. Agriculture in Russia. 2018 [Electronic resource]: Statistical collection. Rosstat – S 29 Moscow (2018). https://www.gks.ru/storage/mediabank/sh_2018.pdf. Accessed 12 Apr 2020
12. Mantino, F.: Rural development in Europe. Policies, institutions and people on the places from the 1970s to the present day. Translation from Ital. I. Khramova, p. 272 p. Business Media of the Sole 24 Ore, Milan (2010)
13. Bondarenko, L.: Conceptual fundamentals of regional policy of rural territories social development and program and target approach to its realization. Econ. Agric. Russia **7**, 60–68 (2019)
14. Hyyrylainen, T.: Governance of local empowerment in Finnish rural policy: collaboration between policy, development and research h Employment Policy Research Center. In: EPRC, Discussion Paper, vol. 4, pp. 81–94. Hirosaki University, Japan (2010)
15. Kotler, F.: Marketing management: analysis, planning, implementation, control. Saint-Petersburg (1998)
16. Petrikov, A.: Problems and prospects of sustainable rural development. The role and place of agricultural and industrial complex in doubling of gross domestic product of Russia: Materials I All-Russian congress of economists and agrarian, Moscow, pp. 92–103 (2005)
17. Chayanov, A.V.: About agricultural cooperation: (Selected chapters and articles). Saratov, p. 176 p. (1989)

Advances in Digital Economics

Technique for Internal Control of Company's Equity

I. N. Bogataya[1] ⓘ, E. M. Evstafyeva[1](✉) ⓘ, and Denis Lavrov[2] ⓘ

[1] Rostov State Economic University, Rostov-on-don, Russian Federation
2982232@mail.ru
[2] Financial University Under the Government of the Russian Federation,
Moscow, Russian Federation

Abstract. The article is aimed at developing the methodology of internal control of equity, considered from the standpoint of accounting and economic approaches through the development of a descriptive structural and logical model of internal control, taking into account the characteristics of equity. In order to achieve this goal, it is necessary: first, to determine the features of equity capital as an object of internal control from the standpoint of accounting and economic approaches; second, to develop and justify a descriptive structural and logical model of internal control of equity. Technique for Internal Control of Company's Equity is designed through the development of a structural and logical model that takes into account the peculiarities of equity capital and considers it from the standpoint of both accounting and economic approaches based on accounting software. The results of the study can be implemented in the formation of an internal control system focused on creating value in commercial organizations. In particular, the developed structural and logical model allows us to systematize and improve such types of support within the internal control system as: regulatory, conceptual, organizational and methodological, information, technological, which contributes to a systematic approach to conducting internal control of equity and creates a basis for the development of its methods in the conditions of digital transformation and uncertainty of the external environment.

Keywords: Internal control · Equity · Accounting approach · Economic approach

1 Introduction

In the modern economic literature devoted to the issues of internal control and audit, due attention is not paid to such an object of accounting as equity capital. The questions connected with the accounting and carrying out of audit of own capital, are disclosed in the works of such scientists as Abeysekera [1], I. A., Alekseeva I. V. [2], Betge Y., Bogataya I. N. [3], Beschetnaya S. V., Van Breda M. F., Evstafieva E. M. [4], Kalyugina I. V. [5], Kyshtymova E. P. [6], Novodvorsky V. D. [7].

These researchers focused more on the methodology of accounting and auditing, as well as the implementation of audit procedures in relation to equity and its individual

T. Antipova (Ed.): ICADS 2021, AISC 1352, pp. 33–44, 2021.
https://doi.org/10.1007/978-3-030-71782-7_4

components. From the perspective of the COSO concept, they comprehensively considered issues related to the assessment of the internal control system in relation to equity. However, approaches to the specifics of internal control of equity capital have not been developed.

The mentioned researchers to a greater extent concentrated their attention on the methodology of accounting and auditing, as well as implementation of auditing procedures in relation to equity and its separate components. From the position of COSO concept they have extensively considered the issues related to the assessment of the internal control system in relation to the equity. However, no approaches to the specifics of internal control of equity have been developed.

In view of the fact that preservation and increase of the value is the most important strategic task of business organizations, the issues should be considered from the position of organization of internal control of the value creation process within the framework of five basic elements of internal control system. Thus to the questions of internal control of own capital, in our opinion, it is necessary to approach comprehensively that assumes realization not only of accounting approach to understanding of own capital, but also its consideration from position of economic approach. It assumes consideration of own capital not only as cost of its assets minus liabilities (cost of net assets), but also as set of means of the organization which are in its ownership, represented from a position of the concept of plurality of the capital by its various kinds (financial, industrial, intellectual, human, social and reputation, natural, etc.) that corresponds to the economic approach.

From the perspective of the conceptual framework of internal control as well as the conceptual framework of enterprise risk management (COSO ERM), strategy plays an important role in the process of value creation in business organizations. For the successful implementation of a strategy aimed at value creation, it is important to take into account the peculiarities characteristic of capital. These peculiarities should be taken into account within the framework of the internal control system. Key features of the own capital are: client orientation, circulation of the capital, use of a wide spectrum of estimations of the own capital and methodical approaches within the limits of various kinds of the accounting, self-growth of cost, social orientation, features of the own capital caused by organizational and legal form of an economic entity; plurality of types of the own capital.

2 Improved Technique for Internal Control of Company's Equity

In the course of research, we developed a structural and logical model of organization and implementation of internal control of equity capital, presented in Fig. 1, which involves consideration of equity capital both from the perspective of accounting and economic approach and taking into account the features characteristic of equity capital in conditions of digital transformation.

The rationale for the logic of the organization and implementation of internal control of equity capital was carried out on the basis of highlighting the different types of collateral used, such as: 1) regulatory and legal; 2) conceptual 3) organizational and methodological; 4) informational; 4) technological.

The basis for organization and realization of internal control of the equity capital is the regulatory and legal support.

Aggregately, the regulatory framework includes acts in the field of:

1) internal control (international (COSO, COSO ERM), national and local or internal);
2) accounting and formation of non-financial reporting (IFRS, ISIR, Russian legislative, regulatory, methodological and organizational acts).

As a feature of the regulatory framework should be noted its focus on international standards and guidelines in the field of accounting, internal control. Besides, approaches to accounting and control provision formation, connected with inclusion of aspects traditionally characteristic for economic approach, connected with risk estimation, search of internal production reserves, strategy of organization, planning of activity, organizational culture and ethical values of organization have considerably expanded. It is necessary to note the general tendency connected with the convergence of the Russian normative base with the international one. The large number of standards of international standards also draws attention, for example, devoted to the evaluation of risks in an organization and the formation of non-financial information.

Conceptual support of internal control of equity capital includes fundamental concepts, as well as general scientific (induction and deduction, analysis and synthesis, modeling, analogy, etc.) and specific methods related to the specifics of equity capital, considered from the standpoint of accounting and economic approaches. The basic principles and methods within the aforementioned approaches may differ significantly. If the accounting approach to a greater extent is aimed at compliance with the requirements of the current legislation and reliability of reporting, within the framework of the economic approach the issues related to efficiency of use of equity capital, process of value creation and use of various types of capital (financial, intellectual, financial, etc.) are in the foreground.

Organizational and methodological support is based on the Conceptual Framework of Internal Control COSO and other acts included in the regulatory framework governing the organization of internal control.

The peculiarity of information support is the use of both external and internal information. At the same time, external sources become important, which is due to expansion of practice of formation of estimated values in the Russian accounting and requires systematic collection of information on them and their periodic review.

Technological support for internal control of equity capital under conditions of digital transformation includes a wide range of software products and services based on the use of digital technologies:

– legal reference systems;
– service programs and services (e.g., the Transparent Business service of the Federal Tax Service, etc.).
– accounting automation programs that involve the use of digital technology;
– engineering programs for management of accounting, control and audit, analytical and strategic processes, including those integrated with external environment sources and involving the use of digital technologies;
– specialized and auxiliary programs, services for control and risk assessment.

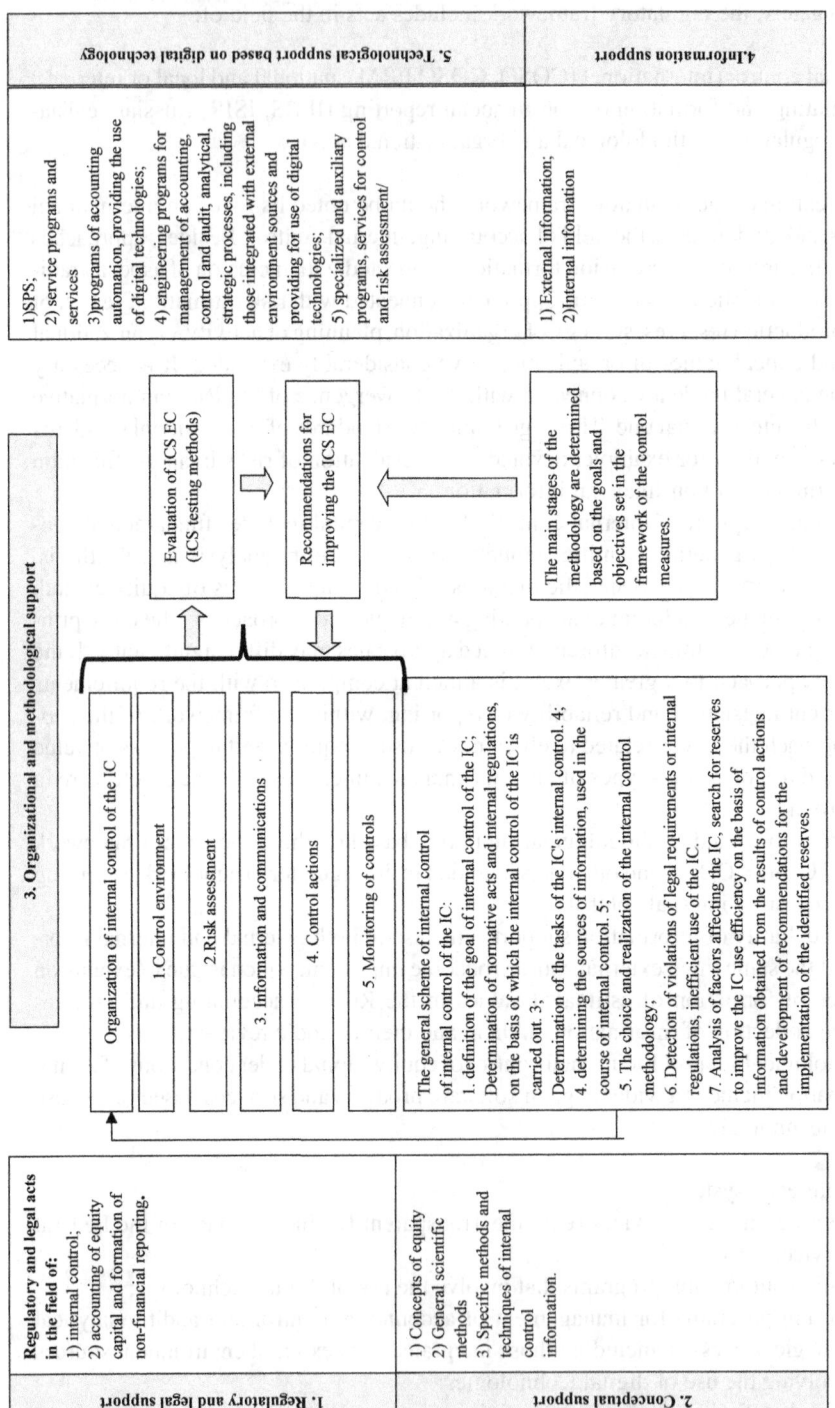

Fig. 1. Descriptive structural-logical scheme of organization and implementation of internal control of equity capital of commercial organizations

The dynamic development of software and hardware support for internal control is connected with the processes of digital transformation, the use of blockchain technologies, the introduction of international and Russian standards in the field of risk assessment and management.

3 Peculiarities of Organizational and Methodological Support of the Internal Control of Equity Capital

The organization and realization of equity's internal control includes 5 basic components.

The first component of the system of internal control - the control environment represents the basis of this system and is directly connected with strategy of the organization. The strategy of the commercial organization, as a rule, is aimed at creation of value and is concretized in the strategies of lower hierarchical levels assuming management of various kinds of capitals (financial, industrial, intellectual, human, social and reputation, natural, etc.). Creation of value is possible only when strategies will be focused on creation of value for clients, realization of effective personnel policy assuming rational use of human and intellectual capital, receipt of profit by investors.

Creating value is impossible without creating and maintaining a culture of integrity and ethical behavior in a commercial organization by top management. Table 1 shows the analysis of the influence of the main elements of the control environment on equity capital.

Value creation is directly linked to risk assessment, which is the second component of the internal control system. In this regard, it is necessary to assess the risks associated with the decline in the value of equity. In our opinion, at the first stage it is advisable to conduct a PEST-analysis, which allows to identify external and internal factors that have a significant impact on the organization, as well as SWOT-analysis, which allows to identify weaknesses and strengths of the organization, as well as opportunities and threats. In addition, it is necessary to form a register of risks, assess their significance, probability of occurrence and make decisions on their minimization to the level of risk appetite. It is important to identify the risks that make it difficult to achieve the strategy aimed at creating value. Thus risks can be connected not only with distortions of various kinds of reporting, but also inefficient use of various kinds of capital that, eventually, will lead to decrease in value of net assets of the commercial organization.

The identification and assessment of risks of material misstatement in accordance with ISA should be conducted at the level of:

1) financial statements, but this approach is also applicable to other types of reporting (e.g., management reporting);
2) Assumptions regarding the types of transactions, account balances and disclosures that are important to external users of statements (primarily investors).

Particular attention should be paid to assessing the risk of dishonest actions, as well as the risks that lead to the inefficient use of various types of capital, which almost always has a negative impact on the value creation process. Since the external and internal environment of an organization is volatile, it is necessary to periodically review the risks. In

Table 1. Analysis of the influence of the main elements of the control environment for capital

Elements of the control environment	Interpretation of equity	
	Equity as the total value of its assets less liabilities (net asset value (accounting approach)	Capital as the totality of funds of an organization owned or controlled by it (economic approach)
Communicating the principle of integrity and ethical values to employees	Compliance with ethical principles contributes to value growth	The risk of unfair practices, inefficient use of various types of capital is reduced
Commitment to competence	It promotes value growth through sound management decisions	It reduces the likelihood of inefficient management decisions on the use of different types of capital
Participation of persons responsible for corporate governance	It increases interest, objectivity of control over the value creation process at its various stages	It reduces the likelihood of inefficient management decisions on the use of various types of capital
Philosophy and management style of leadership	Assumes the development of the organization's strategy aimed at increasing the value with a focus on customers, the profit of investors	Assumes, as part of the organization's strategy aimed at increasing the value, to have a strategy for using different types of capital, primarily human, intellectual
Organizational structure	The efficient organizational structure makes it possible to trace the process of value creation by responsibility centers	The effective organizational structure makes it possible to trace the processes of using different types of capital and evaluate their effectiveness
Distribution of authority and responsibility	It reduces the risks of dishonest actions	It reduces the risks of unfair actions in relation to the use of different types of capital, promotes the rational use of human and intellectual capital
Personnel policies and practices	The effective human resources policy opens up new opportunities with respect to the growth of equity capital based on intellectual capital	The effective human resources policies and practices promote the rational use of almost all types of capital, especially human, intellectual capital

Source: Compiled by the author on the basis of [8].

the COVID-19 pandemic, risk review is mandatory in the internal control assessment. Such revision should be based on the analysis of information from external sources (industry and economic journals, information from mass media, analytical reports, statistical data, etc.) and internal sources represented by the forms of management reporting used in a commercial organization. Both financial and non-financial information is used. Risk assessment involves the use of data, both financial and non-financial reporting for past periods, and forecasting information.

As part of the third component of the internal control system - information and communications, it is necessary to organize an accounting system aimed not only at compliance with current legislation, the formation of reliable financial reporting, but also at the development and improvement of management accounting methods, as well as the formation of non-financial and management reporting, allowing the disclosure of information about the value creation process. This will require: firstly, the definition of economic transactions in the greatest degree, influencing the growth of the value, secondly, the use of information technologies, thirdly, developed communications in terms of interaction with management and those responsible for corporate governance in making management decisions aimed at increasing the value and the presence of feedback.

Informing in the commercial organization can be carried out on the basis of system of the internal regulations concerning policies in the field of internal control, risk management, formation of various types of the reporting, etc.

It should be taken into account that the selected elements of accounting policy for accounting (financial) purposes may lead to information asymmetry and financial accounting data may require adjustments to make management decisions aimed at creating value. In organizations that form the accounting policy for accounting (management) purposes, the elements of this policy are aimed at the formation of relevant information. Changes in accounting policies should be reflected in the financial statements retrospectively, with the exception of those organizations that use simplified methods of accounting.

The information system associated with various types of reporting, formed in a commercial organization, involves initiating, recording, processing and summarizing the organization's operations (as well as events and conditions), ensuring accounting of relevant assets, liabilities and equity; timely correction of identified errors; fixing the facts of bypassing controls; transferring information from transaction processing systems to the main register; recording information that is significant for the reflection in the reporting, ensuring the collection, recording, processing At the moment Russia widely uses cloud accounting based on SaaS concept (1C: Accounting, Info-Predprinitelstvo, Buchsoft "Moye Delo", Kontur, Nebo, etc.), as well as mobile accounting applications. And mobile accounting applications (e.g., Regulatory, Invoicing, Mobile Bank, etc.), which, on the one hand, requires the digital skills of personnel, and, on the other hand, increased attention to cybersecurity and digital trust.

The fourth element of the internal control system - control actions - should be built with focus on prerequisites of reporting, both financial and managerial.

The main areas of validation of reporting prerequisites with respect to equity are presented in Table 2.

Table 2. Main directions of confirmation of the prerequisites of reporting on equity capital

Prerequisites	Capital as net asset value (NAV)[a]	Capital as the sum of capitals
Assumptions about the types of transactions and events, as well as the appropriate disclosures for the audited period		
Availability	FEA related to equity disclosed in different types of statements are relevant to the organization	FEA related to different types of capital disclosed in different types of reporting are relevant to the organization
Completeness	FEA related to equity as disclosed in different types of statements are reported comprehensively	FEA related to different types of capital disclosed in different types of statements are reflected comprehensively
Accuracy	FEA related to equity disclosed in different types of statements are correctly evaluated and described	FEA related to various types of capital disclosed in various types of statements are correctly evaluated and described
Timeliness of recognition	FEA related to equity disclosed in different types of statements are reflected in the proper reporting period	FEA related to different types of equity disclosed in different types of statements are reflected in the proper reporting period
Classification	FEA related to equity, disclosed in different types of statements, are reflected in the appropriate accounts	FEA related to different types of capital disclosed in different types of statements are reflected in the appropriate accounts
Submission	FEA related to equity disclosed in the various types of statements are properly grouped or disaggregated and clearly described, and the related disclosures are relevant and understandable in the context of the requirements of the applicable financial reporting framework	FEA related to the various types of capital disclosed in the various types of statements are properly grouped or disaggregated and clearly described, and the related disclosures are appropriate and understandable in the context of the requirements of the applicable financial reporting framework
Assumptions about account balances and related disclosures at the end of the period:		

(*continued*)

Table 2. (*continued*)

Prerequisites	Capital as net asset value (NAV)[a]	Capital as the sum of capitals
Existence	Assets, liabilities related to equity do exist	Assets, liabilities related to various types of capital do exist;
Rights and obligations	The organization owns or controls the rights to the assets, and the liabilities represent the legal obligations of the organization;	The organization owns or controls the rights to the assets, and the liabilities represent the legal obligations of the organization;
Completeness	All assets, liabilities and equity interests required to be recorded have been accounted for and all relevant disclosures required to be included in the financial statements have been included	All assets, liabilities and equity interests required to be required to be recorded have been recorded and all relevant disclosures required to be included in the financial statements have been included
Accuracy, measurement, and allocation	Assets, liabilities, and equity interests are included in the financial statements at appropriate amounts, all appropriate measurement or allocation adjustments are properly recorded, and appropriate disclosures are properly valued or described	Assets, liabilities and equity interests are included in the financial statements in the appropriate amounts, all appropriate measurement or allocation adjustments are properly recorded, and related disclosures are properly evaluated or described
Classification	Assets, liabilities, and equity interests are recorded in the appropriate accounts	Assets, liabilities, and equity interests are recorded in the appropriate accounts
Submission	Assets, liabilities and equity interests are appropriately grouped or disaggregated and unambiguously described, and the related disclosures are appropriate and understandable within the context of the applicable financial reporting framework	Assets, liabilities and equity interests are appropriately grouped or disaggregated and unambiguously described, and the related disclosures are appropriate and understandable within the context of the applicable financial reporting framework

[a]the fact of economic activity (FEA).
Source: Compiled by the author on the basis of [8]

The main types of control actions and their impact on equity, considered from the standpoint of accounting and economic approaches, are presented in Table 3.

Table 3. The main control actions and their focus on equity capital, considered from the standpoint of accounting and economic approaches.

The aggregate areas of control actions highlighted in ISA 315	A focus on capital as a	
	NAV	The sum of the different capitals
Performance reviews	Analysis of the value creation process	Analysis of the use of different types of capital
Information processing	Accuracy control calculations	Accuracy control calculations
Physical controls	Detect or prevent the diversion of assets and subsequent NAV reduction	Aimed at preventing asset theft
Segregation of duties	Prevent bad faith actions leading to NAV depreciation	Legality of transactions with different types of capital

The fifth element of the internal control system - monitoring of controls - allows to evaluate the effectiveness of control procedures used in relation to equity, considered from the standpoint of accounting and economic approaches, as well as to develop recommendations for their improvement.

Internal auditors, and in small and medium-sized businesses, employees performing similar functions, perform the functions of monitoring the means of control of the organization, based on both internal and external information, assess the effectiveness of the internal control and make recommendations for its improvement [8, 9].

Methodological support of internal control over own capital is a set of methods and techniques used both at the organization of internal control and at implementation of control actions.

It is possible to distinguish a group of methods aimed at internal control evaluation in relation to own capital, which is considered from the standpoint of accounting and economic approaches, and a group of methods of control actions inspection, aimed at receiving information about compliance with applicable legislation and regulations, reliability of different types of reports, which disclose information about own capital and main types of capital, evaluation of efficiency of own capital use and value creation process, as well as about the effectiveness of the control actions.

These groups of techniques can be used by internal control and internal audit services of large organizations, external auditors and consultants, audit commission.

The first group of techniques can be implemented on the basis of internal control testing in respect of equity capital.

The second group of techniques is focused on classical external audit procedures, as well as operational audit of the equity. At the same time, within the framework of this technique it is necessary to:

1. to determine the internal control objectives (for example, compliance with the requirements of the current legislation, search for reserves with regard to the use of equity capital, etc.);
2. to determine the normative acts and internal regulations, on the basis of which the internal control of the internal control is carried out, as they may significantly differ depending on the purpose of such control;
3. to define the tasks within the scope of the set goal;
4. to define the sources of information used during the internal control;
5. To make a choice and implement the internal control methodology;
6. To reveal violations of legal requirements or internal regulations, inefficient use of various types of capital.
7. To analyze the factors influencing the own capital, to search for reserves to increase the efficiency of use of the internal control and various types of capital on the basis of the information received by results of performance of control actions and to work out recommendations on implementation of the revealed reserves.

4 Conclusion

Therefore, in the course of research the descriptive structural and logical model of the organization and implementation of internal control of equity capital in the commercial organization, which includes the following types of support: regulatory, conceptual, organizational and methodological, informational and technological support on the basis of digital technologies was formed. This model reveals the general scheme of internal control of IC considered both from the position of accounting and from the position of economic approach. This scheme includes: 1. Determination of the aim of the internal control of the IC; 2. Determination of the statutory acts and internal regulations, on the basis of which the internal control of the IC is realized; 3. Determination of the tasks of the internal control of the IC; 4. Determination of the information sources, used during the internal control; 5. Selection and realization of the internal control methods; 6. Detection of violations of the legislation or internal regulations. 7. Analysis of factors influencing the shareholders' equity, search of reserves to increase the efficiency of SK and various kinds of capital on the basis of information received by the results of control operations, and working out recommendations on implementation of the reserves revealed, which will allow to estimate the legality of operations with the shareholders' equity, reliability of financial and non-financial information, and increase the efficiency of SK and value creation process in a commercial organization.

References

1. Abeysekera, I.A.: Framework to audit intellectual capital. Intellectual Capital and Human Resources as Objects of Accounting and Control, p. 352 (2001). https://citeseerx.ist.psu.edu/viewdoc/download?doi=10.1.1.456.5040&rep=rep1&type=pdf
2. Alekseeva I.V., Evstafieva E.M., Charaeva M.V., Mosentseva, V.A.: Development path of agricultural companies' internal control system. Int. J. Econ. Bus. Adm. **VII**(1) (2019). https://www.scopus.com/sourceid/21100890303. Research Studies, vol. XX, no. 3B (2019)

3. Bogataya, I.N., Kislaya, I.A., Kuznetsova, M.A., Krohicheva, G.E., Kuznetsova, M.A.: Intellectual capital and human resources as objects of accounting and control. Int. J. Econ. Bus. Adm. **VII**(2), 343–353 (2019)
4. Evstafyeva, E.M., Charaeva, M.V., Paleev, A.V., Krokhicheva, G.E.: Business processes' reengineering in the context of increasing the firm's value. Eur. Res. Stud. J. **XXI**(1), 125–135 (2018). https://www.ersj.eu/index.php?option=com_content&task=view&id=1212
5. Kalyugina, I.V.: Development of accounting and audit of the equity capital of agricultural organizations: author's abstract.... Ph.D. Thesis: 08.00.12. Voronezh-2007, 27c (2017)
6. Kyshtymova, E., Parushina, N., Lytneva, N., Polyanin, A., Plotnikov, V.: The value of the company and transformation of its evaluation under the influence of informatization. In: Proceedings of the 32nd International Business Information Management Association Conference, IBIMA 2018 - Vision 2020: Sustainable Economic Development and Application of Innovation Management from Regional Expansion to Global Growth, pp. 4395–4407 (2018)
7. Novodvorsky, V.D., Marin, V.V.: Accounting of own capital, 159 p. Ekonomist, Moscow (2004)
8. ISA 315 "Identification and assessment of the risks of material misstatement by Studying the organization and its environment" (introduced in the Russian Federation by Order of the Ministry of Finance of the Russian Federation of 09.01.2019 N 2n)
9. Evstafyeva, E.M., Charaeva, M.V., Paleev, A.V., Krokhicheva, G.E.: Business processes' reengineering in the context of increasing the firm's value. Eur. Res. Stud. **XXI**(1), 125–135 (2018). https://www.ersj.eu/index.php?option=com_content&task=view&id=1212

Analysis of the Balance Sheet Liquidity in the Digital Age

Liudmyla Lakhtionova[(✉)] [iD]

National Aviation University, 1, Liubomyr Husar Avenue, Kyiv 03067, Ukraine
Ludmilala@i.ua

Abstract. The article examines the concept of balance sheet liquidity and proposes its improvement in modern financial analysis, which will facilitate the procedure of assessing the risks and prospects for improving the solvency and liquidity
of the enterprise in the age of digital technologies. The opinions of well-known
world scientists-analysts on the essence of balance sheet liquidity are reflected,
debatable issues are revealed and proposals for their elimination are developed.
The author's improved notion of the essence of balance sheet liquidity in the
modern financial analysis is given. The concept of balance sheet liquidity limit
is proposed. The method of determining the absolutely liquid type of balance
has been improved. For the first time its dependence on a kind of activity of the
enterprise is established. According to the balance sheet (statement of financial
position) of State Enterprise of the Service Air Traffic of Ukraine as an example of
the implementation of the proposed theoretical and methodological provisions for
the analysis of balance sheet liquidity is shown. The proposed forms of analytical
tables will contribute to the unification of analytical calculations and increase the
clarity and informativeness of analytical conclusions on the liquidity of the balance in terms of digital technologies. The results of the study are useful for users
in order to make the right management decisions in the digital age.

Keywords: Digitization of financial analysis · Balance sheet · Balance sheet
liquidity

MSC Codes: 5.02

JEL Codes: C51 · D21 · G32 · M41

1 Introduction

It has become commonplace that the fourth industrial revolution is accelerating radical
change around the world in all spheres of human life. This is due to the spread of
attempts to actually use artificial intelligence, the widespread use of robotics, the mass
introduction of digital computer technology. The use of digital computer technology is
becoming the norm, part of the efficiency of human resources.

The break in the technological paradigm reveals the latest opportunities and prospects
for the world community. However, the same break in the technological paradigm poses

© The Author(s), under exclusive license to Springer Nature Switzerland AG 2021
T. Antipova (Ed.): ICADS 2021, AISC 1352, pp. 45–57, 2021.
https://doi.org/10.1007/978-3-030-71782-7_5

new serious challenges for humanity, associated with the transformation of business entities, the acquisition of small companies, the emergence of megastructures - huge international and global corporations and more.

The fourth industrial revolution changed the format of the world and the way of life.

The cosmic speed of the introduction of digital technologies in the activities of mankind leads to changes in the composition and quality of responsibilities and functions in the professions of accountant-analyst, financial analyst, auditor, financial manager and many other professions. At the same time, some professions disappear and others appear.

Over time, the categorical apparatus of the sciences changes, new terms appear, and existing ones are improved. The categorical apparatus of financial analysis is changing and improving. Financial analysis is a science that exists on the basis of numerous calculations. Therefore, its theory, methodology, technology and organization are significantly influenced by the fourth industrial revolution and digital computer technology.

Such changes also apply to the concept of liquidity of the balance sheet of the enterprise.

The analysis of the financial condition of the enterprise according to the balance sheet was considered by many scientists.

Thus, the issue of construction of financial statements (including the balance sheet), the use of its data in the assessment (analysis) of the financial condition of the enterprise are considered by scientists from different countries: 1) USA: L.A. Bernstain (2002) [1], T.P. Carlin (2001) [2], A.R. McMeen (2001) [2], Conrad Karlberg (2008, 2015, 2016, 2019) [3–6], R.C. Higgins (2007) [7], Charles H. Gibson (2009, 2013) [8, 9], Lyn M. Fraser, Aileen Ormiston (2010, 2013, 2016) [10], Walter Aerts, Peter Walton (2013, 2017)) [11, 12], Joanne V. Flood (2015) [13]; Mihir A. Desai (2019) [14]; 2) France: B. Kolass (1997) [15], Jacques Richard [16]; Krishna G. Palepu, Paul M. Healy (2013) [17]; 3) British scientists (London): McKenzie Wendy (2003, 2006) [18], Thomas Padberg (2017) [19], D. Stone (1998), K. Hitcheeng (1998) [20]; P. Atrill (2007, 2015) [21, 22], E. McLaney (2007, 2015, 2017) [21–23]; 4) Ireland: Walsh Ciaran (2001) [24]; 5) Latvia: V. Paupa. (2008) R. Sneidere (2008) [25]; 6) Iraq: Saoud Chayed Mashkoor Alamry (2019) [26]; 7) India: T.S. Grewa H. S. Grewal G. S. Grewa, R. K. Khosla (2019) [27]; 10) Russia: А.Д. Sheremet, E. A. Kozeltseva (2020) [28]; Vit. V. Kovalev, V.V. Kovalev (2015, 2016) [29, 30]; O.V. Efimova (2014) [31–33]; O. Rozhnova (2019) [32, 33]; S. Grishkina, V. Sidneva, Y. Shcherbinina, Dubinina (2019) [34]; 11) Belarus: G.V. Savitskaya (2017, 2013) [35, 36] and others.

However, many issues are unresolved and debatable. The balance sheet is a picture of the financial condition of the enterprise. The concept of balance sheet liquidity is a rather debatable issue. Accordingly, the methodology and organization of the balance sheet liquidity analysis deserve serious discussion and solution. The urgency of these issues is growing in the context of globalization of the economy against the background of the fourth industrial revolution, which accelerates the development of digital technologies in the methodology and organization of financial analysis at the micro and macro levels.

The purpose of the article is to develop proposals for improving the concept and analysis of balance sheet liquidity in modern financial analysis of the digital age.

The objectives of the article are: research of existing definitions and disclosure of debatable issues on the concept of balance sheet liquidity, development of proposals to improve the concept of balance sheet liquidity, proposals to improve the classification of groups of assets and liabilities to analyze the liquidity of the balance sheet and enterprise; giving an example of the implementation of the proposals on the concept and analysis of the balance sheet liquidity on the example of State Enterprise of the Service Air Traffic of Ukraine.

2 Research Method

The study is used: 1) general scientific methods: historical and logical approach, induction, deduction, analysis and synthesis; 2) special techniques: balance, reporting, spreadsheet, generalization.

3 Research Results

Prominent world scientist-analyst A.D. Sheremet (Russia) notes that "the need to analyze the liquidity of the balance sheet arises in market conditions due to increasing financial constraints and the need to assess the creditworthiness of the organization" [28, p. 124]. By definition, A.D. Sheremet: "Liquidity of the balance sheet is defined as the degree of coverage of liabilities of the organization with its assets, the term of conversion into cash corresponds to the maturity of liabilities" [28, p. 124]. Similar definitions are given by analysts G.V Savitskay (Belarus) [35, 36]. S.I. Krylov (Russia) points out that "along with the concept of liquidity of the organization there is the concept of liquidity of its balance sheet. In relation to the balance sheet of the organization, liquidity is the speed of sale of the organization's assets in order to convert them into cash to cover short-term liabilities" [37, pp. 42–43]. L.I. Ivanova and A.S. Bobyleva (Russia) included in the book "Analysis of financial statements" intact section "Analysis of the liquidity of the balance sheet and creditworthiness of the organization" [38, pp. 113–149]. S.V. Dybal, M.A. Dybal (Russia) in the book "Financial analysis: theory and practice" contains a paragraph "Grouping of balance sheet items to assess the liquidity of the balance sheet" [39, pp. 111–117]. V.I. Barilenko, S.I. Kuznetsov, L.K. Plotnikova, O.V. Kairo emphasizes that "the purpose of the liquidity balance analysis is to assess the ability of the company without any disruption of the normal business cycle to make urgent payments at the expense of the corresponding urgency of income from the sale of values [40, p. 27].

However, many analysts do not support the existence in the categorical apparatus of financial analysis and analysis of financial statements of the concept of balance sheet liquidity. Yes, Vit. V. Kovalev and V.V. Kovalev (Russia) points out that... Mixing the concepts of "liquidity" and "solvency" can be seen even in the works of famous scientists and analysts. Thus, the founder of accounting analysis N.R. Weizman wrote "An enterprise whose balance sheet is liquid is always able to pay its obligations, in other words, it is in readiness" [N.R. Weizman 1924, pp. 162 [41]]. The absolute failure of this interpretation is obvious [42, p. 498].

Moreover, today among scientists around the world there is no consensus on the concepts of solvency of the enterprise, liquidity of the enterprise, liquidity of assets, balance sheet liquidity.

Thus, the concept of balance sheet liquidity, its right to exist and improve in modern financial analysis is quite urgent. It becomes especially relevant in the era of digital technologies and digitalization of financial analysis procedures.

It should be noted that in the economic literature of some countries, as a rule, the concept of balance sheet liquidity is not used at all. What can not be said about Russia, Ukraine, Belarus and a number of other post-Soviet countries. In these countries, this concept is widely used in the theory and practice of business analysis, financial analysis, financial management, financial reporting analysis and others. This is due to the fact that in some countries the terms "Balance Sheet" and/or "Statement of Financial Position" are used for financial reporting, while in other countries only the term "Statement of Financial Position" is used but the term "Balance Sheet" is not used in the financial statements. The second option is typical of countries that apply International Financial Reporting Standards (IFRS), International Accounting Standards (IAS) [43] and Generally Accepted Accounting Principles (US GAAP). That's why there is logical that the term "balance sheet liquidity" is not used in these countries.

There are many controversial issues among the existing concepts of balance sheet liquidity. For example, the degree of coverage may vary and the coverage of certain groups of liabilities by certain assets will not always indicate the liquidity of the balance sheet. Non-current assets cannot be classified as such assets. The concepts of enterprise solvency and balance sheet liquidity are often confused. Some definitions are rather superficial and inaccurate, while others do not fully reflect the essence of this concept.

It is considered appropriate to clarify and provide the following concept of this economic category: balance sheet liquidity is the ratio of certain groups of assets depending on the degree of their transformation into cash to the corresponding certain groups of liabilities depending on the maturity of debts, if these terms coincide with the terms of conversion of certain groups of assets in cash, namely - current assets must be greater than or equal to long-term and current liabilities, and non-current assets must be less than or equal to equity.

Justification of this point of view:

- firstly, when assessing the liquidity of the balance sheet is a comparison of certain groups of assets depending on the degree of their transformation into cash and the corresponding certain groups of liabilities depending on the maturity of debts. Therefore, the term "degree of coverage" is replaced by the term "ratio" with the subsequent disclosure of the definition of liquidity of the balance sheet and it will more accurately reflect the essence of this concept and the analytical process (technology) of liquidity assessment.
- secondly, the degree of coverage of liabilities by assets (certain groups may be covered and some may not, there may be different combinations of them) will depend on the type of liquidity of the balance sheet, but this is not the essence of its liquidity as a category of financial analysis. The degree of coverage of certain types of debt by certain assets determines the type of liquidity of the balance sheet, ie acts as a classification feature;
- thirdly, in the definition of liquidity of the balance sheet we must first talk about groups of assets that can be converted into money, and then about certain groups of liabilities that must be repaid, so this concept first focuses on this;

- fourthly, the comparison of certain groups of assets and liabilities with each other is based on a time criterion, namely - the term of transformation of certain groups assets must coincide with the maturity of liabilities, but this has been stated in previous personal research;
- fifthly, the definition reflects the main condition of liquidity of the balance sheet - this is when all current assets are greater than long-term and current liabilities together, and all non-current assets are less than the total amount of equity;
- sixthly, the definition of the minimum liquidity condition is the equality of current assets with long-term and current liabilities and the equality of non-current assets with equity.

Thus, the proposed definition of balance sheet liquidity more accurately reflects the essence of this concept and reveals at the same time the technology of its analysis and determination of the type of liquidity.

The next controversial issue is the grouping of assets and liabilities to analyze the liquidity of the balance sheet and the enterprise. The current classification also deserves further improvement.

In 2013 in Ukraine there were further changes in the legal and regulatory framework for accounting. The structure and content of the balance sheet (statement of financial position) is subject to significant changes. This is due to the requirements of approximation to IFRS [43] and compliance with the requirements of EU Directive № 2013/34/EU [44], accelerating the process of European integration, digital economy, globalization of economic processes and harmonization of financial reporting. In order to bring the norms of national legislation to the provisions of EU Directive № 2013/34/EU [44] in 2017, the Law of Ukraine "On Amendments to the Law of Ukraine" On Accounting and Financial Reporting in Ukraine [45]. The role of the financial statements, including the balance sheet (statement of financial position), has significantly increased. The growing influence of the fourth industrial revolution caused new requirements for the implementation of financial and analytical calculations. All this led to further research in the methodology, methodology and organization of liquidity analysis of the balance sheet.

Today the issue needs further clarification. The proposals concern the delimitation of the group of non-current assets and equity components (Table 1).

The total amount of non-current assets should be divided into non-current financial and non-current non-financial assets, as these groups have different specifics of their origin and will differ in the level of liquidity. Similarly, the articles of equity of the enterprise have different origins. Unpaid and withdrawn capital are not included in the currency of the first section of the liabilities of the Ukrainian balance sheet. Therefore, it is also advisable to divide them into two separate groups. This detail of the grouping of assets and liabilities will help to better understand the level of liquidity of the balance sheet and the reasons for its changes.

It was found that the type of absolutely liquid balance will depend not only on the degree of compliance of the amount of cash (subject to conversion of certain groups of assets into money) to the size of certain groups of debt, but also on the type of activity and status of entities.

Table 1. Proposals for grouping balance sheet assets and liabilities for liquidity analysis.

Assets		Liabilities	
Group name	Group composition	Group name	Group composition
1. Absolutely liquid assets (A_1)	Cash	1. Most-term liabilities (P_1)	Current liabilities on settlements, other current liabilities
2. Highly liquid assets (A_2)	Cash equivalents, current financial investments	2. Term liabilities (P_2)	Current accounts payable for goods, works, services and promissory notes issued
3. Accelerated realizable assets (A_3)	Current receivables, short-term promissory notes received, other current assets	3. Short-term liabilities (P_3)	Short-term bank credits and current payables under long-term liabilities
4. Quickly realizable assets (A_4)	Finished goods, goods, non-current assets and disposal groups intended for sale	4. Long-term liabilities (P_4)	Long-term bank credits, long-term financial liabilities, deferred tax liabilities
5. Slowly realizable assets (A_5)	Inventories, current biological assets, unfinished production	5. Other long-term liabilities (P_5)	Other long-term liabilities
6. Financial assets that are difficult to be realized (A_6)	Non-current financial assets	6. Unpaid and withdrawn capital (P_6)	Unpaid capital, withdrawn capital
7. Non-financial assets that are difficult to be realized (A_7)	Non-current non-financial assets	7. Permanent liabilities (P_7)	Equity (registered (share) capital, capital in revaluations, additional capital, reserve capital retained earnings (uncovered loss)

(Source: author's elaboration).

It is advisable to mathematically describe the types of liquid balance sheet, illiquid balance sheet, balance sheet liquidity limits, absolute balance sheet liquidity (Table 2).

New is the establishment of the formula of absolute liquidity of the balance sheet for those enterprises that do not have stocks. That is, the type of absolute liquidity of the balance sheet will be influenced by the type of activity of the enterprise.

Thus, the type of balance sheet liquidity will still depend on the type of activity of economic entities, which is not taken into account in the existing methodology of

Table 2. Types of liquid balance sheet, illiquid balance sheet, balance sheet liquidity limits, absolute balance sheet liquidity.

Types of liquidity (illiquidity) of the balance sheet	Terms	Types of enterprise activities	Mathematical description
Type of liquid balance	Current assets must be greater than or equal to long-term and current liabilities, and non-current assets must be less than or equal to equity	All activities	$A1 + A2 + A3 + A4 + A5$ $\geq$ $P_1 + P_2 + P_3 + P_4 + P_5$ $A_6 + A_7 \leq P_6 + P_7$
Type of illiquid balance	Current assets are less than current and non-current liabilities, and non-current assets more equity	All activities	$A1 + A2 + A3 + A4 + A5$ $<$ $< P_1 + P_2 + P_3 + P_4 + P_5$ $A_6 + A_7 > P_6 + P_7$
Limit (critical point) of balance sheet liquidity	Non-current assets exceeds the value of equity and, accordingly, the value of long-term and current liabilities exceeds the value of current assets	All activities	$A1 + A2 + A3 + A4 + A5$ $=$ $P_1 + P_2 + P_3 + P_4 + P_5$ $A_6 + A_7 = P_6 + P_7$
Absolutely liquid type of balance sheet	Current assets must be greater than or equal to long-term and current liabilities, and non-current assets must be less to equity	a) production; b) construction; c) trade; d) both production (construction) and trade; e) both production and construction; g) parallel provision of services simultaneously with the above types of activities	$A_1 + A_2 \geq P_1 + P_2$ $A_3 \geq P_3$ $A_4 + A_5 \geq P_4 + P_5$ $A_6 + A_7 < P_6 + P_7$
Absolutely liquid type of balance sheet	Current assets must be greater than or equal to long-term and current liabilities, inventories are greater than or equal to zero, non-current assets must be less to equity	Provision of services	$A_1 + A_2 \geq P_1 + P_2$ $A_3 \geq P_3 + P_4 + P_5$ $A_4 + A_5 \geq 0$ $A_6 + A_7 < P_6 + P_7$

(Source: author's elaboration).

balance sheet liquidity analysis. According to the current method of analysis, no matter how well a given economic entity works, the balance will never be completely liquid. In such a situation, the quick liquidity ratio will always be equal to the total liquidity ratio, because stocks are zero, so it will always negatively characterize the liquidity of such entities: if the total liquidity ratio meets the established limits, the quick liquidity ratio will be too high. Meets the established limits, the total liquidity ratio will be too low.

According to the proposed grouping of assets according to the degree of decrease in their liquidity and classification of liabilities according to the degree of growth of their maturity, the type of absolute liquidity of the balance sheet has the following ratio (for enterprises with no inventories):

$$A_1 + A_2 \geq P_1 + P_2$$

$$A_3 \geq P_3 + P_4 + P_5$$
$$A_4 + A_5 \geq 0$$
$$A_6 + A_7 < P_6 + P_7$$

Thus, the method of conducting liquidity analysis depends on what the entity is doing.

We will show an example of the implementation of the proposals on the theory and methodology of balance sheet liquidity analysis on the example State Enterprise of the Service Air Traffics of Ukraine [46].

It should be noted that the work of modern scientists is devoted to the study of airlines. Thus, Rigas Doganis reveals in some detail the issues of economics and airline marketing [47]. Padberg Thomas covers the analysis of airline financial statements [48, 49].

The grouping of assets and liabilities of the balance sheet of State Enterprise of the Service Air Traffic of Ukraine for 2019 is shown in Table 3.

We compare current assets with current and long-term liabilities, and non-current assets with equity.

At the beginning of 2019:

Current assets: 1 639 107 + 788 458 + 213 + 45 511 = 2 473 289 (thousand UAH).

Current liabilities and long-term liabilities:

194 703 + 50 569 + 174 316 + 65 008 = 484 569 (thousand UAH).

Non-current assets: 17 681 + 4 388 171 = 4 405 852 (thousand UAH).

Equity: 6 394 545 thousand UAH.

2 473 289 thousand UAH > 484 569 thousand UAH.

4 405 852 thousand UAH < 6 394 545 thousand UAH.

Thus, at the beginning of 2019 the balance was liquid.

At the end of 2019:

Current assets: 494 224 + 692 068 + 232 + 35 743 = 1 222 267 (thousand UAH).

Current liabilities and long-term liabilities:

253 508 + 49 528 + 24 312 + 134 850 + 2 180 = 464 378 (thousand UAH).

Non-current assets: 209 627 + 4 380 650 = 4 590 277 (thousand UAH).

Equity: UAH 5 348 166 thousand.

1 222 267 thousand UAH > 464 378 thousand UAH.

4 590 277 thousand UAH < 5 348 166 thousand UAH.

Thus, at the end of 2019, the balance is also liquid.

Both at the beginning and at the end of the year, the ratio was observed, which indicates the liquidity of the balance sheet:

$$A_1 + A_2 + A_3 + A_4 + A_5 \geq P_1 + P_2 + P_3 + P_4 + P_5$$
$$A_6 + A_7 \leq P_6 + P_7$$

Let's determine whether the balance is absolutely liquid.

At the beginning of 2019 :

1 639 107 > 245 272;   788 458 > 82444;

2 834 182 > 0;   4 405 852 < 6 394 545

Table 3. Analysis of the liquidity of the balance of State Enterprise of the Service Air Traffic of Ukraine for 2019, thousand UAH.

Assets	At the beginning of the year	At the end of the year	Liabilities	At the beginning of the year	At the end of the year	Surplus or lack of payments	
						At the beginning of the year	At the end of the year
Absolutely liquid assets (A_1)	1 639 107	494 224	Most-term liabilities (P_1)	194 703	253 508	*	*
Highly liquid assets (A_2)	-	–	Term liabilities (P_2)	50 569	49 528	*	*
Total A_1 + A_2	1 639 107	494 224	Total P_1 + P_2	245 272	303 036	1 393 835	191 188
*	*	*	Short-term liabilities (P_3)	–	24 312	*	*
*	*	*	Long-term liabilities (P_4)	174 316	134 850	*	*
*	*	*	Other long-term liabilities (P_5)	65 008	2 180	*	*
Accelerated realizable assets (A_3)	788 458	692 068	Total P_3 + P_4 + P_5	82 444	161 342	706 014	530 726
Quickly realizable assets (A_4)	213	232	*	*	*	*	*
Slowly realizable assets (A_5)	45 511	35 743	*	*	*	*	*
Total A_4 + A_5	2 834 182	728 043	*	0	0	2 834 182	728 043

(*continued*)

Table 3. (*continued*)

Assets	At the beginning of the year	At the end of the year	Liabilities	At the beginning of the year	At the end of the year	Surplus or lack of payments	
						At the beginning of the year	At the end of the year
Financial assets that are difficult to be realized (A_6)	17 681	209 627	Unpaid and withdrawn capital (P_6)	-	-	*	*
Non-financial assets that are difficult to be realized (A_7)	4 388 171	4 380 650	Permanent liabilities (P_7)	6 394 545	5 348 166	*	*
Total A_6 + A_7	4 405 852	4 590 277	Total P_6 + P_7	6 394 545	5 348 166	−1988 693	− 757 889

(Source: Financial statement Audit Report from the official website at https://uksatse.ua/index.php?act=Part&CODE=383# [46]).

At the end of 2019 :

$$494\,224 > 303\,036; \quad 692\,068 > 161\,342;$$
$$28\,043 > 0; \quad 4\,590277 < 5\,348\,166$$

The balance of State Enterprise of the Service Air Traffic of Ukraine both at the beginning and at the end for 2019 is absolutely liquid, as the inequality is fulfilled:

$$A_1 + A_2 \geq P_1 + P_2; A_3 \geq P_3 + P_4 + P_5 ; A_4 + A_5 \geq 0; A_6 + A_7 < P_6 + P_7.$$

The activity of the airport in terms of liquidity balance can be assessed positively.

4   Conclusion

Thus, in the process of studying the liquidity of the balance sheet came to the following conclusions:

1) a new concept of balance sheet liquidity was provided, namely: balance sheet liquidity is the ratio of certain groups of assets depending on the degree of their transformation into cash to the corresponding certain groups of liabilities depending on the maturity of debts, if these terms coincide with the terms of conversion of certain

groups of assets in cash, namely - current assets must be greater than or equal to long-term and current liabilities, and non-current assets must be less than or equal to equity;

2) the grouping of assets and liabilities to assess the level of the balance sheet liquidity of (absolutely liquid or not) has been clarified;

3) the method of determining the absolutely liquid type of balance has been improved. For the first time its dependence on a kind of activity of the enterprise is established;

4) the proposed forms of tables will promote the unification of calculations, increase the clarity and informativeness of analytical conclusions on the liquidity of the balance sheet in terms of digital technologies.

Scientific Novelty: The developed methodology of balance liquidity analysis and its mathematical description will be easily carried out in the conditions of digital technologies and further implementation of artificial intelligence of companies.

Theoretical and Practical Significance of the Research: The developed methodology of balance liquidity analysis makes it possible to monitor the effectiveness of the company's asset and liability management strategies, assessment of prospects for the development of its activities, forecast the onset of liquidity crisis and insolvency. The proposed forms of analytical tables will contribute to the unification of the management document of the «Report on liquidity and solvency», it will provide comparability of analytical information about their changes in decision-making by interested external and internal users of financial information.

Socio-Economic Effect that Arises as a Result of the Implementation of Scientific Results: The obtained scientific results allow to increase the information culture of documenting changes of balance liquidity analysis in the process of managing the company, to increase clarity and informativeness of analytical conclusions about liquidity, significantly reduce the complexity of analytical work of management staff of the business entity, especially in digital technology. The results of the research are useful for users in order to make the right management decisions in the digital age.

Prospects for Further Researches: The issues of further development of the composition and methods of determining the absolute and relative liquidity of the enterprise will be the subject of further research.

References

1. Bernstain, L.A.: Analiz finansovoy otchyotnosti: teoriya, praktika i interpretatsiya (Financial Statement Analysis: Theory, Practice and Interpretation). Financy i Statistika, Moscow (2002)
2. Karlin, T.R., Makmin, A.R.: Analiz finansovykh otchyotov (na osnove GAAP) (Financial Report Analysis (Based on GAAP)). INFRA-M, Moscow (2001)
3. Carlberg, C.: Analiz finansovoy otchyotnosti s ispolzovaniyem Excel (Financial Statement Analysis Using Excel). Dialektika, Kiev (2019)
4. Carlberg, C.: Regressionny analiz v Microsoft Excel (Regression Analysis Microsoft Excel). Wiliams, Moscow (2016)

5. Carlberg, C.: Biznes-analiz s pomoshchyu Microsoft Excel (+ CD-ROM) (Business Analysis with Microsoft Excel). Wiliams, Moscow (2008)
6. Carlberg, C.: Biznes-analiz s ispolzovaniyem Excel (Business Analysis: Microsoft Excel 2010). Wiliams, Moscow (2015)
7. Higgins, R.S.: Finansovy analiz: instrument dlya prinyatiya biznes-resheniy (Financial Analysis: a Tool for Business Decision Making). OOO "I.D. Wiliams", Moscow (2007)
8. Gibson, C.H.: Financial Statement Analysis, 13th edn. Cengage Leaning EMEA, Boston (2013)
9. Gibson, C.H.: Finance analysis. In: Financial Reporting & Analysis. Eleventh edition. University of Toledo, Toledo, USA (2009)
10. Fraser, L.M., Ormiston, A.: Understanding of Financial Statements, 11th edn. Pearson, New York (2016)
11. Aerts, W., Watlon, P.: Global Financial Accounting and Reporting: Principles and Analysis, 3rd edn. Cengage Leaning EMEA, Boston (2013)
12. Aerts, W., Watlon, P.: Global Financial Accounting and Reporting: Principles and Analysis, 4th edn. Cengage leaning EMEA, Boston (2017)
13. Flood, J.M.: Wiley GAAP 2015: Interpretation and Application of Generally Accepted Accounting Principles. Wiley, New York (2015)
14. Desai, M.A.: How Finance Works: The HBR Guide to Thinking Smart About the Numbers. Harvard Business Review Press, Boston (2019)
15. Kolass, B.: Upravleniye finansovoy deyatelnostyu predpriyatiya. Problemy, kontseptsii i metody (Managing the Financial Activity of an Enterprise. Problems, Conceptions and Methods). Finansy, UNITI, Moscow (1997)
16. Zhak, R.: Audit i analiz khozyaystvennoy deyatel'nosti predpriyatiy (Audit and Analysis of Economic Activities of Enterprises). Audit, UNITY, Moscow (1997)
17. Palepu, K.G., Healy, P.M.: Business Analysis Valuation Using Financial Statements, 5th edn. Cengage Leaning EMEA, Boston (2013)
18. Wendy, M.W.: Ispolzovaniye i interpretatsiya finansovoy otchyotnosti (Financial Statement Usage and Interpretation). Balans-Klub, Biznes Buks, Dnepropetrovsk (2006)
19. Padberg, T.: How to Analyse Bank Financial Statements: A Concise Practical Guide for Analysts and Investors. Harriman House, UK (2017)
20. Stone, D., Hitching, К.: Buchgalterskiy otchyot i finansovy analiz (Accounting and Financial Analysis). "Sirin", "Biznes-Inform", Moscow (1998)
21. Atrill, P., McLaney, E.: Accounting and Finance: An Introduction, 8th edn. Pearson, London (2015)
22. Atrill, P., McLaney, E.: Accounting and Finance for Non-Specialists, 9th edn. Pearson Education Limited, London (2015)
23. Eddie, M.: Business Finance: Theory and Practice, 11th edn. Pearson, London (2017)
24. Ciaran, W.: Kliuchovi finansovi pokaznyky. Analiz ta upravlinnia rozvytkom pidpryiemstva (Key Financial Indexes. Analysis and Management of Enterprise Development). Vseuvyto; Naukova Dumka, Kyiv (2001)
25. Paupa, V., Sneidere, R.: Uzdevumu krajums finansu analize. Baltimoras konsultaciju centres, Riga (2008)
26. Alamry, S.C.M.: Overview of financial statements analysis. In: Analysis of Financial Statements, Al-Alalamia for printing and Designs Sammawa, Iraq (2019)
27. Grewal, T.S., Grewal, H.S, Grewal, G.S., Khosla, R.K.: Analysis of Financial Statements. Sultan Chand & Sons Private Limited, New Delhi (2019)
28. Sheremet, A.D., Kotheltceva, E.A.: Finansovy analiz (Financial Analysis). Ekonomicheskiy fakul'tet MGU imeni M.V. Lomonosova, Moscow (2020)
29. Kovalev, V.V., Kovalev, V.V.: Analiz balansa, ili kak ponimat' balans (Balance Analysis, or How to Understand Balance). Prospekt, Moscow (2015)

30. Kovalev, V.V., Kovalev, V.V.: Analiz balansa (Analysis of Balance). Prospekt, Moscow (2016)
31. Yefimova, O.V.: Finansovyy analiz: sovremennyy instrumentariy dlya prinyatiya ekonomich-eskikh resheniy (Financial Analysis: Modern Tools for Making Economic). Izd-vo Omega-L, Moscow (2014)
32. Efimova, O., Rozhnova, O., Gorodetskaya, O.: XBRL as a tool for integrating financial and non-financial reporting. In: Advances in Intelligent Systems and Computing, vol. 1114, pp. 135–147 (2020)
33. Efimova, O., Rozhnova, O.: The corporate reporting development in the digital economy. In: Advances in Intelligent Systems and Computing, vol. 850, pp. 71–80 (2019)
34. Grishkina, S., Sidneva, V., Shcherbinina, Y., Dubinina, G.: Comparability of financial report-ing under different tax regimes. In: Advances in Intelligent Systems and Computing, vol. 850, pp. 88–93 (2019)
35. Savitskaya, G.V.: Kompleksnyy analiz khozyaystvennoy deyatel'nosti predpriyatiya (Com-prehensive Analysis of the Company's Economic Activities). INFRA-M, Moscow (2017)
36. Savitskaya, G.V.: Kompleksnyy analiz khozyaystvennoy deyatel'nosti predpriyatiya (Com-prehensive analysis of the company's economic activities). INFRA-M, Moscow (2013)
37. Krylov, S.I.: Finansovyy analiz (Financial Analysis). Izd-vo Ural. un-ta, Yekaterinburg (2016)
38. Ivanova, L.I., Bobyleva, A.S.: Analiz finansovoy otchetnosti (Analysis of Financial State-ments). KNORUS, Moscow (2018)
39. Dybal, S.V., Dybal, M.A.: Finansovyy analiz: teoriya i praktika (Financial Analysis: Theory and Practice). KNORUS, Moscow (2019)
40. Barilenko, V.I., Kuznetsov, S.I., Plotnikova, L.K., Kayro, O.V.: Analiz finansovoy otchetnosti (Analysis of Financial Statements). KNORUS, Moscow (2016)
41. Veytsman, N.R.: Schetnyy analiz: Metody issledovaniya deyatel'nosti torgovogo predpriy-atiya po dannym yego bukhgalterii (Counting analysis: methods for researching the activities of a trading enterprise according to its accounting data). Moscow (1924)
42. Kovalev, V.V., Kovalev, V.V.: Analiz balansa, ili kak ponimat' balans (Balance analysis, or how to understand balance). Prospekt, Moscow (2012)
43. Mizhnarodni standarty finansovoi zvitnosti (International Financial Reporting Standards). https://zakon.rada.gov.ua/laws/main/929_010. Accessed 15 July 2020
44. Directive on the annual financial statements, consolidated financial statements and related reports of certain types of undertakings of the European parliament and of the council of 26 June 2013 N 2013/34/EU. https://eur-lex.europa.eu/legal-content/EN/ALL/?uri=CELEX% 3A32013L0034. Accessed 09 Feb 2020
45. Ukrainy, Z.: Pro vnesennia zmin do Zakonu Ukrainy "Pro buchgalterskyi oblik ta finansovu zvitnist v Ukraini" № 2164-VIII vid 5.10.2017 r. (The Law of Ukraine "On Changes to the Law of Ukraine "On Accounting and Financial Reporting in Ukraine") № 2164-VIII of 5 10 2017. https://zakon.rada.gov.ua/laws/show/2164-19. Accessed 09 Feb 2020
46. Zvit nezalezhnoho audytora. Financial statement Audit Report. https://uksatse.ua/doc/aud_ report_2019.pdf. https://uksatse.ua/index.php?act=Part&CODE=383#. Accessed 21 Dec 2020
47. Doganis, R.: Flying Off Course: Airline Economics and Marketing. Routledge, UK (2019)
48. Thomas, P.: Financial Statement Analysis of Airlines. Paphos, Cyprus (2017)
49. Thomas, P.: Bilanzanalyse von Fluggesellschaften. Independently published (2017)

The Prosperity Pyramid as a Consequence of the Low Level of Financial Culture of the Population

Irina Korostelkina(✉) ⓘ, Oksana Fokina ⓘ, Natalia Varaksa ⓘ,
and Marina Vasilyeva ⓘ

Orel State University, Naugorskoe Highway, 40, 302020 Orel, Russian Federation
cakyra_04@mail.ru

Abstract. In 2019, the concept of a financial pyramid turned 100 years old, but it still remains a phenomenon in the system of economic, social and legal relations. Today, as before, the «survivability» of financial pyramids is due to the thirst for profit, but in the end, only those who have a higher level of financial literacy can quench this thirst. At the same time, financial literacy is only an element, one of the conditions for the formation of financial culture. Therefore, it cannot be argued that a high level of financial literacy will always reflect the financial development and high financial culture of society. The purpose of this article is to study the phenomenon of the financial pyramid, the foundations of its prosperity, namely, the low financial culture of society with the application of the best international management practices and evaluate the measures taken by States to reduce them. General and specific research methods, methods and tools for graphical interpretation, comparative analysis, and related changes are used as methodological tools for this study. The article substantiates the direct relationship between financial literacy, financial culture and financial pyramids, assesses the level of financial literacy of the population by country, provides an overview of the types and types of financial pyramids, and highlights public policy measures to reduce fraudulent schemes and increase the level of financial literacy.

Keywords: Financial pyramid · Responsibility · Trust · Financial culture · Financial literacy

JEL classification: G18 · G28 · К42

1 Introduction

Most people in the world consider financial stability to be one of their most important goals and make efforts to achieve this state of their personal budget. There are many ways to achieve financial stability, but, in General, their essence is to increase the individual's income and exceed the total amount of income over expenses. The reliability of financial instruments that generate household income also plays an important role. All financial

© The Author(s), under exclusive license to Springer Nature Switzerland AG 2021
T. Antipova (Ed.): ICADS 2021, AISC 1352, pp. 58–69, 2021.
https://doi.org/10.1007/978-3-030-71782-7_6

instruments differ in the level of profitability, liquidity, the amount of start-up capital and the level of security of funds placement.

Capital accumulation is a complex process that requires constant cash injections, control, and time. But not everyone knows this. In most developed countries, where this type of activity is common, people pay enough attention to their financial literacy. This makes the investment process more efficient and secure. The share of fraud in developed countries is quite low due to the fact that investors' expectations coincide with the real picture. Unfortunately, the same cannot be said for less developed countries and countries with low financial literacy. In accordance with international practice, financial literacy is defined as the ability of individuals to manage their own funds and make effective decisions in the financial sphere. The organization for economic cooperation and development (OECD) links financial literacy to financial education, i.e. a system of knowledge about financial products, skills in determining financial risks, and skills to improve personal well-being [1]. Financial authorities in the UK similarly assess the essence of an individual's financial competence: to live within their means, monitor the state of finances, plan future income and expenses, especially retirement, choose the right financial products and understand financial issues [2].

Silina S.N. and StupinV.Yu. in the framework of their research [3] define financial literacy of the population as one of the main components of financial culture, including being a quantitative evaluation indicator, that is, a measurable indicator. Financial literacy of Russians is regularly measured according to the OECD methodology, using the financial literacy index, which consists of private indices of knowledge, financial behavior (skills) and attitude to Finance (attitudes) (Fig. 1). Based on the financial literacy index, an annual rating of Russian regions in terms of financial literacy is compiled. First, the index of financial literacy was designed in the Analytical center NAFI the framework of the project «Enhancing financial literacy and developing financial education in the Russian Federation» in 2018 [4].

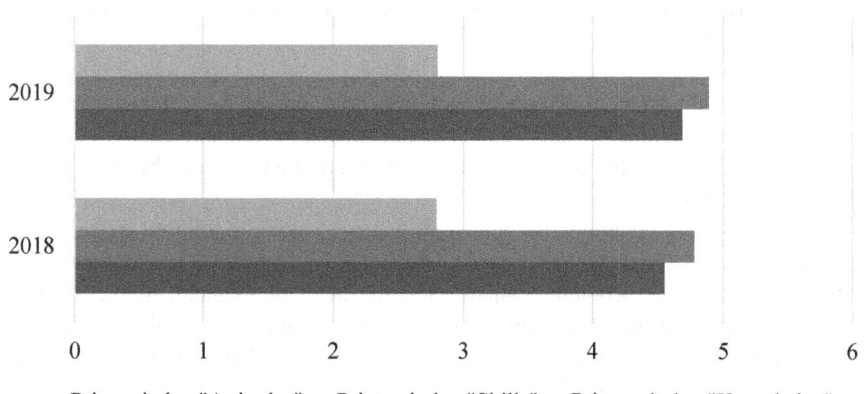

■ Private index "Attitudes"   ■ Private index "Skills"   ■ Private index "Knowledge"

Fig. 1. Private indexes «Knowledge», «Skills», «Attitudes» of the integral indicator of financial literacy (compiled according to [5])

Overall financial literacy, estimated using this methodology for 2018, in the Russian Federation is 12.2 points [5]. Taking a value from 1 to 21 points, the Russian Federation is an average indicator, but in comparison with the CIS countries (Belarus – 13.4 points, Kazakhstan – 13.1 points) it has a lower value. For 2019, this indicator was 12.37 points. According to this indicator, Russia ranks 9th among the G20 countries (Fig. 2).

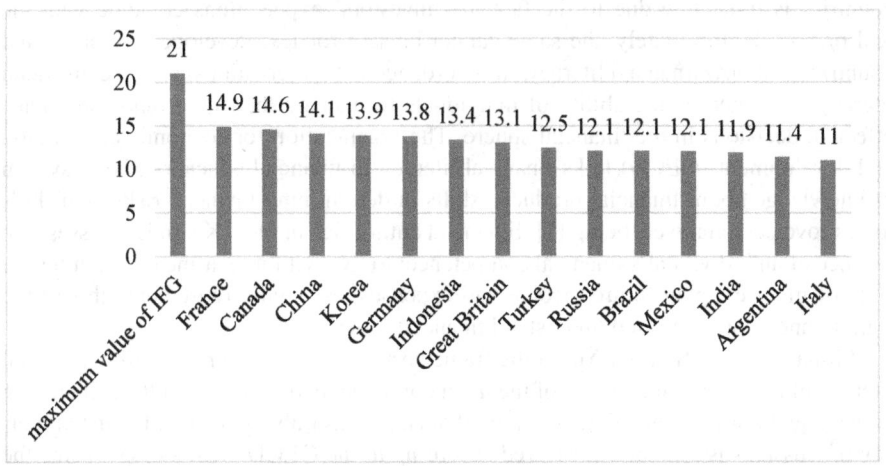

Fig. 2. Dynamics of the financial literacy index for G20 countries (compiled according to [4])

According to a study by the NAFI Analytical center [4], the main problems of insufficient financial literacy are:

– lack of understanding of the relationship between risk and profitability;
– lack of information security, as well as involvement in fraudulent activities with Bank cards and accounts.

According to the study, the financial stability of the family is estimated at 45% [6]. Such studies and their analysis show that less than 30% of respondents can find out about the pyramid scheme and fraudulent activities, and more than 34% can become a victim of financial pyramids.

Thus, the modern economy, the information space, including digital space, creates new challenges and threats to the security of households [7]. Scientists and practitioners are trying to solve the problem of increasing the level of financial literacy through mass educational and training events in the banking sector, in the field of insurance, investment, taxation, personal Finance and cyber security. Therefore, the study of the competence approach to financial literacy, objective prerequisites for the emergence of financial pyramids and the involvement of society in them is important and necessary at the present stage of economic development and ensuring not only financial, but also information security of the individual.

The purpose of the study is to study fraudulent schemes operating in the world and in Russia, as well as the experience of combating low levels of financial literacy, on the basis of which the types and types of financial pyramids, their structure and functioning features are determined. The dynamics of countering financial pyramids from the legal side is revealed, as well as the need for society in a financially literate population is justified.

2 Materials and Methods

The presented research is based on General scientific methodological tools. The relevance of the topic, the results and conclusions were justified using the methods of scientific search, classification, comparison and generalization of information. In the course of the research, methods of analysis, comparison, analogy, deduction, systematization and grouping of information were used to formulate conclusions based on the results of the study. The categorical apparatus of financial culture and the theoretical overview of financial pyramids were studied using the tools of graphical interpretation and comparative analysis. The study of the main indicators that characterize financial culture and financial literacy was conducted on the basis of a competence-based approach and presented using comparison methods. Types and types of financial pyramids are grouped using methods of mental modeling, generalization, and the principle of mutual communication. Public policy measures to reduce fraudulent schemes and improve financial literacy are disclosed through a descriptive method and content analysis.

3 Discussion

Applying a competency-based approach to the study of the concepts of «financial literacy» and «financial culture» is not only justified, but also necessary. However, it seems to us that financial literacy is the Foundation of knowledge and skills for the development of financial culture, and skills and attitudes based on knowledge and basic skills form a person's financial behavior, which is based on individual psychological characteristics of the individual.

The competency - based approach to financial literacy defines the knowledge and understanding formed through little experience that is applied when choosing and using a limited number of financial instruments in the process of managing personal financial resources. Lack of knowledge makes it impossible to acquire and develop even basic skills. Therefore, it is financial literacy that determines the formation of a citizen's personal financial attitudes and a particular type of financial behavior (savings or irrational). At the same time, all types of behavior and types of financial decisions made by citizens are covered by the financial culture (Fig. 3).

In this regard, it is very important to systematically and methodically improve the personal financial culture of each person in order to see the positive development of the financial culture of society as a whole. However, many people confuse cause and effect in the described concepts and actions, believing that the financial culture is formed due to a high level of labor income and savings. That is why, against the background of a low financial culture, financial pyramids attract high rates and promises of easy capital increase. Considering financial culture as a system of moral norms and rules of economic activity, we should also talk about the low culture of the organizers of financial pyramids with a sufficient level of financial literacy.

As a result, we observe a low level of financial culture in society, on the one hand, due to financial illiteracy, and on the other hand, due to the violation of social and moral values, which in both cases does not allow us to develop an adequate model of financial behavior.

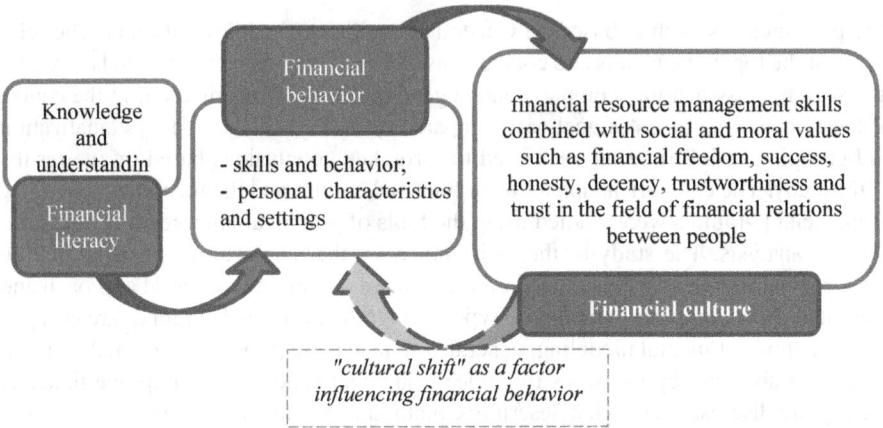

Fig. 3. The definition of a financial culture on the basis of competence approach (compiled by the authors)

On the example of the Russian Federation, it can be seen that the majority of the population in the early 90's of the XX century did not have the skills of rational distribution of funds and financial planning, did not know what to do with the vouchers that they were issued by the state. Only a small percentage of people had a sufficient level of financial literacy, in the hands of which, subsequently, the main part of material resources was concentrated [8]. During the economic restructuring, the population with an unformed financial culture suffered great financial losses due to interaction with dubious organizations, financial pyramids, and inattention to the object of investment.

A pyramid scheme is a kind of fraud system, since it is a way to generate income for those who invented it. And, like any other crime, its creation is criminally punishable (article 159 of the criminal code «Fraud» and article 172.2 of the criminal code «Organization of activities to attract funds and (or other property)» when attracting funds in large and especially large amounts) or entails administrative responsibility under article 14.62 of the administrative Code of the Russian Federation «Activities to attract funds and (or) other property», if the actions do not contain a criminal offense, including for advertising financial pyramids.

Despite its wide popularity, both in Economics and law, and in the public consciousness, the phenomenon of the «financial pyramid» from the scientific and legal points of view has not yet acquired clear methodological boundaries and stable justifications. A unified approach to understanding this scientific category has not yet developed, despite the fact that in 2020 it was 100 years since the creation of a modern-style financial pyramid by an American of Italian origin Charles Ponzi [9] (the essence of the scheme was to exchange postage stamps from the United States to other European countries in order to get more postage stamps, and then sell these postage stamps, make a profit, and divide them among depositors. The pyramid's yield was 50%). However, the first financial pyramid is considered to be the pyramid of J. Law, Minister of France, who in 1717 created the «Company of the Indies» project to raise funds for the development of French-owned colonies along the Mississippi river [10].

To date, taking into account the expert opinion, there are several approaches to the essence of the financial pyramid, each of which reflects the essence of the phenomenon, giving the greatest weight to one or another of its aspects (crime, irrational behavior, etc.). Despite the diversity in the interpretation and perception of the concept of a financial pyramid, its economic meaning and essence remain unchanged – the source of income for primary investors is monetary investments from subsequent investors participating in the financial pyramid. With this in mind, the problem of this phenomenon is also indicated – the organization is recognized as a financial pyramid after the fact, when the situation has already taken on a scale, and law enforcement agencies have accepted dozens of applications from defrauded depositors.

Financial pyramids are built either on a step-by-step basis, or on the principle of accumulating all funds from the organizer (Table 1). The first prototype of the pyramid (Ponzi pyramid) was single-stage. Examples of such pyramids are the MMM S. Mavrodi (1994), the pyramid of iPhones (2011), the project of financing AIDS medicines by B. Tannenbaum (2009).

Table 1. Types of financial pyramids

View	Feature	Source
Single-level	Accumulation of funds from the organizer	The current contributions of the participants
Multilevel	The principle of network marketing	
Matrix	A more complex version of the multi-level structure	Initial payments, inflated cost of goods

Multi-level financial pyramids are currently the most widespread in the world. These include MMM S. Mavrodi (2011), binary options, Talk Fusion, which uses the principles of network marketing (2005). In the modern economy, matrix schemes are also gaining momentum, including on the Internet, since they have a long-term nature [11].

The creation and operation of financial pyramids is prohibited in many countries: in the UAE and China, the penalty is the death penalty, and in Russia - criminal liability [12]. There are also known fraudulent projects by B. Madoff (1960), Stanford International Bank by A. Stanford (2009), L&G by K. Nami (2000), YingkouDonghua Trading V. Zhendong (2002–2005), and others [10] whose creators were prosecuted.

In everyday life, it is customary to identify the concepts of «financial pyramid» and«fraud». However, this is not always correct: it happens that quite law-abiding commercial organizations turn into a pyramid against their will. Or as it was with the next pyramid MMM-2011, when S. Mavrodi honestly told all his so-called investors that he was building a pyramid. And this time he could not be accused of fraud, which is always built on deception. Therefore, in order to eliminate this issue and give it certainty, article 172.2 was introduced in the criminal code of the Russian Federation in 2016, which significantly facilitated the activities of law enforcement agencies in the fight against this phenomenon [13]. But the article introduced did not give an exact definition of the concept of «financial pyramid», describing this scheme very carefully, with reservations and clarifications. The reasons for this are seen in the fact that financial pyramids are disguised as a variety of business types: it can be a travel company, or a credit and financial organization. Illegal mechanisms for ensuring high profitability can even hide under the guise of organizations that extract minerals, rare metals, and jewelry [14].

In this regard, it is more correct to consider a financial or investment pyramid as a business scheme that is only occasionally launched forcibly and briefly, but in most cases, of course, with the aim of obtaining benefits and profit at the expense of ordinary citizens.The so-called HYIP projects (from the English. HYIP-High Yield Investment Program - that is, high-yield investment programs). At the same time, it is worth noting that HYIP investors (people who understand that they carry their money into the pyramid and do it purposefully) clearly differ in the level of financial culture from citizens who fall into the pyramid by deception. HYIP investors create communities where they exchange ideas and strategies for such earnings, assess risks, and discuss modern best-selling financial books.

Financial pyramids were criminally prosecuted, due to the fact that they are certainly dangerous for citizens and for society as a whole. In the hope of improving their financial situation in the network of scammers fall masses of poor and frankly poor, as well as socially unprotected citizens. Sooner or later, the flow of newly received funds, through which the organizer of the pyramid pays interest to citizens who already participate in it, runs out. And then most of the citizens who invested their money in the hope of getting a solid profit, do not receive any profit, or the amount originally invested. For the citizen, this is fraught with ruin, and for society - instability.

However, according to experts, not only naive and trusting people fall for the bait of financial fraudsters. In the case of the «Cashbury» pyramid (the largest financial pyramid in recent years, whose activities were stopped at the end of 2018 by the Bank of Russia), the bet was made on the creative class – «Cashbury» positioned itself as the flagship of p2p lending. Also often fall into the trap of pensioners. At the same time, scammers know that often people aged and with savings migrate to the South, so financial pyramids are created there. If the pyramid format seems successful to the creators, it also applies to the North of the country.

It is alarming that the creation of pyramids in Russia is still practiced, and the number of victims is growing every year. The title of the largest financial pyramid in 2019 is claimed by AirBitClub, which attracted money from citizens to invest in cryptocurrency. And since citizens have heard that you can make a lot of money on cryptocurrencies quickly, but few people understand the whole process, the legend of investing in cryptocurrency is successfully working to attract new customers. The founders of AirBitClub managed to collect more than 500 million rubles [15].

4 Results

In General, after studying the market of financial pyramids, we can distinguish the following types:

1. Investment-a person invests money and waits for interest without doing anything. Such pyramids are also called HYIP projects. And since the origins of the pyramid are most often professional investors, they are well aware and understand what a HYIP is and what stages of development such companies exist. Like investors in the stock market, HYIP investors do not invest all their money in a single pyramid, they diversify their portfolio.

   All HYIPs sooner or later collapse and this is inherent in their very system. And a novice investor entering such a pyramid should not hope that he will be among the first and will be able to make a huge profit. By the time a novice investor learns about the pyramid, it will already include professional investors who are waiting for such newcomers to make money on them.

   Famous HYIP pyramids are Roy Club (2019), Hermes Management (2014), Mizes (2018), Amir Capital (2018) [16].

2. Classic pyramid scheme or mutual assistance cash register. Money in such projects is transferred from one participant to another. This type of financial pyramid scheme was not relevant for a long time (from 2016 to 2020), but in 2019 it was revived again.

3. Masked pyramid scheme – a classic pyramid scheme with the addition of a masking element, such as unnecessary products or products that can only be received by some participants.

In 2019, the Bank of Russia found 237 organizations with signs of financial pyramids in Russia, which is significantly more than in 2018 and 2017 (Fig. 4). Using a polynomial series, the damage forecast for 2019 is more than 6.5 billion rub. In total, the regulator has identified more than 900 probable pyramids in five years [17]. According to the observations of the Bank of Russia, pyramid organizers are most often activated in the first quarter, as citizens have more money after annual bonuses. In the second quarter, they are closed after collecting money. This fact once again confirms the big difference in the financial culture of pyramid organizers and people who carry money there, often hard-earned.

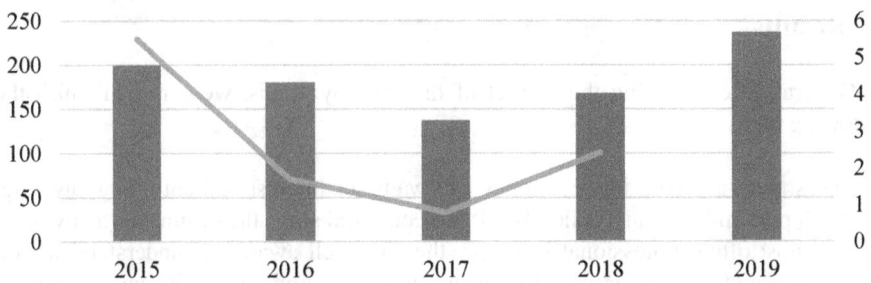

Number of organizations identified by the Bank of Russia that show signs of financial pyramids, units

The established amount of damage from the activities of financial pyramids, billion rubles

Fig. 4. Dynamics of indicators that characterize the state's counteraction to financial pyramids (compiled according to [17])

In Russia in 2019, 161 criminal cases were opened as part of the fight against financial pyramids. More than 77 thousand people were recognized as victims, which is 12.6% more than in 2018, when there were 143 cases in criminal proceedings [14].

This situation occurs due to the low financial literacy of the population, which depends on General education, age and level of well-being. Taking advantage of the lack of public knowledge in this area, as well as the availability and widespread use of sophisticated banking and financial mechanisms, unreliable financial organizations earn high profits. Using aggressive marketing and advertising, the scammers are easy to arouse the desire for profit. And until citizens understand that if the declared profitability of a company is significantly higher than inflation or the key rate of the Bank of Russia and this is associated with a huge risk, and they still carry their money to such companies, financial pyramids will always have fuel and a reason to exist.

The Bank of Russia believes that this dynamic is a confirmation of the development of the system of countering financial pyramids, but at the same time it speaks about the constant adaptation of such organizations to changing realities. The Bank of Russia is really coming up with new ways to deal with fraudsters, and since the second half of 2018, the regulator has launched a special robot on the Internet that searches for illegal financial schemes [14].

Thus, public policy measures to reduce fraudulent schemes and improve financial literacy should be aimed at:

– systematic and coordinated work of all stakeholders, which are public authorities, universities, vocational education institutions, schools, aimed at developing and implementing educational programs to improve financial literacy with more active participation of the business community in this process;
– creating elements of institutional infrastructure in the regions that support regional initiatives to implement best practices in reducing fraudulent schemes and improving the level of financial literacy of the population, as well as tax culture, which together ensures the sustainable development of the state [18];

– development of a system of indicators for assessing the level of financial literacy and the effectiveness of measures aimed at improving this level, which allow identifying problem areas at an early stage for taking operational regulatory measures;
– creation and maintenance of unified channels for the dissemination of information resources to improve financial literacy of various target groups and financial advice to the population in order to explain the risks associated with a particular financial product.

All over the world, Governments form various bodies to improve financial literacy, in particular, the US Government has created Financial Literacu and Education commissi - FLES, a specialized Internet resource, as well as the Advisory Council on financial literacy under the President of the United States [19]. In the UK, financial education is handled by the financial conduct Authority (FSA) [20]. Developed countries in the fight against financial pyramids attach particular importance to improving the level of financial literacy through the implementation of the functions of coordination, monitoring, analysis, control and regulation of educational and educational activities.

Russia needs to implement the best European and global practices in implementing financial education programs among different target audiences (students, schoolchildren, pensioners, and adults). In Russia, the process of improving the level of financial literacy, aimed at creating a system of financial education and information in the field of consumer protection of financial services, has been implemented since 2017 (with the adoption of the strategy for improving financial literacy in the Russian Federation for 2017–2023, dated September 25, 2017, № 2039-R) [21]. Public organizations in this field (for example, the Association for the development of financial literacy, the national center for financial literacy), information and educational Internet resources (ABC of financial literacy, Financial culture) have been created in Russia to broadcast the best world practices. The activities of the macro – regulator, the Bank of Russia, play a significant role in improving financial literacy and developing mechanisms to counteract financial pyramids. The measures of the state policy will enhance the individual responsibility of each person for having taken financial decisions and to protect its interests in obtaining financial services, as well as to promote a financially literate citizen that is a basic requirement to improve the level and quality of life of the population.

The fight against financial pyramids must continue and, first of all, it is necessary to ensure the inevitability of the act for this action, since the organizers of the schemes count on impunity, on the fact that they have the opportunity to illegally collect money and not risk anything. The Bank of Russia, in cooperation with law enforcement agencies, has all the tools to make such illegal activities less profitable and therefore less economically attractive, for example, to create more obstacles by blocking sites, forcing providers to check information and advertising that fraudsters actively place and by these actions increase their expenditure items and increase the risk of criminal prosecution for them.

And, of course, the most important direction in the fight against financial pyramids should be to train people to financial prudence against the background of their financial literacy training. The only way for a person to avoid the pitfalls of financial pyramids, and for the Russian society and economy - to eradicate these pyramids – is to raise the level of financial culture. It takes time, desire, and effort to raise the level of financial culture, both personal and social.

Now, pyramid projects are a response to the needs of people, to their desire to live well. The creators of financial pyramids have learned to adapt and offer «products» with the growing literacy. In addition, the organization of training courses that promise to improve financial literacy [22] and teach how to work in the financial market, thus attracting investors to the financial pyramid, has become a new area of activity for fraudsters.

5 Conclusion

Financial literacy of the population is characterized by knowledge, skills and attitude to Finance. The level of financial literacy in the world is determined by the financial literacy index, which was first calculated in the Russian Federation based on private indicators in 2018 by the NAFI Analytical center. According to the study, Russia's place (9th) among the G20 countries was determined. The obtained value of the level of financial literacy allowed us to identify problems and threats to household security. This circumstance allowed us to propose the use of a competence approach to improve the financial literacy of the population and the financial culture of society, as well as to translate it to the problem of the spread of financial pyramids. Based on a retrospective of the development of financial pyramids and the analysis of modern experience of their creation and mass participation in them, we have structured their types, proposed public policy measures to reduce fraudulent schemes and increase the level of financial literacy. Russia is actively working to protect the interests of citizens in obtaining financial services. Currently, state, educational and public organizations, including volunteers, actively conduct activities that contribute to the formation of financially competent behavior of citizens, ultimately ensuring an increase in the level and quality of life.

References

1. The concept of basic knowledge and skills in financial literacy for adults OECD IFRS (Electronic resource). https://fingramota.econ.msu.ru/sys/raw.php?o=1519&p=attachment
2. Brislavka, E.A., Zelentsova, V.A., Demidov, D.N.: Improving financial literacy: international experience and Russian practice (Electronic resource). https://ur-consul.ru/Bibli/Povyshyen iye-finansovoyi-gramotnosti-nasyelyeniya-myezhdunarodnyyi-opyt-i-rossiyiskaya-praktika. html
3. Silina, S.N., Stupin, V.Yu.: From financial literacy to financial culture. Bulletin of the Baltic Federal University named after I. Kant. Series: Philol. Pedagogy, Psychol. **4**, 71–79 (2017)
4. Financial literacy rating of Russian regions-2018 (Electronic resource). https://nafi.ru/pro jects/finansy/reyting-finansovoy-gramotnosti-regionov-rossii-2018
5. Financial literacy levels in Eurasia. OECD analytical report 2018 (Electronic resource). https://www.oecd.org/education/financial-education-cis.htm
6. Financial literacy rating of Russian regions (Electronic resource). https://www.karta.vashif inancy.ru/
7. Korostelkina, I.A., Popova, L.V., Dedkova, E.G., Korostelkin, M.M.: Information risks and threats of the digital economy of the XXI century: objective prerequisites and management mechanisms. In: Advances in Intelligent Systems and Computing. AISC, vol. 1114, pp. 185–194 (2020)

8. Fahitov, A.I., Nasibullin, R.T.: Problems of forming the financial culture of the Russian population through the prism of sociological research. Bull. TOGU **2**, 235–244 (2010)
9. Timchenko, D.: History of financial pyramids (Electronic resource). https://zen.yandex.ru/media/mmetrics/istoriia-finansovyh-piramid-5c19d639b6a0da00aac86398
10. Bachurka, S.: The largest financial pyramids in Russia and a broad. (Electronic resource). https://zen.yandex.ru/media/id/5dc83f884052283fcdc469ba/krupneishie-finansovye-piramidy-v-rossii-i-za-rube-5dc98d322535bb706a2f733c
11. What is a pyramid scheme? (Electronic resource). https://journal.tinkoff.ru/wiki/pyramid/
12. Sokolov, M.: Financial pyramid - what it is: definition, main types, signs of financial pyramids (Electronic resource). https://feelwave.ru/biznes-terminy/finansovaya-piramida-chto-eto-takoe
13. Mubarakov, I.I., Sagdullin, D.F., Ehlakova, E.A.: Financial pyramid: definition, history, fraud. Forum Young Sci. **11–2**, 120–127 (2018)
14. Cherunkova O.: «Fear and greed»: who invested in pyramid. Newspaper, December 2019. [Электронный ресурс] – Режим доступа. https://www.gazeta.ru/business/2019/12/23/128 79674.shtml
15. Kaledina, A.: And the ship does not sail: the Central Bank told about the new tricks of the creators of the pyramids. News, February 2020 (Electronic resource). https://iz.ru/980148/anna-kaledina/i-korabl-ne-plyvet-v-tcb-rasskazali-o-novykh-ulovkakh-sozdatelei-piramid
16. List of pyramid schemes in 2020. Russia (Electronic resource). https://besuccess.ru/spisok-finansovyx-piramid-2015
17. Annual report of the Bank of Russia for 2017, for 2018 (Electronic resource). https://cbr.ru/Collection/Collection/File/19699/ar_2018.pdf
18. Korostelkina, I.A., Dedkova, E.G., Varaksa, N.G., Korostelkin, M.M.: Models of tax relations: improving the tax culture and discipline of tax payers in the interests of sustainable development. In: E3S Web of Conferences (BTSES-2020), vol. 159, p. 06014 (2020)
19. Financial Literacy and Education Commission (Electronic resource). https://home.treasury.gov/policy-issues/consumer-policy/financial-literacy-and-education-commission
20. Financial Conduct Authority (Electronic resource). https://www.fca.org.uk/
21. Strategy for improving financial literacy in the Russian Federation for 2017–2023 dated September 25, 2017 № 2039-R (Electronic resource). http://static.government.ru/media/files/uQZdLRrkPLAdEVdaBsQrk505szCcL4PA.pdf
22. Tretyak, A.: The number of financial pyramids in Russia has reached a record (Electronic resource). https://www.vedomosti.ru/finance/articles/2020/02/04/822206-finansovih-piramid

Organizational Capacity Assessment Model for Digital Transformation

Zhanna Mingaleva[1]([✉]) [iD], Alina Kostyreva[1] [iD], Elena Shironina[1] [iD], and Kirill Dvinskikh[1,2] [iD]

[1] Perm National Research Polytechnic University, Perm 614990, Russian Federation
mingal1@pstu.ru
[2] LUKOIL-PERM, Perm 614990, Russian Federation

Abstract. The aim of the research is to theoretically substantiate the need for an in-depth assessment of the current organizational potential of a company in order to successfully conduct digital transformation. The study is based on a method developed by the authors for assessing the current organizational potential in accordance with the requirements and characteristics of the digitalization process. The method includes an extended classification of management problems in conditions of increasing uncertainty and the rate of change in the business environment, a methodology for assessing (calculating quantitative values) the main parameters of the Digital Transformation Snake model (DTS model), a graphical representation of a digital transformation model based on improving the organization's personnel structure. The DTS model includes the identification of two zones of organizational impact on the workforce of the organization: the zone of restructuring of the workforce and the zone of personnel development. The conclusion is made about the use of various technologies for improving the organization's personnel to achieve a successful digital transformation. Technologies are distributed across the designated areas of organizational impact.

Keywords: Organizational potential · Digital transformation · DTS model · Organizational change · Personnel development · Change management

1 Introduction

The World Economic Forum's report, "Digital Transformation: Powering the Great Reset", noted that the COVID-19 pandemic is a watershed moment for digital business transformation [1]. The rules for success have changed and are increasingly based on the possibilities of digital models to create new value and experience. At the moment, there are new opportunities to implement digital transformation, which corresponds to the changing role of business: to be a long-term tool for creating value for all stakeholders.

Laurent-Pierre Baculard, Laurent Colombani, Virginie Flam, Ouriel Lancry and Elizabeth Spaulding of Bain & Company have determined that digital leaders in every industry are vastly superior to their competitors, and digital transformation can deliver significant results [2]. By comparing the financial performance of companies with varying degrees of digital complexity, Bain & Company found that digital leaders' revenues

© The Author(s), under exclusive license to Springer Nature Switzerland AG 2021
T. Antipova (Ed.): ICADS 2021, AISC 1352, pp. 70–80, 2021.
https://doi.org/10.1007/978-3-030-71782-7_7

grew more than twice as fast as revenues of companies lagging in the digital space, and digital leaders' profits are nearly twice as often as the profit of lagging companies.

At the same time, the introduction of digital technologies is changing the scale of employment and the structure of the workforce, including in leading innovative companies [3]. Y. Harari noted the hopeless obsolescence by 2050 of the idea of "lifelong employment" and the concept of "profession for life", as well as the emergence of "extra" people class because of the lack of work, education or weak psychological stability [4]. According to Bain & Company forecasts the rate of labor movement in the next decade could be two or three times higher than during other major transformation periods in modern history [5].

The World Economic Forum report "The Future of Jobs" indicates that by 2025, the working hours of people will be equal to the working hours performed by machines and algorithms, there will be problems with the employment of workers in different industries and geographic regions. Research by the Future of Jobs Survey showed "that by 2025, 85 million jobs may be displaced by a shift in the division of labour between humans and machines" [6, p. 4]. Bain & Company predicts that automation could eliminate from 20% to 25% of existing jobs by the end of the 2020s, hitting middle and low-income workers the hardest [5].

2   Theoretical Background

Digital transformation is not a new phenomenon for society and system of government [7–9], economy and business [10–12], transport [13] and all. However, as stated in the World Economic Forum's report Digital Transformation: Powering the Great Reset many attempts have failed [1]. In the scientific literature this is explained by the influence of a number of factors and phenomena [14, 15].

The first is the level of readiness of companies and enterprises for digital transformation. Bain & Company surveyed 1,000 companies around the world to assess their readiness for digital transformation. And two points were highlighted: the return on digital transformation can be impressively high, but the probability of success is unfortunately low [2]. Bain & Company survey showed that only 5% of the companies involved in the digital transformation have reached or surpassed planned expectations (for comparison, it is 12% for conventional changes). 75% of companies experienced a decrease in value and mediocre results [2]. Thus, the researchers conclude that digital transformation is significantly more complex than traditional change management programs.

The reason for this phenomenon is the sheer difficulty of using technology to increase company's speed and large-scale training. Even in situations where some traditional companies are successful in implementing innovations in product development, not many of them are able to quickly embrace digital technology to solve problems and improve organizational productivity [3, 16, 17].

Another reason for the failure of the digitization process is with startups. Startups companies typically start the digital transformation process with a set of isolated initiatives, which are aimed at the most problematic point. However, these companies find it difficult to translate these initiatives into complete solutions and opportunities that could have a significant impact on the economy of the entire company. More mature companies are better at clustering digital initiatives around shared strategic goals and are starting to focus on improving individual support processes, primarily IT. However, these initiatives are also localized and not distributed throughout the company without broad organizational impact. As the researchers explain there are many organizations that are responding to digital change in certain ways, but the overall digital transformation is being held back by legacy systems [2, 18].

The positive experience of digital transformation shows the following.

Digital leaders link their strategic ambitions to the innate abilities and behaviors they will need to achieve their ambitions. At the beginning of transformation leaders translate their strategy into a clear set of digital initiatives. Then they invest heavily in fundamental changes in their way of working, communication, organizational culture. This allows them to take digital initiatives both faster and on a larger scale. The conclusion that many researchers draw is that digital transformation implies changes at all levels of the organization [2].

Ronald Heifetz and Marty Linsky highlighted technical and adaptive problems [19, 20]. Technical challenges require the use of honed skills to solve well-defined problems. Adaptive problems are confusing, timeless and unclear. In many cases the question is difficult to formulate and the answer is even more difficult to find. Chris Ertel and Lisa Kay Solomon point out that the most important strategic challenges organizations face today are adaptive challenges [21].

It is necessary to assess the current organizational capacity for change to drive digital transformation across the entire company.

3 Research Method

In order to assess the current organizational potential of a particular organization for digital transformation we have developed a special research method. For our method we used the concept and hierarchy of integrated systems [22] and the Heifetz classification [19, 20, 23] to which we added another class of problems - complex problems. The characteristics of the complex problems that we introduce into the analysis method as

Table 1. Comparative characteristics of technical, complex and adaptive problems

Criterion	Technical problems	Complex problems	Adaptive problems
Problem formulation	Clear	(a) A sufficiently clear problem, or (b) a wide variety of clearly formulated problems	Fuzzy, indefinite
Initial conditions	Initial data, goal, results, resources required are known	A lot of initial data, there is uncertainty in the goals, results, resources required	Not installed
Decision	Correct	Optimal	Better-worse ratio
Solution methods	Application of known methods	Application of known methods	There are no ready-made methods
Terms of decision	Definable	Definable	Not definable

Source: compiled by the authors

well as the characteristics of the technical and adaptive problems of Heifetz are presented in Table 1.

The classification of problems given in Table 1 was used to determine such a parameter for assessing the current organizational potential of a particular organization for digital transformation as the complexity of the problems being solved.

The second parameter of assessment is the parameter "participation in value creation". For these parameters (the complexity of the problems being solved and the participation in the creation of value) the following levels were set: high, medium and low. The assessment is carried out according to the point method.

The complexity of professional activity is determined by the following components: level of education, training, complexity of the work performed, type of activity. Evaluation characteristics and the number of points for determining the complexity of the problems being solved are shown in Table 2.

Table 2. Assessment of the complexity of the problems being solved

Component	Evaluation characteristics and number of points		
	Low	Medium	High
Number of points	*0–1*	*2–3*	*4–5*
The level of education	Secondary specialized, secondary professional	Higher (bachelor's degree, specialty)	Higher (master's degree, academic degree)
Training	Enough corporate training, development of skills within the company	Knowledge that is necessary only in their field of professional activity study of regulatory documents, training as needed	Availability of knowledge, skills in several areas, training to acquire knowledge, abilities, skills
The complexity of the work performed	Simple functions (work strictly according to instructions, constant repetition of the same type of actions)	Complex functions (knowledge of regulatory documents is required)	Complex heterogeneous functions (work related to solving issues in several areas)
Kind of activity	Information technology activities	Analytical and constructive activity	Organizational and administrative activities

Source: compiled by the authors

The author's method assumes the summation of the scores for all the components shown in Table 2. The total score shows the group and the complexity of the problems being solved. The following rating scale is established:

- from 0 and less than 6 - solving technical problems,
- from 6 and less than 14 - solving complex problems,
- from 14 to 20 inclusive - solving adaptive problems.

Table 3 presents the characteristics for evaluating the components of the parameter "participation in value creation".

Table 3. Assessment of participation in value creation

Component	Evaluation characteristics and number of points		
	Low	Medium	High
Number of points	0–1	2–3	4–5
Responsibility for the result of activities	Responsibility only for the result of activities	Responsibility for the result of their unit	Responsibility for the result of several departments, for the organization as a whole
Material responsibility	Responsibility for work equipment and labor tools	Work related to the possibility of material damage, accidents, equipment breakdowns, its downtime, damage to raw materials, semi-finished products, defective finished product	Disposal of financial and material assets
Involvement of an employee in business processes	Auxiliary process, involves the maintenance of the main process	Main process	The main process with interaction with the client, work at the customer's facility
The importance of involvement in business processes,	The results of employees' activities do not affect the results of other employees. The results of employees' activities affect the performance of labor functions by other employees within only one department	The results of employees' activities affect the performance of labor functions by other employees within only one department The impact is simultaneous, not permanent	The results of the employees' activities affect the results of the activities of employees of several departments, on the organization as a whole. The influence happens all the time

Source: compiled by the authors

By summing up the scores for all components, you can determine the degree of participation in value creation. The following rating scale is established:

- from 0 and less than 6 - indirect participation,
- from 6 and less than 14 - direct,
- from 14 to 20 inclusive - immediate.

Based on the assessment of two key parameters (the complexity of the problems being solved and participation in the creation of value), we propose the following model for assessing the current organizational potential of a particular organization for digital transformation. The model is shown in Fig. 1.

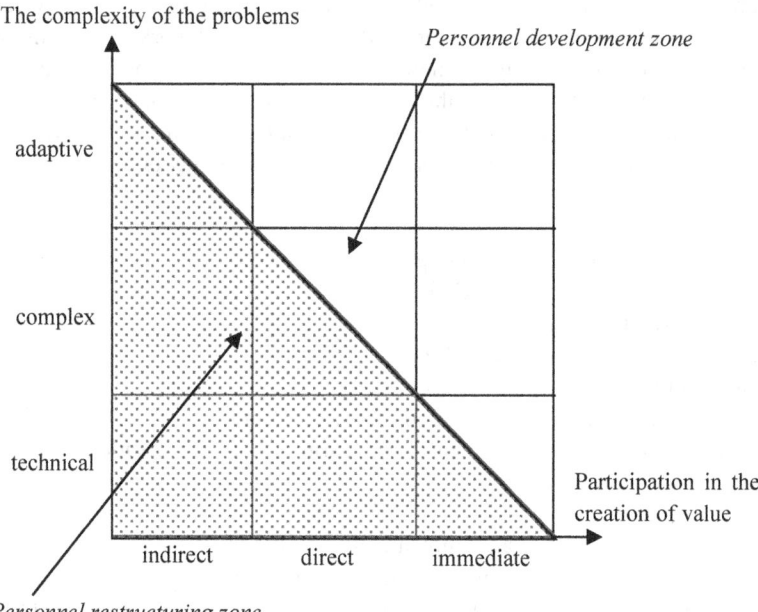

Fig. 1. Model for assessing organizational potential for digital transformation. Source: compiled by the authors

Further, in the research process the main measures were identified to change the organizational potential in each of the two designated areas for the digital transformation of the organization.

4   Research Results

Figure 1 shows the model we have developed for assessing the organizational potential for digital transformation.

The model identifies 2 areas of impact on the company's personnel in terms of the company's readiness for digital transformation:

1. Personnel restructuring zone
2. Personnel development zone

The organizational arrangements for creating the conditions for a successful digital transformation in these zones differ.

The personnel restructuring zone as a basis for digitalization includes the process of automating the functions performed [24]. The automation changes the way we work, the interaction between different employees and the movement of resources in a company. But automation also doesn't just improve the status quo; automation involves rethinking/reengineering business processes, how a company works, how a company's employees work. Automation, firstly, is impossible without revising labor functions and secondly, without reducing labor functions and accordingly, reducing excess staff. Thus, labor functions in the field of administrative, clerical, analytical services with repetitive tasks or tasks based on standardized procedures and work algorithms are easily automated. Automation begins with the alignment of organizational capabilities and implementation requirements, goes through all stages of business planning and only then is completed with deployment and implementation affecting the entire company and covering the entire value creation process. We are talking about such technologies as:

- Process discovery and process mining
- Optical character recognition (OCR)
- Low-code/no-code automation
- Business process management (BPM)
- Workflow automation
- Robotic process or desktop automation (RPA or RDA)
- Artificial intelligence, including machine learning and conversational AI (natural language processing/generation)

The process of developing and implementing digital initiatives to make fundamental changes in the way of working, communication, organizational culture within the personnel restructuring zone includes 5 sequential stages of gradual transition from simpler problems (technical) to more complex (adaptive) ones while maintaining the level of personnel participation in value creation companies at the levels of "mediated" and "direct".

The trajectory of movement of the complete transformation process in both zones is shown in Fig. 2.

When the level of personnel qualifications becomes sufficient to solve adaptive problems, it becomes possible to move to the personnel development zone (5th stage, transition from point 5 to point 6 in Fig. 2).

The personnel development area is characterized by:

- development of critical thinking,
- solving complex and non-standard problems,
- self-management skills.

The complexity of the problems

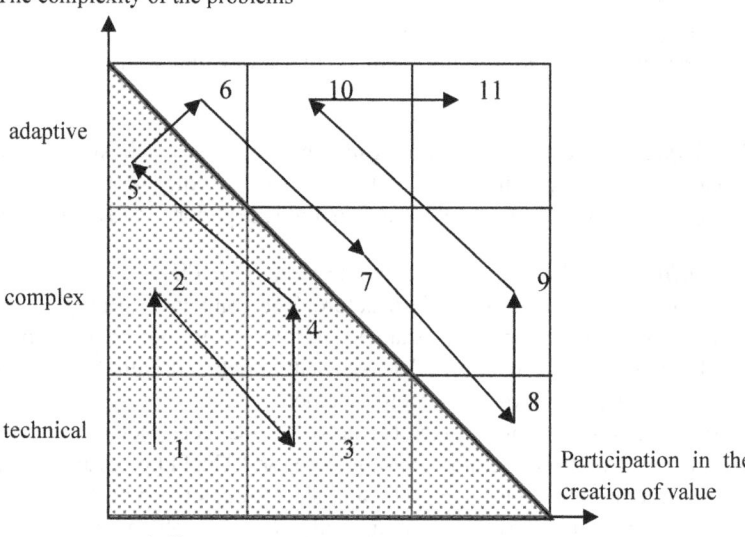

Fig. 2. Model "Snake of digital transformation" based on the assessment of labor functions. Source: compiled by the authors

The basis for the implementation of activities within this zone is digitization [19]. At the same time, the digitization is not a single stand-alone project, it is a corporate ecosystem that unites people, processes, technologies, support services and includes:

- Analytics,
- Business Innovation,
- Craftsmaster/Expert Network,
- Design,
- Digital Marketing,
- Enterprise Technology.

Analytics. Analytics enables to turn data into a competitive asset, gain insights and make smarter, faster decisions. Analytics turns data into a source of value - from customer service to employee interactions.

Business Innovation. This is not just about innovations and innovation activity but also about an entire ecosystem of people-generators of ideas, entrepreneurs, company leaders, venture capitalists, innovation experts, product managers seeking to forge mutually beneficial relationships. This allows to develop innovative strategies, optimal operating models, create and scale new products or entire companies.

Craftsmaster/Expert Network. Combining the best tools, technologies and their carriers: service providers, external experts from different sectors, geographic regions, different specializations, which opens up access to new opportunities and the emergence of synergies.

Design. The mandatory is rapid prototyping, usability testing, human-centeredness, customer journey mapping. Design nowadays involves interdisciplinarity, attention to user needs, special testing and training methods that allow quickly bring a product to the market.

Digital Marketing. Currently, we are talking not only about data analysis, systems integration, creativity, working with media, but also about working in a team with very different people with different personal value structures, experience and areas of interest.

Enterprise Technology. A company's corporate strategy is directly influenced by its technology strategy and technology infrastructure. The company's strategic ambitions must be aligned with the technology strategy and only then goes business planning. Systems and architecture are being modernized with a detailed digital transformation roadmap, modern operating models and agile methods.

A corporate ecosystem that includes the above elements cannot function without new competencies. The World Economic Forum's report "The Future of Jobs" indicates that according to employers the main skills and skill groups that will be increasingly in demand by 2025 include: critical thinking and analysis, problem-solving, skills in self-management such as active learning, resilience, stress tolerance and flexibility.

5   Conclusions

Digital transformation involves innovative changes introduced throughout the entire system of an organization's activities, at all its levels. For the successful implementation of digital transformation, a preliminary assessment of the current organizational potential and further monitoring of the effectiveness of changes are required. We have developed an assessment method and toolkit that meets the requirements and features of digitalization.

The method assumes, firstly, the differentiation of personnel depending on the problems to be solved - grouping and assessing the possibility of transition from technical problems (requiring the use of honed skills to solve well-defined tasks) to solving complex problems (with a large number of diverse tasks with a lot of initial data and limitations in resources, terms) and the final output to the formation of capabilities and competencies for solving adaptive problems (which do not have a clear statement and a known solution method). Secondly, our proposed method includes assessing the level of employee participation in creating value for the customer: from indirect participation (involvement in an auxiliary business process and limited impact on the performance of labor functions by other employees) to direct (involvement in the main business process and impact on performance of labor functions by other employees) and direct participation (performance of work at the customer's facility).

The results of the assessment serve as the basis for building the Digital Transformation Snake model (DTS model). The model includes a clear identification of two zones of organizational impact on the workforce of an organization to achieve digital transformation: a zone for restructuring the workforce and a zone for personnel development.

The workforce restructuring area relies on process automation. At the same time, the automation not only improves the situation, but also involves rethinking/reengineering business processes. The automation is impossible without revising labor functions, reducing redundant labor functions and, accordingly, reducing excess personnel.

The personnel development zone is based on digitalization. Digitalization is not a single stand-alone project. It is a corporate ecosystem that brings together people, processes, technology and support services and includes Analytics, Business Innovation, Craftsmaster/Expert Network, Design, Digital Marketing, Enterprise Technology. This corporate ecosystem cannot function without personnel mastering new skills: developing critical thinking, solving complex and non-standard problems, and self-management skills.

The DTS model allows you to develop a set of organizational and management measures for the successful implementation of the digital transformation of an organization.

Further research involves a more detailed analysis and identification of specific groups of personnel regarding their job functions, distribution on the matrix for assessing organizational potential and making specific management decisions regarding the restructuring/development of these groups in the process of digital transformation of the organization.

Acknowledgment. The work is a part of base portion of the state task of the Ministry of Education and Science of the Russian Federation to Perm National Research Polytechnic University (topic № FSNM-2020-0026).

References

1. World Economic Forum, Digital Transformation: Powering the Great Reset. https://www.wef orum.org/reports/digital-transformation-powering-the-great-reset. Accessed 10 Nov 2020
2. Baculard, L.-P., Colombani, L., Flam, V., Lancry, O., Spaulding, E.: Orchestrating a Successful Digital Transformation (2017). https://www.bain.com/insights/orchestrating-a-succes sful-digital-transformation. Accessed 09 Nov 2020
3. Danilina, E.I., Mingaleva, Z.A., Malikova, Y.I.: Strategic personnel management within innovational development of companies. J. Adv. Res. Law Econ. VII **5**(19), 1004–1013 (2016)
4. Harari, Y.N.: 21 Lessons for the 21st Century, 1st edn. Random House, New York City (2018)
5. Harris, K., Kimson, A., Schwedel, A.: Labor 2030: The Collision of Demographics, Automation and Inequality. https://www.bain.com/insights/labor-2030-the-collision-of-demograph ics-automation-and-inequality. Accessed 09 Nov 2020
6. The Future of Jobs Report 2020. World Economic Forum. Geneva, Switzerland (2020)
7. Antipova, T.: Organizational model for public sector auditing bearing in mind time factor. In: Rocha, Á., Abreu, A., de Carvalho, J., Liberato, D., González, E., Liberato, P. (eds.) Advances in Tourism, Technology and Smart Systems. Smart Innovation, Systems and Technologies, vol. 171, pp. 297–305. Springer, Singapore (2020). https://doi.org/10.1007/978-981-15-2024-2_27
8. Tsvirko, S.: Fuzzy logic approach for evaluation of sovereign wealth funds' management. In: Antipova, T., Rocha, A. (eds.) Digital Science, DSIC18 2018. Advances in Intelligent Systems and Computing, vol. 850. Springer, Cham (2019). https://doi.org/10.1007/978-3-030-02351-5_37

9. Voskanyan, Y., Shikina, I., Kidalov, F., Andreeva, O., Makhovskaya, T.: Impact of macro factors on effectiveness of implementation of medical care safety management system. In: Antipova, T. (eds.) Integrated Science in Digital Age 2020. ICIS 2020. Lecture Notes in Networks and Systems, vol 136. Springer, Cham (2021). https://doi.org/10.1007/978-3-030-49264-9_31

10. Bashminov, A., Mingaleva, Z.: The use of digital technologies for the modernization of the management system of organizations. ICIS 2019. Lecture Notes in Networks and Systems, vol.78. Springer, Cham (2020). https://doi.org/10.1007/978-3-030-22493-6_19

11. Ha, K.M.: Digital business leadership: digital transformation, business model innovation, agile organization change management. R&D Manag. **50**(2), 171–172 (2020)

12. Silva, E., Lopes, P.: A study on market intelligence: the professional, the applicability of information technologies to innovate and gain competitive advantage. J. Digit. Sci. **1**(1), 51–62 (2019)

13. Rozhkova, N., Rozhkova, D., Blinova, U.: An overview of aspects of autonomous vehicles' development in digital era. In: Antipova, T. (eds.) Integrated Science in Digital Age 2020, ICIS 2020. Lecture Notes in Networks and Systems, vol. 136. Springer, Cham (2021). https://doi.org/10.1007/978-3-030-49264-9_28

14. Parry, W., Kirsch, C., Carey, P., Shaw, D.: Empirical development of a model of performance drivers in organizational change projects. J. Change Manag. **14**(1), 99–125 (2014)

15. Rosca, D., Banica L., Sirbu, M.: Building successful information systems – a key for successful organization. In: Annals of "Dunarea de Jos" University of Galati Fascicle I – 2010. Economics and Applied Informatics, vol. 2, pp. 101–108 (2010)

16. Antipova, T.: Digital view on the financial statements' consolidation in russian public sector. In: Antipova, T., Rocha, Á. (eds.) Information Technology Science, MOSITS 2017. Advances in Intelligent Systems and Computing, vol. 724. Springer, Cham (2018). https://doi.org/10.1007/978-3-319-74980-8_12

17. Mingaleva, Z., Deputatova, L., Starkov, Y.: Values and norms in the modern organization as the basis for innovative development. Int. J. Appl. Bus. Econ. Res. **14**(10), 124–133 (2016)

18. Barata, J., Da Cunha, P.R.: Towards a business process quality culture: from high-level guidelines to grassroots actions. In: Proceedings of the 23rd International Conference on Information Systems Development, Varaždin, Croatia, pp. 6–13 (2014)

19. Heifetz, R., Linsky, M.: Leadership on the Line: Staying Alive through the Dangers of Leading. Harvard Business Review Press, Brighton (2002)

20. Hiefetz, R., Linsky, M., Grashow, A.: The Practice of Adaptive Leadership: Tools and Tactics for Changing Your Organization and the World. Harvard Business Review Press, Brighton (2009)

21. Ertel, Ch., Solomon, L.K.: Moments of Impact: How to Design Strategic Conversations That Accelerate Change. Simon & Schuster, New York (2014)

22. Jørgensen, T.H., Remmen, A., Mellado, M.D.: Integrated management systems – three different levels of integration. J. Clean. Prod. **14**(8), 713–722 (2006)

23. Senderek, R., Stich, V.: Work-based-learning in the digital age. In: Proceedings of the 7th International Joint Conference on Knowledge Discovery. Knowledge Engineering and Knowledge Management (KMIS), vol. 3, pp. 268–273 (2015)

24. Perez-Arostegui, M.N., Benitez-Amado, J., Tamayo-Torres, J.: Information technology enabled quality performance: an exploratory study. Indian Manag. Data Syst. **112**(3), 502–518 (2012)

Digitalization as a Key Aspect of the Development of Social Infrastructure in Rural Areas

Larisa Akifieva(✉) ⓘ, Mikhail Polyakovⓘ, Natalia Sutyaginaⓘ,
Alexander Mansurovⓘ, and Olga Shaminaⓘ

Nizhny Novgorod State Engineering and Economic University, Knyaginino 606340, Russia
laraakif@mail.ru

Abstract. The article considers digitalization as a key aspect of the development of social infrastructure in rural areas. The main role is defined by one of the most promising sectors of social infrastructure in rural areas - housing and communal services. Digitalization in the sphere of housing and communal services allows solving many pressing problems of the industry, as well as pressing socio-economic problems, simplifying communication between numerous subjects and objects of the housing and communal services industry, improving the quality of social housing and communal services, and increasing the productivity of the industry. The article defines the main directions of digitalization of housing and communal services. The tasks and problems of digitalization of housing and communal services are highlighted, and ways to solve these problems are proposed.

Keywords: Digitalization · Rural areas · Social infrastructure · Housing and communal sphere

1 Introduction

Digitalization is a new trend in all areas of the global economy. The main documents on digitalization in Russia are "The Strategy for the development of the information society in the Russian Federation for 2017–2030" and "The Digital economy of the Russian Federation" program [1, 2].

Digitalization has also affected rural areas, but according to the Russian Ministry of Agriculture, its level in rural areas is about 10%, growing by 2–3% every year.

Digital transformation of agriculture is the transformation of economic activity in agriculture through the introduction of digital tools-technologies and platform solutions designed for generating, processing, in-depth analysis and translation of analysis results in the form of numerical information about objects and subjects of the agricultural economy for the subsequent adoption of sound management decisions, ensuring a technological breakthrough in the agro-industrial complex and achieving significant (at times) productivity growth at agricultural enterprises [3].

T. Antipova (Ed.): ICADS 2021, AISC 1352, pp. 81–90, 2021.
https://doi.org/10.1007/978-3-030-71782-7_8

The draft national program "Digital economy" and the departmental project for 2019–2024 of the "Digital agriculture" plans to increase productivity, to reduce the share of material costs in the unit cost of agricultural products, reduce costs for business and administration [4].

Today, the basic regulatory, logistical, technological and organizational conditions for active use of information technologies in practice of municipal management was formed as a result of administrative reform, which is reflected in the presence of municipalities, and many municipal enterprises official, regularly supported and updated Internet sites and the ability to receive many municipal services in electronic form, work multifunctional domain services, organization of electronic document management and interdepartmental interaction [5].

Nevertheless, the expediency of optimizing the activities of local authorities in terms of the use of information technologies is emphasized by Esipov S. V., Sergeychuk A. V. [6], Salabutin A. V. [7], Zaitseva M. V., Loginova A. A. [8].

The main problem of implementing digital technologies in rural areas is the lack of the necessary basic information infrastructure that provides rural residents with access to information resources [9, p. 1297].

At the moment, a draft Federal law "On amendments to the Federal law "On communications" has been prepared, which provides for the following key changes:

– refusal of services using collective use points (CUP);
– installation of access points in localities with a population of 100–500 people.

As of the beginning of 2019, 73.2 households have broadband Internet access. Up to 97% of the state will provide broadband access to the Internet by 2024.

Connecting remote and sparsely populated areas to broadband access as of 2019 is shown in Fig. 1.

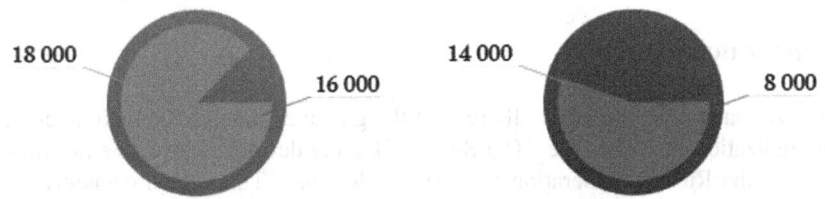

Localities with 500-10, 000 inhabitants: out of 18,000, there are already 16,000   Localities with 250-500 inhabitants: out of 14 000, there are already 8000

Fig. 1. Connecting remote and sparsely populated areas to broadband access as of 2019 [10].

Provalenova N. V. highlighted the following obstacles that hinder the penetration of information technologies in rural areas:

– low income of the rural population, which does not allow paying for expensive equipment and making monthly payments for telecommunication services;

- lack or insufficient development of digital skills in the rural population, which makes it impossible to use all the advantages of the Internet network;
- the lack of available services and on-premises applications;
- high cost of creating information infrastructure in rural areas, with low investment attractiveness [9].

Kuratova L. A. believes that the information infrastructure performs certain functions that determine its role in the development of rural areas:

- provides an opportunity for rural residents to actively participate in public life, providing access to information about current events;
- provides remote provision of services in the field of education, health, public services, etc.;
- integrates municipalities into a single socio-economic system [12].

The works of Knyazkova V. S. [13], Dubinina M. G. [14], Lukovtseva A. K. [15], Kazakova N. V. [16], Murashova N. V. [17] and others are devoted to the study of digital transformation heterogeneity in time and territorial implementations, as well as to the study of effectiveness of digitalization and informatization processes for rural economies.

1.1 Digitalization of One of the Most Promising Branches of Social Infrastructure in Rural Areas – Housing and Communal Services

The quality of life of rural residents differs significantly from the urban standard of living, especially in terms of social amenities and infrastructure development, which attracts young people to cities, and ultimately the shortage of the staff, threatens the further development of rural areas.

Undoubtedly, the recognition of social infrastructure as an important condition for the development of rural areas, which guarantees the necessary level and quality of life in rural areas, is the main factor in the efficiency of agricultural production.

Under the social infrastructure of rural territories, N. V. Provalenova understands a complex interconnected complex of economic activities aimed at meeting material and spiritual needs, as well as creating conditions for the safe life of the rural population, ensuring the reproduction of highly qualified personnel for agriculture and sustainable development of rural territories [18, p. 129].

One of the most promising sectors of social infrastructure in rural areas is the housing and utilities sector, where various digital technologies are currently being introduced to improve the development of social infrastructure.

The most significant areas of digitalization in the housing and utilities sector are (Fig. 2).

1. Federal state information system of housing and communal services (hereinafter GIS HCS), established on the basis of Federal law "On state information system of housing and communal services" according to which from 1 July 2017, the suppliers

are obliged to place in the system information of the metering instruments, resource levels, etc.

Information system for managers of companies "Domoport" in accordance with the requirements of the program "Digital economy of the Russian Federation" and FL-209 provides all the facilities to meet driving-organizations of the Russian legislation and posting information in a GIS utilities.

The level of readiness of homeowners' associations, management companies and resource-supplying organizations to implement digital technologies is increasing, as well as assessment of the opportunities and potential of the housing and utilities market. There is an objective need to change the reforms in the housing sector in the period of digital transformation of the economy and assessing their impact on the development of housing services and activities of ICT companies.

The main objectives of the GIS utilities are: the formation of a convenient socio-oriented content in the housing sector for citizens to obtain in one place all reliable information; monitoring the actual state of transactions between participants of the housing sector; the formation of the unified register of management companies and other organizations of housing and communal complex; control activities of managing organizations; maintaining information about regional targeted programs for capital repairs of apartment buildings, regional capital repair programs, short-term plans for the implementation of regional capital repair programs, receiving reports on the implementation of these programs and plans; improving the efficiency of interaction between departmental information systems, information systems of housing and utilities market participants, etc. [19, p. 177].

This system is currently not fully operational, as the system is constantly being updated, improved, supplemented, and depends on the effective work of responsible suppliers.

In addition, in our opinion, it is advisable to present information from the State Information System of Housing and Communal Services in the context of urban and rural settlements in order to optimally organize the process of making managerial decisions to reduce the difference in the level of comfortable living of the population.

2. Crowd technologies are quite effective tools that are actively used by many commercial organizations, which allow us to consider the possibility of their application in the management of housing and communal services. Crowd technologies are technologies of interaction with the public that allow solving various tasks related to increasing social activity of citizens and involving them in self-government processes.

The idea of co-financing certain works and programs is not new, despite the fact that these concepts have entered the professional lexicon relatively recently. This is due to the development of the digital economy and the Internet technologies, as fundraising are carried out using special crowd-funding platforms-developed Internet sites. Crowd technologies are very actively used abroad, in Russia this mechanism does not work as much as we would like, the reason, among other things, is the lack of legislative

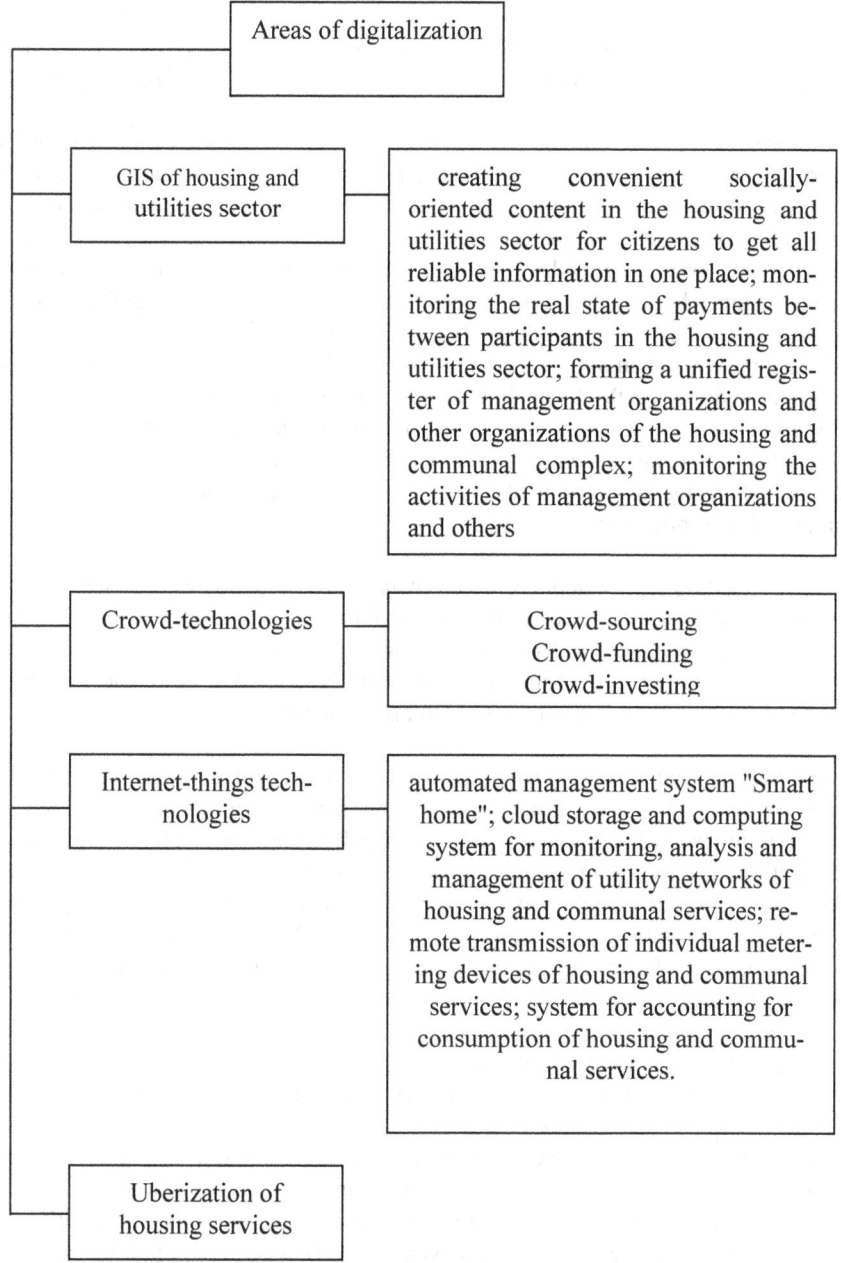

Fig. 2. Significant areas of digitalization in the housing and utilities sector

regulation of the process. Mostly, crowd-funding platforms in Russia are used to finance creative projects, for example, to create a movie, computer game, etc.

In the housing and utilities sector, it is advisable to use crowd technologies as a tool for interacting with the population and developing startups and innovations in the industry. In addition, a crowd-funding platform can function as a launching pad for public-private partnerships.

The process of crowd-funding is usually implemented through a crowd-funding platform on the Internet and gives each user the right to get acquainted with existing projects, assist those who like them, activates the desire to get original content for its financing, and in addition, come up with and implement their own idea and form on the basis of the project, in order to subsequently acquire funding from users for its implementation. Thus, it can act as a specially created platform or a site for a project, which organizes a variety of crowd-funding projects [20, p. 12].

But the main problem here is that the housing and communal services management bodies have to first of all think not about the introduction of information technologies in this area, but about improving the efficiency of their activities, profitability, the problem of the quality of housing and communal services, etc. In almost most municipalities, the problems of housing and communal services are not solved due to the presence of low incomes and corruption in this area [21].

3. Internet of things technologies, which include: automated control system "Smart home"; cloud storage and computing system for monitoring, Analytics and management of engineering networks of housing and communal services; system for remote transmission of readings of individual metering devices of housing and communal services; system for accounting for consumption of housing and communal services.

Creating a unified information environment for control and supervision in the housing and utilities sector, drivers and barriers to the development of the IoT ecosystem (Internet of things) and the industrial Internet in public utilities, creating infrastructure for collecting and storing information, using block-chain technologies in housing and utilities, unmanned and intelligent robotic systems, and much more form a platform for digital transformation of housing and utilities. In this regard, there is a need for active participation of representatives of state regulatory bodies, regional and municipal authorities, housing and utilities – service, operating companies and resource suppliers, Telecom operators, system integrators, IoT-startups, manufacturers of infocommunication equipment and solutions, communication departments and services, information technology and information security from various sectors of the economy, consultants and industry experts, and industry media.

IoT is necessary both for the things themselves (sensors, devices, actuators), and for platforms that manage production and administrative processes. In this regard, the Internet of things market is rapidly developing, and in several vectors at once. With fairly simple products, such as smart metering devices, the development of platforms and their interconnection into complexes of industry and inter-industry solutions gradually begins. In addition, new technologies are emerging, standards for the Internet of things are being formed, and this path will be quite long.

The Smart housing and utilities platform allows you to accept and process applications and requests as quickly as possible, take into account resource consumption, work on accruals and payments, provide services and individualize the information of residents up to the address by name. Communication with the management company becomes available from anywhere in the world.

Figure 3 shows a diagram on the market volume of dispatching and automation systems, where the ASCEM is an automated system for commercial electricity metering, which fully provides remote collection of information from specially equipped metering devices and transmits this information to the collection point. ASCAHE – an automated system for commercial accounting of heat energy; ASCGM is an automated system of commercial metering of gas.

As you can see, in the forecast version for the end of 2019 and in the future for 2020, there will be an increase in the use of automated data transfers. The state policy implemented in the country also contributes to this to a certain extent.

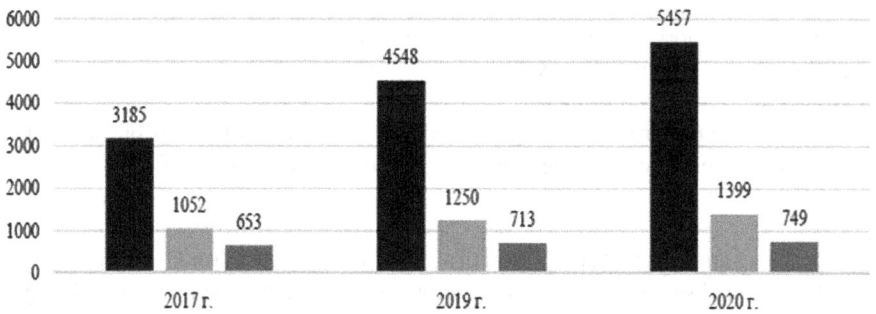

Fig. 3. Market size of dispatching and automation systems [22]

4. Uberization of housing services. Uberization means using computer platforms to conduct direct transactions between service providers and their customers. There is also a broader understanding: uberization is the use of services by market participants that coordinate work and exclude intermediaries from the service – consumer chain. In any case, this phenomenon is based on a certain peer-to-peer platform (information and communication technology), which serves as the basis for the formation of a stable small-or medium-scale economic system [23, p. 53].

Despite the current trends in the development of digitalization, the percentage of enterprises in the housing and utilities sector that seek information solutions is very small across the entire industry.

According to Rosstat, only more than half of the total number of organizations in 2019 used digital solutions. Some businesses do not seek digitalization due to lack of funds for this and unwillingness to reduce payment according to standards.

Digitalization in the housing and utilities sector increases the level of transparency, and most housing and utilities companies are not interested in this. Resource-supplying companies that want to use digital technologies to reduce their debt to them do not have direct access to residents and funds for equipment modifications. The solution to this problem remains with the state. It may require only smart meters to be installed starting from a certain date.

2 Results

The main problem with the introduction of digital services in housing and utilities is related to human risks and is rather at the psychological level. In addition, owners and managers are not always ready to change the paradigm in their minds; this is difficult even for fairly advanced people. A big question is also with the staff, which in many cases will have to be massively reduced or retrained. In the era of digitalization, housing and utilities companies face a number of challenges: excessive regulatory pressure, rapid development of technologies, increasing competition, and difficult market conditions. But these problems are solved not with the help of digitization.

Approaches to managing the housing and utilities sector, taking into account digital transformation, should assume the following:

1. Regulatory authorities, first of all, determine the rules for the operation of private enterprises for automating services.
2. Protect the rights of consumers.
3. Guarantee the performance of high-quality public services with a visible end result.
4. Regularly reconstruct and upgrade municipal infrastructure-tour of the new digital technologies.

Thus, being an important component of social infrastructure of rural territories, housing and communal services is one of the factors of sustainable development of the agricultural sector, which not only contributes to the improvement of rural areas, improving the quality of life of the rural population, but also provides the improvement of technological processes in agricultural production, increases productivity, which ultimately has a positive effect on economic efficiency of agriculture in general.

Thus, in the context of digitalization of the housing sector, it is necessary to focus on the following tasks:

– strengthen the role of the state in the development of the digital economy and the digital transformation of the utilities, increase the level of readiness of private housing services management companies and resource supplying organizations to the introduction of digital technology, to assess the capability and capacity of the utilities market;
– to activate the process of forming public-private partnership for the development of the digital economy in housing and communal services, aimed at implementing the national project "Creating a comfortable urban environment" within the framework of the digital economy development program and its impact on the digital transformation of housing and communal services;
– to step up work on the implementation and further development of GIS for housing and utilities, as well as the creation of a unified information environment for control and supervision in the housing and utilities sector;
– to step up work on creating infrastructure for collecting and storing information, using block-chain technologies in housing and utilities, and implementing bespoke and intelligent robotic systems;

- analyze the state and prospects of using renewable energy sources in housing and utilities heating systems, develop intelligent decision support systems in emergency and crisis situations that ensure information and cyber security of housing and utilities, form digital platforms for intelligent management of energy savings in the housing and utilities sector;
- to form a unified information and analytical system for monitoring and controlling the city's housing and utilities sector to implement the "Safe city" concept. Use GLONASS/GPS navigation systems in IOT/M_2M housing and utilities systems.

References

1. Decree of the President of the Russian Federation № 203 of May 9, 2018 "On the strategy for the development of the information society in the Russian Federation for 2017–2030". https://kremlin.ru/acts/bank/41919. Accessed 20 Sept 2020
2. The Digital economy of the Russian Federation program (Decree of the Government of the Russian Federation No. 1632-R of July 28, 2017). https://government.ru/docs/28653/. Accessed 20 Sept 2020
3. Frolova, O.A., Shamin, A.E., Shavandina, I.V., Kutaeva, T.N.: Smart village. problems and prospects in Russia. In: Proceedings of the International Conference «The 2018 International Conference on Digital Science», DSIC 2019: Digital Science 2019, pp. 400–486 (2019)
4. Digitalization of agriculture in Russia: steps, results, plans. https://gpsgeometer.ru/blog/tsifro vizatsiya-selskogo-hozyajstva-v-rossii-etapy-itogi-plany. Accessed 8 Nov 2020
5. Akifieva, L.V., Bolshakova, Y.A., Ilicheva, O.V., Kuchin, S.V., Belousova, O.A.: Problems of use of information technologies for management of social and economic development of the municipal unit. In: Antipova, T., Rocha, Á. (eds.) Digital Science 2019. DSIC 2019. Advances in Intelligent Systems and Computing, vol. 1114, pp. 470–479. Springer, Cham (2019). https://doi.org/10.1007/978-3-030-37737-3_40
6. Esipov, S.V., Sergeychuk, A.V.: Optimization of activities of local self-management authorities to provide municipal services in conditions of digitalization. St. Petersburg Econ. J. **2** (2019). https://cyberleninka.ru/article/n/optimizatsiya-deyatelnosti-organov-mestnogo-samoupravleniya-po-predostavleniyu-munitsipalnyh-uslug-v-usloviyah-tsifrovizatsii
7. Salabutin, A.V.: Issues of digitalization of municipal government: trends and problems. Sci. Without Borders **5**(45) (2020). https://cyberleninka.ru/article/n/voprosy-tsifrovizatsii-munits ipalnogo-upravleniya-tendentsii-i-problemy
8. Zaitseva, M.V., Loginova, A.A.: The role of municipal employees in the development of the territory. Int. J. Humanit. Nat. Sci. 1–2 (2019). https://cyberleninka.ru/article/n/rol-munitsipa lnyh-sluzhaschih-v-razvitii-territorii
9. Provalenova, N.V., Shamin, A.E.: Digitalization as an instrument of the development of social infrastructure of the rural areas. Econ. Entrepreneursh **10**(111), 1297–1303 (2019)
10. Ministry of digital development, communications and mass media of the Russian Federation. https://digital.gov.ru/uploaded/2019results/home.html. Accessed 12 Nov 2020
11. Kasimova, Zh.V., Kasimov, A.A.: Digital transformation of rural territories. Bull. NGIEI **8**(111), 117–126 (2020). https://doi.org/10.24411/2227-9407-2020-10079
12. Kuratova, L.A.: The peculiarities of informational infrastructure of the rural areas. Bull. KRAGSiU. Ser.: Theory Pract. Manag. **15**(20), 40–46 (2015)
13. Belyatskaya, T.N., Knyazkova, V.S.: Digital divide in modern information society. Econ. Sci. Today **10**, 209–217 (2019)

14. Dubinina, M.G.: Uneven development of digital economy in federal districts of Russia. Manag. Sci. Scientometr. **3**(14), 368–399 (2019)
15. Lukovtseva, A.K.: Economic inequality and its impact on the development of national innovation system. Actual Probl. Law **4**(68), 61–75 (2020)
16. Bannikov, S.A., Zhiltsov, S.A., Kazakova, N.V.: Digitalization trends and causes of digital divide in rural areas. Bull. NGIEI **11**(114), 137–149 (2020). https://doi.org/10.24411/2227-9407-2020-10112
17. Kovalenko, E.G., Murashova, N.V.: Digitalization and informatization as factors of sustainable development of territories. Bull. NGIEI **10**(113), 119–128. https://doi.org/10.24411/2227-9407-2020-10100
18. Provalenova, N.V.: The role and place of housing and communal services in the social infrastructure of rural areas. Sci. Bus. **1**(103), 128–130 (2020)
19. Potapenko, O.S.: Use of modern digital technologies in quality management of housing and utilities services. In: Collection of Articles of the I all-Russian Scientific and Practical Conference: Challenges of the Digital Economy: Conditions, Key Institutions, Infrastructure, pp. 176–180 (2018)
20. Efimova, T.B.: Housing and utilities as one of the areas of crowd-sourcing in municipal management. In: Efimova, T.B., Kolotilina, M.A., Levina, Yu.V., Kuznetsova, E.Yu. (eds.) Scientific Forum: Technical and Physical and Mathematical Sciences: Collection of Articles Based on the Materials of the XVI International Scientific and Practical Conference, vol. 6, no. 16, pp. 11–19 (2018)
21. Akifieva, L., Polyakov, M., Sutyagina, N., Zvereva, I., Zhdankina, I.: Use of, information management technologies in housing and communal utilities. In: Antipova, T. (eds.) Integrated Science in Digital Age 2020. ICIS 2020. Lecture Notes in Networks and Systems, vol. 136, pp. 369–374. Springer, Cham (2020). https://doi.org/10.1007/978-3-030-49264-9_33
22. Krylova, A.: Digitalization of housing and utilities services: pilot's time. IKSMedia. https://www.iksmedia.ru/articles/5501597-Czifrovizaciya-ZHKX-vremya-pilotov.html
23. Terelyansky, P.V., Zyabkin, A.S.: Digital technologies and development of new services based on the management company in the field of housing and communal services. E-Management **1**, 50–58 (2020)

Coronavirus Pandemic as a Challenge in Debt Sphere

Svetlana Tsvirko[✉] [iD]

Financial University Under the Government of the Russian Federation,
Leningradskiy prospekt 49, Moscow 125993, Russian Federation
s_ts@mail.ru

Abstract. The aim of the study is to carry out analysis of the coronavirus pandemic as a challenge in debt sphere and to develop forecasts regarding debts in the global economy, as well as some recommendations, connected with debt management. The trend in debts' dynamics such as simultaneous increase in volume of advanced and developing countries' debts is analyzed. Risks in debt sphere are detected, including contingent liabilities, increases in debt servicing costs in developing countries and problems of refinancing. Vicious circle of different types of debt (public and private) is revealed. The paper considers the modern problems with debt relief. It was shown that current situation in terms of the prospects of the world economy, structure of debts and specifics related to creditors complicates the process of debt restructuring. Some recommendations for more efficient management of debts' presented in the paper are as follows: economic growth as main provision for deleveraging; increase in accountability and transparency of all types of debt; full coverage of internal and external risks; activization of the usage of new financial instruments; improvements in regulation, aimed at the participation of private creditors in debt settlement.

Keywords: Pandemic · Crisis · Public debt · Private debt · Debt relief · Management · Efficiency

1 Introduction

Current situation with COVID-19 and quarantine measures carried out by most countries of the world and the widespread reduction in economic activity is unprecedented. There is no experience in creating conditions for recovery from a shock of such magnitude. Both developed and developing countries urgently need significant funding to sustain economic activity, especially jobs. Ant-crisis measures undertaken by the governments of different states lead to increase in borrowing and, accordingly, public debt. Debt overhang usually slows investment and hinders economic growth for years, all this reduces the well-being of the population. So, public debt management is one of the most important directions of economic policy and a constant search of its efficiency is one of the purposes of the government in each state.

There is significant amount of publications devoted to the problem of debt in the global economy. Among the modern authors, who contributed to the creation of the

© The Author(s), under exclusive license to Springer Nature Switzerland AG 2021
T. Antipova (Ed.): ICADS 2021, AISC 1352, pp. 91–102, 2021.
https://doi.org/10.1007/978-3-030-71782-7_9

foundations of debt management we should name the following researchers: C. Reinhart [1–3]; K. Rogoff [1–4, 7, 8]; V. Reinhart [3]; A. Abbas, A. Pienkowski [4]; S. G. Cecchetti, M. Mohanty and F. Zampolli [5], M.A. Kose, P. Nagle, F. Ohnsorge and N. Sugawara [6]. At the same time, a number of issues related to the latest trends in the debt sphere remain not revealed and are of scientific interest. There are many publications devoted to methods of debt management, developed by J. Bulow with co-authors [7, 8]; S. Claessens with co-authors [9]; B. Clements with co-authors [10]; P. Krugman with co-authors [9, 11]. The research gap is connected with the fact that many recommendations for prudent debt management and restructuring of debts in case of problems with repayments were formulated for the conditions that are quite different from current situation in terms of the prospects of the world economy, structure of debts and specifics related to creditors.

There is a fast-growing list of literature devoted to the studies of the COVID-19 epidemic and its impact on national and global economics. For instance, T. Antipova reveals coronavirus pandemic as Black Swan event for the world [12]. C. Hevia and P.A. Neumeyer show COVID19 as a perfect storm for emerging economies [13]. But it should be noted that we are at the early stage of understanding of the pandemic effects and corresponding ant-crisis measures in the global economy.

The purpose of this study is to consider the coronavirus pandemic as a challenge in debt sphere, to develop analysis regarding the volume and structure of debts in the global economy and to formulate some recommendations.

The findings of this research can be useful for the financial authorities that deal with debt management (Ministries of Finance, Central Banks, Debt Agencies, etc.) and participants of the financial markets.

2 Methodology of Research

The methodological basis of the research are system approach, fundamental provisions of economic theory and the theory of international economic relations. We applied methods of mathematical modelling, comparative analysis, expert assessments. Mathematical modelling (Eq. (1)) helped to speculate the idea about possibility to reduce the volume of public debt. The method of comparative analysis was used to determine the differences between debt situation in advanced and developing countries currently; as well as to reveal the differences between present situation and previous periods of debts' restructuring.

Statistical data of international economic and financial organizations (International Monetary Fund, World Bank), for instance, Global Financial Stability Report by International Monetary Fund [14] and International Debt Statistics by the World Bank [26] as well as publications of foreign [1–11] and domestic authors [20, 24] were used as sources of information.

3 Overview of the Situation in the Global Economy and Global Finance in 2020

Global economy is currently experiencing a combination of a severe health crisis, a deep recession in many countries and regions, and unprecedented social consequences for all countries. The governments of many countries impose mitigation and lockdown policies that limit social interactions to stop the spread of COVID-19. These measures have been effective in taming the spread of the disease but caused significant economic costs.

The COVID-19 pandemic has negatively affected a considerable number of sectors of the economy and threatened the existence of a large percentage of companies around the world. The situation in many economies around the world is characterized by a significant decline in business activity. Contraction in global economic activity is unprecedented, with global growth projected at –4.4% for year 2020, according to the October 2020 World Economic Outlook. Both advanced and developing economies are affected, with more than 85% of countries around the world expected to see subzero growth in 2020 [14, p. 1].

The most significant losses in the COVID-19 situation were incurred by service sector [15]. Deterioration is noted in the wholesale and retail trade, the tourism industry and the restaurant business, sectors providing other services. Small and medium-sized businesses have difficulties. Significant decrease in the consumption of energy resources is forecasted. There are sharp declines in car travel, flights; as well as lower electricity and heat consumption in restaurants and offices. There is a threat of massive bankruptcies in the passenger transport, tourism and hospitality industries.

A significant decrease in production is observed in industrial sectors focused on investment demand (mechanical engineering, production of electronic and electrical devices, etc.). Some segments of mechanical engineering will also be challenged. For example, airlines are unlikely to make new orders for aircraft in the near future. Overall, companies in many industries will need to rethink their strategies and business models.

Additional negative factors were the changes in labor markets caused by the pandemic, associated with an increase in underemployment and a decrease in income of the population, and a decrease in the inflow of remittances in some developing countries. As a result, domestic consumer demand declined; there is a "cooling" of credit activity and a general decline in the indicators of enterprises.

In general, the following factors are negatively affecting national economies: weakening domestic and external demand, measures of social isolation and limited communication between states, uncertainty about the prospects for economic development, deteriorating business sentiment, as well as disruption of value chains. Many governments were forced to make fiscal transfers to support consumption during the lockdowns [15].

Developing countries have been hit by COVID-19 significantly more than advanced countries. It can be explained by overreliance on revenue from foreign trade and international tourism. Such countries are also more financially vulnerable and faced significant capital outflow. As it is revealed by IMF experts, the extraordinary level and speed of portfolio outflows from February to April 2020 created serious disruptions for emerging markets [14, p. 20]. Major economies have been able to implement massive monetary and fiscal ant-crisis measures, that are not affordable to other countries. For emerging markets limited policy space can prevent necessary anti-crisis policies.

4   Analysis of Debts' Dynamics and Structure

By objective reasons the modern world economy can't develop without movement of borrowed funds between national economies, as well as between their entities. The global economy at the present stage is characterized by reliance on debt. Debt can be classified as public and private as well as it can be external and internal.

Public debt arises when government spending begins to exceed revenues and budget deficits become chronic; at the same time, the coverage of the budget deficit is carried out not by monetary emission methods, but by borrowing.

The growing public debt, on the one hand, can be associated with the implementation of state regulation of the economy, aimed at ensuring progressive shifts in the structure of the economy, but, on the other hand, in many cases, it is a reflection of the crisis processes taking place in the country and requires the adoption of stabilization measures. As it is stressed by one of the most famous researches in the sphere of debt, World Bank Chief Economist C. Reinhart: «Debt is what enables governments to have extra resources they need to invest in health systems, education, or infrastructure... If you have a debt problem, all those ambitions suffer. That's why it's important to get the debt onto sustainable ground as quickly as possible. We can't afford another lost decade» [16].

Debt problems often become the most difficult economic problems to deal with. A sharp increase in the growth of government debt is always caused by an economic downturn, a budget deficit, and the inability of the government of a country to keep the dynamics of the financial situation under control. Unsustainable debt makes it difficult to reform the economy, slows down economic growth, and has a negative impact on the state's position in the world community.

Dynamics of general government debt in advanced economies and in emerging economies during the period 1880–2020 years are revealed in Fig. 1.

According to IMF, even prior to the pandemic, public and private debt were already high and rising in most countries, reaching 225% of GDP in 2019, 30% points above the level prevailing before the global financial crisis. Global public debt rose faster over the period, standing at 83% of GDP in 2019. In 2020, global general government debt is estimated to make an unprecedented increase up to almost 100% of GDP [17, p. ix]. We can see nearly vertical part of the curve, that shows the jump of public debt both in advanced and emerging markets economies. The patterns of such increase are similar to the conditions of the Global Financial Crisis (especially in advanced countries) and, unfortunately, it is similar to the circumstances of the World War I and World War II. The major drivers of such significant growth of the public debts are as follows: sharp contraction in economic activity and decline in revenues to the budgets; increase in public expenditures; rises in the primary deficits and as consequence – necessity to increase borrowing.

As for low and middle-income countries, additionally to problems, caused by increased expenditures and decreased revenues of the budgets, they faced another challenge, that was connected with high levels of capital flight from their financial markets. This in turn has led to devaluation of the national currencies, aggravated the burden of external debt and identified the problem of debt refinancing.

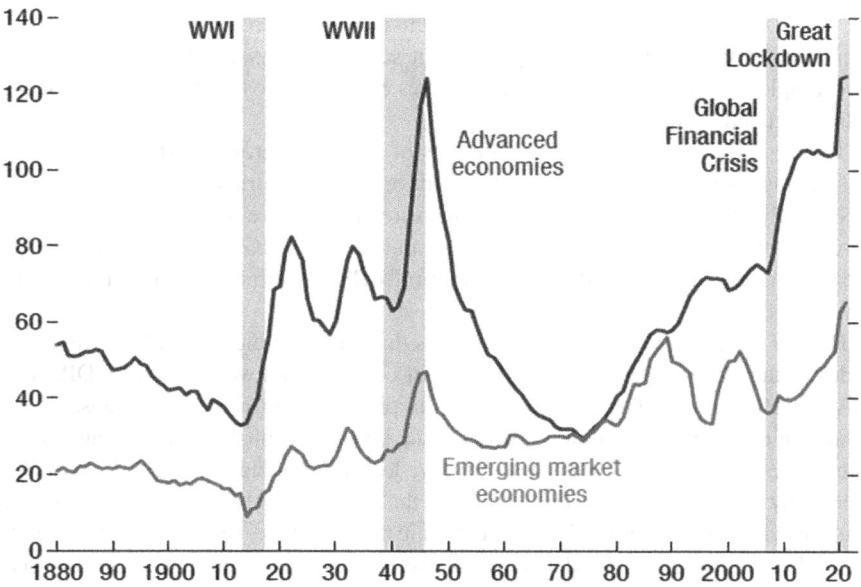

Fig. 1. Historical Patterns of General Government Debt, percent of GDP. Source: [17, p. 2].

Advanced economies show higher rates of public debt to GDP in comparison with emerging markets economies. As it can be seen from Fig. 1, in 1970–1990 the ratios of general government debt to GDP in % for advanced and developing countries were rather close. In previous several decades the problem of over-indebtedness has been observed mainly in poorest countries and emerging markets. In recent years the debt of the leading developed countries, whose politicians are faced with the need for significant fiscal consolidation, has spun out of control.

A simplified model for public debt can be applied to calculate the indicator of public debt to GDP:

$$(D/G)_t = \left[\left[\frac{1 + i_t}{1 + g_t} \right] (D/G)_{t-1} \right] - Pb_t, \tag{1}$$

where $(D/G)_t$ is the ratio of public debt to GDP in the future period, g_t is the forecast of growth in nominal GDP, i_t is the interest rate on government obligations, $(D/G)_{(t-1)}$ is the current debt to GDP, Pb_t is the positive primary balance, that is, the fiscal balance minus interest payments.

Equation (1) stresses the role of the growth rates for GDP, interest rates on government obligations, current ratio of public debt to GDP and the primary balance as factors influencing the debt volume relative to GDP. Already accumulated significant debt, low growth rates of projected GDP and deficit of the budget will lead to further increase of the public debt to GDP ratio. As for the current interest rates, we can forecast positive influence of this factor on the situation with public debt. In several previous years low interest rates have been observed on the global capital market. Thus, even if government debt increased, interest payments increased slightly. In 2020, as anti-crisis measures the Federal Reserve System in the US and the European Central Bank have continued to rely on a policy of quantitative easing, respectively, low interest rates, although earlier planned to finish carrying out this policy. But such decisions also have shortcomings. For example, they lead to negative interest rates on bank deposits, that increases banks' costs and cuts their profits.

Taking into account the current forecasts of economic growth and budget deficits, public debts in the majority of advanced and developing countries will continue to grow. Accordingly, there is a need to develop measures to stabilize and make an attempt to decrease public debt. To reduce the ratio of public debt to GDP, a significant budget surplus is required, that is sufficient to pay off debt, or an increase in nominal GDP growth.

History offers some examples of countries that have significantly reduced their debt burden. After World War II, the public debt in Great Britain was 238% of GDP, in the USA - 121% of GDP. These figures were sharply reduced (in the US because of two decades of considerable GDP growth and in the UK as a result of spending cuts). There were successful deleveraging episodes of Sweden and Finland in the 1990s. As it was revealed by the experts of McKinsey based on the experience of Sweden and Finland, the critical markers of progress before the economic recovery takes off are the following: the financial sector is stabilized and lending volumes are rising; structural reforms have been implemented; credible medium-term public deficit reduction plans are in place; exports are growing; private investment has resumed; and the housing market is stabilized, with residential construction reviving [18]. In the recent past, Canada reduced its public debt from 91% of GDP in 1995 to 51% of GDP in 2007, thanks to strong growth in the global economy and commodity exports. So, the circumstances of the deleveraging process were different; but it is clear that this process is politically challenging [19].

An important aspect of public debt and its management is the problem of contingent liabilities. Contingent liabilities are obligations that are not explicitly recorded on government balance sheets and that arise in the event of a particular discrete situation, such as a crisis [17, p. 55]. Based on the experience of different countries, S. Storchak has classified basic forms of the contingent liabilities: 1) state guarantees and state insurance; 2) pension obligations for future periods; 3) commitments of regional and local authorities; 4) obligations under the public-private partnership; 5) obligations of enterprises with state participation; 6) commitments to maintain the stability of the financial system [20, p. 253]. There has been a steady increase in contingent obligations in the world economy. Underestimation of the volume of public debt can be connected with the fact, that there is no complete comprehensive accounting of all contingent liabilities accumulated by governments. This in its turn leads to the lack of the authorities' ability to manage contingent liabilities and emerging risks in the sphere of debt.

Another type of the debt is private debt, that can include the debts of banks, corporates as well as debt of households. Nonfinancial corporate and household debt has trended upward for two decades, reaching almost 150% of GDP in 2019 and exceeding public debt by a large margin in most Group of Twenty countries. There was significant increase in corporates' borrowings in the first half of 2020, with $5.4 trillion secured by companies across the globe, including $3.9 trillion since the start of March 2020 [17, p. 23]. There can be risks for the public finances because of the nexus between private and public debt. The interconnectedness of different segments of debts and possible contagion are revealed in Fig. 2.

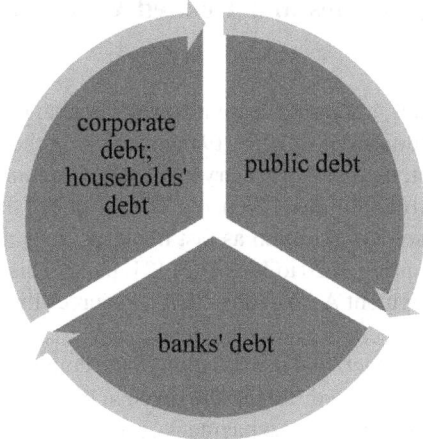

Fig. 2. Vicious circle of different types of debt. Source: compiled by the author.

Let's discuss the possible links between public debt, banks' debt and the debts of corporates and households.

1st stage. In the situation of economic slowdown and limited fiscal space the sovereigns start to face complexities with their debts. In case of the concerns, connected with credit-worthiness of the governments, there can be problems for banks, that hold corresponding governments' obligations. That was the case for example in 2011–2012 in euro area, when banks faced mark-to-market losses on their holdings of governmental bonds of some European countries. In this scenario there is usually a decrease in the demand for sovereign as well as banks' bonds and consequently increase in required yields on them. Sovereigns and banks face problems with servicing of their debts and with refinancing. Under such conditions banks' funding costs increase and are transferred to corporates and households through higher interest rates on loans.

2nd stage. This increases the probability of nonperforming loans from the side of corporations and households. The situations is often accompanied by further economic slowdown. Companies and individuals have lower incomes, that in its turn decreases tax revenues to the budget. Banks face losses. The problem of contingent liabilities begins to manifest. As an extreme there can be bankruptcies of financial institutions and non-financial firms.

3rd stage. Public sector funds are used to rescue falling banks and companies that have strategic importance for the economy of the country. As a result it increases the expenditures of the state budget and the overall situation worsens. The government is forced to further increase the debt or implement austerity measures.

Under current conditions vulnerabilities in public or private debt can exacerbate each other through sovereign-corporate linkages.

5 Debt Relief Approaches and Related Contradictions and Problems

Different types of countries, not just developing countries, can face temporary financing challenges under the conditions of COVID-19 pandemic. For these purposes there is the lending capacity of the International Monetary Fund in the amount of $1 trillion, of which about one-fourth is already committed [17, p. ix]. The volume of financial resources from the World Bank comprises $14.5 billion as fast track facilities and $160 billion in the 15 months starting from spring 2020 to June 2021 [21, p. ix]. This includes $50 billion of new International Development Association (IDA) resources through grants and highly concessional loans.

Taking into account the fact, that the situation in the countries differs, the measures to mitigate debt problem are also different: the revision of the loan payment schedules, introduction of a moratorium on interest payments, provision of short-term or long-term capital.

As part of the plan for post-crisis economic recovery, debt relief measures are required. The International Monetary Fund, the World Bank, regional development banks and other international organizations support debt relief programs to help countries get through the COVID-19 epidemic. For instance, responding to a call from the World Bank and the International Monetary Fund, the G20 endorsed the Debt Service Suspension Initiative (DSSI) in April 2020 to help up to 73 of the poorest countries manage the impact of the COVID-19 pandemic [16]. It is a temporarily suspension of the payments by poor countries to repay official bilateral debt. Access to grants and concessional financing is necessary for some countries with low-income developing states.

Contradictions and problems in debt restructuring are connected with the structure of debts in terms of creditors and forms of borrowings. Over some previous years when low interest rates were observed on the global capital market many developing countries were able to increase their borrowing in the form of bonds. It helped to raise important financing; at the same time debt servicing costs have grown significantly. What complicates the management of the debts is the fact that the share of commercial creditors has risen: 69% of developing countries' debt is now owned to commercial creditors compared to 41% in 2010 [22, p. 98]. It means that initiatives by international organizations with suspension of debt payments will not cover all problematic debts and debtor countries. The G20 has called on private creditors to participate in moratorium for debt payments but as of September 2020 there has been no private participation [22, p. 98].

6 Discussion

Current conditions of high and simultaneous increases in financing in different countries lead to actualization of the debt problem. For countries with unsustainable debt, options for debt restructuring must be considered. C. Arellano, Y. B. and G. P. Mihalache show that debt relief programs can have profound positive effects: debt relief supports consumption, can reduce the severity of the debt crisis, and can save lives [23]. In this context, their research paper suggests that the recent debt relief policies promoted by the International Monetary Fund and other international organizations, are right on target to combat the costs associated with COVID-19. We do not fully share the optimism demonstrated for debt relief.

Unfortunately, the prerequisites and necessary conditions for active debt cancellation have not been created. As it was proved by E.A. Zvonova [24, p. 15], there is a high probability that globalism, especially hyperglobalization, will give up positions, the world economy will return to a bloc confrontation. The dominant trends in the future will be regionalization and protectionism. In this situation, some creditors will be reluctant to restructure the debts owed to them and to indulge debtors. The possibilities of debt relief will depend on many factors, including the probability of repayment of the remaining part of the debt, types of the creditors (official or private) and their incentives. That is why detailed statistical information on all of the aspects of the debts is of crucial importance.

There are changes in the world economy, structure of debts and specifics related to creditors that can make management of the debts more complex.

The difficulty in the work of international organizations, including the IMF, World Bank, regional development banks is due to the fact that an increase in allocated funds, along with a moratorium on debt repayment, will put the international financial institutions themselves in a difficult financial situation. The volatility of payments from middle-income borrowing countries can reduce aid to the poorest countries. Despite all the difficulties of the multilateral cooperation, it is necessary in the sphere of debt restructuring.

The conducted research does not claim to fully highlight the key problems of debt management. As areas for further investigation we can name the following topics: approaches to increase accountability and transparency of debt management, including new technologies, including Distributed Ledger Technology (DLT); detailed analysis of the methods of debt restructuring and accumulated experience of different counties in this sphere in previous decades; practice of some countries, that successfully conducted deleveraging in their economies.

7 Conclusion and Recommendations

There is some kind of consensus about the fact, that the lockdown policies are useful for alleviating the health crisis, caused by COVID-19. But such measures carry large economic costs that can finally lead to debt crisis. Both advanced and developing countries deal with the problem of debt nowadays. But the ability of different countries to respond to pandemic crisis and challenges in debt sphere is different.

We can make some general and some more specific recommendation in connection to debt management.

1. As for general recommendations, the main provision for successful deleveraging is economic growth. Governments should prepare conditions and stimulus for activisation. There should be measures to support safe and successful reopening; programs

to increase employment and economic activity; facilitation of the transformation to a post-pandemic economy that should be more resilient, inclusive and greener. Different countries can have their own paths to deleveraging.

2. To improve the quality of decision-making, it is necessary to increase accountability of debt. As it was highlighted by T. Antipova [25], the role of accounting increases under the conditions of the financial crisis. For example, in response to an urgent need for greater debt transparency, the latest edition of the International Debt Statistics (IDS) report [26] provides more detailed and more disaggregated data on external debt than ever before in its near 70-year history (including breakdowns of what each borrowing country owes to official and private creditors in each creditor country, and the expected month-by-month debt-service payments) [16]. Both advanced and developing countries should improve accounting and prudent management of contingent liabilities.

3. It is required to enhance transparency in the public management sector in many countries. It is necessary to take into account that, for example, scenario forecasts perform an important function of forming the expectations of society and economic entities [27].

4. Debt management should be more holistic, comprehensive, with an understanding of existing relationships between different types of debt and the full coverage of internal and external risks.

As for some more specific recommendations, the following should be mentioned:

1. The variety of financial instruments should be increased. For example, already now developing countries are more actively using complex debt instruments, such as the central bank and currency swap arrangements that represent loans from other central banks.

2. Additionally to traditional, conventional bonds as a main tool that is used to borrow, it is advisable to attract financial resources by new types of obligations: sustainability bonds, green bonds, social bonds, vaccine bonds. It can be recommended for the potential investors to pay attention to the opportunities of Environmental, Social, Corporate Governance (ESG) investments.

3. There should be improvements in legislation connected with restructuring of the debts. The changes in the structure of creditors brought new challenges that should be addressed. Possible solutions aimed at participation of private creditors in debt settlement include incentives provided by Financial Stability Board, Bank for International Settlements and other regulators in the form of lessening of some requirements; or, contrary, compulsion to suspend debt payments, even in the absence of a force majeure clause in bond contracts. Legal measures should be aimed at preventing the predatory behavior of some creditors, among them are limits on amounts that can be obtained from the sale of government bonds, that were purchased at a significant discount; counteracting to blocking by predatory funds of decision-making in connection with debt restructuring.

Measures taken within debt management should be aimed at borrowing at favorable conditions, rationally allocating attracted financial resources and effective usage of them; monitoring of the volume and structure of the debt and countering the negative consequences of the crisis events.

Acknowledgements. The article was written based on the results of the research carried out at the expense of budget funds, which were provided to the Financial University as part of the state contract.

References

1. Reinhart, C., Rogoff, K.: This Time Is Different. Eight Centuries of Financial Folly. Princeton University Press, Princeton (2009)
2. Reinhart, C., Rogoff, K.: This time is different: a panoramic view of eight centuries of financial crises. Ann. Econ. Finance **15**(2), 215–268 (2014)
3. Reinhart, C., Reinhart, V., Rogoff, K.: Dealing with debt. J. Int. Econ. **96**(Supplement 1), S43-55 (2015)
4. Abbas, A., Pienkowski, A., Rogoff, K.: Sovereign Debt: A Guide for Economists and Practitioners. Oxford University Press, New York (2019)
5. Cecchetti, S.G., Mohanty, M., Zampolli, F.: The real effects of debt. Working papers № 352 (2011). www.bis.org/publ/work352.htm
6. Kose, M.A., Nagle, P., Ohnsorge, F., Sugawara, N.: Global Waves of Debt: Causes and Consequences. Advance Edition. World Bank, Washington, DC. License: Creative Commons Attribution CC BY 3.0 IGO (2020)
7. Bulow, J., Rogoff, K.: The buyback boondoggle. Brookings Pap. Econ. Act. **2**, 675–704 (1988)
8. Bulow, J., Rogoff, K.: Sovereign debt: is to forgive to forget? Am. Econ. Rev. **79**(1), 43–50 (1989)
9. Claessens, S., Diwan, I., Froot, K., Krugman, P.: Market-based debt reduction for developing countries. Principles and Prospects, The World Bank (1992). 62 p.
10. Clements, B., Bhattacharya, R., Nguyen, T.Q.: Can debt relief boost growth in poor countries? IMF Economic Issues, № 34. International Monetary Fund (2005)
11. Krugman, P.R.: Market-based debt-related reduction schemes. NBER Working paper № 2587 (1988)
12. Antipova, T.: Coronavirus pandemic as black swan event. In: Antipova, T. (ed.) Integrated Science in Digital Age 2020. Lecture Notes in Networks and Systems, vol. 136, pp. 356–366. Springer, Cham (2020). https://doi.org/10.1007/978-3-030-49264-9
13. Hevia, C., Neumeyer, P.A.: A perfect storm: COVID-19 in emerging economies. https://voxeu. org/article/perfect-storm-covid-19-emerging-economies. Accessed 1 Nov 2020
14. Global Financial Stability Report: Bridge to Recovery. International Monetary Fund. International Monetary Fund, October 2020. https://www.imf.org/en/Publications/GFSR. Accessed 1 Nov 2020
15. EDB Macroreview. Outlook update/April 2020. https://eabr.org/upload/iblock/881/EABR_M acroview_04_2020_EN_web_2.pdf. Accessed 25 June 2020
16. Debt Burden of Least Developed Countries Continues to Climb to a Record $744 Billion in 2019. https://www.worldbank.org/en/news/press-release/2020/10/12/debt-burden-of-least-developed-countries-continues-to-climb-to-a-record-744-billion-in-2019. Accessed 1 Nov 2020
17. Fiscal Monitor. Policies for Recovery. International Monetary Fund, October 2020. https:// www.imf.org/en/Publications/FM/Issues/2020/09/30/october-2020-fiscal-monitor. Accessed 1 Nov 2020

18. Debt and deleveraging: Uneven progress on the path to growth. Mckinsey & Company, January 2012. https://www.mckinsey.com/~/media/McKinsey/Featured%20Insights/Glo bal%20Capital%20Markets/Uneven%20progress%20on%20the%20path%20to%20growth/ MGI_Debt_and_deleveraging_Uneven_progress_to_growth_Report.pdf. Accessed 25 Oct 2020

19. Debt and (not much) deleveraging. Mckinsey & Company, February 2015. https://www. mckinsey.com/~/media/McKinsey/Featured%20Insights/Employment%20and%20Growth/ Debt%20and%20not%20much%20deleveraging/MGI%20Debt%20and%20not%20much% 20deleveragingFullreportFebruary2015.pdf. Accessed 25 Oct 2020

20. Storchak, S.: Contingent liabilities. AST, Zebra E, Moscow (2009). 443 p. (in Russian)

21. Saving Lives, Scaling-up Impact and Getting Back on Track. World Bank Group COVID-19 Crisis Response Approach Paper, June 2020. https://documents1.worldbank.org/curated/ en/136631594937150795/pdf/World-Bank-Group-COVID-19-Crisis-Response-Approach-Paper-Saving-Lives-Scaling-up-Impact-and-Getting-Back-on-Track.pdf. Accessed 25 July 2020

22. Financing for Development in the Era of COVID-19 and Beyond. UN, September 2020. https://www.un.org/sites/un2.un.org/files/part_ii-_detailed_menu_of_options_fin ancing_for_development_covid19.pdf. Accessed 1 Nov 2020

23. Arellano, C., Bai, Y., Mihalache, G.P.: Deadly debt crises: COVID-19 in emerging markets. NBER Working Paper No. 27275. https://www.nber.org/system/files/working_papers/w27 275/w27275.pdf. Accessed 25 Oct 2020

24. Zvonova, E.A.: Transformation of the world economy and the pandemic. Econ. Taxes Law 13(4), 6–19 (2020). https://doi.org/10.26794/1999-849X-2020-13-4-6-19. (in Russian)

25. Antipova, T.: The impact of financial crisis on the role of public sector accountants. In: Antipova, T., Rocha A. (ed.) Information Technology Science. Advances in Intelligent Systems and Computing, vol. 724, pp. 208–314. Springer, Cham (2018). https://doi.org/10.1007/ 978-3-319-74980-8_19

26. International Debt Statistics 2021. World Bank. https://openknowledge.worldbank.org/bitstr eam/handle/10986/34588/9781464816109.pdf. Accessed 1 Nov 2020

27. Tsvirko, S.: Informational technologies for the efficiency of public debt management in Russia. Advances in Intelligent Systems and Computing, vol. 724, pp. 104–113 (2018). https:// doi.org/10.1007/978-3-319-74980-8_10

Advances in Digital Education

The Use of Technology on Inclusive Education in Brazil - A Discussion on the Teacher and Student Views

Łukasz Tomczyk[1]([⊠]) [iD], Valéria Farinazzo Martins[2] [iD],
Cibelle Albuquerque de la Higuera Amato[2] [iD], Maria Amelia Eliseo[2] [iD],
and Ismar Frango Silveira[2] [iD]

[1] Pedagogical University of Cracow, Ingardena 4, Kraków, Poland
`lukasz.tomczyk@up.krakow.pl`
[2] Mackenzie Presbyterian University, R. da Consolação, 930 - Consolação, São Paulo, Brazil

Abstract. Among the various strands of discussion on Inclusive Education, a recurring topic deals precisely with the perception of the main educational actors involved in this process, namely, teachers and students. In this sense, the present article brings an analysis carried out based on two samples: one from undergraduate students in Pedagogy and the other from teachers working in the classroom. The results obtained through this research show that there is still a lack of knowledge about the use of ICTs for Inclusive Education (technology and methodologies) by students and teachers of Pedagogy in Brazil.

Keywords: Inclusion · Including education · Digital technologies · Brazil

1 Introduction

According to the Demographic Census carried out by the Instituto Brasileiro de Geografia e Estatística - IBGE (Brazilian Institute of Geography and Statistics) (IBGE 2010), in 2010, about 24% of the Brazilian population declared to have some disability. Given this, in 2015, Brazil approved Law No. 13,146, centered on social inclusion and citizenship (Estatuto da Pessoa com Deficiência-Statute for Persons with Disabilities) (Brasil 2015), which aims to guarantee and promote, equally, the exercise of rights and freedoms for the disabled person.

The current understanding of disability should be based on an inclusive model, as well as assistive or support technology, products and services aimed at people with disabilities, to propose solutions, methodologies, devices that reduce the person's limitations and the physical and social environment (Scatolim et al. 2017).

In the scope of school inclusion, this same law obliges schools to accommodate students with disabilities in regular education and to adopt the necessary adaptation measures, without any financial burden being passed on to the fees. Recent perspectives consider the use of Information and Communication Technologies (ICTs) in the classroom and, at the same time, offer a methodological transformation in teaching through hybrid education (that is, combining classroom and distance elements).

© The Author(s), under exclusive license to Springer Nature Switzerland AG 2021
T. Antipova (Ed.): ICADS 2021, AISC 1352, pp. 105–115, 2021.
https://doi.org/10.1007/978-3-030-71782-7_10

Through a systematic literature review by Amiel and Oliveira (2018), the proposals identified to reduce the problem of inclusion and accessibility in schools were the incorporation of assistive technology for physical education teachers, training of foreign language teachers in ICTs, experiments for math teachers; training for teachers of kindergarten, training of teachers in free software to reduce social inequality. Among the means of communication adopted, uses of laptops, programming languages, educational software, mobile devices, social networks, Internet publishing tools were identified. There was greater coverage both in training offered to public education networks and specific audience by government programs.

About the knowledge generated by the investigated studies, the authors emphasize that the most recurrent problems for training are the lack of teacher preparation for the use of technology, poor school infrastructure, lack of time and support. According to the studies, there is a tendency for teachers to replicate traditional methodologies in classroom practices. The mobilization of knowledge about technology does not directly produce a transformation in pedagogical practices, as they are also impacted by the working conditions of teachers (Amato et al. 2019; Rodrigues 2012; Medeiros and Queiroz 2018; Rebelo and Kassar 2017).

In Brazil, the educational, cultural and social demands placed on schools are increasing. The population is growing in diversity, and educational authorities often suggest reforms that pressure teachers to provide adequate instruction to students. In this scenario, it is necessary to understand the two prisms of education: the teacher and the student. It is necessary to understand how teachers are using ICTs in the school environment, thinking about the digital skills necessary to do so. On the other hand, it is essential to consider updated and constant training for students who will become teachers, preparing them in a theoretical and practical way, both in terms of content and psychological training to support pedagogical decisions.

In light of this, this research sought to verify the perception of teachers working in the classroom and Pedagogy students about the situation of Inclusive Education in Brazil and the ICTs usage to promote inclusion and accessibility.

This article is structured as follows. Section 2 presents the concepts of Inclusive Education. In Sect. 3, works related to this area are discussed. Section 4 presents the methodology of this research. The results are presented in Sect. 5 and the conclusions in Sect. 6.

2   Inclusive Education and Information and Communication Technology

The current understanding of education is connected with equal opportunities, upbringing to respect for otherness, supporting both pupils with deficits and those who are significantly out of the way (gifted pupils). All educational and upbringing activities related to equalisation of opportunities are classified as inclusion. Inclusion may be implemented intentionally, among others, with the use of pre-imposed concepts and methodological solutions. Most often these are activities established by expert institutions, or representing disadvantaged groups or in the course of activities coming from government administration institutions. The second type of inclusion activities takes place from the bottom up. It is the teachers - educating through everyday, intentional actions that create environments conducive to equal opportunities, both among pupils with deficits and gifted pupils. Education as a science oriented towards human development includes in its mission, expressed through research, expertises, metatheoretical analyses, a component devoted to support. It is precisely the wise co-existence, the perception of deficits and the creation of situations that eliminate unwanted states that are elementary foundations of inclusion (Dimitrova-Radojichich and Chichevska-Jovanova 2014). Inclusion education is a permanent part of the research of scientists focused on special pedagogy, but also on seemingly distant sub-disciplines, such as media pedagogy.

The relationship of media pedagogy with human deficits, very often perceived as a disability, is multidimensional and difficult. By definition, ICTs, which are very often used as effective teaching aids, are understood instrumentally. ICT is treated primarily in pedagogical studies as a tool for remote learning, educational gadgets that increase engagement, or increase attractiveness in the educational process. However, there are a number of research results showing that digital technologies can be effective tools in the educational inclusion process. In this case, digital technologies are not only becoming tools supporting psychosocial functioning of selected groups with disabilities, but also create new hybrid solutions ensuring increased quality of life (Plichta 2013). ICT for people with disabilities are not only ordinary tools for learning, entertainment and other basic e-services. ICT can be a source of positive change. Access to ICT in the context of educational inclusion should therefore be one of the fundamental rights of people with deficits (Plichta 2017). The challenges of ICT integration and educational inclusion currently only go beyond the classic problems of equipment availability and are related to modern media pedagogy, which should incorporate both the paradigm of opportunity and risk of media pedagogy (Pyżalski et al. 2019).

One of the basic conditions for effective integration with ICT is the proper preparation of the teaching staff. Awareness of technical solutions, the ability to integrate new media into teaching and upbringing, the ability to recognize the needs of people with deficits and connect them to the potential of ICT is a challenge for contemporary educators. Digital competencies, i.e. knowledge of how ICT works and the level of their effective service, become a barrier. All of this causes that there are still different groups of educators using ICT to work with students with special educational needs. Inclusion through ICT therefore divides the teachers into four elementary groups: techno-optimist, techno-realist, techno-pessimist and techno-ignorant (Tomczyk et al. 2015). Each group has different attitudes towards using media in the work with people in need of support, different levels of digital competence, different ways of experimenting with new technologies (websites and devices). Currently, a particular challenge is to adequately prepare the staff implementing inclusion processes by equipping them with specialist knowledge and digital competences. Of course, this requirement does not apply to all special educators, because both in methodological studies (good case studies) and research results assigned to special pedagogy and media, there are many valuable projects which are characterized by high effectiveness of combining knowledge about digital inclusion with effective use of digital media.

3 Related Works

There is a large number of studies on the theme of inclusive education, both nationally and internationally. Many researches indicate that inclusive education is beneficial for all students with and without disabilities. The record that included students develop better skills and learning, have higher attendance rates, fewer behavioral problems and tend to follow their studies up to higher education have been shown as relevant results. Concerning students without disabilities who study in inclusive environments, the research highlights the socio-emotional benefits, such as developing less prejudiced opinions (Hehir et al. 2016; Brasil 2010; Andrade and Ferreira 2016).

In relation to the pedagogical proposal, several studies recommend Universal Design for Learning (UDL) as a proposal that meets the student diversity, as pointed out in the literature review carried out by (Ribeiro and Amato 2018), in which 35 works about UDL are presented. Researches in several countries show that teachers agree with inclusive education in the theoretical proposition, but do not know how to apply it in the classroom routine. Several authors who addressed the use of ICTs on inclusive education also point out this perspective. Professionals recognize the importance but question their own capacity for using it in the classroom (Ramberg and Watkins 2020; Budiyanto et al. 2020; Dovigo 2020).

Some studies indicate the gains that the inclusive educational system provides for the socialization of all students. With regard to students with disabilities, studies show the benefits of developing and maintaining positive relationships with peers of the same age group and interests (Hehir et al. 2016).

Barreto and Oliveira (2018) bring a study on the perception of Mathematics teachers about Inclusive Education in a city in the State of Rio Grande do Sul. Basso and Campos (2019) present and discuss the students' perception of their initial formation for the inclusion of students with disabilities. Two hundred thirty-six students from the last year of the Science Degree courses (face-to-face learning), namely Biology, Physics and Chemistry, from the state public universities of the state of São Paulo, were part of the study.

The studies no longer discuss the proposal for Inclusive Education, but how to enable the Brazilian educational system to be effectively inclusive. In this context, the use of ICTs becomes a great ally and researches that broaden and deepen knowledge about the relationship between inclusive education and ICTs usage is a necessity.

4 Methodology

The study aimed to investigate the knowledge that students and teachers in the field of Pedagogy have about ICTs usage in inclusive education. In order to achieve this goal, it tries to understand the importance of ICTs for these two groups, to then relate their views on the use of ICTs in Inclusive Education. A questionnaire was designed for them, with questions regarding ICTs usage in Education and Inclusive Education. This survey is an excerpt from a larger survey that addressed both more general ICTs and Inclusive Education issues. The full tool has ten questions (with subscale) related to the following issues: ICT usage style, self-evaluation of digital competences, experiences with e-learning, assessment of ICT usage in private life and educational processes. An international team consisting of educators and IT specialists created the tool.

Table 1 shows only the issues related to inclusion. Questions 1 to 10 had answer options on the 5-point Likert scale (from 1 as strongly agree to 5 as strongly disagree). Questions 4 and 5 present answer options by frequency of occurrence.

Table 1. Questionnaire applied to students and teachers

Use of digital technologies To what extent do you agree or disagree with the following statements?
1. Digital technologies have positively changed our lives
2. It is necessary to use digital technologies in the learning and teaching process
3. Digital teaching resources are better than non-digital learning resources
4. The use of digital technologies by the teacher has a positive impact on student learning
5. The use of digital technologies by the teacher has a positive effect on the student's motivation
6. The use of digital technologies by the teacher has a positive effect on student's engagement
7. The use of digital technologies by the teacher has a positive effect on student satisfaction
Inclusive education To what extent do you agree or disagree with the following statements?
8. I think that student/learner differences should be considered as an essential aspect of human development in any conceptualization of learning
9. I think that all educators must believe that they are qualified/capable of teaching all students
10. I think that all educators should continually develop new creative ways of working with others
Information and communication technology as a tool to support learning How often do you use the following techniques as a teacher or student in the last year of undergraduate?
11. Special ICT (Information and Communication Technology) tools to support teaching and learning for people who are deaf, blind or physically disabled
12. Method to support digitally excluded people (for example, the elderly, migrants)

5 Results

Responses to the application of the questionnaire were collected from a hundred and six Pedagogy students and a hundred and four teachers, online, spread throughout Brazil. This questionnaire was made available from August to September 2019 on the Internet, through Google Forms, without identifying the participants, using a sample for convenience to teachers and students.

Regarding teachers, 59.61% were women and 40.39% men. As for the workplace, 43.14% of teachers worked at public schools, 48.04% at private schools and 8.82% at public and private schools. Most teachers had a bachelor's degree in education (36.54%), specialization (25.97%), 14.43% had master's degrees, 17.30% doctorate, and 5.76% completed high school in teacher training.

In the students' group, 55.67% were women, 41.50% men, and 2.83% answered the other option. Most of the participants in this group (62.63%) attended the Pedagogy course, and 31.31% attended high school. 6.06% were in other levels of training (specialization and master's).

The first statement in Chart 1 brings a fundamental consideration to the learning and inclusion process, the diversity of human development and the learning process. The vast majority of participants, 94.34% of students and 89.43% of teachers agreed with this perspective.

With regard to the impact of digital technologies on daily life, the group of teachers was more affected than the group of students: 83% of teachers agree with the statement "Digital technologies have positively changed our lives" (question 1, Table 1), while among students, responses with a positive bias totaled 74%. More than pointing out a criticism from the students, this result is possibly due to the fact that the majority of this group grew up in environments already endowed with technological resources, while the majority of teachers experienced the changes imposed by the ubiquity of technologies digital. Figure 1 shows these data.

This information is quite interesting because, according to Soares-Leite and Nascimento-Ribeiro (2012), among the problems for the use of ICT in Education in Brazil, there are two directly related to teachers: lack of continuing education by teachers and insecurity in the use of ICT in the classroom context.

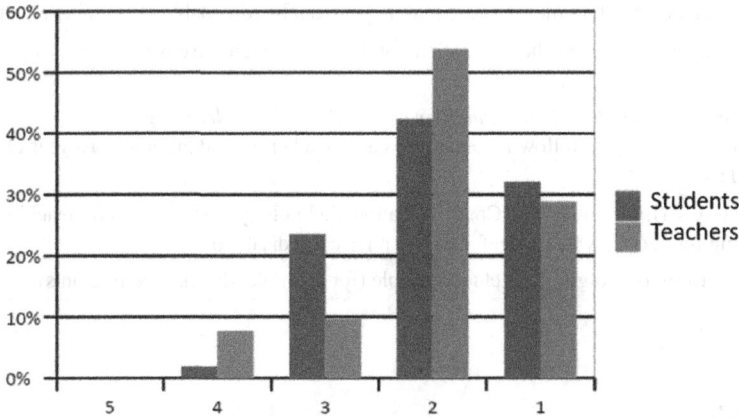

Fig. 1. Perception of teachers and students regarding the impacts of ICTs.

However, when analyzing question 2, regarding the use of technologies for teaching and learning situations, there is a slightly stronger positive bias (85% versus 82%) on the part of teachers than of students, as can be seen in Fig. 2.

In this sense, the analysis of question 3 can shed light on the process (Fig. 3): when answering the question "Digital teaching resources are better than non-digital resources for learning", there is an important degree of disagreement on the part of the students (37%), even higher than that of teachers (31%) - they have a much higher degree of agreement with this statement (37%) than that of students (20%).

As pointed out by Soares-Leite and do Nascimento-Ribeiro (2012), Brazil faces many problems with the inclusion of ICTs in Education, such as the lack of public policies for the realization of this inclusion, lack of ICTs inserted in the curricula of courses aimed at teacher training, lack of school structure to support digital inclusion, among others. Thus, the disagreement on this point, by teachers and students, may be strongly linked to the lack of knowledge of the potential of the use of ICTs in the teaching-learning process.

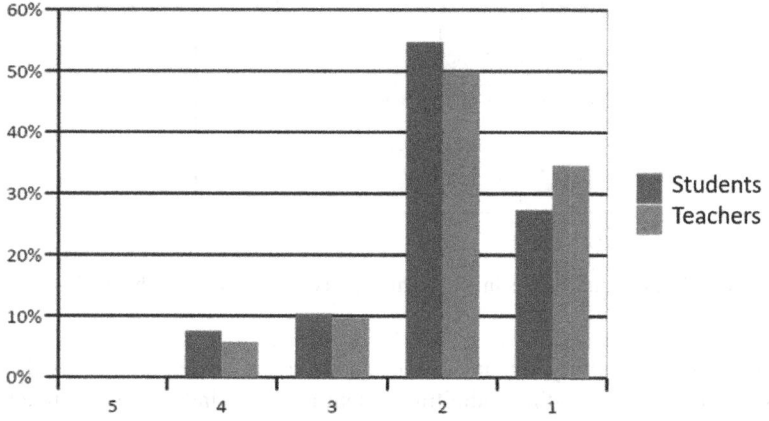

Fig. 2. Perception of students and teachers regarding the use of technologies in teaching

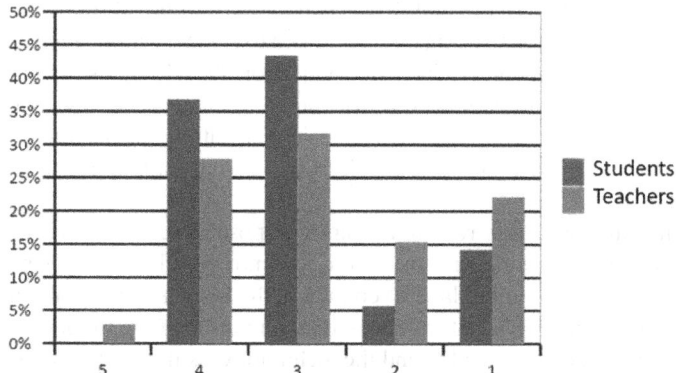

Fig. 3. Comparative analysis of digital resources versus non-digital resources

Questions 4 to 7 refer to four dimensions of learning, namely: the learning process itself, the motivation, engagement and satisfaction of the students with regard to the use of technologies in the teaching and learning processes. Teachers and students tend to have positive responses to the four dimensions, with emphasis on learning itself, assessed as being positively impacted by technologies by 90% of teachers and 85% of students. The satisfaction dimension is the one with the lowest positive perception among students, mainly, who rate 77% (teachers have 79% of positive responses in this dimension). Figure 4 shows such results.

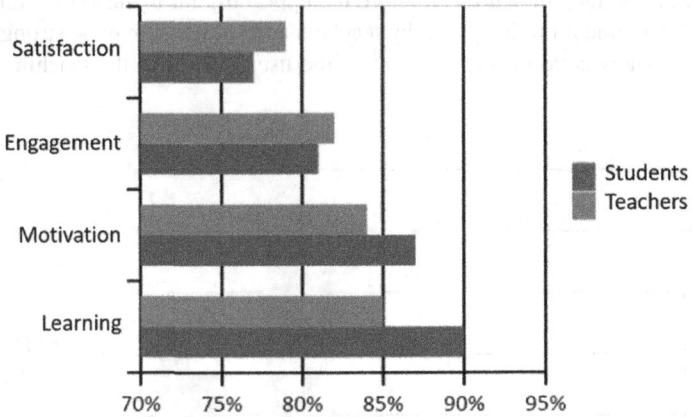

Fig. 4. Four dimensions in the learning process with digital technologies

Concerning accessibility, the answers to question 11, carried out now with a new Likert scale (from 1 to 4, without admitting a neutral point), indicate that both teachers and students rarely or never use or envision using technologies to support people with different disabilities, as can be seen in Fig. 5 - 69% of students and 64% of teachers. Thus, an interesting fact is that although teachers and students are aware that ICTs have positively changed everyday life and that it is necessary to use ICTs in the teaching-learning process (in addition to being directly linked to student's motivation, satisfaction and engagement), when it comes to Inclusive Education, they do not use them for support people with disabilities. As addressed by Bazom et al. (2018) and Souza and Pletsch (2017), there is a lack of public policies for Inclusive Education, teacher training curricula that need to address these issues and a lack of continuing training for teachers in the Brazilian context.

Due to the sample of this research is not composed exclusively of teachers and Pedagogy students who work with inclusion, it is not possible to say whether these negative numbers are due to the lack of contact with students with disabilities or lack of knowledge of technologies that can be used as support learning to such an audience. However, the discussion extends beyond the technology, as the interviewees also point out the absence or lack of knowledge of methodologies for this (question 12), which incurs the need for action in this regard in the teachers' training. Based on the data collected, it was noted that 64% of students and 69% of teachers pointing out the absence of methodologies for working with digital technologies in the learning of people with disabilities or belonging to vulnerable groups, such as the elderly and migrants. In any case, these data reflect an attention point in the teachers' preparation for performing with this group of students.

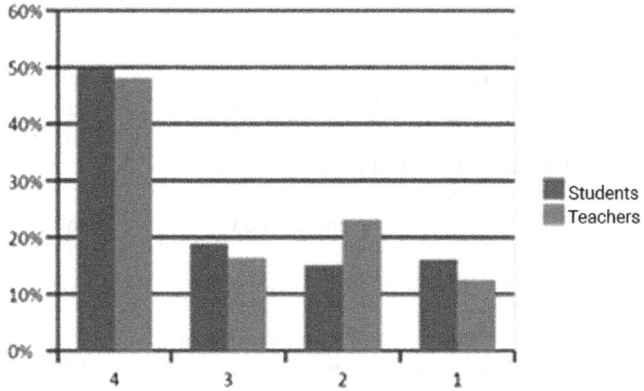

Fig. 5. Use of digital technologies as support to students with disabilities.

6   Conclusions

This article presented an exploratory survey carried out with a hundred and four teachers and a hundred and six Pedagogy students spread throughout Brazil, regarding the conceptions about the use of digital technologies in the teaching and learning processes, as well as the perception and knowledge in the ICTs usage to support the learning of people with disabilities or belonging to groups with some degree of vulnerability.

In Brazil, the right to attend traditional teaching schools is guaranteed to students with disabilities. However, as pointed out by Bazon et al. (2018), Amato et al. (2019) and Souza and Pletsch (2017), and corroborated by this sample, neither teachers nor pedagogy students, as a whole, are prepared for the Inclusive Education mission. On the other hand, it is perceived through data from the larger research that, about the use of technologies for teaching and learning situations, there is a slightly stronger positive bias of teachers concerning students (85% versus 82%) (Blind review). In addition to the lack of technological knowledge to promote Inclusive Education, the methodological aspect is quite accentuated, leading to the reflection on the need of improvement and dissemination of teaching techniques, strategies and good practices concerning the use of digital technologies in teaching processes and learning by people with disabilities or belonging to groups that suffer different forms and degrees of exclusion.

Future work points towards expanding the scope of the research on the one side, increasing the significance of the sample; and on the other side, deepen the survey, carrying it out with specific groups, such as teachers who work with students with disabilities or even students belonging to different groups with some degree of vulnerability.

Acknowledgments. This work was supported by the ERANET-LAC project which has received funding from the European Union Seventh Framework Programme. Project Smart Ecosystem for Learning and Inclusion – ERANet17/ICT-0076SELI. In Poland the project was supported by the National Centre for Research and Development (NCBiR). This research was developed also with the support from FAPESP (Fundação de Amparo à Pesquisa do Estado de São Paulo, Proex (Programa de Excelência Acadêmica) and CAPES-Brasil (Coordenação de Aperfeiçoamento de Pessoal de Nível Superior).

References

Amato, C.A.D.L.H., Silveira, I.F., Eliseo, M.A., Martins, V.F.: ICT in education fostering inclusion: the Brazilian context. In: Tomczyk, Ł., Oyelere, S.S. (ed.) ICT for Learning and Inclusion in Latin America and Europe: Case Study from Countries: Bolivia, Brazil, Cuba, Dominican Republic, Ecuador, Finland, Poland, Turkey, Uruguay, 1 edn, vol. 1, pp. 35–54. Pedagogical University of Cracow, Cracow (2019)

Amiel, T., Oliveira, T.P.: A formação docente em serviço para e sobre tecnologia: Uma revisão sistemática. Rede IEB, out (2018)

Andrade, L.B.P., Ferreira, G.S.: A educação inclusiva no contexto da educação infantil. Serviço Soc. Realidade **25**(2), (2016)

Barreto, A.B., Oliveira, D.L.D.: Visão do professor de matemática do município de Canoas em relação à educação inclusiva. Encontro Regional de Ensino de Ciências (2: 2018: Porto Alegre, RS). Anais do II Encontro Regional de Ensino de Ciências: formação do professor e o ensino de ciências. UFRGS, Porto Alegre (2018)

Basso, S.P.S., Campos, L.M.L.: Licenciaturas em Ciências e Educação Inclusiva: a visão dos/as licenciandos/as (Science licensing courses and Inclusive Education: the graduate's vision). Revista Eletrônica de Educação **13**(2), 554–571 (2019)

Bazon, F.V.M., Furlan, E.G.M., Faria, P.C.D., Lozano, D., Gomes, C.: Formação de formadores e suas significações para a educação inclusiva. Educação e Pesquisa 44 (2018)

Brasil, Ministério da Educação. Secretaria de Educação Especial: Marcos Políticos-Legais da Educação Especial na perspectiva da Educação Inclusiva. Secretaria de Educação especial, Brasília (2010). https://portal.mec.gov.br/index.php?option=com_docman&view=download&alias=6726-marcos-politicos-legais&Itemid=30192. Accessed Mar 2020

Brasil: Lei Nº 13.146 Lei Brasileira de Inclusão da Pessoa com Deficiência (Estatuto da Pessoa com Deficiência), Casa Civil, julho de 2015 (2015). https://www.planalto.gov.br/CCIVIL_03/_Ato2015-2018/2015/Lei/L13146.htm. Accessed June 2019

Budiyanto, Sheehy, K., Kaye, H., Rofiah, K.: Indonesian educators' knowledge and beliefs about teaching children with autism. Athens J. Educ. **7**(1), 77–98 (2020)

Dimitrova-Radojichich, D., Chichevska-Jovanova, N.: Parents attitude: inclusive education of children with disability. Int. J. Cogn. Res. Sci. Eng. Educ. (IJCRSEE) **2**(1), 13–17 (2014)

Dovigo, F.: Through the eyes of inclusion: an evaluation of video analysis as a reflective tool for student teachers within special education. Eur. J. Teach. Educ. **43**(1), 110–126 (2020)

Hehir, T., Freeman, B., Lamoreau, R., Borquaye, Y.B.: Os benefícios da educação inclusiva para estudantes com e sem deficiência. Instituto Alana ABT Associates (2016)

IBGE: Instituto Brasileiro de Geografia e Estatística. Censo Demográfico de 2010, IBGE (2010). https://censo2010.ibge.gov.br/. Accessed June 2019

Medeiros, M.M., Queiroz, M.J.: Tics na educação: o uso de software livre na promoção da acessibilidade. Revista Brasileira da Educação Profissional e Tecnológica **1**(14), 6875 (2018)

Plichta, P.: Młodzi użytkownicy nowych mediów z niepełnosprawnością intelektualną. Między korzyściami i zagrożeniami. Dziecko krzywdzone. Teoria, badania, praktyka **12**(1), (2013)

Plichta, P.: Socjalizacja i wychowanie dzieci i młodzieży z niepełnosprawnością intelektualną w erze cyfrowej. Wydawnictwo Adam Marszałek, Toruń (2017)

Pyżalski, J., Zdrodowska, A., Tomczyk, Ł., Abramczuk, K.: Polskie badanie EU Kids Online 2018. Najważniejsze wyniki i wnioski. UAM, Poznań (2019)

Ramberg, J., Watkins, A.: Exploring inclusive education across Europe: some insights from the European Agency Statistics on Inclusive Education. FIRE Forum Int. Res. Educ. **6**(1), 85–101 (2020)

Rebelo, A.S., Kassar, M.D.C.M.: Escolarização dos alunos da educação especial na política de educação inclusiva no Brasil. Inclusão Social **11**(1), (2017)

Ribeiro, G.R.D.P.S., Amato, C.A.H.: Análise da utilização do Desenho Universal para Aprendizagem. Cadernos de Pós-Graduação em Distúrbios do Desenvolvimento **18**(2), (2018)

Rodrigues, D.: As tecnologias de informação e comunicação em tempo de educação inclusiva. In: Giroto, C.R.M., Omote, S. (eds.) As tecnologias nas práticas pedagógicas inclusivas. Cultura Acadêmica editor, pp. 25–40 (2012)

Scatolim, R.L., dos Santos, J.E.G., da Cruz Landim, P., de Toledo, T.G., Fermino, S.C.M., Cardozo, D., Sanches, R.S.: Legislação e tecnologias assistivas: Aspectos que asseguram a acessibilidade das pessoas com deficiências. InFor **2**(1), 227–248 (2017)

Soares-Leite, W.S., do Nascimento-Ribeiro, C.A.: A inclusão das TICs na educação brasileira: problemas e desafios. Magis. Revista Internacional de Investigación en Educación **5**(10), 173–187 (2012)

Souza, F.F.D., Pletsch, M.D.: A relação entre as diretrizes do Sistema das Nações Unidas (ONU) e as políticas de Educação Inclusiva no Brasil. Ensaio: Avaliação e Políticas Públicas em Educação **25**(97), 831–853 (2017)

Tomczyk, Ł, Szotkowski, R., Fabiś, A., Wąsiński, A., Chudý, Š, Neumeister, P.: Selected aspects of conditions in the use of new media as an important part of the training of teachers in the Czech Republic and Poland - differences, risks and threats. Educ. Inf. Technol. **22**(3), 747–767 (2015). https://doi.org/10.1007/s10639-015-9455-8

Modernization of Higher Education in Russia: New Challenges and Approaches

Irina Zhdankina$^{(\boxtimes)}$ ⓘ, Natalia Ignatieva ⓘ, Darya Bykova ⓘ, and Yulia Sysoeva ⓘ

Nizhny Novgorod State Engineering and Economic University, Knyaginino 606340, Russia
irka-zh@mail.ru

Abstract. From the beginning of the 21st century the Russian education system is in the process of transformation. Higher education has undergone the greatest changes within the framework of Russia's integration into the Bologna process. Nowadays, priority projects are being realized to increase the openness, accessibility and competitiveness of Russian higher education at the local and international levels. Nevertheless, there is a period of stagnation of modernization caused by the instability of progressive development and the progress of reforms. According to the authors' opinion, this situation is caused by a significant lag in the level of educational activities of universities. The article analyzes the rating presence of Russian educational institutions, analyzes the introduction of modern trends that exist in the global educational space. The authors draw conclusions about the discrepancy between educational activities in Russian universities of modern educational standards, which do not have a synergistic effect that would allow universities to reach a new level in the realization of programs and competitiveness.

Keywords: Academic mobility · Competence approach · Monitoring the effectiveness · Educational activity · E-learning · Educational export

1 Introduction

The last decades have become a challenge for the Russian education system as a whole. At the same time, the management of the education system has undergone significant changes: there has been a division of ministries responsible for problems related to different levels of the education system. National Projects "Education" and "Development of the export potential of the Russian education system" are priority development projects, which involve improving the quality, accessibility, openness and attractiveness of higher education programs at the local and international levels. However, in the process of modernization of education, there is stagnation, as well as an insufficient level of educational activity of higher education institutions.

The competitiveness of the Russian educational services market remains at a low level. Despite actively developing international, academic, economic and social and cultural ties, the increasing citation in world scientific publications, it is not easy for Russian educational organizations to take high positions in the ranking of the world's leading

T. Antipova (Ed.): ICADS 2021, AISC 1352, pp. 116–125, 2021.
https://doi.org/10.1007/978-3-030-71782-7_11

universities. Therefore, increasing the competitive advantages of Russian educational organizations is becoming of national importance.

It should be noted that the current trend of internationalization and commercialization of higher education is developing. The first process is primarily related to the development and deepening of international relations, increasing the level of international academic mobility, participation in scientific and educational programs. The commercialization of education is due to the increased role of paid services provided by educational organizations. Commercialization is taking place against the background of an annual increase in funding for the higher education sector.

The general combination of factors that actively function and influence the market conditions and processes taking place on it changes the configuration of the competitive environment to a certain extent, but in general its main parameters will remain in the near future.

The entry of Russian educational organizations into the world educational space can be achieved with the help of state funding and clearly formulated strategy for the development of education and science.

Thus, the global nature of modern educational space sets the task of increasing the international competitiveness of the Russian higher education system.

2 Methodology

The article uses statistical data on the modern state of indicators of the number, equipment and staff potential of Russian universities. Data on monitoring the effectiveness of educational organizations are analyzed. The article presents data from international rating agencies on the presence of Russian educational institutions among the best in the global educational space. The ratio of the competence content of training areas in accordance with the two-stage structure of higher education is given, as well as the opinions of members of the scientific community regarding trends in the modern educational environment.

3 Discussion

The government has set ambitious goals to increase the competitiveness and export of educational services at the international level. In terms of higher education, this project has a close correlation with joining the Bologna process in 2003. Russian universities have undergone qualitative changes in the structure of higher education over the past years. So, instead of specialty courses, Bachelor's, Master's and Doctoral courses have appeared.

Beginning with 2004 the introduction of state educational standards (SES) has begun, and the requirements for the development and implementation of educational programs have changed. Modernization of educational standards continues to this day. Nowadays, these processes are supported at the state level through the introduction of new state projects. Thus, the Presidential Council for strategic development and priority projects approved the priority project "Development of the export potential of the Russian education system" (May 30, 2017). This document supposes the designing and improvement of the regulatory legal framework, as well as creating conditions for increasing the attractiveness of educational programs [1]. Special attention is also paid to the problems of online learning and the introduction of educational products in a foreign (English) language. The National Project "Education", approved on January 1, 2019 for the period until December 31, 2024, "involves the realization of 4 main directions of development of the education system: upgrading its content, creating the necessary modern infrastructure, training relevant professional staff, their retraining and advanced training, as well as creating the most effective management mechanisms in this area" [2]. Thus, despite the intensive transformation process, Russian education needs a number of changes related to increasing the openness, accessibility, attractiveness and competitiveness of the educational sector. This is confirmed by both international and local ratings. For example, two leading Russian universities – Lomonosov Moscow State University and St Petersburg University –are not leaders in world rankings, such as ARWU, QS World University Rankings, Thomson Reuters (according to 2019) [3–5] (Table 1).

Table 1. Rating positions of Lomonosov Moscow State University and St Petersburg University

University	Rating position		
	ARWU	QS world university rankings	Thomson reuters
Lomonosov Moscow State University	87	90	199
St Petersburg University	301–400	235	501–600

Source: according ARWU, QS World University Rankings, Thomson Reuters

As for the situation at the local level, the effectiveness monitoring of universities reveals the same problems from year to year. The rating was developed by the Main Information and Computing Center of the Ministry of Education and Science of Russia. It is based on statistical indicators, and the basic ones are:

1) Educational activity;
2) Scientific activity;
3) International cooperation;
4) Financial and economic activity;
5) Staff salary;
6) Extra indicator (infrastructure) [6].

The following examples of four Russian universities show the organization's position in terms of key indicators in comparison with the threshold values. These institutions belong to different federal districts of the Russian Federation, including the best educational institutions presented in the different world rankings (Fig. 1, Fig. 2).

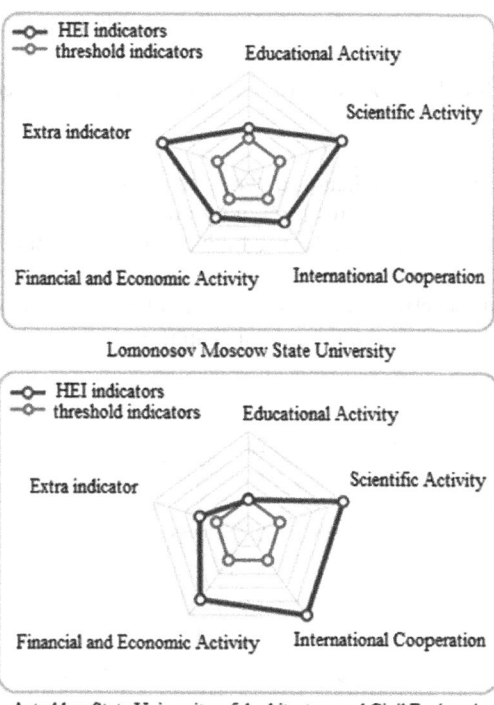

Fig. 1. Efficiency monitoring indicators of Lomonosov State University and Astrakhan State University of Architecture and Civil Engineering. Source: https://indicators.miccedu.ru/monitoring/ [6]

The first indicator is formed taking into account 15 groups of statistical data, the most important among them are:

- average score on the Unified State Exam (USE) among bachelor applicants;
- the number of students participating in educational and scientific competitions of the all-Russian and international level;
- number of students accepted for post-graduate studies programs;
- the number of students from other universities enrolled in advanced training and retraining programs, etc. Such data, according to the authors, depend essentially on the organization of teaching activities, the relevance and level of teaching of disciplines.

The diagrams show (Fig. 1, Fig. 2) that educational institutions do not significantly exceed the threshold indicator of educational activity.

At the same time, the thresholds for international and research activities may be significantly exceeded. Weakly developed financial activity, in our opinion, is a direct consequence of low indicators of educational activity (due to attracting a small number of applicants who are ready to study for a fee). There is a tendency for educational activity of higher education institutions to lag significantly behind other types. It also causes low competitiveness of universities.

The problem of low level of educational activity can be caused by a number of reasons:

– insufficient education level applicants entering universities;
– low threshold values of USE scores for admission to universities;
– insufficient using of the world's advanced educational technologies to improve the quality of students' education;
– insufficient correlation between the ecosystems of universities (absence of systematic work of scientific, practical, and international spheres within one university).

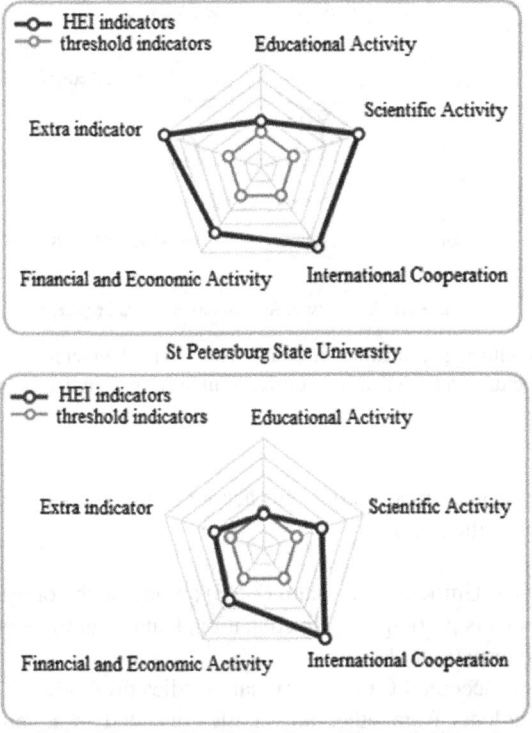

Fig. 2. Efficiency monitoring indicators of St Petersburg State University and Nizhny Novgorod State Engineering and Economic University. Source: https://indicators.miccedu.ru/monitoring/ [6]

Optimization of educational activity, in our opinion, should be carried out taking into account the trends in modern world education, shown in Fig. 3.

These processes are interconnected and actual in global educational environment and require to the reality of the modern labour market. Besides they correlate with aims set before the Russian universities in the National Project "Education".

SMART-education is a synergy of methods, technologies and processes that are linked with the content of training. It can also be defined as "a model that gives an idea of a given process, its goals and functions, and also demonstrates the place and role of a person in it" [7, 52]. It is the interconnection of all elements that allows us ensuring high-quality realization of educational programs, compliance with the principles of openness, accessibility, continuity, and flexibility.

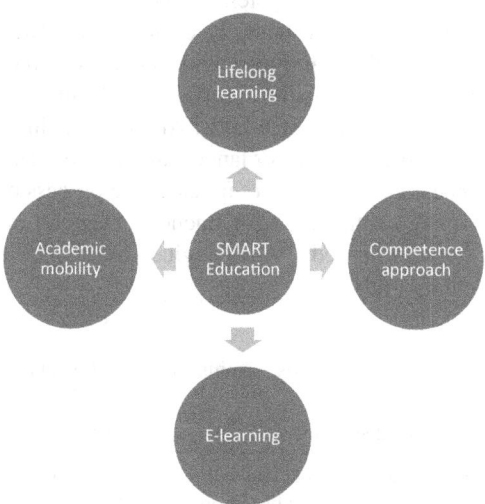

Fig. 3. Trends realized in modern world education. Source: compiled by the authors

For instance, the competence approach should be realized in accordance with the concept of continuing education. This should help to update the training content, expected results, and approaches to quality control. It is necessary to ensure smooth transition from one level of higher education to another, as well as further access to additional education or professional retraining programs. Thus, the student will add his "piggy bank" of his competencies, and the employer will be able to determine which of them need to be formed by the employee. Also, the criteria for evaluating learning outcomes should be unified to ensure academic mobility both within and outside the country of study [8, 218]. In this perspective, academic mobility contributes to the formation of communicative competencies, the ability to adapt to different conditions, and getting of new educational experience [9, 2232].

In the same time, the competence approach requires the introduction of certain techniques and training technologies that would be able to form various types of competencies. In modern education, you can't do without a set of standard tasks for automating certain skills. Interactivity, interdisciplinarity, and a variety of training tools are necessary, since a modern specialist must adapt to new tasks in a rapidly changing professional environment [10, 38].

Continuity of education, its flexibility and openness are ensured through the introduction of digital educational technologies and the implementation of the e-learning concept

[11, 24]. Digitalization of education contributes to the expansion of the range of used training tools, types of interactive work, and implements the principles of accessibility, visibility, and consistency. The use of online platforms for broadcasting, performing creative tasks and games, and co-writing lectures and notes allows using of such activities as online broadcasts, creating projects online, co-writing lectures and notes online, and serves as one of the most important learning goals – dissemination of information [12, 83].

Within the framework of a two-stage system of higher education, continuity of skills, professional actions and competencies should be ensured. The third-generation Federal State Educational Standards (FSES) aim to provide students with professional competencies (PC), general professional competencies (GPC) and multi-purpose competencies (MC) based on professional goals and standards. At the same time, the PCs are determined by the educational institution in accordance with the chosen professional standard and are formulated independently, which complicates the process of internal academic mobility. Using the example of the FSES for Bachelor Degree Level [13] and Master Degree Level [14] in the field of study "Agro engineering", we will consider whether the MCs are complementary (Table 2).

Table 2. General professional competences forming in the field of study "Agro engineering"

Forming competence	Bachelor degree level	Master degree level
GPC-1	Able to solve typical tasks of professional activity based on knowledge of the basic laws of mathematical and natural sciences with the use of information and communication technologies	Able to analyze modern problems of science and production, solve development problems in the field of professional activity and (or) organization
GPC-2	Able to use normative law acts and arrange special documentation in professional activity	Able to pass professional knowledge with the use of modern pedagogical techniques
GPC-3	Able to create and support safe conditions of arranging production processes	Able to use knowledge of problem solving methods in the development of new technologies in professional activity
GPC-4	Able to realize modern techniques and apply them in professional activity	Able to conduct research, analyze results, and prepare reporting documents
GPC-5	Able to participate in experimental research in professional activity	Able to carry out a feasibility study of professional activity projects
GPC-6	Able to use fundamental economic knowledge and define economic efficiency in professional activity	Able to manage teams and organize production processes

Comparison of the above mentioned competencies shows only partial continuity and correlation of the formed competencies. It is obvious that new competencies at the Master Degree level appear, for example, pedagogical, but the competencies formed at the Bachelor Degree level of creating safe conditions in professional activities and using regulatory documentation are not formed in any way. The sequence of formation of competencies related to scientific activities, the use of existing and new technologies, and the economic aspect of professional activity is also not entirely obvious. However, the list of these competencies shows that the future specialist will be able to form them only in the conditions of using interactive techniques, various training tools, and digital technologies; have self-learning skills. Accordingly, there is a question about the availability of sufficient documentation, staff and technical base for training such a specialist in Russian universities.

Statistic data show that public expenses on education accounts for 10.8% of the total, while only 1.6% is spent for higher education. In addition, the total number of educational institutions providing training of higher education programs has decreased from 1,068 units in 2005 to 741 units in 2019. The number of teaching staff with academic degrees and academic titles has also been decreased, which is associated with stricter requirements for obtaining them. Despite the fact that there is an increase in computerization of higher education institutions, there are only 23.4 computers per 100 students [15]. Can such trends promote the growth of competitiveness and attractiveness of Russian higher education?

4   Conclusion

Educational activity in Russian universities does not fully satisfy the requirements of modern educational standards, and do not have a synergistic effect that would allow universities reaching a new qualitative level in the implementation of programs and competitiveness. It is necessary to create a new culture of learning, starting from the school level. Reforms in the Russian education system, in our opinion, are fragmentary, and the transition to the new generation of standards is not progressive and not universal. This can be explained by the following peculiarities:

1. The specific of the realization of the National Project "Education", which firstly provides support for pre-school, school and additional education. For example, it is supposed that by 2024, only 20% of students will have completed individual (online) courses in organizations providing the quality of training of students of the world level. Russia's place in the world's top-500 global university rankings should shift from 17 to 10. The Project does not pay enough attention to academic mobility, both internal and external.
2. The absence of consolidated efforts and networking. The trend of internal academic mobility is extremely unpopular, the agreements of network cooperation are rarely signed, joint interdisciplinary or online courses are not created, and there are not enough events for the exchange of methodological and pedagogical experience.
3. The local peculiarities of educational institutions: location, technical equipment and staff potential.

However, it cannot be said that Russian higher education is in a state of crisis. The phase of its development can be defined as an incomplete transition to qualitative changes caused by the instability of progressive development. A special role in the stagnation is played by a significant lag in the level of educational activity of higher education institutions along with other indicators of their activity. The introduction of SMART-learning elements [16], the modernization of the content of academic disciplines, and the organization of network cooperation can cause the increasing of the accessibility, openness, and competitiveness of Russian education.

References

1. PASSPORT of priority project Development of the export potential of the Russian education system approved 30 May 2017. https://static.government.ru/media/files/DkOXerfvAnLv0vF KJ59ZeqTC7ycla5HV.pdf
2. National project Education till 2024. https://edu.gov.ru/national-project
3. Academic ranking of worldwide universities. https://www.shanghairanking.com/ARW U2019.html
4. QS World University Ranking. https://www.topuniversities.com/university-rankings/world-university-rankings/2019
5. The World University Rankings. https://www.timeshighereducation.com/world-university-rankings/2019/world-ranking#!/page/0/length/25/sort_by/rank/sort_order/asc/cols/stats
6. Informational and analytical materials on the results of monitoring. https://indicators.mic cedu.ru/monitoring/2019/index.php?m=vpo
7. Ardashkin, I.B., Surovtsev, V.A.: SMART-education as a new educational paradigm: proetcontra. Bull. Tomsk State Univ. Philos. Sociol. Politol. **54**, 51–61 (2020)
8. Lomakina, G.R.: Competence approach as pragmatic oriented approach to results of higher education. Theory Pract. Soc. Dev. **12**, 217–220 (2012)
9. Todorescu, L., Greculescu, A., Dragomir, G.M.: The Bologna process. Academic mobility in Romanian technical higher education. Proc. Soc. Behav. Sci. **47**, 2229–2233 (2012)
10. Bolatbaeva, Z.T.: Competence approach in the sphere of higher professional education. Sci. World. **2**(6), 37–38 (2014). https://www.elibrary.ru/download/elibrary_21671446_53222634. pdf
11. Barul, H.K.: Revolutionizing Modern Education through meaningful E-learning implementation. IGI Global, USA (2016). https://books.google.ru/books?hl=en&lr=&id=ClWiDA AAQBAJ&oi=fnd&pg=PP1&dq=modern+education&ots=srABoq78OG&sig=HhIuSdiv0 fGoU-k42EQ1k0A7t1M&redir_esc=y#v=onepage&q=modern%20education&f=false
12. Nancy, W., Parimala, A., Merlin, M., Livingston, L.: Advanced teaching pedagogy as innovative approach in modern education system. Proc. Comput. Sci. **172**, 382–388 (2020). https:// doi.org/10.1016/j.procs.2020.05.059
13. Order № 813 on approval of the Federal state educational standard for higher education – Bachelor Degree in the field of study 35.04.06 "Agro engineering", 23 August 2017. http:// www.edu.ru/file/docs/2017/08/m813.pdf#page=3
14. Order № 709 on approval of the Federal state educational standard for higher education – Master Degree in the field of study 35.04.06 "Agro engineering", 26 July 2017 http://www. edu.ru/file/docs/2017/07/m709.pdf#page=3

15. Bondarenko, N.V., Gokhberg, L.M., Kovaleva N.V.: Education in numbers: 2019: brief statistical book. National scientific university Higher School of Economics, Moscow (2019). https://www.hse.ru/data/2019/08/12/1483728373/oc2019.PDF
16. Ignatieva, N., Zhdankina, I., Bykova, D., Sysoeva, Yu.: Using smart technologies at the classes of foreign languages at a non-linguistic university. In: Antipova, T. (ed.) Integrated Science in Digital Age 2020. Lecture Notes in Networks and Systems, vol. 136, Springer, Switzerland (2020). https://doi.org/10.1007/978-3-030-49264-9_21

Public Universities' Performance Evaluation

Tatiana Antipova[✉] [iD]

Institute of Certified Specialists, Perm, Russia
antipovatatianav@gmail.com

Abstract. Public University has state ownership and/or receives public funds through a federal/national or subnational government. As the budgetary entity, Public University spends budget money and every citizen demands greater understanding of where their tax money goes and spends. The purpose of this paper is to study performance evaluation of Public Universities complying with Public Sector laws/rules. This paper suggests that the construction process of performance measurement should be aligned with outputs and outcomes of budget policy. Performance evaluation of Public University as a whole organization is considered in the three aspects: design/creating, evaluating, and challenging/monitoring. Subsequently, author aims to investigate the latest achievement in evaluation, monitoring, and challenges for public universities' KPI system. Texts and contents from different sources such as financial annual reports and budgetary entities' websites provided insightful and findings. In addition, the research result lightens the most important drawbacks and challenges facing during the public universities performance measurement.

Keywords: Public university · Public sector key performance indicators (PSKPIs) · Performance evaluation · Budgetary entity · COVID-19 impact · Effectiveness in public sector · Budget money · Budget policy · Supreme audit institution

1 Introduction

Public University is a university that is in state ownership or receives public funds through a federal/national or subnational government, as opposed to a private university. So Public University is subject to the laws of the public sector and is a budgetary entity. As the budgetary entity Public University expenses budget money and every citizen demands greater understanding of where their tax money goes and spends [1]. An increasing social interest in effectiveness of resources management and in efficiency of operations in public management has triggered a change in perception of the roles and tasks of analysis in this group of institutions [2].

Such public organizations as public universities should be good citizens, fulfil social roles and tackle social problems, thus obtaining social recognition by engaging actively in dialogue with stakeholders. In this vein, public universities are responsible for the significant influence over a large population of future leaders. This influence is not limited to teaching and research, but it extends to include the need and ability to maintain

T. Antipova (Ed.): ICADS 2021, AISC 1352, pp. 126–137, 2021.
https://doi.org/10.1007/978-3-030-71782-7_12

its long-term performance better than its competitors. Public universities started gearing their key performance indicators (KPIs) towards social responsibility by voluntarily integrating social activities into their core policy and social performance measurements [3].

The public universities' performance can be measured by the quality, effectiveness, productivity, efficiency, innovation and quality of work life. Improving the performance of an educational institution is expected to generate good output, in which many preparations should be taken into account including enhancing the quality of facilities, infrastructure, education/teaching, research and social community service [4].

Performance evaluation in academia has traditionally been somewhat biased towards research indicators and it seems that this orientation is becoming even more profound. A growing number of reports document that universities striving for government funding, research grants and high rankings have adopted strategies including recruitment and reward systems that favour academics with top publications or the potential to secure such publications [5].

Indicators of CSR studies have been embedded with the existing CSR studies, for example, "CSR disclosure and organizational attraction", "motivation of CSR adoption", "internal CSR practice", "commitment of CSR education", "integration of CSR into teaching and research", and "standalone or embedded CSR subjects", were mainly implicated in practice, and have not been fully explained in the literature. In addition, to minimize the ongoing uncertainty of COVID-19, Public Universities should be collectively engaged with stakeholders for long-term value creation, and decision-making by governing bodies should focus on sustainable development as an integral part of society [3].

Achieving the quality of education is achieved through the existence of a mechanism that clarifies the policy that the institution should follow from the efficiency of administrative organization and the provision of high-level training systems for the educational and administrative staff. The quality of education is the totality of the attributes and characteristics that relate to the educational service and can meet the needs of students. [6]. Beyond recognizing the importance of openness, transparency, and quality, institutions and funders should work together to enable the establishment of local resources that assist and support researchers in fostering these values [7].

Performance evaluation of Public University as a whole organization is considered in three aspects: design/creating of KPI system, evaluating, and changing as shown in following parts of this work.

2   Materials and Methodology

There is a very large body of literature on key performance indicators system. Our search for "key performance indicators in public universities" in Taylor & Francis databases gave us 371,088 results on 06.01.2021. Despite the extensive literature on this topic, there are relatively few evidences that the contributions of governance dimensions to public sector performance actually determine the quality with which particular governance system function requires careful, systematic and replicate measurement. In implementing an analysis process, this study was aimed to provide sufficient transparency and reproducibility to allow for both replication, critique and alternative approaches to this analysis similarly to this analysis in works [6, 8–10, 12–15, 18, 22, 29, 30, 32].

In the case of agonistic, for example, this type of behavior can occur because the teacher tries to show anger on his face to control the class or because he cannot clearly express his ideas to make himself understood by the students. In the case of submitting, the teacher tries to keep their attention (pressure from the class) under control by means of humility and resignation, which at the time can be similar to sadness. The important thing is that it is possible to determine the emotional behavior of the teacher by the emotional vectors assigned to a time window, and to know what recommendations to provide to deal with a specific topic during the teaching-learning process and thereby feed a recommender system [10].

To evaluate the results of Public University activity there is needed to design Key Performance Indicators system.

3 Design of KPI System for Public Universities

The current performance management system takes account of multi-dimensional and multi-layered performance objectives and pursues realization of various values including not only efficiency and effectiveness, which are important factors in management activities, but also equity and fairness of the public services. Thus, performance evaluation is conducted by using both qualitative and quantitative measures at the individual level as well as the organizational level in the aspect of both short and long periods of time. It is emphasized to construct a comprehensive system that encompasses strategic objectives and crisis management to provide the necessary public services to the people at all times [11–13].

Many researches consider KPI for evaluation of quality education/graduation, e.g. [28, 29]. In addition, there is a removal from the types of boredom and general frustration, as well as unjustified fatigue resulting from the constant pressure of quality requests in a disturbing manner, which is called emotional stress. The faculty member faces several requirements between the basic tasks of the teaching function, the quality requirements, and professional development on the one hand, as well as the general depletion of mental, muscular, and psychological capabilities. 88% of the study sample confirmed that, despite this, they are cooperating with colleagues in order to develop academic curricula and develop university work, despite bearing the intense pressure burdened with quality requirements. Many faculty members suffer from the lack of clarity in the university's mechanism to take care of the opinions and observations of employees, as well as 76% suffer from severe pressure because of direct interaction [6].

But this study aims to evaluate Public university as a whole organization as a part of budget/government system in the other words as public sector subject. In this aspect some of scientists offer four groups of indicators for public universities: 1) personnel, 2) material-technical, 3) economic, and 4) pedagogical content to comprehensively inspect the process from the perspective of an organization activity [8].

As the result of our previous researches [1, 2, 4, 5], we stated that the most important KPI of Universities' activity as a whole might be divided on three groups of indicators: 1) indicators that determine the financial position of the universities, 2) indicators related to the publications activity of teaching staff/faculties, 3) quantitative indicators counted staff, student, area of buildings and amount of tax. Each group of indicators consists of several units as shows in Table 1.

Table 1. The most important KPI of Universities' activity

Financial position	Publications activity of faculties	Quantitative indicators
Cash leftover	Web of Science/Scopus Publications per 100 Teachers	Total number of students
Total expenditure	Number of citations per 100 Teachers	Area of educational and laboratory buildings
Total revenue	Number of citations from Web of Science/Scopus per 100 Teachers	Tax
Value of fixed assets		The total number of teaching staff
Budget funding		Total number of employees
Revenue from paid services		
Average salary of employees		

Using KPI system from Table 1 we ranged chosen 11 Russian Universities and study results [2] show that used factor analysis for ranking is behind the constellations of performance measures. Our ranking was compared with the data available on Internet: the leaders were the same. The significance of the factor analysis should be enhanced with the increase in the number of the objects and the period of reporting.

For a more comprehensive evaluation of universities, we need a comprehensive system, which is discussed in the next paragraph.

4 Performance Evaluating of Public University

Performance measurement is only one of the management tools. The basis of this position is simple: if the organization is complex and consists of many different components, evaluation of this activity should be complex and multifaceted. It has been said that the public sector cannot improve what it does not measure, especially given the significant reduction in government funding with increased community demand for quality services. Public sector key performance indicators (PSKPIs) need to be based on a comprehensive system of indicators [12].

Logic model of the construction and implementation of a rational and optimal PSKPI system must consist of following elements: (1) gather historical data from the organization, (2) organize and prepare the final database, (3) ascertain and define the numbers of strategic perspectives and performance indicators connected to those, (4) assemble the cause-and-effect link between all strategic perspectives and, lastly, (5) employ and operate this management tool for long-term vision. This logic sequence for Public University's performance evaluation is shown on Fig. 1.

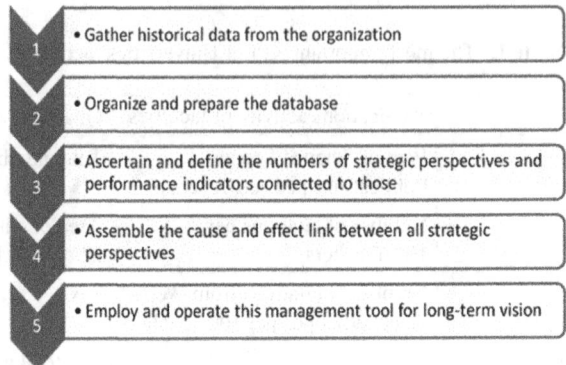

Fig. 1. Logic sequence of performance evaluating (adopted from [11]).

Most budgetary entities need a clear, precise statement of purpose and a description of the work assignment, in conjunction with readily identifiable responsibility for the quality of their work. In this case, one of the main effectiveness criteria should be an assessment of how well they manage to achieve outcomes and/or outputs according to New Public Management (NPM).

In internal performance evaluation it is reasonable to use auditors' methodology to evaluate Public University activity to avoid revealing the misuse, fraud, violent, etc. by Supreme Audit Institutions (SAI) auditors during budgetary control [15, 21, 22].

To manage Public University evaluating results we need to use of modern digital technologies but digital evaluating is not just a technology-based effort. It involves changing the expectations of what is included within an evaluation, and adjusting managers' knowledge, skills and abilities. This is especially true in terms of implementation of results-based budgeting, as well as openness and transparency in the sphere of Public Sector evaluating [1, 12, 15, 17, 20, 22]. The formation of an information database in the field of Public Sector evaluating results will allow for the recording, monitoring and analysis of inefficiencies and take measures to optimize budget expenditures. This database will provide the opportunity to hold videoconferences, implement electronic document management, work with documents, create and use an electronic library, conduct training, store information of the and other control and accounting agencies and implement other services.

It would be great if the results of Public Sector evaluating were taken into account for budgeting of the next period. In the future, it is possible to link the definition of the size of the budget subsidy depends on KPI achievement [12, 22]. To achieve it needs to be ensured that KPI system reflects Public University activities fair, precise, and accuracy. Block-scheme for evaluating of Public University's KPI system is shown on Fig. 2.

The use of Block-scheme on Fig. 2 would help to analyze the result of budgetary funds using of Public University, find ways to solve the problem of rational use of budget resources and optimize fiscal relations in Public Sector unit.

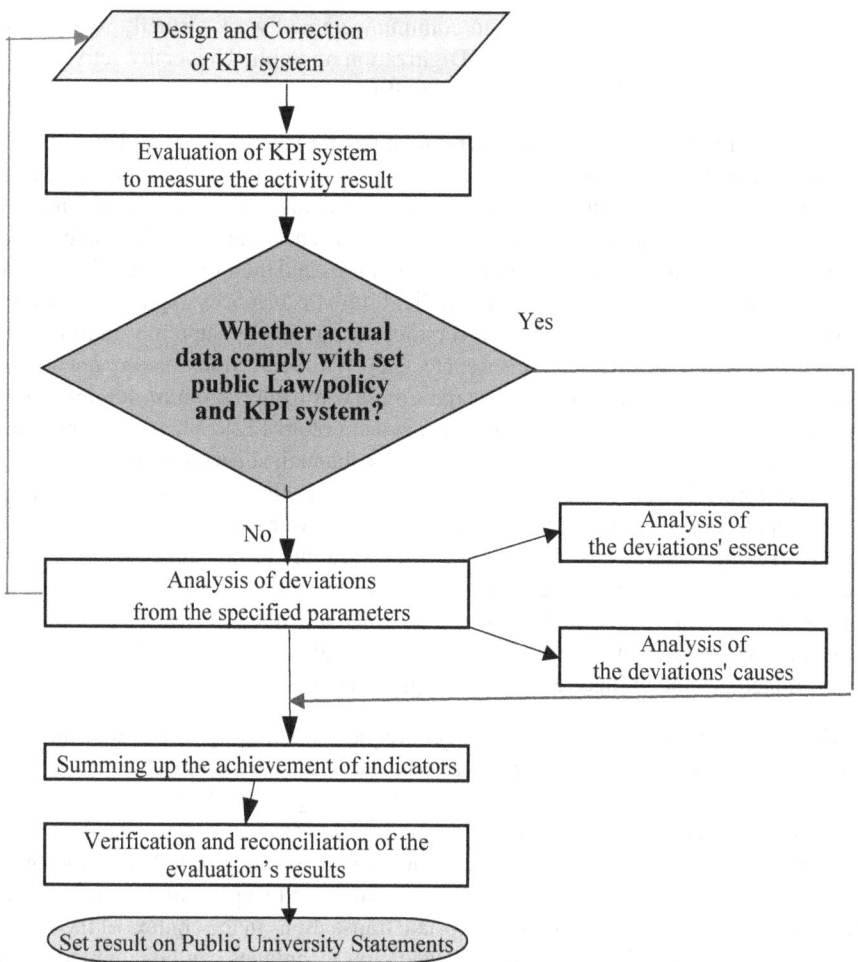

Fig. 2. Block-scheme for evaluating of Public University's KPI system (author's modification of Fig. 1 in [20]).

Due to experts' opinion, the future trend of Public University evaluating [20]: Digitalization; Preliminary; Continuity; Analysis.

Digitalization. The digital revolution is changing the reality surrounding universities and changing the universities themselves. The connection of university changes with the total digitalization of the economy and society has become a generally accepted idea. The modern realities of the increasingly rapid introduction of digital learning have led to the creation of various digital environments at universities: a virtual educational environment, a mobile learning environment, an adaptive and intuitive environment [11, 14, 16–19, 24, 28, 29, 31].

The key for the future is learning to communicate and work virtually with Public University. More details about impact of Digitization on Public University activities are considered in works [9–12, 14, 16–19, 24, 28–30].

Preliminary. Public University managers must ensure reasonable confidence in preventing or detecting suspicious acquisitions in a timely manner at the expense of budgetary funds. To enable predictive modeling, data sets must comprise transactions considered suspicious and the outcomes of investigations. There can be used simple model with Excel function as in [29]. Combining observations and their outcomes allows managers to begin to build the link needed to predict future occurrences of purchase misuse. It should not be forgotten that Public Universities are audited by Supreme Audit Institutions which reveal frauds, violent, misuse, etc. [21]. While rule violations are not always indicative of misuse, they are an effective and simple way to alert program administrators to aberrant behavior especially using blockchain technology [22]. Whether a public sector transaction supervision system is automated or manual, supervisors should explore ways to integrate business rules and rule violations into the Public Sector evaluating process depends on time factor [29]. Wherein an effective notification system operates over the Treasury central server, delivers event messaging to predefined employees in "real time", bearing in mind time factor [29], as the event occurs, and is sent directly to the employees and their smart devices. This level of event notification ensures that the people who need to know about an incident are made aware in a timely manner and fosters immediate and unified response as required [12].

Continuity. Continuous evaluating involves frequent monitoring throughout the year to ensure that transactions are captured properly and are flowing correctly to the income statement. During Digital Budgetary Transactions Surfing flows of transactions are continuously monitored, identifying transactions that match certain pre-determined integrity constraints and, in the event of a constrain violation, alert the Public University manager and copy the transaction data to a file. A natural first step in implementing data-driven techniques is to determine to appropriate transactions in the context of the Public University's day-to-day life [20]. Digital evaluation techniques can be combined with this.

Analysis. Data analytics should be the foundation of Public University evaluating. Exploratory analysis and trending allow managers to identify patterns and detect anomalous behavior. Basic statistics like mean, standard deviation and skew, along with commonly accepted "tests", help identify transactions that are unusual. Such outliers and anomalies should raise red flags with supervisors and indicate the need for further investigation. Using existing technology, managers examine Public University transactions to spot trends. For example, a customer who's previous on time payment suddenly slows to 10 days late may signal a risk of default.

5   Challenges for Public Universities' KPI System

Historically, researchers have used many quantitative and qualitative variables to evaluate the performance of KPI system implementations, ranging from return on investment to quality of user decision-making. Academic institutions, such as public universities, often have knowledge creation as a final goal, adding to the complexity of the KPI performance measurement challenge [23].

However, the application of quality in a manner that leads to a set of challenges facing the educational institutions, such as waste of human resources and misdirection leads to what is known as institutional burnout, and to meet these challenges had to be the proper and comprehensive application of the concept of total quality management to improve quality levels and enable the organization to excellence [6].

One problem cited is the cost in terms of data collection and analysis: many public universities lack this specific expertise and must employ consultants to assist. A further challenge is the difficulty with attributing performance to such specific budgetary entity as Public University. What is more, any performance evaluation system must take into account the fact that performance is the result of effective activity. So, the one of the practical drawbacks in the evaluation models lies in the problem of linking performance indicators that objectively must be reflecting dimensions of budget funding. Consequently, budgetary entities are often afraid to report bad news in case it affects their future funding. It must be born in mind that lack of differentiation in performance evaluation is also a very drawback factor [1].

In addition, unless experimental methods are used to isolate a research group and a control group, attributing any outcome in a beneficiary's life to a specific budgetary entity intervention, will also be challenged. However, the need remains for budgetary entities to show the difference they make in their communities, to be clear about the outcomes they are working towards, and to use performance frameworks to utilise scarce resources effectively. For this reason, it is important to understand how social welfare fit into the conventional set of performance indicator distinctions.

Well-known that in recent (2019/2020, 2020/2021 learning years) reality most meaningful unprecedented phenomenon on the sphere of education is COVID-19 pandemic [25, 26, 31]. The outbreak of the pandemic is a global concern for universities, professors, students and public policy makers around the world. Reducing the funding of public universities will have medium and long-term effects and also puts us at a disadvantage for the new norm [27].

Because of the Covid-19, and the pandemic restrictions, we all have had to face situations that we have never imagined before. Every day we live in special circumstances that we have to understand and cope with, as best as possible. Teachers suddenly have to be not only educators but most of all, humans, that must take care especially of the needs, feelings and response to stress of their students. The educational system worldwide has not been prepared for such a sudden change, but both teachers and students have managed to cope, often with unexpected success, with this radical transformation. We start by emphasizing some obvious benefits of distance education brought beside and in opposition to the classic/face-to-face education [24]. At the present time, learning processes are being developed in online models, thereby also allowing flexibility of the process [10].

The impact of COVID-19 on the quality and evaluation of higher education has already been covered quite extensively in the recent literature [22–28].

Teachers and students are the main factors affecting the quality of online education. Due to the lack of information technology and insufficient information technology literacy of many teachers, online education implemented during the epidemic period copied the traditional education mode to the online environment, resulting in low educational effect [31]. As a survey result, the most meaningful barrier factor in teaching-learning process were mentioned: Social interactions (78% of respondents); Learner motivation (70% of respondents); Technical problems (50% of respondents); Personal perception (30% of respondents); Language skills (25% of respondents) [28].

6 Conclusions

This study was considered Performance evaluation of Public University as a whole organization in the three aspects: design/creating, evaluating, and challenging/monitoring. Subsequently, there were state the latest achievement in evaluation, monitoring, and challenges for public universities' KPI system.

Distance education is a difficile task both for teachers and students. As long the necessity to keep education online as long the danger of generations with many gaps in education and certain subjects and topics. On the other hand, there are so many challenges in front of a qualitative distance education and the effects of the lack of social interactions, beside Distance Teaching-Learning-Evaluation Triad drawbacks, are about to be known and revealed during the following years [24].

It is obvious that digitalization of education involves the use of mobile and Internet technologies by students, expanding the horizons of their knowledge, making them limitless. The productive use of digital technologies, the inclusion of students into an independent search, the selection of information, and participation in project activities form the competences of the 21st century.

It is clear that the benefits outweigh the cons. Technology can be a very effective tool, but it is just a tool. Technologies are not intended to replace the teacher, rather, the idea is to create a learning environment that will allow you to switch the organization of the educational process from one-actor theater to cooperation and interaction.

Digitization has no doubt changed our education system, but we cannot say that it has diminished the value of traditional classroom learning. The best part about the digitization of education in the 21st century is that it is combined with the aspects of both; classroom learning and online learning methods. Walking hand in hand both act as a support system to each other, which gives a stronghold to our modern students. The implementation process can upset someone, annoy, take a lot of time and effort, but ultimately technology can "open doors" to new experiences, discoveries, ways of learning and cooperation of students and teachers [17].

The issues of pedagogical design and development of courses were not put on the agenda for the transition to distance learning. This was a consequence of force majeure circumstances that forced universities and teachers to mobilize all available resources and make a breakthrough in the massive emergency introduction of distance learning technologies.

However, during the process professors improved the quality of online learning tools, which certainly takes much longer. In general, we can conclude that Russian education system coped with challenges and continued an educational process without failures in accordance with the schedule. Its format has changed, but not its quality. Moreover, improvements become a part of day-to day professors' work [24].

Introducing new approaches and techniques in Public Sector evaluating has been a challenging yet immensely rewarding process. By improving the impact of evaluation, Public Universities can steadily move closer to better public governance. Embracing digital Public Universities evaluating is a necessary investment to move higher education to new and evolving techniques by making full use of current and emerging technologies. We are in an information age and the exponential growth of data brings both challenges and opportunities to overhaul traditional sampling-based educational approaches and fully leverage technology. Public University evaluation provides a window to view trends, issues, and relationships across a wider expanse of data, and provide more meaningful and insightful observations to Public Sector leaders and stakeholders for improving Public Sector performance [20]. To do this, we should reform the education system and mechanism, introduce rules and regulations on socialized coordinated supply of educational resources and services, standardize and guide the participation of diverse subjects in online education [32].

The COVID-19 pandemic has a dramatic Impact that presents an unprecedented challenge to public educational system. Now is the time for global solidarity and support, especially with the most vulnerable in our societies, particularly in the emerging and developing world. Only together can we overcome the intertwined social and economic impacts of the pandemic and prevent its escalation into a protracted humanitarian catastrophe, with the potential loss of already achieved development gains [31].

References

1. Antipova, T., Antipov, A.: Performance measurement in the Russian public sector: drawbacks and challenges. In: Bourmistrov, A., Khodachek, I., Aleksandrov, E. (eds.) Budget Developments in Russia's Regions: New Norms, Practices and Challenges, pp. 113–130 (2017)
2. Antipova, T., Shestakova, L., Melnik, M., Computer-assisted factor analysis of university performance indicators. In: 2016 11th Iberian Conference on Information Systems and Technologies (CISTI), Las Palmas, pp. 1–5 (2016). https://doi.org/10.1109/CISTI.2016.752 1605.
3. CSR implication and disclosure in higher education: uncovered points. results from a systematic literature review and agenda for future research (2021). Sustainability **13**(2), 525. https://dx.doi.org.aucklandlibraries.idm.oclc.org/10.3390/su13020525
4. Mariina, E., Tjahjadi, B.: Strategic management accounting and university performance: a critical review. Acad. Strategic Manage. J. **19**(2), 1–5 (2020). https://0-search.proquest.com.www.elgar.govt.nz/scholarly-journals/strategic-management-accounting-university/docview/2414423340/se-2?accountid=40858
5. Cadez, S., Dimovski, V., Groff, M.: Research, teaching and performance evaluation in academia: the salience of quality. Stud. High. Educ. **42**(8), 1455–1473 (2017). https://doi.org/10.1080/03075079.2015.1104659

6. Abdelnabi, H.A.I., Abobaker, F.A., ElBadawey, K.M.: The relationship between ineffective methods of applying quality management and functional combustion in Arab universities. Centre for Business & Economic Research, London (2020). https://0-search.proquest.com. www.elgar.govt.nz/conference-papers-proceedings/relationship-between-ineffective-met hods-applying/docview/2471512261/se-2?accountid=40858

7. Noémie, A.B., Pinxten, W.: Advancing science or advancing careers? Researchers' success opinions on indicators. Cold Spring Harbor Laboratory Press, Cold Spring Harbor (2020). https://dx.doi.org.aucklandlibraries.idm.oclc.org/10.1101/2020.06.22.165654

8. Štrangfeldová, J., Štefanišinová, N.: Value for money in organizations providing public education services and how to measure it. Nase Gospodarstvo: NG **66**(2), 62–70 (2020). https://dx.doi.org.aucklandlibraries.idm.oclc.org/10.2478/ngoe-2020-0012

9. Chun-Kai, H., Neylon, C., Hosking, R., Montgomery, L., Wilson, K., Ozaygen, A., Brookes-Kenworthy, C.: Evaluating institutional open access performance: methodology, challenges and assessment. Cold Spring Harbor Laboratory Press, Cold Spring Harbor (2020). http://dx.doi.org.aucklandlibraries.idm.oclc.org/10.1101/2020.03.19.998336

10. Arias, S.A., Moreno-Ger, P., Verdu, E.: Hierarchical clustering to identify emotional human behavior in online classes: the teacher's point of view. In: ICCS 2020, LNNS, Vol. 186, pp.269–275. Springer (2021). https://doi.org/10.1007/978-3-030-66093-2_26

11. Antipova, T., Yuhertiana, I.: Indonesian and Russian public sector accounting education: difficulties, challenges, and effectiveness. In: Gaol, F.L., Hutagalung, F. (eds.) The Role of Service in the Tourism & Hospitality Industry. CRC Press (2015). eBook ISBN 978-1-315-68852-7. https://doi.org/10.1201/b18238-15

12. Antipova, T.: Human-computer interaction in the public sector performance evaluation analysis. In: Rocha, Á., Correia, A., Adeli, H., Reis, L., Costanzo, S. (eds.) Recent Advances in Information Systems and Technologies. WorldCIST 2017. Advances in Intelligent Systems and Computing, vol. 570. Springer, Cham (2017). https://doi.org/10.1007/978-3-319-56538-5_56.

13. Park, S.: Performance evaluation in public institutions in Korea. Historical evolution and challenges for the future. Public Management in Korea, p. 37 (2018). https://doi.org/10.4324/9781351061384-7

14. Tsvirko, S.: Informational technologies for the efficiency of public debt management in Russia. In: MosITS 2017. AISC, vol. 724, pp. 104–113 (2018). https://doi.org/10.1007/978-3-319-74980-8_10

15. Antipova, T.: Public sector performance auditing. In: 2019 14th Iberian Conference on Information Systems and Technologies (CISTI), Coimbra, Portugal, pp. 1–6 (2019). https://doi.org/10.23919/CISTI.2019.8760933.

16. Rozhkova, D., Rozhkova, N., Blinova, U.: Digital universities in Russia: prospects and problems. In: Antipova T., Rocha Á. (eds) Digital Science 2019. DSIC 2019. Advances in Intelligent Systems and Computing, vol 1114, pp. 252–262 Springer, Cham (2020). https://doi.org/10.1007/978-3-030-37737-3_23.

17. Bilyalova, A.: Integration of digital technologies into education. J. Digit. Art Hum. **1**(2), 20–33 (2020). https://doi.org/10.33847/2712-8148.1.2_2

18. Rosca, S., Riurean, S., Leba, M., Ionica, A.: An educational model of graduation project for students at automation and computer engineering. J. Digit. Sci. **1**(1), 34–42 (2019). https://doi.org/10.33847/2686-8296.1.1_4

19. Rozhkova, D., Rozhkova, N., Blinova, U.: Digital universities in Russia: digitization with extra speed. J. Digit. Sci. **2**(1), 76–81 (2020). https://doi.org/10.33847/2686-8296.2.1_7

20. Antipova, T.: Digital public sector auditing: a look into the future. Qual. Access Success **20**(Supplement S1), 441–446 (2019). https://www.srac.ro/calitatea/en/arhiva/supliment/2019/Q-asContents_Vol.20_S1_January-2019.pdf

21. Antipova, T.: Fraud prevention by government auditors. In: 12th Iberian Conference on Information Systems and Technologies (CISTI), pp. 1–6 (2017). https://doi.org/10.23919/CISTI.2017.7976024

22. Antipova, T.: Using blockchain technology for government auditing. In: 13th Iberian Conference on Information Systems and Technologies (CISTI), Caceres, pp. 1–6 (2018). https://doi.org/10.23919/CISTI.2018.8399439

23. Kleist, V.F., Williams, L., Graham Peace, A.: A performance evaluation framework for a public university knowledge management system. J. Comput. Inf. Syst. **44**(3), 9–16 (2004). https://doi.org/10.1080/08874417.2004.11647577

24. Antipova, T., Riurean, I.P., Riurean, S.: Is distance teaching-learning-evaluation triad a form of digital art? J. Digit. Art Hum. **1**(2), 03–19 (2020). https://doi.org/10.33847/2712-8148.1.2_1

25. Coronavirus disease 2019 (COVID-19). Situation Reports. https://www.who.int/docs/default-source/coronaviruse/situation-reports/

26. Antipova, T.: Coronavirus pandemic as black swan event. In: Antipova, T. (ed.) Integrated Science in Digital Age 2020. ICIS 2020. Lecture Notes in Networks and Systems, vol. 136, pp. 356–366. Springer, Cham (2021). https://doi.org/10.1007/978-3-030-49264-9_32

27. Clery, A., Molina, L., Linzán, S., Muirragui, V., Zambrano-Maridueña, R., Córdova, A.: Importance of university communication in times of pandemic. J. Digit. Art Hum. **1**(2), 42–49 (2020). https://doi.org/10.33847/2712-8148.1.2_4

28. Rozhkova, D., Rozhkova, N.: COVID-19 and E-learning: challenges for Russian professors. Journal of Digit. Art Hum. **1**(2), 34–41 (2020). https://doi.org/10.33847/2712-8148.1.2_3

29. Antipova, T.: Organizational model for public sector auditing bearing in mind time factor. In: Rocha, Á., Abreu, A., de Carvalho, J., Liberato, D., González, E., Liberato, P. (eds.) Advances in Tourism, Technology and Smart Systems. Smart Innovation, Systems and Technologies, vol. 171, pp. 297–305. Springer, Singapore (2020). https://doi.org/10.1007/978-981-15-2024-2_27

30. Ross, G., Liechtenstein, V.: Management of financial bubbles as control technology of digital economy. In: Antipova, T., Rocha, Á. (eds.) Information Technology Science. MOSITS 2017. Advances in Intelligent Systems and Computing, vol. 724. Springer, Cham (2018). https://doi.org/10.1007/978-3-319-74980-8_9

31. Antipova, T.: Digital view on COVID-19 impact. In: Antipova, T. (eds.) Comprehensible Science. ICCS 2020. Lecture Notes in Networks and Systems, vol. 186, pp. 155–164. Springer, Cham (2021). https://doi.org/10.1007/978-3-030-66093-2_15

32. Xue, E., Li, J., Xu, L.: Online education action for defeating COVID-19 in China: an analysis of the system, mechanism and mode. Educ. Philos. Theory (2020). https://doi.org/10.1080/00131857.2020.1821188

Visualization Lecture in the Digital Educational Process at the University

Alena Guznova[1]([⊠]) [iD], Olga Belousova[1], Maria Arkhipova[2] [iD],
Alexander Martyanichev[1] [iD], and Valery Polyakov[1] [iD]

[1] Nizhny Novgorod State Engineering and Economic University, Knyaginino, Russia
ilichevalga@yandex.ru

[2] Minin Nizhny Novgorod State Pedagogical University, Nizhny Novgorod, Russia

Abstract. Information and communications technology appears to be the only way to achieve educational goals, meeting needs of the modern generation. The purpose of the research is to analyze a visualization lecture as an effective form of learning in the digital educational process. The work is devoted to the study of modern visual methods. The effectiveness of a visualization lecture as a form of conducting classes is confirmed by the results of testing students' knowledge, questionnaires. We conducted a study on the use of visualization and its significance for the development of digital education. For this purpose, 2 groups were selected: the first of them continued to study according to the usual program, and the second became experimental. In the latter, visualization lectures were held during the semester, i.e. using presentations, tables, graphs, clusters, video lectures, and other materials both from the teacher and prepared by students.

Keywords: A visualization lecture · The digital educational process · The method

1 Introduction

The COVID-19 pandemic has triggered deep changes in all spheres of life, education systems being no exception. Teachers faced a myriad of challenges having no choice but to rapidly find new approaches inspiring students to adapt to the new reality, therefore improving the quality of education.

Global informatization naturally affects the education process, impelling teachers to identify current trends and introduce pedagogical technologies into classroom environment with the aim to achieve beneficial learning results. Information technology in the educational process includes computer training programs (electronic textbooks, simulators, testing systems); telecommunications tools, such as e-mail, teleconferences, communication networks; digital libraries. The latest technologies integrated into teaching can bridge the generation gap existing between teachers and students, encouraging desirable response and bringing beneficial educational perspectives.

Information and communications technology appears to be the only way to achieve educational goals, meeting needs of the modern generation. The traditional Sage-on-the-Stage approach proves its inefficiency, while distant learning appeals to the educational

T. Antipova (Ed.): ICADS 2021, AISC 1352, pp. 138–148, 2021.
https://doi.org/10.1007/978-3-030-71782-7_13

needs of present generation, referred to by various terms as Digital One [1], Digital Natives [2], iGeneration, Gen Tech [3, 4], the Net Generation [5]. Online teaching, being the only way-out in present day lockdown, meets preferences of the new generation cohort, being technologically savvy, their information-age mindset easily adapting to new online classes.

It is possible to boost academic performance by combining traditional means with the latest achievements of science and technology. The ICT is seen as a combination of technical devices and didactic tools used to process and present information [6]. The researchers underline various characteristics of effective teaching by means of information and communications technology, among which social interaction, contextual sensitivity, connection to different sources and data, individuality as a way to adapt the process of teaching to individual students' needs [7], there are no time or place restrictions, high degree of cooperation, autonomy which enables students to control their learning [8].

Numerous and quite convincing examples are demonstrated proving effectiveness of information and communications technology in the pedagogical process, which can be used at the stage of presenting educational material, its drilling and consolidation, control and self-control.

E-teaching and e-learning applications, various online platforms undoubtedly help to endure the education process [9]. E-learning brings societal benefits during the current pandemic. Nevertheless, any educational programs must necessarily be checked for their proper pedagogical effect, pass their own kind of expertise, taking into account the value criteria. It is necessary to take into account possible negative impact of technology on the psychological and emotional state of students and, ultimately, on their health. Infomania, the state of information overload, causes serious health hazards and threat to overall wellbeing. The role of a teacher is to find balance, using the ICT as an effective teaching tool in order to achieve beneficial learning results.

The World Health Organization added gaming to the section on addictive disorders. Digital-/video-gaming is characterized by a person's inability to control time for play with the priority given over other daily activities and interests, which brings significant impairment in personal, social, educational and other areas of life. People who engage in digital-/video-gaming activities are to be alert to the amount of time spent [10].

The role of a teacher is to find balance, using contemporary educational approaches with the ICT viewed as an effective teaching tool improving learning outcomes.

The ICT is seen as a means arousing students' interest, appealing to the emotional sphere being strong motives in any activity, including educational. Interest and positive emotions aroused by effective educational tools in the modern day classroom have a beneficial influence on learning outcomes. Motivated students strive for success by getting solid knowledge, taking initiative and seeking responsibilities in solving problems.

The purpose of information and communications technology is to make the teaching process more intensive and flexible. With the widespread use of the media the transmission of information is carried out both in writing and oral forms. The need to teach a foreign language more people, i.e. to speak and understand it, has also increased with the growth of international political, economic, and cultural ties. The ICT tools are particularly important in teaching languages due to the shift in emphasis on language acquisition as a means of communication.

2   Visualization Lecture in the Digital Educational Process at the University (On the Example of the Discipline «Russian Language and Speech Culture»)

It is known, «… in the course of training it is necessary to form such competences those will make the future graduate competitive in the labour market» [11].

As professor O.A. Frolova writes, «it is possible to make remarkable curricula, to develop strategic tasks but if there is absence of desire to change approaches to digitalization of educational process – nothing will turn out» [12, p. 328].

Everywhere Smart education develops. It means:

S - Self-Directed (self-governed and self-checked);
M - Motivated (motivated);
A - Adaptive (adaptive, flexible);
R - Resourse-enriched (enriched with various resources);
T - Technological.

The concept of Smart education is the flexibility assuming existence of a huge number of sources, different types of multimedia, ability to be adjusted quickly and easily.

Besides, the mixed training (blended learning) is the model of joint educational activity of a teacher and a student or a group on preparation and carrying out educational process providing the most effective achievement of stated purposes is used.

Students gain knowledge on classes and independently online. They prepare cards, presentations, booklets of video.

The results of poll which are carried out to GBOU VO «Nizhny Novgorod State Engineering and Economic University» showed that students would like to see at our university:

– «total rejection of printing editions;
– maximum computerization: online consultations, use of electronic textbooks …»
– the model of joint educational activity of a teacher and a student or a group on preparation and carrying out educational process providing the most effective achievement of stated purposes.

During teaching the discipline «Russian Language and Speech Culture» information technologies are actively used:

1. electronic textbooks;
2. monitoring procedure with attraction of IT technologies and resources of the Internet;
3. information search for reports;
4. watching educational videos;

5. preparation of the presentations;
6. work with video lectures;
7. organizing webinars of professional orientation value;
8. organizing lectures and practical lessons in the form of webinars (students can make record of lessons and once again look through them in out-of-class time, download training material, presentations and to work with them independently in free time);
9. carrying out consultations in the form of webinars (individual and and in groups before examinations);
10. organizing seminars for staff.

Students gain knowledge in classrooms and independently online at home. They prepare cards, presentations, booklets, videos.

The purpose of the research is to analyze visualization lecture as an effective form of learning in the digital educational process. The scientific and practical significance of the article is to review the techniques of visualization in Humanities cycle with the use of remote technologies and techniques on the subject «Russian Language and Speech Culture».

The work is devoted to the study of modern visual methods in the digital educational process. The research describes visualization as an educational method, describes visualization techniques, their practical application during the lecture and analyzes the effectiveness of using visualization lecture as a form of conducting classes confirmed by the results of testing students' knowledge using questionnaires. They say: «It is better to see once than hear a hundred times». Indeed, most of information from the world around us is obtained through vision, that is why the visual channel of perception should be maximally involved in learning. Even in the 20th century, visual teaching methods were widely used, providing for the display of visual AIDS in the classroom to facilitate understanding, memorizing the material and applying it in practice [13–15]. Visual in Latin means «visually perceived». There were significant changes in visual teaching methods due to the growth of information, the development of digital technologies and their application in the educational space in recent decades. Students get more information that is easier to understand and remember visually when presenting structured information [16].

Visualization is widely used in the educational process at all levels, including in higher education. With the development of digitalization visibility has received a new idea: the old posters, maps, visual AIDS have been replaced by computer technologies those allow you to use not only illustration and demonstration, but also multimedia. The teacher relies on the contingent, chooses the optimal form of presentation of the material using computer technology: for example, an interactive table will be effective for one group, a cluster for another group, and so on. While preparing for the lesson, it is necessary to study the material in terms.

Economist Jon Schwabish has formulated 4 principles of data visualization: data clarity; visual noise; graph and text – a single whole; pre-attentive processing [17]. In our opinion, visualization as a teaching method also has a set of principles that contribute to the successful application of the method in the educational process. Among them, it is worth to think about:

- clarity of the material (visual representation of specific information that should be clear to students);
- consistency (compliance with the laws of logic, ways of thinking);
- simplicity (ease of perception);
- optimality (optimal amount of information).

Visualization is a tool for activating students' mental activity, and also has an interactive component: not only the teacher gives ready-made images for visual perception, but also the students themselves are involved in the process of creating «visible» and understandable information that passes into the category of their own knowledge obtained by heuristic means. In the works of a number of researchers, it is said about visual thinking that can create new images that are endowed with a semantic load making information visible [18–20].

It is particularly worth noting that the use of the visualization method in Humanities classes has its own nuances associated with the presence of a large volume of mostly textual information. Students of technical specialties who study Humanities within the framework of educational programs have difficulties in working with text material, so it is necessary to use visibility during teaching.

In our opinion, cluster, infographics, intelligence map, word cloud, time tape (chronograph) are effective visualization techniques in Humanities classes. A cluster is a figurative scheme that explains the essence of a key concept through its associative reproducible elements. The cluster can be based on a picture with cause-and-effect relationships. Another technique that is close to a cluster is the intelligence map. An intelligence map is a way of displaying information in the form of a tree diagram that reflects interrelated ideas and concepts (branches of a tree) coming from the Central one (trunk). Infographics is a visualization technique that reflects information through graphic images, including tables, diagrams, charts, graphs and drawings. Time tape, or chronograph, is a visualization technique that reflects a sequence of events.

These techniques can be used in the course of both lectures and practical classes in Humanities effectively. In the context of the pandemic, all universities were forced to resort to distance learning.

Students need to perceive the teacher through a computer monitor, where the presentation of the lecture material comes to the fore now. The teacher's word is a sound background. Visualization in such conditions is in the first place. Teachers need to convey new material through clear, memorable images, as well as teach students to visualize the material they are working with.

Visualization lecture as an interactive form of conducting classes allows you to use various visualization techniques through digital technologies in the educational process. Visualization lecture helps to transform oral and written information into visual images by folding a large text into a concise visual element, as well as the process of speech development through «deployment»: explaining visual images with complete oral monologues. This type of lecture implements the principle of material availability.

As an example, we will give a methodological development of a class on the subject «Russian language and speech culture» on the topic «Communicative qualities of speech: richness, diversity, expressiveness of speech». The purpose of the lesson is to study the richness and expressiveness of speech as communicative qualities. The main tool of the teacher in the distance learning format is a presentation containing visual images those help to reveal the content of a new topic, as well as tasks those address the independent creation of visual images by students on the topic.

At the beginning of the lesson, the already studied material is systematized by analyzing clusters, infographics, for example, Figure 1 presented below.

This cluster reflects elements that do not meet the standards of the literary language and clog it. Students formulate the concept of «correctness of speech».

The interactive table «Language norms» is aimed at repeating the language norms and their compliance with the rules, and draws a parallel with the sections of linguistics. Students must fill in the gaps in the table in the columns of language norms or rules themselves.

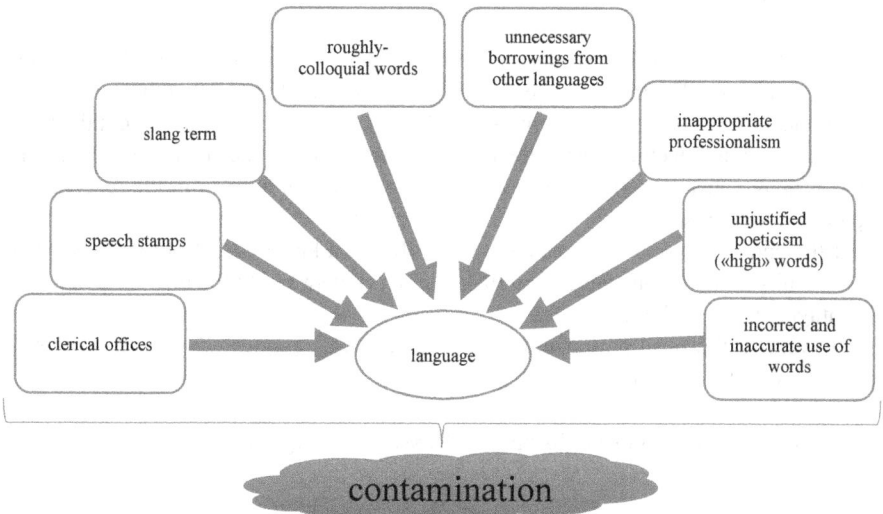

Fig. 1. The correctness of speech

The study of new material begins with analysis of supply artificially created by the linguist L. V. Shcherba: «Glokaya kuzdra shteko budlanula bokra and kudryachit bokryenka». Students are invited to make an illustration of the proposal in a graphic editor or on a piece of paper, take a photo and post it for discussion. We can conclude that even unfamiliar words make it possible to imagine what is being said, to assume the partial belonging of words.

Speaking about the semantic richness of words, the teacher suggests students to work with word clouds, where synonymous series are presented. Students determine the shades of meaning by noting positive or negative connotations, determining the presence or absence of expression in words.

When analyzing the richness and expressiveness of the grammatical structure of the language, students are invited to work with the table (infographic): to compare the verbs «to have read» and «to have been reading».

Students are asked to create a cluster that reflects the concepts those are part of the named communicative quality to formulate the definition of speech expressiveness. The teacher creates a cluster online, adding students' responses to it. In the course of the work, the cluster is proposed to be transformed into an intelligence map, adding the visual image with conditions that affect the expressiveness of a person's speech, as well as recommendations for improving their speech expressiveness.

Repetition of visual and expressive means can be submitted via an infographic (table or diagram) or an intelligence map of students' choice. Thus, a fairly large amount of information is visually presented not only from the position of the teacher, but also from the position of students.

3   Results

We conducted a study on the use of visualization and its significance for the development of digital education. For this purpose, 2 groups were selected: the first of them continued to study according to the usual program, and the second became experimental. In the latter, visualization lectures were held during the semester, i.e. using presentations, tables, graphs, clusters, video lectures, and other materials both from the teacher and prepared by students.

At the end of the semester, students were offered a knowledge test, which showed higher results of the second subgroup-79% of correct answers. In the first subgroup, this percentage was much lower – 57%.

To evaluate the results, the same questionnaire was conducted in both groups. It included the following questions:

1. Did you like the lessons on discipline «Russian language and speech culture»?

   a) Yes
   b) No
   c) Not really

2. Do you remember them?

   a) Yes
   b) No
   c) Not really

3. Will you remember them long?

   a) Yes
   b) No
   c) Not really

4. Would you like to return to these classes?

    a) Yes
    b) No
    c) Not really

The answers to each of the questions were the same, which simplified the assessment system. Based on the responses provided, a comparative table was compiled below (Tables 1 and 2).

Table 1. Knowledge test of the 1st group

The question	The answer «yes», %	The answer «no», %	The answer «not really», %
1. Did you like the lessons on discipline «Russian language and speech culture»?	56	7	37
2. Do you remember them?	49	1	50
3. Will you remember them long?	47	2	51
4. Would you like to return to these classes?	59	3	38

Table 2. Knowledge test of the 2nd group

The question	The answer «yes», %	The answer «no», %	The answer «not really», %
1. Did you like the lessons on discipline «Russian language and speech culture»?	76	6	18
2. Do you remember them?	81	1	18
3. Will you remember them long?	79	2	19
4. Would you like to return to these classes?	72	2	26

Moreover, after finishing the questionnaire the students of the second group were invited to answer the question «Do you think lectures full of presentations, graphs, diagrams, clusters and tables are more effective than regular so to say «dictation» lectures taught with chalk / a marker at the board? Answer «yes»/«no» and say why? We need detailed answers». It is necessary to give a small survey.

The students commented their attitude to the new form of the lecture, i.e. visualization lecture, in the oral form. 3 students of the group said that it was always individual.

One of them underlined: «I understand more when I write under dictation and with a blackboard. It's easier for me. I don't know what's best for the others». The second one commented that «it depends on the material. I better remember graphs and tables, but I think that there will be no special difference». The other student offered to combine classical lecture with the visualization one.

There was one who understood the belonging of visualization lecture more to the distance learning and commented it: «If you talk about distance and full-time learning, then my personal opinion is that full-time is of course better, and the combination of lectures at the blackboard with chalk and also saturated with presentations when you sit at a desk is better. There is a combination of old and new type of learning».

There was a student who said: «I think that visualization lecture is more effective. The information is easier to remember, lessons are more interesting, and generally speaking, the information in color is remembered better and faster than written with chalk or a marker on the board».

Two students underlined the necessity of video materials at the lessons. One of them said «Yes, yes and yes again! Classical lectures are good, but for a good mark of a teacher it would be better, if there were more video materials. And it would be possible for a teacher to get an «Oscar» for the lesson».

The second one decided to characterize the teacher of the discipline «Russian language and speech culture»: «You are a good teacher, you explain everything and help with tasks. But you can add more videos to your presentations and other materials».

There were also some other opinion: «Lectures with presentations and graphs are easier to understand than writing them as it were a dictation. But sometimes you still need optional lectures with a blackboard. At it you can ask certain things on which you have questions during the study and get a high-quality answer. Mostly it is easier with presentations, but you need the blackboard to ask questions and get the answers at a certain time of the lesson. This student even had a desire to offer one decision as if he were a teacher: «That is better to conduct a certain number of classes with presentations and at the end of the last of them to explain the remained questions at the blackboard».

The answers of the students help the teacher again to understand that education is a bilateral process and some so to say «grains» of it we can get not only from the books, but also from our students who can be both the object and the subject of the education.

4 Conclusion

Students of the second group noted the simplicity of presentation of the material and ease of memorization. As you know, the majority of people are visual artists, so it is easier for them to perceive the material systematically, visually, taught in the form of diagrams, tables, presentations, clusters, and video lectures. The students said that they began to understand the discipline more and its importance for the overall development.

Even teachers of other disciplines began to use visualization after student's offer: to draw up diagrams and tables at specialized lectures, as well as to structure knowledge in this way. As well as the teachers of the discipline «Russian language and culture of speech» noted the increase in students' motivation, as well as their colleagues borrowed the experience of conducting visualization lectures.

Appendix

Plan of the lecture visualization

Lesson topic: Communicative qualities of speech: richness and diversity, expressiveness.
Methodological goal: to activate the work of students in the classroom through the use of interactive forms and methods of teaching.
The purpose of the lesson: to continue studying the communicative qualities of speech, to consider the expressiveness of speech.
Tasks:
Educational: to repeat the studied communicative qualities of speech, to study the richness and expressiveness as communicative qualities of speech.
Developing: to develop analytical thinking, competent oral and written speech, the ability to use the communicative qualities of speech.
Educational: to form the ability to work in a team, to develop aesthetic feelings in relation to a language and speech.

Lesson progress

1. Organizational moment: greeting, attendance check (2 min).
2. Entrance control (repetition of the studied material, terms, diagrams, table) (10–12 min).
3. Content of the training material (lecture).
   Visualization lecture «Communicative qualities of speech: richness and diversity of speech, expressive speech» (work with diagrams, fixing the definitions, work with texts, discussion) (60–65 min).
   The richness of speech: the identity of the Russian language; the richness of its vocabulary; semantic richness (polysemy of words, synonyms, idioms); expression; grammatical structure of the language.
   Expressiveness of speech: features of speech structure; conditions of expressiveness; visual and expressive means.
4. Output control (working with the table 5–7 min)
5. Summing up, reflection (2 min).
6. Homework (1 min).

References

1. Stillman, D., Lancaster, L.C.: When generations collide: who they are. Why they crash. How to solve the Generational Puzzle at Work. HarperCollins. 352 p. (2002)
2. Homan, A.: Z is for Generation Z? Innov. Educ. (2015). https://tiie.w3.uvm.edu/blog/who-are-generation-z/#.Wt8DRVw5bIX. Accessed 27 Oct 2015
3. Horovitz, B.: After Gen X, Millennials, what should next generation be? USA Today (2012). https://usatoday30.usatoday.com/money/advertising/story/2012-05-03/naming-the-next-generation/54737518/1. Accessed 5 Apr 2012
4. Reynol, J., Jeanna, M.: Connecting to the net generation: what higher education professionals need to know about today's students. NASPA. 164 p. (2007)
5. Shapira, I.: What comes Next after Generation X? Washington Post (2008). https://www.was hingtonpost.com/wp-dyn/content/article/2008/07/05/AR2008070501599.html. Accessed 6 July 2008
6. Egorova, S.Yu.: the Use of technical learning tools in the classroom in a foreign language. University of the XXI century: Scientific measurement, pp. 106–112 (2011)
7. Klopfer, E.: Augmented Learning: Research and Design of Mobile Educational Games. MIT Press, Cambridge (2008)
8. Lakarnchua, O., Reinders, H.: Implementing mobile language learning with an augmented reality activity. Mod. English Teach. **23**(2), 42–50 (2014)
9. Harsha, R.: Covid -19 Lockdown-challenges to higher education (2020). https://doi.org/10.13140/RG.2.2.16290.25281
10. World health statistics 2018: monitoring health for the SDGs, sustainable development goals. Geneva: World Health Organization (2018)
11. Belousova, O.A., Polyakov, V.M., Mikhailiukov, L.V.: Methods diagnosing terminological culture of students studying in agricultural institutions. Mod. Pedag. Educ. **4**, 109–112 (2019)
12. Shamin, A.E., Frolova, O.A.: University of the future in the digital economy era. Training modern economic personnel to meet new challenges. Nikon Readings, no. 23 (2018)
13. Asimov, E.G., Shchukin, A.N.: New dictionary of methodological terms and concepts (theory and practice of language teaching). Moscow: publishing house IKAR (2009). https://method ological_terms.academic.ru/1047
14. Kalinichenko, A.V.: Interactive electronic didactic tools with cognitive visualization. Sci. Tech. Bull. Inf. Technol. Mech. Opt. **17**(2), 359–364 (2017). https://cyberlemnka.m/artide/n/interaktivnye-elektronnye-didakticheskie-sredstva-s-kognitivnoy-vizualizatsiey
15. Krotova, I., Kamoza, T., Donchenko, N.: Visualization method in the system of innovative training. High. Educ. Russia (4) (2008). https://cyberleninka.ru/article/n/metod-vizualizatsii-v-sisteme-innovatsionnogo-obucheniya
16. Fedosova, O.A., sokolina, E.N.: on the value of visualization of educational information. Probl. Pedag. **3**(35) (2018). https://cyberleninka.ru/article/n/o-znachenii-vizualizatsii-uchebnoy-informatsii
17. Schwabish, J.: Memo on working with visual data. https://infographer.ru/four-principles-of-data-viz/
18. Mindzaev, E.V., Matevosova, J.V.: Visual thinking as a factor of formation of ICT-competence of students of the University. Quest. Mod. Sci. Pract. **1**(32), 155–158 (2011)
19. Reznik, N.A.:Methodological foundations of teaching mathematics in high school using visual thinking development tools: dis. on the map. 500 p. SPb (1997)
20. Dochkin, S.A., Michurina, E.S.: Technology of knowledge visualization as a necessary aspect of training University teachers. Prof. Educ. Russia Abroad **3**(15) (2014). https://cyberleninka.ru/article/n/tehnologii-vizualizatsii-znaniy-kak-neobhodimyy-aspekt-podgotovki-prepodavateley-universiteta

Comparative Analysis of Machine Learning Models for Students' Performance Prediction

Leila Ismail[1](✉) ⓘ, Huned Materwala[1] ⓘ, and Alain Hennebelle[2]

[1] Distributed Computing and Systems Research Laboratory, Department of Computer Science and Software Engineering, College of Information Technology, United Arab Emirates University, Al Ain, Abu Dhabi 15551, United Arab Emirates
`leila@uaeu.ac.ae`
[2] Al Ain, Abu Dhabi, United Arab Emirates

Abstract. Machine learning for education is an emerging discipline where a model is developed based on training data to make predictions on students' performance. The main aim is to identify students who would have difficulty in their learning and to take precautionary measures to help them. In this paper, we conduct a comparative analysis of the most used machine learning classification models in the literature. We evaluate the performance of the models in terms of accuracy, F-measure, and execution time using two real-life education datasets. The performance of the models is data-driven. We give insights into the models' performance and advise on the best model to use accordingly. We believe the results of this paper will be widely used by education professionals for accurate predictions.

Keywords: Artificial intelligence · Classification models · Educational data mining · Educational machine learning · Student performance prediction

1 Introduction

Education plays an important role in the development of a nation. Educational Data Mining (EDM) is an emerging discipline of data analytics where the machine learning approaches are applied to the educational data for extracting insights [1]. The objective is to determine the usefulness of the current education systems, analyze the academic performance of students, and to develop an academic failure prevention plan. Prediction of students' academic performance is crucial to develop strategies for weak learners to improve their overall performance. The performance of students depends on different social, demographic, psychological, and family factors. Machine learning classification models are used to predict the performance of students based on these factors.

The work on machine learning for education in the literature evaluates the performance of the classification models using heterogeneous datasets and evaluation metrics. In this paper, we evaluate and compare the performance of the most used classification models in the literature, Decision Tree (DT), Naïve Bayes (NB), Artificial Neural Network (ANN), Support Vector Machine (SVM), and Random Forest (RF). This is in a

---

A. Hennebelle—Independent Scientist Engineer.

© The Author(s), under exclusive license to Springer Nature Switzerland AG 2021
T. Antipova (Ed.): ICADS 2021, AISC 1352, pp. 149–160, 2021.
https://doi.org/10.1007/978-3-030-71782-7_14

unified setup using student performance dataset – Portuguese [2] and student performance dataset - xAPI [3, 4] datasets. We evaluate the performance of the models in terms of accuracy, F-measure, and execution time. The statistical error measures such as Root Mean Squared Error (RMSE) and Mean Absolute Error (MAE) are not used to evaluate the models. This is because these error measures are used for continuous numeric values. Prediction of students' performance is a classification problem with class labels as prediction results rather than continuous numeric values. consequently, it is very difficult to find the error between the predicted class label (pass for instance) and the actual class label (fail for instance). A low RMSE does not always guarantee low misclassification rate.

The rest of the paper is organized as follows. In Sect. 2, we discuss the related work. The classification models used in this study are explained in Sect. 3 in the context of students' performance prediction. The experimental setup, experiments, and the analysis of the results in terms of accuracy, F-measure, and execution time are described in Sect. 4. The paper is concluded in Sect. 5 along with recommendations and future research directions.

2 Related Work

Work in the literature evaluated the performance of machine learning classification models for students' performance. Table 1 shows the work on the most used classification models in the literature, i.e., DT, NB, ANN, SVM, and RF. These models are evaluated

Table 1. Evaluation of past works on DT, NB, ANN, SVM, and RF models.

Models evaluated	Work	#features	#observations	Accuracy	F-measure	Execution time
DT and NB	[7]	11	279	✓	✗	✗
DT and RF	[8]	28	450	✓	✗	✗
ANN and SVM	[9]	49	127	✗	✗	✗
ANN and RF	[10]	16	480	✓	✓	✗
DT, NB and ANN	[11]	37	60	✓	✗	✗
	[12]	11	666	✓	✗	✗
DT, NB and SVM	[13]	23	776	✗	✓	✗
DT, ANN and RF	[14]	39	32593	✓	✓	✗
DT, SVM and RF	[15]	16	500	✓	✓	✗
DT, NB, ANN and SVM	[16]	19	262	✗	✓	✗
		15	161			
	[17]	16	344	✓	✗	✗
DT, NB, ANN and RF	[18]	33	395	✗	✓	✗
DT, NB, SVM and RF	[19]	19	2459	✓	✗	✗
DT, NB, ANN, SVM and RF	[5]	11	125	✓	✗	✗
	[6]	11	114	✓	✗	✗
	This work	32	1044	✓	✓	✓
		16	480			

using different datasets and evaluation metrics making an objective comparison difficult. Works [5] and [6] compare these models in a unified setup. However, these works use a single dataset with less than 150 students' records for the evaluation of the models which makes it difficult to generalize the results. Moreover, these works evaluate the models in terms of accuracy only. We argue that for an imbalanced dataset, it is important to use F-measure as accuracy alone can be misleading. In this paper, we conduct a comparative analysis between the models under study in terms of both accuracy and F-measure. In addition, we compare the models in terms of execution time.

3   Machine Learning Models For Students' Performance Prediction

In this section, we explain the machine learning classification models under study for students' performance prediction. We take an example of classifying students' marks into three levels: high, medium, and low. However, other classification labels, such as A, B, C, D, and F can be used.

3.1   Decision Tree

Decision Tree (DT) model develops a tree-like structure from the training dataset to define the sequences of decisions and their corresponding outcomes [20]. Each feature in the dataset is represented as a node. The node having the class labels (high, medium, and low) is called a leaf node. At each node, the model traverses down by selecting a particular node. The model selects the node having the maximum value of information gain, i.e., the most informative feature. The value of information gain for a feature F is calculated using Eq. (1).

$$InfoGain_F = H_{High} - H_{High|F} \tag{1}$$

where H_{High} is the base entropy for the high class which is calculated using Eq. (2) and $H_{High|F}$ is the conditional entropy for the high class corresponding to the feature F which is calculated using Eq. (3).

$$H_{High} = \sum_{\forall\, High\, \in\{High, Medium, Low\}} P(High) \log_2 P(High) \tag{2}$$

$$H_{High|F} = \sum_f P(f) H(High|F = f)$$
$$= \sum_{\forall f \in F} P(f) \sum_{\forall\, High\, \in\{High, Medium, Low\}} P(High|f) \log_2 P(High|f) \tag{3}$$

where $P(High)$ is the probability of the number of students in the high class compared to the total number of students in the dataset and f is the set of all values for the feature F.

3.2 Naïve Bayes

Naïve Bayes (NB) is based on the Bayes' theorem that formulates the relationship between the probabilities and conditional probabilities of two features [20]. The objective of NB model is to predict the class for a student with a set of features $(F_1, F_2, ..., F_n)$ from the set of classes, $C = \{High, Medium, Low\}$. This is in a way that the conditional probability $P(High| F_1, F_2, ..., F_n)$ or $P(Medium| F_1, F_2, ..., F_n)$ or $P(Low| F_1, F_2, ..., F_n)$ is the maximum for the class being predicted. The general form of Bayes' theorem for calculating the posterior probability of the high class for a student with a set of features is stated in Eq. (4).

$$P(High|(F_1, F_2, \ldots, F_n) = \frac{P(F_1, F_2, \ldots, F_n|High).P(High)}{P(F_1, F_2, \ldots, F_n)} \tag{4}$$

where $P(F_1, F_2, ..., F_n| High)$ is the likelihood, $P(High)$ is the prior probability of the high class and $P(F_1, F_2, ..., F_n)$ is the prior probability of the feature set.

The NB model extends the Bayes' theorem stated in Eq. (4) based on two assumptions:

1) For a given class (High/medium/low), there is no dependency among the features as stated in Eq. (5)

$$P(F_1, F_2, \ldots, F_n|High) = P(F_1|High).P(F_2|High).\ldots.P(F_n|High) = \prod_{i=1}^{n} P(F_i|High) \tag{5}$$

2) Removing the denominator from Eq. (4) as it remains constant. Consequently, the conditional probability of the high class for a student with a set of features (Eq. (4)) can be expressed as stated in Eq. (6).

$$P(High|(F_1, F_2, \ldots, F_n) = P(High).n \prod_{i=1}^{n} P(F_i|High) \tag{6}$$

3.3 Artificial Neural Network

Artificial Neural Network (ANN) is a network of neurons arranged in layers and each neuron takes in an input and passes the output to the next layer after applying an activation function [21]. ANN is comprised of three different types of layers: 1) input layer that consists of the features for predicting students' performance, 2) hidden layer(s) and 3) output layer that consists of the class labels high, medium and low. A neuron in the network feeds its output to all the other neurons in the next layer with an associated weight. The implementation of ANN in the literature is mostly based on Multilayer Perceptron (MLP) which is a feedforward neural network that utilizes back-propagation for training [22]. Back-propagation is used to tune the weights in the network based on the error rate. After each iteration, the model computes the error for each neuron by comparing the result of the output layer with the actual class labels. The error values are then back-propagated to tune the weights in the hidden layer in a way that the prediction accuracy increases in the next iteration. The output is ANN is calculated using Eq. (7).

$$a = \emptyset(w_i + b) \tag{7}$$

where w is the weight matrix for the features, i is the input vector consisting of the features, b is the vector consisting of the bias values and $\emptyset(.)$ is the activation function. In this paper, the sigmoid activation function is considered.

3.4 Support Vector Machine

Support Vector Machine (SVM) model separates the n-dimensional students' information into high, medium, and low classes. This is by creating a decision boundary, known as hyperplane, for separation. The hyperplane is generated with the help of the data points in each class that are the closest to the data points in the other class. These data points are known as support vectors. The generation of hyperplane is an iterative process where the objective is to find the maximum possible margin between the support vectors of the opposite classes. For a students' performance dataset with a set of F features and $C = \{High, Medium, Low\}$ classes, the hyperplane that separates these classes can be represented using Eq. (8).

$$wF + b = 0$$
$$s.t., \ wF_i + b > 0, \ if \ C_i = +1 \ \textbf{and} \ wF_i + b < 0, \ if \ C_i = -1 \tag{8}$$

where w is normal to the hyperplane and b is the bias.

To find the maximum possible margin for optimal hyperplane, the norm of the margin should be minimized as stated in Eq. (9).

$$\min \frac{1}{2}||w||^2 s.t., \ C_i(w.F_i + b) \geq 1, \forall F_i \tag{9}$$

3.5 Random Forest

Random Forest (RF) is an ensemble technique that uses a set of Decision Trees. Each tree is constructed using a randomly selected sample of the dataset [23]. Each Decision Tree predicts the class for a student (high/medium/low) based on the features and voting is performed on the output of each tree. In RF, each DT model will consider only a randomly selected subset of features at each node for splitting the tree and traversing down. The RF model then decides on the high/medium/low class based on the majority of the votes as stated in Eq. (10).

$$C_{RF}(s) = majority \ vote\{C_n(s)\}_1^N \tag{10}$$

where N represents the number of Decision Trees used.

4 Performance Analysis

In this section, we analyze and compare the performance of the most used classification models for students' performance prediction. We evaluate the models in terms of accuracy, F-measure, and execution time with and without feature selection.

4.1 Experimental Environment

We evaluate the performance of the classification models using two educational datasets: one which includes data about Portuguese students from the UCI repository [2] and one from Kaggle by xAPI [3, 4]. Table 2 shows the specifications of the datasets under experiments. To evaluate the performance of the models under study with feature selection algorithm, we use the Information-Gain Attribute Evaluator algorithm as it is found to be the most accurate among others in education machine learning [24]. We use Weka 3.8 [25] for implementation.

Table 2. Specifications of the datasets used in the experiments.

Dataset	#features	Features	#observations	Class labels
student academic performance dataset – Portuguese	32	a	1044	A B C D F
student academic performance dataset – xAPI	16	b	480	High Medium Low

a - Student's school, gender, age, address, family size, parent's cohabitation status, mother's education, father's education, mother's occupation, father's occupation, reason to choose the school, student's guardian, home to school travel time, study time, failures, extra educational support, family educational support, paid classes, extra-curricular activities, attended nursery school or not, wants to take higher education or not, internet access at home, romantic relationship, quality of family relationships, free time, going out with friends, workday alcohol consumption, weekend alcohol consumption, current health status, number of school absences, first period grade and second period grade.

b - Gender, nationality, place of birth, stage id, grade id, section id, topic, semester, relation, raised hands, visited resources, announcement views, discussion, parents answering survey, parent school satisfaction and student absence.

4.2 Experiments

Data Preprocessing. For all the datasets under study, we first convert the categorical features into numerical ones. For instance, for the feature 'Gender', we replace the categorical values 'Male' and 'Female' with numerical values '0' and '1'. For the student academic performance dataset – Portuguese, we combined the separate data files for the Mathematics and Portuguese language subjects. In addition, we created class labels for the predictive grade response 'G3'. This is by mapping the grades to 5 classes based on the Erasmus grade conversion system [26], i.e., grade A (16–20), grade B (14–15), grade C (12–13), grade D (10–11) and grade F (0–9).

Model Building and Testing. For model building and testing, we use 10-fold cross-validation method to obtain the training and testing datasets respectively. For the SVM model, we implement the linear, polynomial, and Radial Basis Function (RBF) kernels. We measure the accuracy, F-measure, and execution time for each model. The accuracy is calculated using Eq. (11).

$$Accuracy = \frac{TP + TN}{TP + FP + TN + FN} \tag{11}$$

where

TP = True Positive = #observations in positive class that are predicted as positive.
TN = True Negative = #observations in negative class that are predicted as negative.
FP = False Positive = #observations in negative class that are predicted as positive.
FN = False Negative = #observations in positive class that are predicted as negative.

The F-measure is calculated using precision and recall as stated in Eq. (12). The values of precision and recall are calculated using Eqs. (13) and (14) respectively. The execution time is computed by adding the model building and validation times.

$$F - measure = \frac{2(Precision * Recall)}{Precision + Recall} \tag{12}$$

$$Precision = \frac{TP}{TP + FP} \tag{13}$$

$$Recall = \frac{TP}{TP + FN} \tag{14}$$

4.3   Experimental Results Analysis

Figure 1 shows the accuracy and F-measure for the Portuguese dataset without feature selection. It shows that the RF model has the highest accuracy and F-measure values compared to the other models. This is because RF selects a subset of the features randomly to fit the data while the other models use all the features resulting in overfitting and consequently inaccurate prediction. The performance of DT is almost similar to RF.

When applying the feature selection algorithm on the Portuguese dataset, the selected features are the age of the student, mother's education, father's education, mother's occupation, student study time, number of past class failures, plan for higher education, number of school absences and first and second period grades. Figure 2 shows the accuracy and the F-measure with feature selection. It shows that the performance of the models is enhanced by using the selected features. In particular, the accuracy of DT, SVM linear, and SVM polynomial increases by 3.16%, 8.6%, and 4.5% respectively

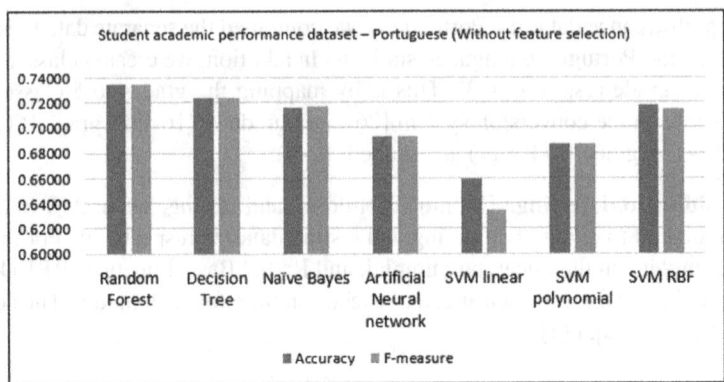

Fig. 1. Accuracy and F-measure of the models for the Portuguese dataset without feature selection.

compared to that without feature selection. Comparing the performance of DT and RF with feature selection, the DT performs better. This is on the contrary to their relative performance without feature selection (Fig. 1). This is because, with feature selection, DT constructs a tree with the significant selected features, whereas RT randomly selects a subset of selected features to construct different trees. Consequently, RF does not always use the most significant feature as the root of the tree.

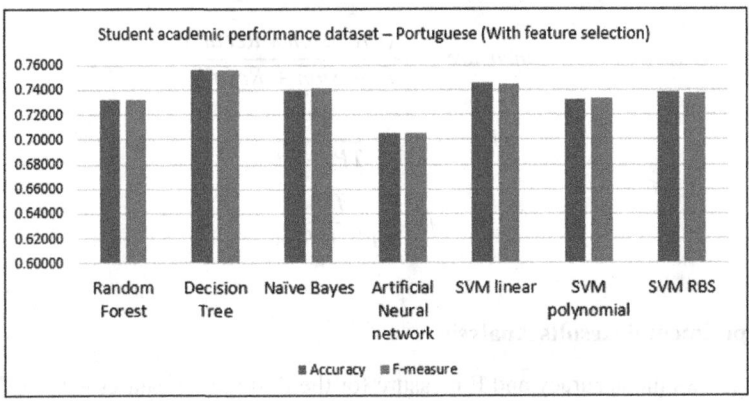

Fig. 2. Accuracy and F-measure of the models for the Portuguese dataset with feature selection.

Figure 3 shows the accuracy and the F-measure for xAPI dataset without feature selection. It shows that SVM Polynomial, SVM linear, and NB models outperforms the other models. The SVM with RBF kernel is having no F-measure value. This is because the model is not able to detect the low and high class labels in the dataset. The reason for this is that the xAPI is a small dataset having only 480 instances out of which 44% of the observations belong to the detected medium class and 26% and 30% of the observations belong to the non-detected low and high classes respectively. Consequently, the SVM

RBF only detects the majority class. With accuracy only as an evaluation metric, this issue cannot be detected.

Applying the feature selection algorithm on the xAPI dataset, our experimental results show that the selected features are Nationality, place of birth of student, the parent responsible for the student, number of times hands raised in class, viewed announcements, and participated in discussions, parent answering a survey, parent satisfaction from school and the number of days the student is absent. Figure 4 shows the accuracy and F-measure for the xAPI with feature selection. It shows that the selected features are not relevant for predicting students' performance as the accuracy of the models under study is not improved significantly. Moreover, the accuracy of the NB, ANN and SVM polynomial models is reduced by 1.25%, 3.95%, and 2.91% respectively after applying feature selection.

Table 3 shows the execution time in seconds for the classification models under study with and without feature selection. It shows that for the Portuguese dataset having 1044 observations, the execution time is reduced after applying feature selection. However, for the xAPI dataset having 480 observations, there is not much reduction in execution time after feature selection.

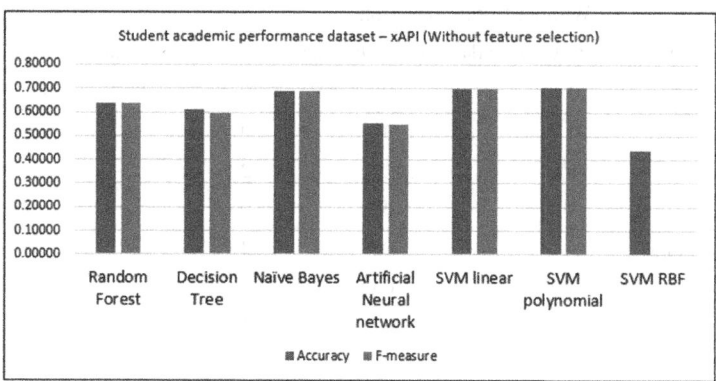

Fig. 3. Accuracy and F-measure of the models for xAPI dataset without feature selection.

In summary, for the Portuguese dataset having 1044 observations, RF and DT outperform the other models under study, whereas for the xAPI dataset having 480 observations, SVM linear, SVM polynomial and NB outperforms the other models. The relative performance of DT and NB models in our experimental results is similar to the one obtained by [7]. Furthermore, the relative performance between ANN and SVM in our experimental environment is the same as in [9], and the DT is performing better than NB, ANN, RF as per [18]. Our experimental results reveal that NB performs better for an imbalanced dataset whereas DT performs better for a balanced dataset. This is also confirmed by the results in the literature [11, 12]. Our results for the xAPI small dataset show that SVM linear outperforms the other models under study, which is consistent with that in the literature when evaluated using small datasets of 125 observations [5] and 114 observations [6]. However, our results are not consistent with the ones obtained

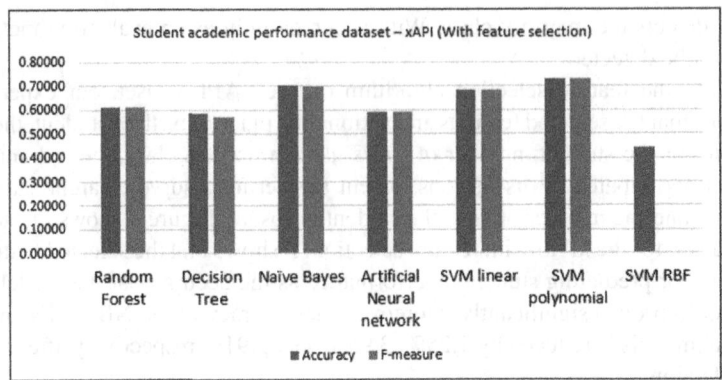

Fig. 4. Accuracy and F-measure of the models for the xAPI dataset with feature selection.

Table 3. Execution time for the models with and without feature selection.

| Classification model | Execution time (seconds) | | | |
| | Without feature selection | | With feature selection | |
	Portuguese dataset	xAPI dataset	Portuguese dataset	xAPI dataset
DT	0	0	0	0
NB	0	0	0	0
ANN	71	203	29	183
SVM Linear	3	0.3	0.1	0.9
SVM Polynomial	5	0.5	4	0.2
SVM RBF	3	0.05	1	0.2
RF	1	0.1	1	2

by [8, 14, 15, 17, 19]. The discrepancy in the results is due to heterogeneity in the features used in these works. The size of the dataset, the features, and the proportion of the classes (balanced or imbalanced data sets) impact the performance of the models.

5   Conclusions

Machine learning is used by education professionals to identify potentially weak performing students from the beginning with the aim of developing strategies and techniques to help those students. Works in the literature have used different machine learning approaches for the prediction of students' performance. However, these works used non-identical datasets and evaluation metrics, making an objective comparison between the developed models difficult. In this paper, we evaluate and compare the performance of the most used machine learning classification models, i.e., DT, NB, ANN, SVM, and RF, for students' performance prediction. We conduct a comparative analysis of the models using two real-life educational datasets. Our experimental results reveal that for a dataset having fewer observations, SVM linear, SVM polynomial, and NB outperforms the other models under study, whereas for a dataset having a large number of observations, DT and RF outperforms the other models under study. However, for generalizing these results, more extensive experiments are required on several balanced and imbalanced datasets.

When selecting a machine learning model, the following requirements should be considered:

1. *Data characteristics:* Our experimental results reveal that the performance of the classification models is data-driven. The SVM linear, SVM polynomial, and NB models are suitable for a dataset with a small number of observations (480 in our study). The RF and DT models are suitable for prediction in the case of a dataset having a greater number of observations (1044 in our study) leading to better accuracy.
2. *Significant features:* While creating the dataset for the prediction of students' performance, it is important to consider features such as the age of the student, parents educational and work background, student study time, number of past class failures, plan for higher education, number of school absences and previous grades. This is because these features improve the prediction accuracy.
3. *Evaluation metrics:* The selection of the evaluation metrics for the analysis of classification models is crucial. Most of the works in the literature use only accuracy. However, in the case of a dataset which is small or has imbalanced classes, accuracy can be misleading. Consequently, F-measure in addition to accuracy should be considered. In addition, RMSE and MAE, often used in the literature, can be misleading. For instance, a model can have a higher error, but it is accurate. Low RMSE does not guarantee a low misclassification rate.

Acknowledgment. This work was funded by the National Water and Energy Center.

References

1. Romero, C., Ventura, S.: Educational data mining and learning analytics: an updated survey. Wiley Interdisc. Rev. Data Min. Knowl. Discov. **10**, e1355 (2020)
2. Cortez, P., Silva, A.M.G.: Using data mining to predict secondary school student performance (2008)
3. Amrieh, E.A., Hamtini, T., Aljarah, I.: Mining educational data to predict student's academic performance using ensemble methods. Int. J. Database Theory Appl. **9**, 119–136 (2016)
4. Amrieh, E.A., Hamtini, T., Aljarah, I.: Preprocessing and analyzing educational data set using X-API for improving student's performance. In: 2015 IEEE Jordan Conference on Applied Electrical Engineering and Computing Technologies (AEECT) (2015)
5. Rimadana, M.R., Kusumawardani, S.S., Santosa, P.I., Erwianda, M.S.F.: Predicting student academic performance using machine learning and time management skill data. In: 2019 International Seminar on Research of Information Technology and Intelligent Systems (ISRITI), pp. 511–515 (2019)

6. López, M.I., Luna, J.M., Romero, C., Ventura, S.: Classification via clustering for predicting final marks based on student participation in forums. In: Proceedings of the 5th International Conference on Educational Data Mining, of EDM 2012, Chania, Greece, pp. 148–151 (2012)
7. Wati, M., Indrawan, W., Widians, J.A., Puspitasari, N.: Data mining for predicting students' learning result. In: 2017 4th International Conference on Computer Applications and Information Processing Technology (CAIPT). IEEE, Kuta Bali (2017)
8. Mehboob, B., Muzamal Liaqat, R., Abbas, N.: Student performance prediction and risk analysis by using data mining approach. J. Intell. Comput. **8**, 49 (2017)
9. Tekin, A.: Early prediction of students' grade point averages at graduation: a data mining approach. Eur. J. Educ. Res. **54**, 207–226 (2014)
10. Almutairi, S., Shaiba, H., Bezbradica, M.: Predicting students' academic performance and main behavioral features using data mining techniques. In: International Conference on Computing, pp. 245–259. Springer, Cham (2019)
11. Mueen, A., Zafar, B., Manzoor, U.: Modeling and predicting students' academic performance using data mining techniques. Int. J. Mod. Educ. Comput. Sci. **8**, 36–42 (2016)
12. Hussain, S., Atallah, R., Kamsin, A., Hazarika, J.: Classification, clustering and association rule mining in educational datasets using data mining tools: a case study. In: CSOC2018 2018: Cybernetics and Algorithms in Intelligent Systems, pp. 196–211 (2018)
13. Daud, A., Aljohani, N.R., Abbasi, R.A., et al.: Predicting student performance using advanced learning analytics. In: Proceedings of the 26th International Conference on World Wide Web Companion, Perth, Australia, pp. 415–421 (2017)
14. Rivas, A., Gonzalez-Briones, A., Hernandez, G., et al.: Artificial neural network analysis of the academic performance of students in virtual learning environments. Neurocomputing **423**, 713–720 (2020)
15. Ajibade, S.-S.M., Ahmad, N.B.B., Shamsuddin, S.M.: Educational data mining: enhancement of student performance model using ensemble methods. In: IOP Conference Series: Materials Science and Engineering (2019)
16. Costa, E.B., Fonseca, B., Santana, M.A., et al.: Evaluating the effectiveness of educational data mining techniques for early prediction of students' academic failure in introductory programming courses. Comput. Hum. Behav. **73**, 247–256 (2017)
17. Kostopoulos, G., Lipitakis, A.-D., Kotsiantis, S., Gravvanis, G.: Predicting student performance in distance higher education using active learning. In: International Conference on Engineering Applications of Neural Networks, pp. 75–86. Springer, Cham (2017)
18. Kiu, C.-C.: Data mining analysis on student's academic performance through exploration of student's background and social activities. In: 2018 Fourth International Conference on Advances in Computing, Communication & Automation (ICACCA). Subang Jaya, Malaysia (2018)
19. Migueis, V.L., Freitas, A., Garcia, P.J., Silva, A.: Early segmentation of students according to their academic performance: a predictive modelling approach. Decis. Support Syst. **115**, 36–51 (2018)
20. EMC Education Services: Data Science and Big Data Analytics: Discovering, Analyzing, Visualizing and Presenting Data. Wiley (2015)
21. Hassoun, M.H.: Fundamentals of artificial neural networks (1995)
22. Ramchoun, H., Idrissi, M.A.J., Ghanou, Y., Ettaouil, M.: Multilayer perceptron: architecture optimization and training. Int. J. Interact. Multimed. Artif. Intell. **4**, 26–30 (2016)
23. Liaw, A., Wiener, M.: Classification and regression by random forest. R news **2**, 18–22 (2002)
24. Ramaswami, M., Bhaskaran, R.: A study on feature selection techniques in educational data mining. J. Comput. **1**, 7–11 (2009)
25. Hall, M., Frank, E., Holmes, G., et al.: The WEKA data mining software: an update. ACM SIGKDD Explor. Newsl. **11**, 10–18 (2009)
26. Erasmus Programme. https://en.wikipedia.org/wiki/Erasmus_Programme. Accessed 18 Dec 2020

The Experience of Ukraine and Kazakhstan of Digitalization Education Under Quarantine Conditions

Elena Yuryevna Klymenko[1](✉) ⓘ and Sholpan Essenbaevna Alpeissova[2] ⓘ

[1] Department of Economics, Entrepreneurship and Social Technologies,
State University of Telecommunications, Solomenskaya Street, 7, Kyiv 03110, Ukraine
`klimenkoelens@gmail.com`
[2] Management Department, S. Seifullin Kazakh Agro Technical University, Zhenis Avenue, 62,
Nur-Sultan 010011, Kazakhstan
`sholpan761@mail.ru`

Abstract. The article aims to study features of digitalization of education in the post-Soviet space under quarantine conditions: the experience of Ukraine and Kazakhstan. While the society of Western Europe starts all the transformations from the personality change, waiting for changes from outside is a characteristic feature for our sociocultural, the post-Soviet area. The existing crisis of social transformations gives an essential stimulus to active formation of a new paradigm of social relations. Both for developed countries, which believed that the onset of a crisis was a problem for the future, so for Ukraine and Kazakhstan, which already live in conditions of combined social systems, the pandemic came as a surprise. Thus, the education must response quickly to such changes, forming new approaches, forms and methods of knowledge of social discourse. The aim of the article is to study new models of educational practices (in particular digitalization) that became the most common in society during the period of crisis associated with quarantine conditions, accompanied by forced self-isolation of citizens and determining roles in online learning, allowing to formulate new strategies for adapting to the conditions of limiting social activity. The authors disclosed its main components, identified key aspects of Internet prospects of studying. Two relatively separate directions of this phenomenon analysis have been analyzed. Special attention has been paid to the research of priorities of digital education.

Keywords: Pandemic · Crisis · Post-Soviet societies · Social transformation · Digital education

1 Global Economic and Society Transformations Amid a Pandemic

1.1 Crisis in Post-Soviet Societies Undergoing Transformation Amid a Pandemic

Fundamental basis of the post-Soviet society development is its combined character conditioned by the processes of system transformation that appeared during fledging years

© The Author(s), under exclusive license to Springer Nature Switzerland AG 2021
T. Antipova (Ed.): ICADS 2021, AISC 1352, pp. 161–172, 2021.
https://doi.org/10.1007/978-3-030-71782-7_15

of Ukraine and Kazakhstan as independent states. They give rise to the strange phenomena in politico-social and economic life, lead to the shift of features of different types of societies. These phenomena look counterintuitive in the view of their contradiction to the logical laws. Combination of different commonly contradictive sources, which can't exist theoretically, but they do exist in a real sense, have combined character, and consist of different by quality components – these are the peculiarities of social changes in modern Ukraine and Kazakhstan.

Considering the crisis as a certain impetus to the development of society, analyzing not only its negative but also positive factors, we tried to analyze which way of adaptation (active or passive) to the fundamentally new living conditions will be chosen by ordinary citizens, people from fundamentally different sociocultural environments of Europe and Asia, who have formed a traumatic Soviet experience.

In recent years, both in Ukrainian and Kazakh societies has developed ways to adapt people to life in crises, among which there are both passive and active models of adaptation. The conditions of quarantine not only limited the opportunities for movement and communication of people, forcing ordinary citizens to abandon traditional social actions and move to rational ones, they sharply reduced the list of possible technologies of personality adaptation, as they changed the temporal-spatial construct of life. The only space that has more or less remained stable and for which time does not matter is the Internet.

In the article, we tried to study the differences between the models of social adaptation to quarantine conditions between the countries of Europe and Asia, which have a common experience in the formation of socio-cultural space in Soviet realities.

In this context to identify the status of the state of modern society in the post-Soviet space one can freely use the notion «transforming». Moreover, under transformation one should understand significant changes of forms and contents of modern social phenomena and processes, as well as global economic transformations. Whereas the transformation can't go with lightening speed – immediately and to the full extent, the society feels the influence of factors of both: its past state and its future one. Therefore, all social transformations of present day have combined character, where qualities and properties of the past overlap the modern times and the future, creating difficult and mixed picture of the social life. This circumstance produces the complex of paradoxes and we are the witnesses of their existence. An example that brightly illustrates such phenomena could be the following: digitalization of education and distance learning.

Thus, our contemporary society lives within the frames of several paradigms at once, that determines formation among people not only of different points of view but also of different mental matrixes and makes an impact on formation of processes of communication between them. The sphere of scientific knowledge is not an exception in this context. Today education is a starting potential in defining the future life path of every person and is notable for constant focus on the adult world. The future of the society depends on the quality of education, conditions under which the process of socializing is taking place – formation of personalities.

In 2020, the crisis became a major problem not only for politicians, civil servants, political scientists, sociologists, but also for most ordinary citizens, as it somehow affected all spheres of life, both Ukrainian and Kazakhstan societies, in particular, and the world community in general. Today, there is almost no person left whose social actions would not be affected by the spread of the terrible disease – COVID-19 and the forced restrictions associated with the introduction of quarantine measures. A specific feature for the post-Soviet society was the situation in which global trends affected the already protracted systemic crisis, which has its roots in the events of the XXI century [1]. A number of crises – political, economic, cultural, which over the past ten years accompanied the lives of Ukrainians and Kazakhstan combined with global, which caused by a pandemic, which created a situation of high threat to national security.

According to the United Nations, more than 6 million Ukrainians are below the poverty line (from 27% to 44%), with one in two families at risk zone and the most vulnerable are single-parent families and families with young children [2].

Due to the pandemic, Kazakhstan is facing the worst economic turmoil in nearly two decades. Prior to this, the country's economy suffered from the fall in oil prices, which affected the stability of the financial sector According to the United Nations, more than 800 thousand people Kazakhstanis are below the poverty line, which amounted to 12.7% [3].

The conditions of quarantine not Ukrainian and Kazakh societies only limited the opportunities for movement and communication of people, forcing ordinary citizens to abandon traditional social actions and move to rational ones, they sharply reduced the list of possible technologies of personality adaptation, as they changed the temporal-spatial construct of life. The only space that has more or less remained stable and for which time does not matter is the Internet.

With an active model of adaptation to fundamentally new social realities, virtual space is used by humanity as an environment directly in which you can find a job, additional opportunities for earnings, new ways of self-realization and gain the necessary knowledge. Otherwise, the virtual environment can serve as a means of passive model of adaptation to life, in times of crisis, by communicating in the Internet community, listening to music, watching movies, reading fiction, reading information sites.

Transformations in the life of society have actualized the issue of mass transition of educational practices into cyberspace. According to UNESCO, 165 countries have introduced quarantine measures through COVID-19, which is why more than 1.5 billion people (91.4%) have switched to distance learning, although according to the Organization for Economic Cooperation (OECD) only 53% of teachers have experience online learning [4]. It is clear that educational institutions have been forced to switch to distance learning, but no less interesting is the experience of involving young and middle-aged people in non-formal education on the Internet, as an option to make the most effective use of quarantined time for self-improvement.

1.2 Analysis of the Main Research of This Problem

Transforming processes in our state have been taking place continuously starting from its formation and it doesn't seem possible to define their end, because they have stable character. The society used to numerous reforms, bypassing from one to another like noise of the sea, adjusted to them and tries to live without taking notice of them. Thus, the factor of variability has become a habit, routine, stimulus for adjustment – combination of incompatible by implication. To describe such state there is a term «oxymoron» in the Greek mythology that is translated like «silly things are ingenious».

Awareness of changes in the spatio-temporal characteristics of social reality in sociology was associated with the concept of forming a new state of world civilization, which W. Beck described as a state of «globality». It is based on the individual's understanding of the impossibility of closing the social space through the commonality of any social relations, which can no longer be processes of national-state scale, because, due to the rapid dissemination of information, inevitably become the property of the world community. Among the most obvious consequences of this condition, W. Beck considers the rapid spread of scientific and educational information [5].

The theory of crises and social change was developed in the first decade of XX century by famous American Sociologist K. Thomas. Crisis, in his opinion, is a phenomenon that «disrupts the normal course of events» and requires a person to form a «new model of behavior» to adapt their actions to the new conditions of changed social reality. The researcher emphasized three main factors, the presence of which contributes to the successful overcoming of the crisis by a social group – the presence of leaders; sufficient level of development of the group, including the availability of the necessary technologies, and most importantly – the desire for progress and flexibility of perception of objective reality [6].

Thus, the study of crises and the study of models of adaptation to them for sociology is not new, but today's realities have necessitated a revision of some aspects, due to the situation in which one crisis is superimposed on several existing ones, complicating the process of individual behavior and strategies of social action, because previously developed methods of adaptation do not work, and social reality requires the creation of new models of adaptation.

Therefore, the mainstream of research is the analysis of the pandemic paradox in education, that lies in, on one side that pedagogues and psychologists recognize the significance and importance of Internet education development in the social space the idea of its inherent worth is emphasized, but, on the other side, sociologists say that educational process is a preparation period for «normal» adult life – socialization. In their turn, the pedagogues, who rank pupils and students by level of knowledge, abilities, and perspectives for future life in general, made no less contribution to the development of the scientific knowledge.

In this context, it is necessary to pay attention to the fact, that crisis, are studied by scientists who are experiencing it themselves, each of them has his own experience and, in the measure of his existence, could be considered to be an expert. At the same time, scientists consider this pandemic from the perspective of present, changing social background. On one side, pedagogues, psychologists and sociologists take in the significance and importance of digital development, emphasize the idea of his inherent worth, but on the other side, they say that study is just a period of socialization, preparation for an adult life.

In this context, referring to the traditions of classical sociology, research prospects could be divided into those, which focus on questions connected with the development of individual education, and those, which consider social – «collective», structural education, namely, constructs and practices typical for many pupils and students. In terms of

P. Berger and T. Luckmann, they went through the process of «habitualization» [7] and became at the end constructs of education space. Such point of view resonates with the scientific theory by S. Bernfeld, who turned his attention to the problem of inconformity of idea about the category of education with realties of child's life. Children, in the scientist's opinion, are recognized by the adult world as a means of pursuing economical, social, political and other goals, and so can't be rightful part of the society [8].

Therefore, the traditional analysis within the theory of functionalism doesn't recognize educational practices as a part of the society and culture adult life, which doesn't allow singling an individual's independent choice of the method of obtaining knowledge.

Interdisciplinary disputes within the mentioned problem also come down to contradiction, concerning the questions of impact of social changes and crisis on children's life and prospects of the society development. Things which are taken by children from their childhood become basis for the development of subsequent generations. Therefore, entering by an individual the definite age stage can't be considered only within structural processes of socialization, as, for example, in the theory by N. Luhmann [9]. Even if we recognize that digital education fits into the context of classical social relations, the need to determine its content and principles that will be applicable to educational practices in a pandemic becomes obvious.

For a long time, education has been studied in the context of a social order, which in the opinion of E. Durkheim, is possible on condition that stable value orientations and patterns of requirements will be impressed to each individual [10]. Moreover, the fact of socialization as adaptation of individual interests and requirements to the social order is a basis for understanding of contrast between an individual and the society in the classical sociology. Thus, educational practices studying in the context of the theory of socialization assumes the analysis of two categories – competent, organized participants of life in the society (adults), and also imperfect participants, because of their incompetence and disorganization (children). Therefore, pupils and students is considered only in the context of knowledge transfer from one generation to another.

However, in the concept by T. Parsons, one of the brightest representatives of the classical sociology, exceptional attention is paid to the studying of peculiarities of adjustment of developing individual requirements to social frames that totally changes the concept of person's adaptation to the social order during the life [11]. This theory became a basis for creating, in the context of evolutionism, the concept of social identity by G. Mead. He reasonably noted that socialization only teaches a person the skill to look at himself from the outside, in terms of social environment, cultural norms and traditions [12]. Therefore, targeted development of the ability to accept roles by an individual forms the basis for the mechanism of cooperation of social groups.

Taking into consideration the above mentioned, one could summarize that studying educational practices in crisis situations in the context of socialization leads to the narrowing of the scientific discourse till the limits of research of little groups (family, collective body) and personal features (adaptation, deviation). E. Klymenko spoke about this in her article [13].

This theoretical base led to the working out of the concepts of social and cultural capital, where a child acts as a subject of social process. So the Nobel prizeman for economy Gary S. Becker stated: «people can provide for old age indirectly, investing money in education of children...» [14]. In other words, the theory of one generation investing into another, which appeared in the end of the twentieth century, became another stimulus for studying the role of education in society. Such approach reduces scientific knowledge to the analysis of the level of relations between parents and children, as a criterion for defining of stability of the social institutes in society.

The specificity of social space is closely related to the specificity of social time, which is the internal dimension of social life, which is forced to rely on the physical rhythms of natural processes. Social time is a quantitative assessment of the path taken by society. It is a measure of the variability of social processes and changes that have historically occurred in people's lives.

2 Transfer of Social Practices from Real Life to the Internet Environment

2.1 Methodological Base of Research

The crisis facing Ukrainian and Kazakh societies has changed the understanding of social space and time, limiting them to the immediate living conditions of specific individuals and shifting communication with society from the real three-dimensional space to the Internet environment. The most significant changes in this context were education, which was forced to radically change the methods of providing educational services, completely transforming the process of formation of knowledge, skills, and abilities, leaving unchanged only the necessary quality of learning outcomes. With the latter, serious problems arose in education before the quarantine. Thus, crisis trends have affected existing social problems.

In our study, quantitative and qualitative systematization of data presented by the media, the results of the monitoring of regulatory documents of legislative bodies and ministries of the executive branch are used, a secondary analysis of sociological studies conducted by other research organizations which in freely available in the free Internet. The methodological basis of the article includes the results of sociological research conducted by the authors.

The empirical basis of the article was the data of the author's sociological studies conducted from April 27 to May 11, 2020 in each country in Ukraine and Kazakhstan. Namely:

1. The results of an online survey of the population of Kyiv and Kyiv region, a total of n = 800 respondents (with the general population 2 million 967 thousand), aged 17 to 50 were interviewed, 46% of men and 54% of women.
2. The results of an online survey of the population of Nur-Sultan and region. A total of n = 400 respondents (with the general population 1 million 107 thousand), aged 17 to 50 were interviewed, 48% of men and 52% of women.

The sample represents the adult population of Ukraine and Kazakhstan according to the main socio-demographic characteristics (gender, age, type of settlement, level of education). The sample is multi-stage, random, at the last stage - quota. The method of collecting information is to fill in a google form upon visiting the Internet platform.

The statistical error does not exceed 2.2%. The coefficient of normal deviation directly depends on the probability with which we extrapolated the results of the sample to the general population in our study, it is equal to 95.45%. Therefore, the sample was defined as n = 800 to n = 400, because in comparison, the general population of Kyiv is twice as large as the general population of Nur-Sultan.

To compare the main indicators, the respondents were conditionally divided into two age groups 17–25 years – young people who are forced to study remotely, and 26–50 years old, those who independently acquire knowledge to improve skills and expand personal worldview.

In Kyiv and Kyiv region respondents have different levels of education: complete general secondary education – 42.9%, higher education – 29.9%, incomplete and basic higher education – 16.1%, vocational education – 5.0%, basic higher, basic (incomplete) – 3.4%, degree – 1.1% (see Fig. 1).

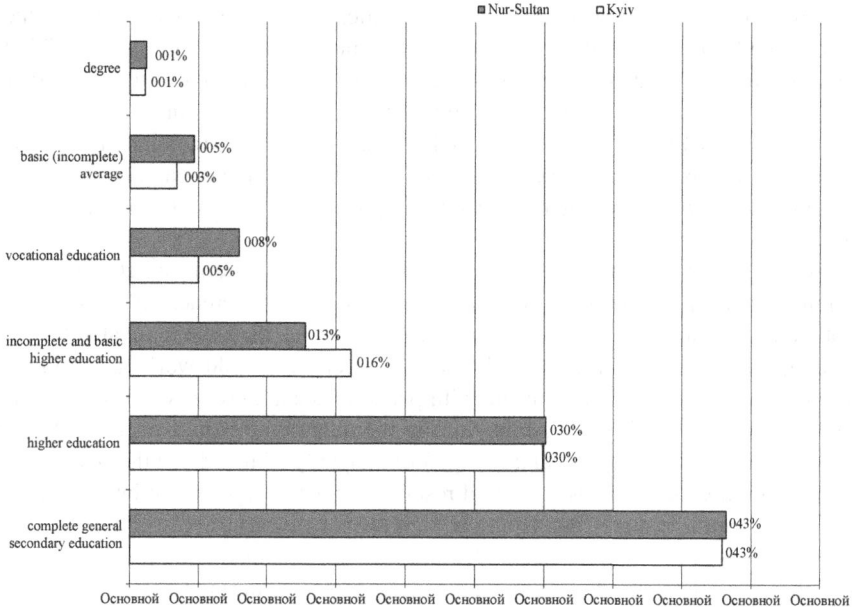

Fig. 1. Respondents have different levels of education in Kyiv and Nur-Sultan.

In Nur-Sultan and region respondents have different levels of education: complete general secondary education – 43.2%, higher education – 30.1%, incomplete and basic higher education – 12.8%, vocational education – 8.0%, basic (incomplete) average – 4.7%, degree – 1.2% (see Fig. 1).

Thus, the quota nature of the sample made it possible to achieve balance and equivalence for the indicated general populations of Kiev and Nur-Sultan (in %).

2.2 Personal Adaptation to Crisis Situations in Conditions of Limited Space

In living conditions, when neither the environment nor the space in which social inter-action takes place over time, a person begins to get used to the surrounding paradoxes of life.

The phenomenon of addiction is one of the passive mechanisms of adaptation of the individual to new conditions and way of life. However, its content can be both positive and negative. Depending on the path of the individual, they will choose forms of social interaction from traditional ways of organizing everyday life (sofa and watching TV, computer, and social networks) to radical changes in life (cosmetic repairs, new hobbies). The results of a sociological survey show that the majority of Ukrainians choose more active forms of adaptation.

To the question «What kind of activity during quarantine/self-isolation did you pre-fer?» (it was possible to choose no more than three options) the answers of the respon-dents were distributed as follows: 49.1% stated that during the quarantine they studied and/or improved their skills; 48.4% paid more attention to sports and improved their physical condition; 46.4% devoted free time to housework (cleaning, repair); 34.2% compensated for forced isolation and social distance by communicating with others, using digital resources (social networks, interest groups on various Internet platforms). Almost every third (30.2%) of the respondents said that during the quarantine period they preferred cooking (cooking), 29.3% had fun (joined computer games, team virtual competitions, browsing the Internet), 18.9% improved their cultural level, and 11.0% engaged in homesteading (see Fig. 2).

The analysis of respondents' responses to the identified patterns of behavior (activ-ities) in the period of new social realities shows some age differences in the choice of adaptation strategies. The age group of 17 to 25 years was the most motivated to study online, apparently to avoid homework and additional household workload, using the time to study in order to communicate with peers on social networks. Thus, according to the survey, 63.0% of young people during the quarantine period were involved in training, 54.2% preferred physical training, 39.3% openly admitted that they established communication with others using digital resources. Such a high rate of involvement in education is primarily due to the requirement of formal educational institutions (schools and universities) to continue distance learning during the quarantine period, and sec-ondly, this socio-demographic group was the least protected during transformations and crises. Therefore, for young people, education has acted as a mechanism for adapting to new social realities.

Respondents of the older socio-demographic age group from 26 to 50 years old noted that during the quarantine period the most popular activities for them were: 59.0% home-work, 37.0% physical training 35.0% cooking (cooking), obviously these two indicators are interrelated and presuppose each other. However, it was interesting that 30.6% of the representatives of this age group actively used their free time to study and improve their skills.

Perhaps such a high level of torture in educational practices is due to high professional demand. According to a sociological study, the most in-demand professionally during the quarantine period were respondents of the oldest socio-demographic group aged 26 to 50 years. Thus, 71.5% of respondents said that they continued their professional

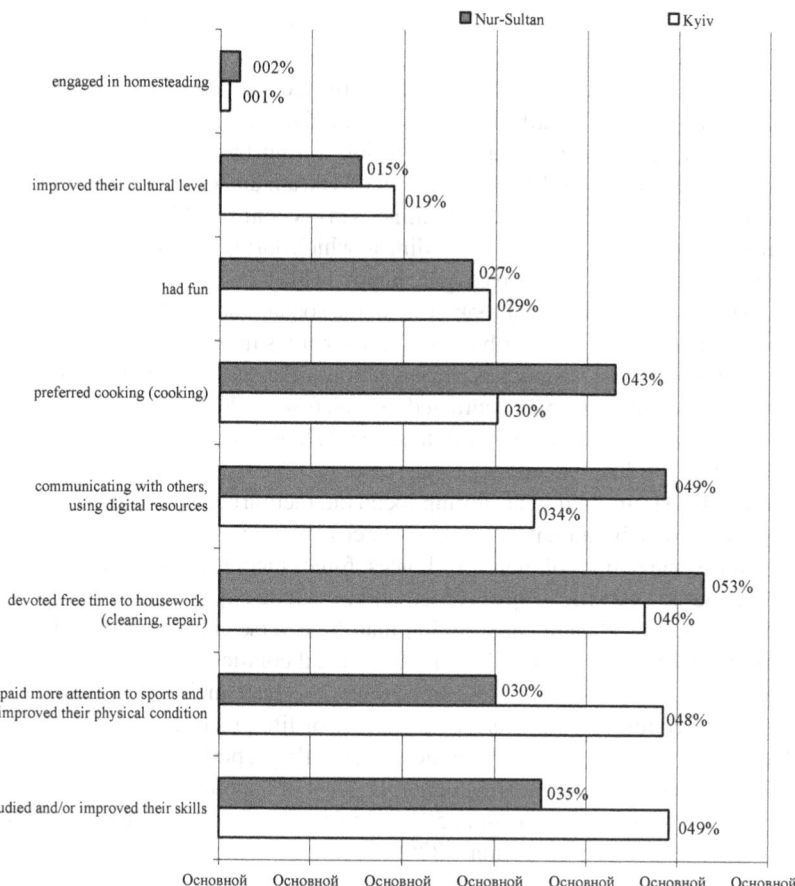

Fig. 2. Comparative analysis of the answers of respondents in Ukraine and Kazakhstan to the question: «What kind of activity during quarantine/self-isolation did you prefer?» (in %).

activities offline or remotely. In the age group of 17 to 25, only 55.8% of young people were able to stay in their jobs, of which 23.3% said they had gone online.

Thus, for most young people, the phenomenon of crisis accustoming, as an organization of individual life in conditions of extremely limited resources, both economic and social, became a trigger for radical changes in forms of activity, which led to a change in perception of social space. Because in the conditions of deficit of practices of social interaction and absence of the developed adaptation strategies the youth directs the basic attention to the preservation of already available potential, and therefore reduces social actions to banal satisfaction of basic needs. All life is centered around recreating yourself physically and having a good time. Therefore, almost 30% of respondents completely ignored the need to participate in distance learning. To the question «Did you find an interesting activity during the quarantine?» 22.7% found it difficult to answer, 36.7% did not find it, but 40.6% were able to take an interest in something (most of them discovered of physical culture).

It is worth noting that there were no significant differences in the answers of age groups to the questions regarding the attitude to quarantine. To the question «Do you like the lifestyle that you formed during quarantine/self-isolation?» only 20.4% of all respondents answered in the affirmative. Interestingly, almost every third part (34.0%) of those respondents who were at a remote job during the quarantine period, noted that they liked this lifestyle. But the majority of Ukrainians (75.3%) said they did not like the lifestyle formed during the quarantine. This is confirmed by studies of the Kyiv International Institute of Sociology, according to which during the quarantine only 45.0% of Ukrainians considered themselves quite happy [15].

Interestingly, the changes that took place in the social practices of Ukrainians during quarantine did not affect the distribution of gender roles in household management. In this area of domestic life, society has chosen actions of an exclusively traditional nature. Thus, 80.0% of surveyed women continued to cook (cook), 70.5% did housework, while 50.0% of men started working on homesteads, and responsibilities were divided in half. Perhaps the latter is due to the desire to increase the time spent in the air.

In times of crisis, the individual during social interaction can not build their life strategies in the usual way, because it lacks the resources: material, social, psychological. This sharply reduces the options of their life choices, forces them to adjust their social behavior, limiting her to short life projects based on immediate opportunities. V. Kryvosheev drew attention to this phenomenon, noting that the reverse side of adaptation during a crisis may be derivation, which will extinguish social conflicts [16].

In this sense, education is not just an effective mechanism for adapting to radically changed living conditions but acts as a stabilizer of life practices. According to KIIS sociological research, the level of happiness is directly proportional to the level of education - the higher the education, the higher the level of happiness. Thus, among people with incomplete secondary education, 50% of respondents consider themselves happy, among people with higher education – 77%.

2.3 Adaptive Function of Digital Education in the Period of Pandemic

As noted above, studying during the quarantine period became the most popular activity for which Ukrainians spent their free time. In general, almost every second – 49.1% of respondents studied during forced self-isolation. One reason for this popularity was the requirement for formal education to continue distance learning. However, 32.9% stated that they joined studying because of being motivated by their desire for personal growth, and 29.5% of respondents joined the studying process for professional growth. Also, it was found that 17.6% of respondents studied because of free time, 12.6% were motivated by the opportunity to join free courses of reputable universities and platforms, which were previously paid and 10.6% admitted that they joined the study out of boredom.

The analysis of motivation for educational practices of different age groups demonstrates the following priorities. 34.4% of young people aged 17 to 25 said that the main motivation for learning was the desire for personal growth, and 29.4% - professional growth. This socio-demographic group has the highest rate of respondents (13.7%) who joined the study out of boredom.

At the same time, almost every third 30.0% of respondents from the senior socio-demographic group joined studying during the quarantine, the same number of those who studied, noted that the main motive was the desire for personal growth; and 28.4% sought professional development, 20% filled their free time in that way.

Thus, the results of the sociological study demonstrate the high potential of education as a mechanism of adaptation to social change in crises, in terms of space-time disorientation and abandonment of traditional forms of social interaction, which is relevant for all age groups of modern Ukrainian society.

In Kazakhstan, the overwhelming majority of respondents rate the effectiveness of online learning below average (42%) and average (41%). Only 12% of the respondents gave a high rating to the new form of education. The overwhelming majority of respondents are convinced that distance learning has an extremely negative impact on the country's education system. The main disadvantage of distance learning is stress for children, teachers, parents. In addition, learning is mainly hampered by technical difficulties and slow Internet connections, as noted by 44% and 40%, respectively. According to 19% of respondents, the main disadvantage of online learning is the low quality of education. 9% of respondents consider the lack of a computer to be the main difficulty. The quality of the Internet connection is one of the main factors in online learning [17].

It is interesting that on the question of what form of education citizens consider the most optimal in the context of the global coronavirus pandemic, the opinions of the respondents were approximately equally divided. 35% believe that people should be given the right to choose a convenient form of education; 33% of Kazakhstanis are for online education, as this is a forced measure; 29% expressed skepticism about the online format. In their opinion, it is better to go to school and university, as before – «illiteracy is worse than the COVID-19».

3 Conclusions

Thus, the results of a sociological study fully confirmed our hypothesis about the significant role of educational practices in the society which for individuals have become a mechanism for adaptation to the crisis in the country and social transformations, which change the spatial and temporal perception of social reality. At the same time, there was a tendency to increase the interest of the older age group in non-formal education and reduce the interest of young people in formal education, which was not ready for such challenges as the transition to the online format. The crisis did not affect most of the everyday life practices of all ages, they remained unchanged, which contributed to the choice of traditional forms of social action during domestic interaction. Women played a leading role in housework and cooking.

The complex of crises that post-Soviet societies are experiencing have become a kind of the impetus for ordinary citizens to improve the mechanisms of social adaptation, one of which under the fundamentally new living conditions was non-formal education. However, the outlined problem field requires further scientific research and more detailed sociological research.

The results of the sociological study demonstrate both in Ukrainian and Kazakh societies the high potential of education as a mechanism of adaptation to social change in crises. In terms of space-time disorientation and abandonment of traditional forms of social interaction, this is relevant for all age groups of modern Ukrainian society and in modern Kazakh society, similar trends are characteristic of young people and middle-aged people. For elderly people, passive forms of adaptation and the use of only entertainment options of the Internet information network are more characteristic.

Potentially relevant for further research, both in Ukraine and in Kazakhstan, will be questions of the peculiarities of social interaction and communication in a non-contact society, a radical change in consumer and leisure practices. The transition of business to online mode will contribute to the already voluntary distancing of workers, for most of whom their own home will become an office, which will contribute to the emergence of many social problems that will require serious scientific understanding in both Europe and Asia.

References

1. Gryzunova, E.A.: Concepts of crisis of social system: the comparative analysis. https://www.academia.edu/3461425/Концепции_кризиса_социальной_системы_сравнительный_анализ. Accessed 16 Mar 2020
2. More than 6 million Ukrainians may find themselves below the poverty line. Homepage. https://business.ua/news/10146-bilshe-6-mln-ukrajintsiv-mozhut-opinitisya-za-mezheyu-bidnosti. Accessed 14 May 2020
3. Due to pandemic, the share of the poor in Kazakhstan may reach 12.7 percent on July. https://news.un.org/ru/story/2020/07/1382661. Accessed 24 Mar 2020
4. Training continues: UNESCO brings together international organizations, civil society and private sector partners into a broad Coalition for Education. https://ru.unesco.org/news/obuchenieprodolzhaetsya-yunesko-obedinyaet-mezhdunarodnye-organizacii-grazhdanskoe-obshchestvo. Accessed 26 Mar 2020
5. Beck, W.: Risk Society. On the way to another Art Nouveau. [Transl. by B. Sedelnik, N. Fedorova], p. 36 (2000)
6. Thomas, K.: The crisis and emergence of scientific theories. Researcher. 1-2, 20–39. https://cyberleninka.ru/article/n/krizis-i-vozniknovenie-nauchnyh-teoriy. Accessed 26 Mar 2020
7. Berger, P.L., Luckmann, T.: The Social Construction of Reality. Society as Objective Reality, p. 89 (1966). http://perflensburg.se/Berger%20social-construction-of-reality.pdf
8. Bernfeld, S.: Sisyphos oder die Grenzen der Erziehung. Frankfurt, p. 36 (1967).
9. Luhmann, N.: Social System. Grundriss einer allg Emeinen theorie Suhrkamp. Das Forderprogramm des Goethe-Institutm Ubersetrungen deutsche Bucher Homepage. https://philosophy.ru/library/luman/luman_soc_sistem.pdf. Accessed 14 Nov 2008
10. Durkheim, E.: Essays on Morals and Education, p. 74 (2005)
11. Parsons, T.: Socialization and interaction process. Sociol. Soc. Philos. 31, 256–266 (1979)
12. Mead, G.H.: Mind, Self and Society: From the Standpoint of a Social Behaviorist. Chicago (1934)
13. Klymenko, E.: Die Schwächsten der Schwachen Ukraine: Kinder im Krieg. Osteuropa. 3–4, 185–191. https://www.zeitschrift-osteuropa.de/hefte/2019/3-4/die-schwaechsten-der-schwachen/. Accessed 01 Apr 2019
14. Becker, G.: Economic Outlook on life: the Nobel lecture Genre Becker December 9, 1992. World economic thought through the ages: 5 t. Worldwide recognition. Lectures of Nobel laureates. Moscow, pp. 688–706 (2004)
15. Panioto, V.: Self-assessment of happiness by the population of Ukraine. https://www.kiis.com.ua/?lang=ukr&cat=reports&id=944&page=1&t=7. Accessed 18 May 2020
16. Krivosheev, V.V.: Short life projects: manifestation of anomie in modern society. Sociol. Res. 3, 58–69 (2009)
17. Quarantine and distance learning exposed the disadvantages of national IT management. https://365info.kz/2020/04/karantin-i-distantsionnoe-obuchenie-obnazhili-minusy-natsionalnogo-it-menedzhmenta. Accessed 18 Apr 2020

The Impact of the Online Project X-Culture on the Development of Students' Emotional Intelligence

Tatiana Baranova[ID], Aleksandra Kobicheva[(⊠)] [ID], and Elena Tokareva[ID]

Peter the Great Saint-Petersburg Polytechnic University, St. Petersburg 195251, Russia
kobicheva92@gmail.com

Abstract. Modern employers choose specialists with a large number of developed skills to perform multifaceted tasks. One of the most in-demand skills right now is emotional intelligence. Currently, universities are striving to solve this problem and develop students' emotional intelligence to improve the qualification of future graduates. In our research, we study the effectiveness of the international online project X-culture and how it influences St. Petersburg Polytechnic University 3rd year students' emotional intelligence development. To obtain the data we used both quantitative and qualitative data. To measure the emotional intelligence (EQ) we decided to rely on the Emotional Competency Inventory (ECI) model, slightly adapted to our experiment. So, we evaluated four clusters - Self-Awareness, Self-Management, Social Awareness, and Relationship Management. The analysis showed that all indicators were improved after the participation in the X-Culture project. The most significant difference was in the results of the following indicators – self-confidence, adaptability, conflict management, inspirational leadership, and collaboration. According to the correlation test, final report results are substantially influenced by all clusters of Emotional Intelligence that confirm the crucial role of indicator and importance of such indicator development.

Keywords: Emotional intelligence · Online project · Virtual team · Digital education · Communication skills

1 Introduction

Higher education fulfills an important social function. At the present stage of society's development, the requirements for the quality of specialists' training and the level of their professionally significant qualities formation are increasing, which are largely laid down at the university stage of vocational training [1]. Emotional orientation and emotional intelligence, as relatively stable personal characteristics of the emotional sphere, have a mediating effect on the success of the professional activity. The importance of the problem of intelligence and non-cognitive attributes of mental development is primarily determined by the role they play in solving a complex of social and individual psychological problems of a person.

© The Author(s), under exclusive license to Springer Nature Switzerland AG 2021
T. Antipova (Ed.): ICADS 2021, AISC 1352, pp. 173–187, 2021.
https://doi.org/10.1007/978-3-030-71782-7_16

Emotional intelligence is a prerequisite for pro-social and other positive behavior, and its development optimizes interpersonal interactions, as evidenced by numerous studies. Emotional intelligence in the broadest sense combines the ability of a person to communicate effectively through understanding the emotions of others and the ability to adapt to their emotional state. Such an ability to control oneself and competently organize interaction turns out to be indispensable for the majority of highly qualified specialists in our time. The growing research interest in issues related to emotional, social competence, and professional self-awareness of future specialists in connection with the success of their activities is due to social trends taking place in society. In this regard, the problem of studying the emotional sphere, communicative, and regulatory features of a future specialist in the context of professionally significant personality traits is one of the central ones.

In this regard, the requirements for the higher education system and the quality of professional training are growing. SPBPU (Peter the Great St.-Petersburg Polytechnic University) conducts different professional trainings for students of various fields of study, implements project activities in the educational process, and organizes various international programs that contribute to the development of communication and social skills that determine emotional intelligence [2–4]. For example, over the past two years, students have been participating in the international online project X-culture, where they have been working in virtual teams for 8 weeks with students from other universities from more than 45 countries. The goal of the project is to solve the case of an international company and write a final report on the chosen case, which is evaluated by the project experts, as well as by the heads of the companies. Communication within the team takes place via Skype, e-mail, and various messengers.

In this paper, we will consider the impact of the international online project X-culture on the development of emotional intelligence in students of 3 years of study, and also clarify the correlation between the competencies of emotional intelligence and the results of project activities.

1.1 Theoretical Background

Definition of Emotional Intelligence. Emotional experiences are an inseparable part of individuals' daily life. Emotions such as pride, anger, and shame influence experiences. They guide behavior, prioritize goals, and they communicate our mood states to others [5]. It is therefore not surprising that emotions influence both the self and the other when individuals are interacting. That is, emotions can be expressed towards others, they can elicit emotions in others, or they can be reactions to the emotions of others [6]. During social interactions, individuals thus not only need to appraise and regulate their own emotions; they also need to keep track of the emotions of their interaction partner to facilitate the interaction and achieve what they want. Some individuals are better at this than others, and part of these individual differences is reflected in emotional intelligence (EQ) [7].

EQ can generally be described as the ability or knowledge to perceive, understand, and manage emotions [8–10]. High-EQ individuals tend to deal with emotions in such a way that their reactions are socially effective which may help them to reach their goals in various life domains. For example, the emotions that a sports coach expresses during a race may motivate and enable an athlete to reach new performance levels. The manifestation (i.e., enactment) of EQ may even influence important life outcomes. To illustrate, EQ is positively associated with a satisfying social life [11, 12], health [13], and job performance [14, 15].

Methods for Measuring Emotional Intelligence. The literature on EQ can roughly be divided in two main approaches that differ in their conceptualization and measurement of EQ [16]. The ability-approach is largely based on the Four-Branch Model of EQ [8, 17]. In this model, EQ is conceptualized as a set of interrelated emotional abilities organized in four branches. The branches consist of (1) the ability to perceive emotions, (2) the ability to use emotions to facilitate thinking, (3) the ability to understand emotions, and (4) the ability to manage emotions to reach (interpersonal) goals. Characteristic of this approach is its measurement with performance-based EQ tests, which is comparable to the way cognitive abilities are measured. In contrast, the trait-approach conceptualizes EQ as a set of emotion-related traits and uses self-report instruments to measure EQ. This approach is more similar to research conducted in the personality field.

There is a relevant limitation in both approaches. Scholars typically use global EQ scores that mask the unique role of self-focused EQ (dealing with emotions of the self) versus other-focused EQ (dealing with the emotions of others). However, recent research suggests that self-focused EQ is particularly relevant to remain healthy, whereas other-focused EQ contributes particularly to social and performance outcomes [18–20]. A further limitation of using global EQ scores is that they do not clearly reveal the different steps involved in processing an emotion. Yet, in real life, an emotion needs to be appraised first before it can be regulated [14].

Emotional Intelligence of Students. We pay special attention to the development of emotional intelligence in students, since adolescents face multiple demands and stressors due to the biological and social changes that take place during this developmental period [21], and also violent and traumatic experiences reach their peak in adolescence [22]. Adolescence is also a time of heightened risk for mental health problems, especially mood and anxiety disorders [23, 24]. EQ involves emotional, social and personal skills, competences, and motivations that help successful adaption [25–27]. In adolescence, emotional expression, recognition and regulation become more effective and sophisticated through increasing flexible planning, creative thinking, and motivation to form multiple social relationships [28, 29]. Ample evidence shows that high EQ is associated with a number of favourable indicators of health, mental health, social relationship, and adaptation to stress and hardships. For instance, a Lithuania follow-up study showed that students with a high EQ showed decrease in depression and anxiety symptoms with time [30], and an multinational study confirmed that a high EQ was positively associated with psychological wellbeing, indicated by high self-esteem, life satisfaction, self-efficacy, and self-acceptance [31].

Intervention studies have shown that teaching people social and emotional intelligence can help increase emotional intelligence [32, 33]. These results also indicate the inclusion of this competence in the curriculum. What's more, several studies show the benefits of emotional intelligence training. For example, young people with high emotional intelligence are more prepared to manage emotions in decision-making related to career choices, demonstrate higher self-confidence and professional commitment [34], and are less likely to drop out of school [35]. In addition, learning enables young people to develop relevant skills and competencies that can enhance their employability, digital adaptation and earning capacity [36, 37].

Online Project X-Culture. Working as a team on one project requires many skills. In our research, we consider an international online project for students. But first of all, it is worth clarifying the concept of project management. According to Knutzon and Bitz, project management is "a set of principles, methods, tools, and techniques for the effective management of objective- oriented work in the context of a specific and unique organizational environment" [38, p. 2]. Kerzner emphasized project management as "the art of creating the illusion that any outcome is the result of a series of predetermined, deliberate acts when in fact it was dumb luck" [39, p. 192]. Havranek expanded on Kerzner's definition. He defined project management as "project management is the art and science of planning, organizing, integrating, directing, and controlling all committed resources – throughout the life of a project – to achieve the predetermined objectives of scope, quality, time, cost, and customer satisfaction" [40, p. 61].

The article by Taras et al. [41] focuses on the experimental component of international business education. An article written by 11 authors from the USA, Poland, Ecuador, Spain and the United Arab Emirates describes an exciting new collaborative consulting project called X Culture. Using state-of-the-art technology, social media and online collaboration tools, students from different schools and countries are challenged to develop real-world consulting projects. The article discusses many of the positive outcomes of students and presents opportunities for other international business teachers.

Another study related to the X-Culture project [42] explores the problems and experiences of X-Culture through the eyes of professors and students and draws attention to the importance of such projects in international business practice, in addition to examining the key factors influencing interculturalism and ICT technologies. The researchers found that many students increased their chances of finding attractive jobs in the labor market and broadened their social and professional ties by participating in the X-Culture International Student Cooperation Project. The problems were mainly related to differences in time, but cultural differences and language barriers often emerged.

2   Materials and Methods

The study involved 3rd-year undergraduate students from Peter the Great St. Petersburg Polytechnic University, who were enrolled in an integrated learning course in 2018 and 2019 (N = 123). Students are 20 years old. The groups consisted of 77 girls and 46 boys. To obtain the data we used both quantitative and qualitative data (Table 1). To measure emotional intelligence (EQ) we decided to rely on the Emotional Competency Inventory (ECI) model. So, we evaluated four clusters - Self-Awareness, Self-Management, Social Awareness, and Relationship Management. To get results on 18 competencies that contribute 4 clusters we conducted 2 online surveys for two groups of students (before and after the online project) and deep interviews with random students (N = 12). Also, we used records of completed peer-evaluation tests in the online project and the results of students' final reports. All results were examined and compared to reveal the effect of the online global project.

Table 1. Methods of data collection.

Results	Sort of data collection	Type of data
Self-awareness	Questionnaire in Moodle online system	Quantitative
	Interview	Qualitative
Self-management	Questionnaire in Moodle online system	Quantitative
	Interview	Qualitative
Social awareness	Peer-evaluation test in online project	Quantitative
Relationship management	Peer-evaluation test in online project	Quantitative
Final report results	Online Project X-culture data base	Quantitative

This paper is based on the following research questions:

1. Does the online project X-culture play an influential role and contribute to students' higher level of Emotional Intelligence (EQ)?
2. Is there a significant influence of Self-Awareness, Self-Management, Social Awareness, and Relationship Management on student's final report results?

For the analysis descriptive statistics, pair-samples Students' t-test and Pearson correlation test were conducted.

3 Results

3.1 Self-awareness

Questionnaire in Moodle Online System. The Self-Awareness cluster constitutes three competencies - emotional awareness, accurate self-assessment and self-confidence. To determine the indicators of students' competencies development we conducted an online survey that included 12 questions on each competency. The online survey was conducted twice, before and after the X-Culture project. The results measured by 10-point Likert scale are presented in Table 2.

The analysis showed that all indicators were improved after the X-Culture project. The most significant difference was in the results of the following indicator – Self-confidence. The significant difference in indicators means that online international project had a great influence on students. Due to the t-value tests the difference in such indicator, as "Emotional awareness" was not significant.

Table 2. Descriptive statistics (students' Self-awareness indicators).

Items	Survey	Results (average mean)	SD	t-value
1. Emotional awareness	Before the project	6,5	0,71	1,31
	After the project	6,6	0,77	
2. Accurate self-assessment	Before the project	6,3	0,8	3,4 **
	After the project	6,8	0,79	
3. Self-confidence	Before the project	6,9	1,01	6,42 ***
	After the project	7,8	0,95	

Note: * p < 0,05; ** p < 0,01; ***p < 0,001

3.2 Self-management

Questionnaire in Moodle Online System. The Self-Management cluster contains six competencies: emotional self-control, transparency, adaptability, achievement, initiative, and optimism. The survey was conducted twice, before and after the X-Culture project, and included 18 questions that defined the level of each competency development. The results measured by the 10-point Likert scale are presented in Table 3.

Table 3. Descriptive statistics (students' Self-Management indicators).

Items	Survey	Results (average mean)	SD	t-value
1. Emotional self-control	Before the project	6,5	0,71	2,31*
	After the project	6,7	0,77	
2. Transparency	Before the project	6,3	0,8	7,9 ***
	After the project	7,3	0,79	
3. Adaptability	Before the project	6,9	1,01	6,42 ***
	After the project	7,8	0,95	
4. Achievement	Before the project	7,3	0,87	1,89
	After the project	7,4	0,88	
5. Initiative	Before the project	7,8	0,94	1,91
	After the project	7,9	0,9	
6. Optimism	Before the project	6,7	0,89	3,2**
	After the project	7,2	1,07	

Note: * p < 0,05; ** p < 0,01; ***p < 0,001

According to the t-value test the difference between such indicators as "Transparency" and "Adaptability" were significant (the significance level was set to 0,001). These results can be explained by the fact that by the end of the project students feel more confident as they know what to expect and to do.

3.3 Students' Interview

For a deeper study of the development of the Self-Awareness and Self-Management competencies we conducted semi-structured interview with random students (N = 12). Interview consisted of 10 questions and further free discussion. Each interview lasted approximately 40 min. The interview was conducted at the end of the project, but before the final grades were given, in order to exclude the influence of the grade on the students' answers. Students' answers were recorded with their voluntary consent. Respondents were asked to answer 10 questions:

1. Describe your emotional state while participating in the project. How did your emotional state change while working on the project?
2. What emotions have you experienced more often?
3. What were the emotional outbursts associated with it?
4. What helped you restore emotional balance?
5. Have you encountered problems controlling your own emotions? If so, how did you solve these problems?
6. How did your participation in the project affect your self-esteem?
7. Did your emotional state affect the productivity of the project?
8. Have you had any conflicts with other project participants?
9. Did you or others in the project show emotional restraint?
10. Do you think that participation in the project has positively influenced your emotional state?

After analyzing the students' answers about their emotional state during the work on the project, it turned out that the majority (79%) of the respondents experienced similar emotions. The most common emotion in students' responses is anxiety. The students mentioned the discomfort caused by anxiety at different stages of the project. After evaluating the students' responses, we compiled a graph showing the number of students experiencing anxiety in a given period of time (see Fig. 1). Since the students did not assess the level of anxiety in numerical terms, we used the parameters - "high level of anxiety", "average level of anxiety" and "low level of anxiety".

The graph shows that all students experienced anxiety. During the interview, such an emotional response was explained - the students had never previously participated in international projects. Consequently, the novelty of the teaching format, the need for active participation, independent decision-making, and communication in a foreign language became the cause of students' anxiety. However, students differ in their resistance to stress, which is reflected in different levels of anxiety among the group. 75% of the respondents felt anxious at the beginning of the project due to the novelty of the training format. In the middle of the project, the values of anxiety depended on the success of work in the project, namely: well-established communication with foreign participants, a developed work plan, active participation in the project. Consequently, those who were actively working in the middle of the project, and mutual understanding between the participants was achieved, experienced minimal anxiety. In turn, those students who were not able to organize their work by the middle of the project were especially worried. Almost all students by the end of the project experienced a moderate level of anxiety

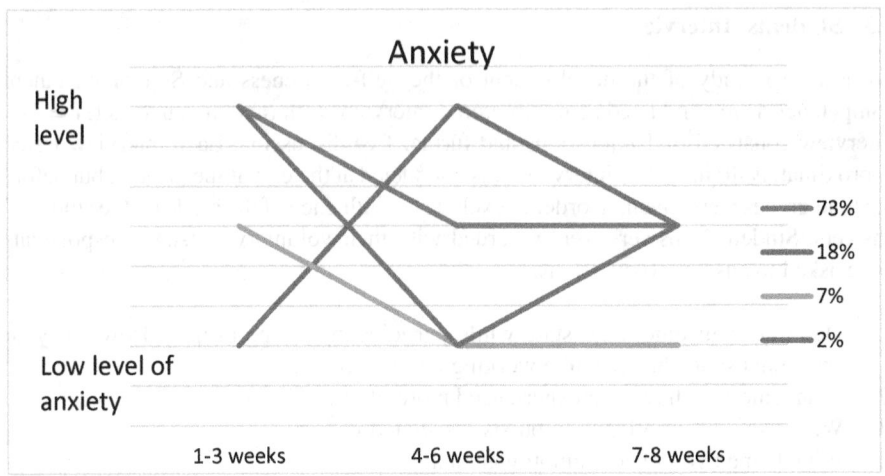

Fig. 1. Anxiety level.

since the main activity of the group was completed, the participants analyzed the findings and summed up the results.

Also, very often the respondents noted happiness. The main reason for the joy was satisfaction from the work done and pleasure from communicating with nice people. In contrast to the anxiety that survey participants experienced at different stages of the project, the majority (66%) noted happiness at the end of the project, which is explained by the completion of the work and the achievement of the result. 24% of students noted the middle of the project as a happy period, as they enjoyed working in a friendly team. 10% of the interviewees noted that they could not appreciate participating in the project and had a deficit of positive emotions.

The respondents also noted such emotions as anger, fear, delight, interest, despondency, irritation, and glee. Most often, negative emotions were associated with conflicts in teams and took place in the middle of the project, during the most active work. One of the respondents noted: "Some of the team members annoyed me because they only criticized the work done". Another student also expressed a similar position: "I was angry when only 1–2 participants from the whole team took the initiative, and the rest did not want to do anything".

Unlike negative ones, students experienced positive emotions more often. At the beginning of the project, many students (56%) were enthusiastic and interested in the proposed project. One student commented: "I liked the idea of participating in an international project so much that I was looking forward to meeting new participants. Although my joy was accompanied by anxiety about the novelty of the training format. But I was happy when the project started, I got to know the team and the fear is gone". 63% of the respondents noted that they experienced pleasure from the work process: "we performed tasks that seemed too difficult for us". At the end of the project, 49% of students noted that they felt mixed feelings: pleasure from the result obtained, but also anxiety due to the assessment of their work by the teacher. 23% of respondents said that the last 2 weeks of the project were the happiest: "we looked at the results of our work, saw what we had achieved, and felt absolutely happy".

The students noted that emotionally, it was the hardest to control their emotions. One student said: "During the project, I learned to restrain my emotions when necessary. I realized that anger and irritation can be overcome in one way – to work. The result is a smoothing out possible dissatisfaction". Many students emphasized the importance of controlling emotions, but they also reported that the ability to understand their own and others' emotions is equally important. "Sometimes I realized only after a few hours that I was angry or rude. At the moment of aggression, I did not fully realize how it looks from the outside, and what the consequences could be. It was a good lesson for me. It seems to me that I have matured." – said, one participant.

All students noted the importance of the emotional background on the productivity of the team as a whole. Show of initiative, optimism, and friendliness promoted active work, effective communication, and focus on results. Some students noted that they did not immediately realize this relationship, which is why they had conflicts and a lack of motivation in the team. However, having improved the atmosphere, they immediately noticed a change for the better and were able to achieve success.

As a result of the interviews, it was revealed that students experienced both negative and positive emotions, had difficulties in recognizing emotions and their control. But by the end of their participation in the project, they began to better control their emotions and determine the influence of the emotional background on the productivity of the team. Also, the work done had a positive effect on the self-confidence of students and increased their self-esteem.

3.4 Social Awareness

Team members evaluated each other every week during participation in the X-Culture project. The idea of these evaluations is to collect valuable data that can be used in various investigations. Also, these evaluations help to detect those students who do not work and contribute to the project and those who should be excluded from the team.

For our research, we use the evaluations of the first and last weeks of the project that determine the following competencies: empathy, organizational awareness and service, and orientation relationship management. The results measured by a 5-point Likert scale are presented in Fig. 2.

The results show the positive trend in developing Social Awareness indicators. According to peer evaluations, y students mostly developed such indicator as service and orientation relationship management. It means that students learned during the project how to behave and interact with their team members successfully.

3.5 Relationship Management

To determine the level of students' Relationship Management development we also use the evaluations of the first and last weeks of the project that determine the following competencies: developing others, inspirational leadership, change catalysts, influence, conflict management, teamwork, and collaboration. The results measured by a 5-point Likert scale are presented in Fig. 3.

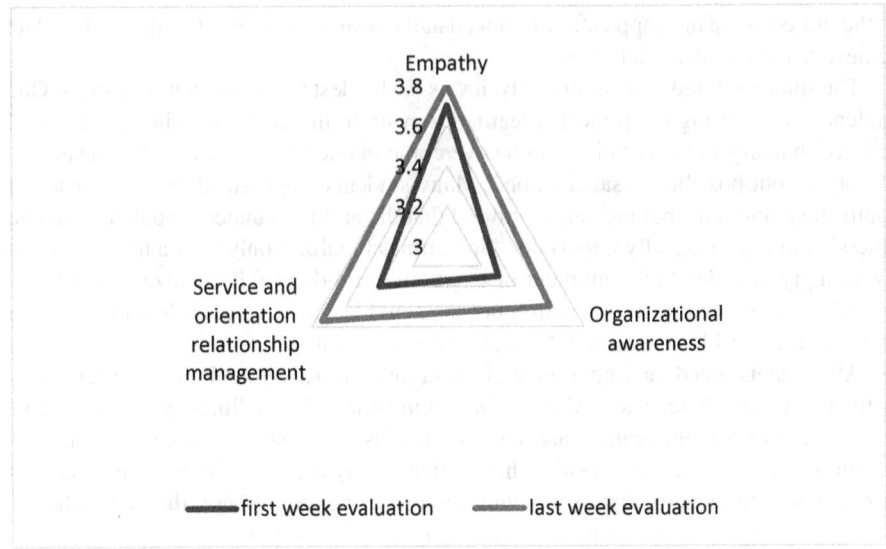

Fig. 2. Social awareness indicators.

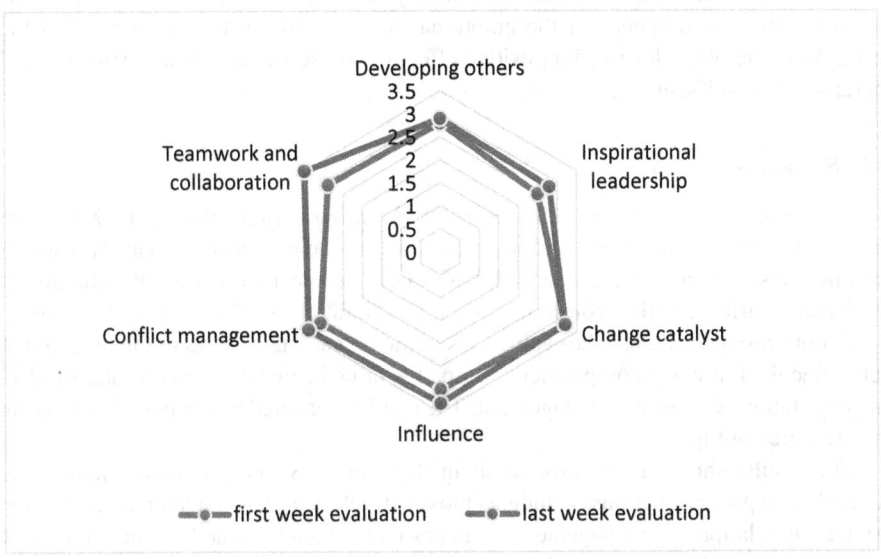

Fig. 3. Relationship management indicators.

According to peer evaluations, students developed significantly the following competencies – conflict management, inspirational leadership, influence and teamwork, and collaboration. The biggest difference was in teamwork and collaboration indicators, that confirms the influence of online project X-Culture especially on this competency.

3.6 Correlation Analysis

We implemented the correlation analysis using the quantitative results of four clusters - Self-Awareness, Self-Management, Social Awareness, and Relationship Management as well as the final report results of all students taken part in the research. The correlation analysis shows the interdependence between four crucial indicators that reflect the level of the students' emotional intelligence development. Our main goal was to confirm the influence of students' EQ level on their final results in international project activity. The indicators we gained are presented in Table 4.

Table 4. The correlation analysis of EQ indicators and students' final report results.

	Self-awareness	Self-management	Social awareness	Relationship management	Final report results
Self-awareness	1				
Self-management	0,24*	1			
Social awareness	0,22*	0,15	1		
Relationship management	0,13	0,18	0,53***	1	
Final report results	0,21*	0,21*	0,37**	0,42***	1

Note: * $p < 0,05$; ** $p < 0,01$; ***$p < 0,001$

According to the results gained the indicators we analyzed have a positive correlation. The strongest relationship is between such indicators as "Relationship Management", "Final report results" and "Social Awareness". The relationship between "Relationship Management" and "Self-awareness" as well as between "Social Awareness" and "Self-Management" is quite weak. We can also note that "Final report results" are substantially influenced by all clusters of Emotional Intelligence that confirm the crucial role of indicator and importance of such indicator development.

4 Discussion

Educational institutions are responding to the modern challenges of the labor market. Since employers are currently looking for sausages with a wide range of developed competencies, emotional intelligence must be developed during university studies. After analyzing scientific works, it was revealed that the development of emotional intelligence is necessary for modern society in connection with changes in communications, methods of group work and methods of solving assigned tasks [8–10]. The relevance of the development of emotional intelligence at an early age, especially in adolescence, is also recognized [23, 24]. The studied scientific works showed us the need to search for the most effective ways to develop emotional intelligence in students, so that after completing their studies, graduates can find decent work, and employers find capable workers. Having considered various types of educational technologies, we chose project activity, since it develops a large number of skills, forming an interconnection between them and making learning complex [38, 42, 43]. It is also worth noting that emotional intelligence has a particular impact on the performance of project activities. We conducted our research on the basis of an international online project X-culture.

To determine the effectiveness of using this project for the development of emotional intelligence we decided to rely on Emotional Competency Inventory (ECI) model and adapted it to our experiment. So, we evaluated four clusters - Self-Awareness, Self-Management, Social Awareness, and Relationship Management. All results were examined and compared to reveal the effect of online global project.

The research has shown that participation in an international online project X-culture contributed to the development of emotional intelligence. Thus, according to the results of the analysis, it can be concluded that students' self-confidence, self-esteem, adaptability to new conditions and optimism have increased. All this is due to the fact that by the end of the project, students felt much more comfortable in group work and project activities, and also felt more confident, as they gained new experience and learned the principles of work. For a deeper analysis, we used students' peer review. According to the data, the students learned how to behave and communicate in the project. Participation in project work had a particular impact on the formation of leadership qualities, teamwork and resolving conflict situations. Thus, the students improved the skills of effective communication, recognition of their own and others' emotions, as well as control over their emotions.

Based on the results of the correlation analysis, it was revealed that the development of all indicators of emotional intelligence had a significant impact on the final result of project activities. We have come to the conclusion that the development of emotional intelligence is appropriate for improving the skills of teamwork and project activities. Consequently, professionals with these skills are more successful in their careers. Thus, we recommend the use of the international online project X-Culture in educational institutions.

4.1 Limitations of the Present Study and Suggestions for Future Research

It should be noted that our study has limitations. Only students from Russia were analyzed, therefore, these students had a similar mentality, which can differ from representatives of other countries and affect the emotional state and perception of the emotions of others. Also, the study was conducted based on the results of two years of participation in the project, which is a short period. The study did not analyze the relationship between the development of emotional intelligence and the gender of the respondents.

In our further research we are going to analyze the cultural peculiarities of X-culture teams and reveal students from which countries work more efficient together.

References

1. Baranova, T., Kobicheva, A., Olkhovik, N., Tokareva, E.: Analysis of the communication competence dynamics in integrated learning. In: Anikina, Z. (ed.) Integrating Engineering Education and Humanities for Global Intercultural Perspectives. IEEHGIP 2020. Lecture Notes in Networks and Systems, vol. 131. Springer, Cham (2020). https://doi.org/10.1007/978-3-030-47415-7_45
2. Baranova, T.A., Kobicheva, A.M., Tokareva, E.Y.: Effects of an integrated learning approach on students' outcomes in St. Petersburg polytechnic university. In: ACM International Conference Proceeding Series, pp. 77–81 (2019). https://doi.org/10.1145/3369199.3369245
3. Kobicheva, A., Safonova, A.: Specialists' training by integration of the high school and business structures. In: E3S Web of Conferences, Vol. 164, p. 12016 (2020). https://doi.org/10.1051/e3sconf/202016412016
4. Baranova, T.A., Kobicheva, A.M., Tokareva, E.Y.: The impact of Erasmus program on intercultural communication skills of students. In: E3S Web of Conferences, vol. 164, p. 12013 (2020). https://doi.org/10.1051/e3sconf/202016412014
5. Frijda, N.H.: Emotions and Action. In: Manstead, A.S.R., Frijda, N., Fischer, A. (eds.) Studies in Emotion and Social Interaction. Feelings and Emotions: The Amsterdam Symposium, pp. 158–173. Cambridge University Press (2004). https://doi.org/10.1017/CBO9780511806582.010
6. Fischer, A.H., Van Kleef, G.A.: Where have all the people gone? A plea for including social interaction in emotion research. Emot. Rev. **2**, 208–211 (2010)
7. Pekaar, K.A., van der Linden, D., Bakker, A.B., Born, M.Ph.: Dynamic self- and other-focused emotional intelligence: a theoretical framework and research agenda. J. Res. Pers. **86** (2020). https://doi.org/10.1016/j.jrp.2020.103958
8. Mayer, J.D., Salovey, P.: What is emotional intelligence? In: Salovey, P., Sluyter, D. (eds.) Emotional Development and Emotional Intelligence: Educational Implications, p. 31. Basic Books, New York (1997)
9. Petrides, K.V.: Ability and trait emotional intelligence. In: Chamorro-Premuzic, T., Furnham, A., von Stumm, S. (eds.) The Blackwell-Wiley Handbook of Individual Differences, pp. 656–678. Wiley, New York (2011)
10. Zeidner, M., Roberts, R.D., Matthews, G.: The science of emotional intelligence: current consensus and controversies. Eur. Psychol. **13**, 64–78 (2008)
11. Lopes, P.N., Brackett, M.A., Nezlek, J.B., Schütz, A., Sellin, I., Salovey, P.: Emotional intelligence and social interaction. Pers. Soc. Psychol. Bull. **30**, 1018–1034 (2004)
12. Schutte, N.S., Malouff, J.M., Bobik, C., Coston, T.D., Greeson, C., Jedlicka, C., Wendorf, G.: Emotional intelligence and interpersonal relations. J. Soc. Psychol. **141**, 523–536 (2001)
13. Martins, A., Ramalho, N., Morin, E.: A comprehensive meta-analysis of the relationship between emotional intelligence and health. Pers. Individ. Differ. **49**, 554–564 (2010)
14. Joseph, D.L., Newman, D.A.: Emotional intelligence: an integrative meta- analysis and cascading model. J. Appl. Psychol. **95**, 54–78 (2010)
15. O'Boyle Jr., E.H., Humphrey, R.H., Pollack, J.M., Hawver, T.H., Story, P.A.: The relation between emotional intelligence and job performance: a meta-analysis. J. Organ. Behav. **31**, 788–818 (2011)
16. Siegling, A.B., Saklofske, D.H., Petrides, K.V.: Measures of ability and trait emotional intelligence. In: Boyle, G.J., Matthews, G., Saklofske, D.H. (eds.) Measures of Personality and Social Psychological Constructs, pp. 381–414. Academic Press, San Diego (2015)

17. Vesely Maillefer, A., Udayar, S., Fiori, M.: Enhancing the prediction of emotionally intelligent behavior: the PAT integrated framework involving trait EI, ability EI, and emotion information processing. Front. Psychol. **9**, 1078 (2018). https://doi.org/10.3389/fpsyg.2018.01078
18. Brasseur, S., Grégoire, J., Bourdu, R., Mikolajczak, M.: The profile of emotional competence (PEC): development and validation of a self-reported measure that fits dimensions of emotional competence theory. PLoS One **8**, e62635 (2013)
19. Mikolajczak, M., Avalosse, H., Vancorenland, S., Verniest, R., Callens, M., Van Broeck, N., Mierop, A.: A nationally representative study of emotional competence and health. Emotion **15**, 653–667 (2015)
20. Pekaar, K.A., Bakker, A.B., Van der Linden, D., Born, M.: Self- and other- focused emotional intelligence: development and validation of the Rotterdam emotional intelligence scale (REIS). Pers. Individ. Differ. **120**, 222–233 (2018)
21. Crone, E.A., Dahl, R.E.: Understanding adolescence as a period of social-affective engagement and goal flexibility. Nat. Rev. Neurosci. **13**(9), 636–650 (2012)
22. Qouta, S., Punamäki, R.L., Montgomery, E., Sarraj, E.: Predictors of psychological distress and positive resources among Palestinian adolescents: trauma, child, and mothering characteristics. Child Abuse Neglect **31**(7), 699–717 (2007)
23. Thapar, A., Collishaw, S., Pine, D.S., Thapar, A.K.: Depression in adolescence. Lancet **379**(9820), 1056–1067 (2012)
24. Song, L.J., Huang, G., Peng, K.Z., Law, K.S., Wong, C., Chen, Z.: The differential effects of general mental ability and emotional intelligence on academic performance and social interactions. Intelligence **38**, 137–143 (2009). https://doi.org/10.1016/j.intell.2009.09.003
25. Afolabi, O.A.: Roles of personality types, emotional intelligence and gender differences on prosocial behavior. Psychol. Thought **6**(1), 124–139 (2013)
26. Salovey, D., Grewal, D.: The science of emotional intelligence. Curr. Dir. Psychol. Sci. **14**, 281–285 (2005)
27. Bar-On, R.: Emotional intelligence: an integral part of positive psychology. S. Afr. J. Psychol. **40**(1), 54–62 (2010)
28. Rajappa, K., Gallagher, M., Miranda, R.: Emotion dysregulation and vulnerability to suicidal ideation and attempts. Cognit. Ther. Res. **36**(6), 833–839 (2012)
29. Davis, S.K., Humphrey, N.: Ability versus trait emotional intelligence. J. Indiv. Differ. **35**, 54–62 (2014)
30. Antinienė, D., Lekavičienė, R.: Psychological and physical well-being of Lithuanian youth: relation to emotional intelligence. Medicina **53**(4), 277–284 (2017)
31. Bhullar, N., Schutte, N.S., Malouff, J.M.: Trait emotional intelligence as a moderator of the relationship between psychological distress and satisfaction with life. Indiv. Differ. Res. **10**(1), 19–26 (2012)
32. Schutte, N.S., Malouff, J.M., Thorsteinsson, E.B.: Increasing emotional intelligence through training: current status and future directions. Int. J. Emot. Educ. **5**(1), 56–72 (2013)
33. Di Fabio, A., Saklofske, D.H.: Emotional intelligence and youth career readiness. In: Keefer, K.V., Parker, J.D.A., Saklofske, D.H. (eds.) Emotional Intelligence in Education: Integrating Research with Practice, pp. 353–375. Springer, Cham (2018)
34. Encinas, J.J., Chauca, M.: Emotional intelligence can make a difference in engineering students under the competency-based education model. Proc. Comput. Sci. **172**, 960–964 (2020). https://doi.org/10.1016/j.procs.2020.05.139
35. Petrides, K.V., Sanchez-Ruiz, M.-J., Siegling, A.B., Saklofske, D.H., Mavroveli, S.: Emotional intelligence as personality: measurement and role of trait emotional intelligence in educational contexts. In: Keefer, K.V., Parker, J.D.A., Saklofske, D.H. (eds.) Emotional Intelligence in Education: Integrating Research with Practice, pp. 49–81. Springer, Cham (2018)

36. Chacaltana, J., Díaz, J.J., Rosas-Shady, D.: Towards a system of continuous training of the labor force in Peru. International Labor Organization (ILO) and Inter-American Development Bank (IDB), Lima, Peru (2015)

37. Kluve, J., Puerto, S., Robalino, D., Romero, J.M., Rother, F., Stöterau, J., Weidenkaff, F., Witte, M.: Interventions to improve the labour market outcomes of youth: a systematic review of training, entrepreneurship promotion, employment services, and subsidized employment interventions. The Campbell Collaboration, 207AD, Oslo, Norway (2017). https://doi.org/10.4073/csr.2017.12

38. Knutson, J., Bitz, I.: Project management: how to plan and manage successful projects. AMACOM Div. American. Mgmt Assn. (1991)

39. Kerzner, H.: Project Management a Systems Approach to Planning, Scheduling, and Controlling, 2nd edn. Van Nostrand Reinhold, New York (1984)

40. Havranek, T.J.: Modern Project Management Techniques for the Environmental Remediation Industry. CRC Press (1998)

41. Taras, V., Bryla, P., Gupta, S., Jimenez, A., Minor, M., Tim, M., Ordeñana, X., Rottig, D., Sarala, R., Zakaria, N., Zdravkovic, S.: Changing the face of international business education: the X-culture project. AIB Insights **12**, 11–17 (2012)

42. Poór, J., Kollar, C., Szira, Z., Taras, V., Varga, E.: central and eastern european experience of the X-culture project in teaching international management and cross-cultural communication. J. Int. Manage. **10**, 5–41 (2018). https://doi.org/10.2478/joim-2018-0001

43. Volodarskaya, E.B., Grishina, A.S., Pechinskaya, L.I.: Virtual learning environment in lexical skills development for active vocabulary expansion in non-language students who learn English. In: 2019 12th International Conference on Developments in eSystems Engineering (DeSE), Kazan, Russia, pp. 388–392 (2019)

Advances in Public Health Care, Hospitals and Rehabilitation

The Nervous System Disorders in COVID-19: From Theory to Practice

Tatyana Zakharycheva[1] (ID), Tatyana Makhovskaya[2] (ID), Alexandra Shirokova[3] (ID), and Irina Shikina[4(✉)] (ID)

[1] Far Eastern State Medical University, Khabarovsk, Russia
dolika@inbox.ru

[2] Central State Medical Academy of the Administrative Office of the President of the Russian Federation, Moscow, Russia
makhovskayat@mail.ru

[3] Diagnostic Centre "Viveya" of the Ministry of Health of the Khabarovsk Territory, Khabarovsk, Russia
a.s.shirokova@mail.ru

[4] Central Research Institute for Organization and Informatization of Medical Care, Ministry of Health of Russia, Moscow, Russia
shikina_irina@mail.ru

Abstract. The article presents a state-of-the-art literature review and personal view on spectrum and pathogenesis of the nervous system lesions in the case of coronavirus infection COVID-19 known to date - from theory, medical and social significance to the main clinical options for lesions of the nervous system in the novel coronavirus infection COVID-19. Basic information is also presented on factors modifying the course and outcome of infectious lesions in the nervous system, mechanisms of their development and possible outcomes scenarios in case of novel coronavirus infection COVID-19. Today's level of knowledge for nervous system damages due to COVID-19 are considered to be life-threatening, and their consequences can have a huge negative impact on the quality of life, especially in those with pre-morbidities.

Keywords: Coronavirus · COVID-19 · Nervous system disorders

1 Introduction

The medical and social significance of infectious lesions of the nervous system is due to a wide range of pathogens, adverse course and outcome of diseases, absence of antiviral therapy and resistance of bacteria to antibacterial agents [1, 2]. The interaction between micro- and macroorganisms due to genetic variability, change of pathogens, and population ageing is also relevant. In present conditions beta-coronavirus SARS-CoV-2, causing "Coronavirus Disease 2019" (COVID-19) has become an extremely dangerous pathogen [3, 4]. Coronavirus infection has been known since the twentieth century as an acute viral disease with weak, predominant lesions of the upper respiratory tract and a

T. Antipova (Ed.): ICADS 2021, AISC 1352, pp. 191–197, 2021.
https://doi.org/10.1007/978-3-030-71782-7_17

good outcome. At the same time, there are reports of finding coronaviruses in the brain of patients with multiple sclerosis [5, 6]. We provide our own view on the spectrum and pathogenesis of nervous system lesions in COVID-19 to our colleagues.

Microorganisms exist in a variety of conditions, living in habitats and biological objects. Therefore, viruses can be treated both as agents of infectious diseases and as factors contributing to the adaptive restructuring of the human body under changing conditions of the external and internal environment. In the latter case, viral diseases and nervous system lesions should be considered as adaptation diseases due to immune defect in humans [7, 8].

It is known that the infection process is determined by several factors such as the properties of a particular pathogen, infection method, the characteristics of the host body, etc. [9–11].

2   Materials and Methods

The article presents a state-of-the-art literature review and personal opinion on infectious lesions of the nervous system problem, known to date, specifically connected with the new novel coronavirus infection COVID-19. Analytical research has been carried out to provide basic information on factors modifying the course and outcome of infectious lesions of the nervous system, mechanisms of their development and possible scenarios of outcomes in the novel coronavirus infection. The researchers independently searched for literature published during the period 1980–2020. Prospective and retrospective observational studies of high methodological quality, analytical journals and original scientific articles were used for the analysis.

3   Results

Many authors have shown that antigens of the main HLA histocompatibility complex determine the intensity of cellular immune reactions [9, 12, 13]. They can be used as genetic markers encoding predisposition to COVID-19 disease and for predicting its course and outcome.

An important role in the pathogenesis of nervous system lesions at COVID-19 may belong to hematoencephalic barrier dysfunction (HBD) due to its pre-morbid characteristics and also due to the "cytokine storm". The pathological process is aggravated by hypoxia, electrolyte imbalance, vitamins and mineral deficiency, intoxications, microtraumas, and microbial invasiveness factors [14].

Viruses can infect the lymphatic system, suppress the immune response and invade the host body [15].

The immune response to different infections may vary significantly. Data on the duration and intensity of immunity concerning SARS-CoV-2 are only accumulating. It should be assumed that in pre-morbid healthy individuals, specific post-infection immunity should be long and strong. In latent (inapparent) and erased forms of infection, as well as in weakened patients' immunity may not be sustained, and re-infection may occur [16, 17]. Cross-immunity to other members of the coronavirus family has not been found [18].

There are two types of central nervous system infectious lesions. Neuroinfections or primary lesions are mainly caused by viruses that are traumatic to nerve cells and irreversibly damage them. Such diseases are characterized by heavy current and residual symptoms. Secondary (post-infection or para-infection) lesions are relatively benign, immune-mediated with vessel involvement, demyelination, more often reversible neuronal lesions [19, 20].

It is known that the leading role in the pathogenesis of COVID-19 is played by type 2 angiotensin-converting enzyme (ACE2), for which beta-coronavirus SARS-CoV-2 has high affinity. The binding of S1-protein coronavirus and ACE2 is the key stage of virus entry into the cell [21–23]. The high expression of AFP2 in the brain determines the tropism of the virus to neurons and glia. This determines the potential neurotropism of SARS-CoV-2 and ability of coronaviruses to cause neuronal death [24, 25]. According to Chinese researchers, 36.4% of patients with COVID-19 had neurological disorders. Their spectrum included anosmia and ageusia, viral meningitis, encephalitis, stroke, acute hemorrhagic necrotizing encephalopathy; post-infection acute disseminated encephalomyelitis, post-infection stem encephalitis, transverse myelitis, Guillain-Barre syndrome, myositis [26].

In connection with the data presented above, we can assume several possible types of nervous system damage at COVID-19.

Firstly, the development of diseases resulting from direct exposure of SARS-CoV-2 beta-coronavirus to the nervous system are following: meningitis, encephalitis, myelitis, encephalomyelitis, including subacute sclerosing panencephalitis as a result of the persistence of viral infection in patients with congenital or acquired immune system defects. The disease outcome in this scenario will be unfavorable.

Secondly, secondary meningitis in patients with purulent-septic complications occurring both with a relevant clinical presentation of purulent meningitis and with focal symptoms due to cerebrovascular complications.

Thirdly, secondary post- and para infection of the central nervous system lesions (acute disseminated acute encephalomyelitis, acute hemorrhagic leukoencephalitis).

Fourthly, acute inflammatory demyelinating (axonal) polyneuropathy (Guillain-Barre syndrome).

The uncontrolled release of endogenous inflammatory mediators, insufficient mechanisms limiting the damaging effect of cytokines, as well as metabolic and microcirculatory disorders in tissues are considered to be the main reasons for the development of organ system disorders in severe forms of COVID-19. The most frequent and severe complications of COVID-19 are acute respiratory distress syndrome and secondary bacterial infections that require intensive therapy with a ventilator. Patients might develop infectious toxic shock, acute damage to the kidneys and heart, and liver dysfunction [18]. Therefore, patients with severe forms of COVID-19 have a high probability of developing critical illness polyneuropathy (CIP) [13, 14]. We observed this type of polyneuropathy in a patient with severe flu A (H1N1) [27, 28].

Long-term clinical observations in the foci of tick-borne encephalitis indicate a particular severity of the disease in elderly and senile people. In this age group, the proportion of focal lesions of the nervous system was 42.3%. Disorders of vital functions were observed in 80.1% of cases. Lethality reached 21.2%. Adverse outcomes were promoted by changes in nervous, cardiovascular, respiratory, immune and other systems caused by ageing [9]. Therefore, patients with COVID-19 of old age, suffering from immunosuppressive diseases, diabetes mellitus, chronic cardiovascular, pulmonary and neurological pathology require special attention [29, 30]. Infections, including respiratory diseases, can serve as triggers for acute vascular disorders. Chronic obstructive pulmonary disease (COPD) is an independent risk factor for cardiovascular disease, and cough associated with increased chest pressure can cause a thromboembolic stroke or dissection of carotid arteries [31, 32].

Hypoxia (acute and chronic) due to respiratory disorders also harms the central nervous system and may cause the development of acute hypoxic encephalopathy or dysfunction of the limbic-reticular complex and the appearance of vegetative disorders, neurosis-like disorders and focal neurological symptoms [33].

The infectious process development is characterized by phases. With COVID-19 there are at least three or four phases: 1) incubation; 2) clinical-symptomatic; 3) convalescence period; 4) distant. The latter period may manifest as asthenic and vegetative syndromes, as well as disorders of motor, cognitive and mental functions. Neurological consequences of COVID-19, in turn, according to existing classifications [1, 33, 34], can be subdivided into residual and progressive.

It is generally recognized that all the highest forms of human behavior are associated with the vital functions of nerve cells that synthesize catecholamines: neurotransmitters noradrenaline, serotonin, dopamine. In patients with neuroinfections at the peak of the infection process, as well as in the period of early and late convalescence period, asthenia, anxiety-depressive and vegetative disorders, sleep disorders are detected. Their severity correlates with clinical forms and disease severity [9, 35–37], which is explained by damage to brain cell structures [38–41]. Study of brains' neurotransmitter systems allows not only to understand the pathogenesis of symptomatology in nervous system lesions including infectious ones but also to make pharmacological interventions of such disorders [42, 43].

The medical and social rehabilitation of COVID-19 rehabilitation centers is one of the important state tasks. Rehabilitation measures imply medication use - general tonic, vasoactive, nootropic, neurotrophic, adaptogenic, vitamin, antidepressants and anticonvulsants [44–47]. In the group of elderly and senile patients with cognitive disorders progressing, during the disease the use of dopamine agonists becomes relevant. The first-choice medicine is non-ergoline dopamine receptor agonist (piribedil) due to more favorable side effects profile [48–51].

A special group that requires close medical supervision are patients suffering from autoimmune inflammatory diseases of the nervous system such as multiple sclerosis, Guillain-Barre syndrome, myasthenia gravis and others. They have a high risk of exacerbation during or after COVID-19 [52].

4 Conclusion

The nervous system lesions due to COVID-19 are life-threatening, and their consequences can have a huge negative impact on a person's life especially pre-morbidly burdened.

At COVID-19 several variants of nervous system lesions are possible: direct effect of SARS-CoV-2 beta-coronavirus on nerve cells with the persistence of the pathogen in the body; intermediate - with the development of secondary purulent meningitis and cerebrovascular complications; autoimmune demyelination of the central and/or peripheral nervous system; development of stroke, acute hypoxic encephalopathy or decompensation of chronic circulatory failure. COVID-19 is highly likely to exacerbate pre-existing conditions in patients suffering from autoimmune diseases of the nervous system.

Special COVID-19 severity and the prospect of its frequent adverse outcomes calls for the expediency of in-depth study of clinical manifestations of the disease, improvement of methods of prevention and early diagnosis of neurological complications as well as their rational pathogenetic treatment.

References

1. Gusev, E.I., Konovalov, A.N., Skvortsova, V.I.: Neurology: national guidelines. GEOTAR-Media **1**, 880 (2018)
2. Sandakov, Y.P., Kochubey, A.V.: The activities concerning improvement of dispensary observation. Probl. sotsial'noi gigieny zdravookhraneniia i istorii meditsiny **26**(6), 428–431 (2018). https://doi.org/10.32687/0869-866x-2018-26-6-428-431
3. Antipova, T.: Coronavirus pandemic as black swan event. In International Conference on Integrated Science, vol. 136, pp. 356–366 (2020). https://doi.org/10.1007/978-3-030-49264-9_32
4. Mirskikh, I., Mingaleva, Z., Kuranov, V., Matseeva, S.: Digitization of medicine in russia: mainstream development and potential. In: International Conference on Integrated Science, pp. 337–345. Springer, Cham (2020). https://doi.org/10.1007/978-3-030-49264-9_30
5. Lobzin, Y.V., Pilipenko, V.V., Gromyko, Y.N.: Meningitis and encephalitis. Monograph. Foliant, St. Petersburg, pp. 2–54 (2006)
6. Lobzin Y.V., Kazantsev A.P.: Guide to Infectious Diseases, p. 736. Phoenix (1997)
7. Umanskiĭ, K.G.: Viral neuroinfections (problem of adaptive pathology). Review of the literature. Zhurnal nevropatologii i psikhiatrii imeni SS Korsakova **80**(8), 1235 (1980)
8. Umansky, K.G.: Ubiquitous nature of viruses and presumption of innocence. Arkh. Patol. **42**(10), 76–81 (1980)
9. Zakharycheva, T.A.: Tick-borne encephalitis in the Khabarovsk Territory: yesterday, today, tomorrow, p. 248 (2014)
10. Shirokova, A.S., Zakharycheva, T.A., Fleishman, M.Y., Obukhova, G.G.: Clinical and immunological features of adolescents - convalescents of enteroviral meningitis. Bull. Neurol. Psychiatry Neurosurg. 29–34 (2018)
11. Zaretskaya, Y.M.: Clinical immunogenetics, pp. 65–66 (1983)
12. Iyerusalimskiy, A.P.: Tick-borne encephalitis. A guide for doctors, p. 360 (2001)
13. Tsinserling, V.A., Chukhlovina, M.L.: Infectious lesions of the nervous system: issues of epidemiology, pathogenesis and diagnostics: a guide for doctors of multidisciplinary hospitals. ELBI-SPb. p. 448 (2005)
14. Somova, L.M.: Pathology of neuroinfections caused by viruses of the tick-borne encephalitis complex: monograph-atlas. LLC SYNTERIA, p. 360 (2018)
15. Shikina, I.B.: The maintenance of security of the elderly and gerontic patients in the hospital conditions. Probl. sotsialnoĭ gigieny zdravookhraneniia i istorii meditsiny **6**, 44 (2007)
16. Armashevskaya, O.V., Ivanova, M.A., Chuchalina, L.Y.: Age features of pathology of women in the peri - and postmenopausal period. Adv. Gerontol. **30**, 363–367 (2017)

17. Zhmerenetskiy, K.V., Sazonova, E.N., Voronina, N.V., et al.: COVID-19: only scientific facts. Far Eastern Med. J. 5–22 (2020)
18. Levin, O.S.: Polyneuropathies: clinical guidelines. Med. Inf. Agency 480 (2016)
19. Umansky, K.G.: Where do the controversial issues of neuroviral diseases lead? S-info 72 (1993)
20. Jia, H.P., Look, D.C., Shi, L., et al.: ACE2 receptor expression and severe acute respiratory syndrome coronavirus infection depend on differentiation of human airway epithelia. J. Virol. **23**, 14614–14621 (2005). https://doi.org/10.1128/JVI.79.23.14614-14621.2005
21. Tang, X., Wu, C., Li, X., et al.: On the origin and continuing evolution of SARS-CoV-2. Natl. Sci. Rev. (2020). https://doi.org/10.1038/s41586-020-2169-0
22. Tortorici M.A., Veesler, D.: Structural insights into coronavirus entry. Adv. Virus Res. 93–116 (2019). https://doi.org/10.1016/bs.aivir.2019.08.002
23. Baig, A.M., Khaleeq, A., Ali, U., Syeda, H.: Evidence of the COVID-19 virus targeting the CNS: tissue distribution, host–virus interaction, and proposed neurotropic mechanisms. ACS Chem. Neurosci. **11**(7), 995–998 (2020). https://doi.org/10.1021/acschemneuro.0c00122
24. Netland, J., Meyerholz, D.K., Moore, S., Cassell, M., Perlman, S.: Severe acute respiratory syndrome coronavirus infection causes neuronal death in the absence of encephalitis in mice transgenic for human ACE2. J. Virol. **82**(15), 7264–7275 (2008). https://doi.org/10.1128/JVI.00737-08
25. Mao, L., Wang, M., Chen, S., He, Q., Chang, J., Hong, C., Zhou, Y., Wang, D., Miao, X., Hu, Y., Li, Y.: Neurological manifestations of hospitalized patients with COVID-19 in Wuhan, China: a retrospective case series study (2020). https://doi.org/10.1101/2020.02.22.20026500
26. Lemann-Horn, F., Ludolph, A.: Treatment of nervous system diseases. MEDpress-Inform. 528 (2005)
27. Zakharycheva, T.A., Menshikov, A.B., Frolova, M.A., et al.: Damage to the nervous system in influenza A (H1N1). Far Eastern J. Infect. Pathol. 43–48 (2012)
28. Suslina, T.S., Gulevskaya, M.Y., Maximova, V.A., Morgunov: Cerebral circulation disorders: diagnosis, treatment, prevention. MEDpress-inform. 536 (2016)
29. Buzin, V.N., Mikhailova, Y.V., Chukhriyenko, I.Y., Buzina, T.S., Shikina, I.B., Mikhailov, A.Y.: Russian healthcare through the eyes of the population: dynamics of satisfaction over the past 14 years (2006–2019): review of sociological studies. Prev. Med. **23**(3), 42–47 (2019). https://doi.org/10.17116/profmed20202303142
30. Chudnovsky, V.M., Makhovskaya, T.G., Mayor, A.Y., Yusupov, V.I., Nevozhai, V.I., Kiselyov, A.Y., Shikina, I.B.: The role of blood foaming in the mechanism of endovasal laser ablation. Phlebology 261–269 (2018). https://doi.org/10.17116/flebo201812041261
31. Tsiskaridze, A., Lindgren, A., Qureshi, A.: Iatrogenic stroke: a manual for practitioners. GEOTAR-Media, p. 432 (2019). [Russian]
32. Amelina, O.A., et.al.: Clinical neurology with the basics of medical and social expertise: a guide for doctors. SPb: Medline-Media, p. 594 (2006)
33. Shtock, V.N., Levin, O.S.: Handbook for the formulation of the clinical diagnosis of diseases of the nervous system. Med. Inf. Agency 520 (2019)
34. Sumlivaya, O.N., Vorobieva, N.N., Karakulova, Y.V.: Postinfectious syndrome in convalescents of ixodic tick-borne borreliosis. J. Infectol. 27–32 (2014)
35. Sumlivaya, O.N., Vorobieva, N.N., Karakulova, Y.V.: Postinfectious asthenia in convalescents after tick-borne encephalitis and methods of its relief. Perm Med. J. 41–48 (2017). https://doi.org/10.17816/pmj34541-48
36. Shirokova, A.S.: Cognitive functions in adolescents who have undergone enteroviral meningitis. Scientific and practical conference of neurologists. In: XX All-Russian Conference "Neuroimmunology. Multiple sclerosis", p. 103 (2015)

37. Alekseeva, L.A., Skripchenko, N.V., Bessonova, T.V., et al.: Markers of damage to glial neurons in cerebrospinal fluid in children with meningitis. Clin. Lab. Diagnost. 204–210 (2017)
38. Skripchenko, N.V., Shirokova, A.S.: Neuron-specific enolase and S100 protein are biomarkers of brain damage. State of the issue and clinical application. Neurosurg. Neurol. Children 4(50), 16–25 (2016)
39. Sumlivaya, O.N.: The relationship between changes in serotonin and cytokines in patients with tick-borne encephalitis. Perm Med. J. 32, 68–73 (2015)
40. Lins, H., Wallesch, C.W., Wunderlich, M.T.: Sequential analyses of neurobiochemical markers of cerebral damage in cerebrospinal fluid and serum in Cnervous system infection. Acta Neurol. Scand. 112, 303–308 (2005)
41. Sumlivaya, O.N., Vorobieva, N.N., Karakulova, Y.V.: Participation of the serotoninergic system in the formation of cerebrasthenic syndrome in tick-borne encephalitis and its correction. Pract. Med. 7(83), 68–71 (2014)
42. Shirokova, A.S., Skripchenko, N.V., Zakharycheva, T.A., Fleishman, M.Y.: Pharmaco-correction of asthenovegetative disorders in children and adolescents - convalescents of enteroviral meningitis. Bull. Neurol. Psychiatry Neurosurg. 9(104), 48–51 (2018)
43. Goldblat, Y.V.: Medical and social rehabilitation in neurology. Polytechnic 607 (2015)
44. Lobzin, Y.V., Skripchenko, N.V.: Modern approaches to the diagnosis, therapy and prevention of infectious diseases in children. Sci. works 904 (2013)
45. Blinova, U., Rozhkova, D., Rozhkova, N.: Management accounting of innovation costs. Vestnik Univ. (1), 43–48 (2018). [Russian]. https://doi.org/10.26425/1816-4277-2018-1-43-48
46. Rozhkova, N., Blinova, U., Rozhkova, D.: The concept of management accounting based on the information technologies application. Inf. Technol. Sci. (2018). https://doi.org/10.1007/978-3-319-74980-8_8
47. Moroz, E.V., Zakharycheva, T.A., Antonyuk, M.V.: Clinical experience of using piribedil in chronic cerebrovascular disease with cognitive impairment. Neurol. Neuropsychiatry Psychosomat. 11(4), 100–103 (2019). https://doi.org/10.14412/2074-2711-2019-4-100-103
48. Parfenov, V.A.: Vascular cognitive impairment and chronic cerebral ischemia (discirculatory encephalopathy). Neurol. Neuropsychiatry Psychosomat. 11(3), 61–67 (2019). https://doi.org/10.14412/2074-2711-2019-3S-61-67
49. Nagaraja, D., Jayashree, S.: Randomized study of the dopamine receptor agonist piribedil in the treatment of mild cognitive impairment. Am. J. Psychiatry 158(9), 1517–1519 (2001). https://doi.org/10.1176/appi.ajp.158.9.1517
50. Perez-Lloret, S., Rascol, O.: Piribedil for the treatment of motor and non-motor symptoms of Parkinson disease. CNS Drugs 30(8), 703–717 (2016). https://doi.org/10.1007/s40263-016-0360-5
51. Moskalev, A.V.: Autoimmune diseases: diagnosis and treatment: a guide for doctors. GEOTAR-Media 224 (2017)
52. Muravyeva, A., Mikhaylova, Y., Shikina, I.: Organizational arrangements for medical assistance to patients with a new coronavirus infection Covid-19 in Stavropol Krai. Current problems of healthcare and medical statistics, p. 4 (2020). https://doi.org/10.24411/2312-2935-2020-00120

Diagnosis of Diabetic Retinopathy Using Data Mining Classification Techniques

Ana Abreu[1] , Diana Ferreira[2] , Cristiana Neto[2] , António Abelha[2] ,
and José Machado[2(✉)]

[1] University of Minho, Campus of Gualtar, Braga, Portugal
`a80276@alunos.uminho.pt`
[2] Algoritmi Research Center, University of Minho, Campus of Gualtar, Braga, Portugal
`{diana.ferreira,cristiana.neto}@algoritmi.uminho.pt`, `{abelha, jmac}@di.uminho.pt`

Abstract. Diabetic retinopathy is one of the complications of diabetes that affects the small vessels of the retina, being the main cause of blindness in adults. An early detection of this disease is essential, as it can prevent blindness as well as other irreversible harmful outcomes. This article attempts to develop a data mining model capable of identifying diabetic retinopathy in patients based on features extracted from eye fundus images. The data mining process was carried out in the RapidMiner software and followed the CRISP-DM methodology. In particular, classification models were built by combining different scenarios, algorithms, and sampling methods. The data mining model which performed best achieved an accuracy of 76.90%, a precision of 85.92%, and a sensitivity of 67.40%, using the Logistic Regression algorithm and Split Validation as the sampling method.

Keywords: Healthcare · Diabetic retinopathy · Data mining · CRISP-DM · Classification

1 Introduction

Despite being an old method, Data Mining (DM) has been increasingly used in recent years [1]. DM is the process of exploring large amounts of data in search of consistent patterns, such as association rules or time sequences, to detect systematic relationships between variables [2]. It is formed by a set of tools and techniques that through the use of learning or classification algorithms based on statistics are able to explore sets of data and to extract or help to identify patterns in the data, thus assisting in the discovery of knowledge [3]. In DM, knowledge can be presented by tools in several ways such as groupings, hypotheses, rules, decision trees or graphs [1].

The use of information technologies allows the automatization of processes for extracting data that promote the attainment of interesting knowledge, which means the elimination of manual tasks and an easier extraction of data directly from electronic records [3]. DM has applications in multiple fields, such as science and research. Data

© The Author(s), under exclusive license to Springer Nature Switzerland AG 2021
T. Antipova (Ed.): ICADS 2021, AISC 1352, pp. 198–209, 2021.
https://doi.org/10.1007/978-3-030-71782-7_18

related to healthcare environments tend to be large and complex in nature, which corresponds to the ideal data type for the implementation of DM models [2]. In fact, the interest in the use of DM by healthcare organizations has been increasing, as it has proven to be a very effective method to improve operational efficiency, contributing to a high level of quality of healthcare. However, it is not very easy to apply DM models to health data, since these data have several limitations, for instance the difficult access to them due to privacy policies [4].

Diabetes is a chronic long-term condition that occurs when the body does not make enough insulin or is unable to use it. Diabetic retinopathy (DR) is one of the complications of diabetes that affects the small vessels of the retina, the region of the eye responsible for the formation of images sent to the brain [5]. When diabetes is not controlled, hyperglycemia triggers several changes in the body that, among other damages, leads to dysfunction of the retinal vessels [6]. As the disease progresses, these vessels become incontinent and release blood fluid into the retinal space or the vitreous, causing vision problems [6].

DR is not only dependent on blood glucose values, but also on other factors such as arterial hypertension, smoking habits, and genetic factors [7]. DR usually affects both eyes and, if not early diagnosed and treated, can lead to irreversible blindness [8, 9].

DR can be classified as non-proliferative or proliferative [10]. Non-proliferative DR is the least advanced stage of the disease. In this phase, microaneurysms (MAs), hemorrhages and blocked blood vessels can be found, causing several areas of the retina to be without blood supply, which contains oxygen and nutrients [11]. If the macula is not affected, this stage of DR may show no symptoms [10]. However, when macular edema is present, vision may appear blurred and the risk of visual loss or blindness increases significantly [5]. Proliferative DR is the most advanced stage of the disease, characterized by the appearance of new blood vessels on the surface of the retina [10]. The main cause of the formation of new vessels is the occlusion of blood vessels in the retina, called ischemia, with impediment of adequate blood flow [12]. Neovessels grow along the retina without causing any symptoms or loss of vision. However, they can breach and release blood, causing severe vision loss and even blindness [5].

The comparison between a normal eye and an eye with DR can be found in Fig. 1.

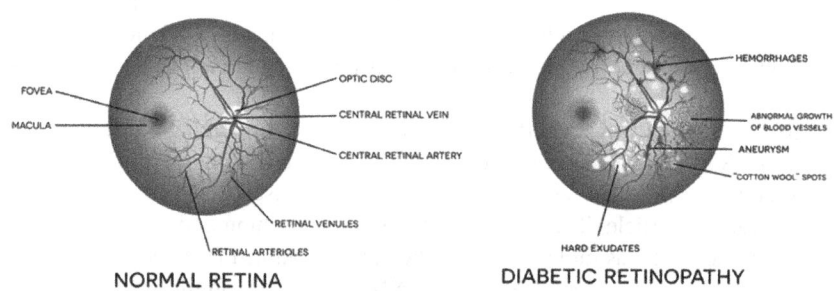

Fig. 1. Comparison between a normal eye and an eye with DR [13].

DR is the most common microvascular complication of diabetes, being the main cause of blindness in adults aged 20 to 74 years old [10]. The number of people with DR is growing higher day by day. It is estimated that the number will rise from 171 million in 2000 to 366 million in 2030 [14]. In order to prevent this, regular eye examinations are essential to detect DR and to allow treatments to start as early as possible, when the chances of controlling the disease are greater. DM can be very useful, as it can provide important information and allow to draw new conclusions about the detection of this disease. Several DM methodologies have been proposed to discover data knowledge, such as the CRISP-DM (Standard Industrial Process for Data Mining), the one adopted in this work.

Hence, this article focuses on the prediction of DR and in the analysis of the performance of different Machine Learning (ML) algorithms, based on the attributes present in the dataset under study. It is intended to use these data as the training data to build a model capable of predicting whether new cases have DR or not.

The remainder of the paper is organized as follows: the next section presents an overview of the existing studies carried out within the scope of the topic addressed in this paper; Sect. 3 presents the entire DM process implemented, referring to how each phase was developed and the results obtained; Sect. 4 refers to an analysis of the results obtained and finally, Sect. 5 presents the main conclusions of this work and prospects for future work.

2 Related Work

The identification of DR and the detection of features such as exudates and blood vessels on fundus images have been studied for many years. There have been multiple studies regarding the detection of DR using different DM techniques.

In 2009, a study made a series of experiments on feature selection and exudates classification and compared the best Naive Bayes (NB) and Support Vector Machine (SVM) classifiers to a baseline Nearest Neighbour classifier using the best feature sets from both classifiers. The authors found that the NB and SVM classifiers performed better than the Nearest Neighbour classifier [15].

In 2014, K. R. Ananthapadmanabhan and G. Parthiban studied the possibility of early detection of this disease with different DM techniques, using the RapidMiner tool. They used NB and SVM to predict the early detection of DR and found NB algorithm to be 83.37% accurate. This methodology has also shown that the DM model developed helps to retrieve useful correlation even from attributes that are not direct indicators of the target class [16].

In 2018, a studied was carried out with the objective of making a decision on the presence of DR by applying ensemble of ML classifying algorithms on features extracted from the output of different retinal images. The authors used exactly the same dataset that is used in this article. They concluded that, after training and testing the model, the accuracy they got was quite similar to the values obtained in the studies presented before, but the algorithm Neural Networks (NNET) was providing a higher accuracy rate for predicting DR, with an accuracy of 75% [17].

3 Data Mining Process

CRISP-DM provides a structured approach to plan a DM project. CRISP-DM remains the standard methodology for carrying out DM projects, because it proves to be less costly, more reliable, more repeatable, and faster, while simultaneously providing flexibility and customization [2]. The CRISP-DM methodology outlines the steps involved in performing data science activities from business needs to deployment, but more importantly defines a framework that allows iterations through all the phases [3]. In real world applications, the iterative nature allows constant improvement via backtracking to previous tasks and repeating certain actions [18]. CRISP-DM consists of six major phases, as shown in Fig. 2.

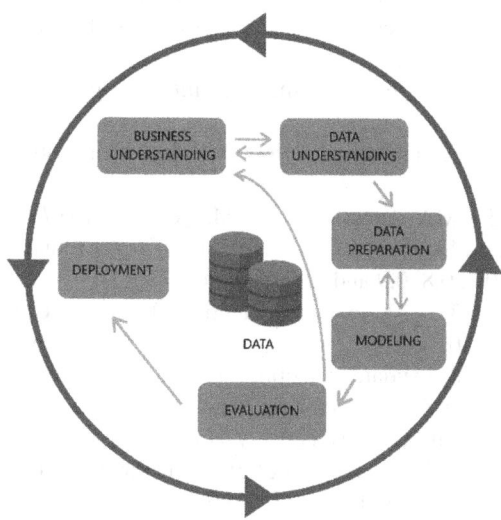

Fig. 2. Lifecycle of the CRISP-DM methodology [2].

3.1 Business Understanding

As data related to health environments tend to be large and complex in nature, it is important to implement DM models to streamline diseases' detection process, thus avoiding irreversible consequences [19].

The business goal of this work is to classify cases with DR, based on the attributes present in the dataset under study. It is intended that these data are used as training data to build a model capable of predicting whether new cases have DR and those that do not. To classify cases with DR, the objective is the "class" attribute. Since the main cause of blindness in adults is due to DR, it is essential that its detection is done as soon as possible.

3.2 Data Understanding

Having already set the business goal, the next phase involved analysing the data available for mining and also its quality.

The dataset consists of 1151 instances and 19 features considered relevant in the identification of cases with DR [20]. These features were extracted from the Messidor (Methods to Evaluate Segmentation and Indexing Techniques in the field of Retinal Ophthalmology) database in order to predict whether new cases have DR or not. The Messidor database, which contains hundreds of eye fundus images, has been publicly distributed since 2008 [21].

The features of this dataset can represent a detected lesion, a descriptive feature of an anatomical part of the eye or a feature at the image level.

As the names of the attributes were barely noticeable, they were changed in order to make observation and interpretation easier. The attributes of the dataset are as follows:

- quality_asses - Binary quality assessment result (0 = poor quality, 1 = sufficient quality).
- pre_screen - Binary pre-screening result (0 = lack of retinal abnormality, 1 = severe retinal abnormality).
- MA_detection_0.5, MA_detection_0.6, MA_detection_0.7, MA_detection_0.8, MA_detection_0.9, MA_detection_1 - Number of MAs found at alpha confidence levels = 0.5, 0.6, 0.7, 0.8, 0.9 and 1, respectively.
- exudate_detection_0.3, exudate_detection_0.4, exudate_detection_0.5, exudate_detection_0.6, exudate_detection_0.7, exudate_detection_0.8, exudate_detection_0.9, exudate_detection_1 - Contains the same information as attributes 2 to 7 for exudates at alpha confidence levels = 0.3, 0.4, 0.5, 0.6, 0.7, 0.8, 0.9 and 1, respectively. However, as the exudates are represented by a set of dots, and not by the number of pixels that make up the lesions, these characteristics are normalized by dividing the number of lesions by the diameter of the ROI (region of interest) to compensate for different image sizes.
- euc_dist - The Euclidean distance from the center of the macula to the center of the optic disc to provide important information about the patient's condition. This feature is also normalized with the diameter of the ROI (region of interest).
- diam_opt_dis - Diameter of the optical disc.
- AM/FM - The binary result of the classification based on amplitude-frequency modulation (AM / FM).
- Class - It can be of type "0" or "1", where "0" means that there are no signs of DR and "1" means that the patient has signs of DR.

To classify cases with DR, the objective is the "class" attribute. All attributes of the dataset are of the numeric type, except for the Class which is of the nominal type.

Figure 3 presents the statistical distribution of the class attribute. This figure shows that there are more patients with DR than without, in this case 611 of the 1151 instances, which corresponds to a percentage of 53.1%.

Also, it was noted that there were no missing values for any instance of the dataset.

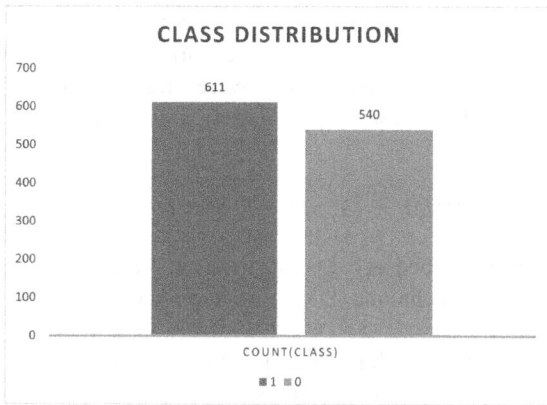

Fig. 3. Class distribution of the target attribute

3.3 Data Preparation

The third stage of this methodology is to select and clean data, depending on the business goals. This is the most relevant phase of the CRISP-DM process, since a proper preparation of the data can greatly improve the performance of a DM model. A cleaning process was performed to preserve the consistency of the data and also to remove any incomplete information.

As it was mentioned before, this dataset has no missing values, so there was no need to remove or replace any incomplete information. However, regarding the data consistency, the *Detect Outliers LOF* operator was used to detect discrepancies. This operator identifies outliers in the given dataset based on Local Outlier Factors (LOF). The LOF is based on a concept of a local density, where locality is given by the k nearest neighbors, whose distance is used to estimate the density. By comparing the local density of an object to the local densities of its neighbors, one can identify regions of similar density, and points that have a substantially lower density than their neighbors [22]. Since this dataset has 1151 instances, the sample is large enough that the discrepant data can be removed without substantial loss of statistical power. A threshold equal to 2 was defined and it was found that there were 8 instances with an outlier value greater than 2. Thus, these instances were removed from the dataset.

In order to discover which are the most relevant attributes to achieve the intended objective, several operators were used to analyze which attributes had the highest weights in the forecast of the target attribute. The operators used were *Weight by Chi Squared Statistic, Weight by Correlation, W-GainRatioAttributeEval, Weight by Information Gain Ratio, Weight by Gini Index* and *W-InfoGainAttributeEval*. The most relevant attributes shown by all these operators are the attributes associated with the number of MAs.

3.4 Modeling

After the Data Preparation stage was completed, the data was finally prepared to feed the models. All experiments were conducted using the RapidMiner software and various modeling techniques were selected and applied. Then, their parameters were calibrated to the most suitable values. In addition, various scenarios were created to test and verify the quality and validity of the model. In this way, several Data Mining Models (DMM) were created, which are based on six parameters, as it is presented in the following equation:

$$DMM = \{A, S, DMT, SM, DA, T\} \tag{1}$$

Each DMM can be described as belonging to an approach (A), being composed by a scenario (S), a data mining technique (DMT), a sampling method (SM), a data approach (DA), and a target (T).

All models were built according to a classification approach (A). There was no need to perform oversampling or undersampling (DA) since the class distribution was already balanced. Through the *Compare ROCs* operator and taking into account what was exposed in the Related Work section, the best six DMT found for this case were the AutoMLP, Perceptron (P), SVM, JRip, Neural Net (NN) and Logistic Regression (LR).

The sampling methods (SM) implemented were Cross Validation (CV) with 10 folds and 7 folds and Split Validation (SV) with 30% and 20% of data used for testing. A single target (T) was considered and used to determine whether the patient has the disease or not, which is the *class* attribute.

Regarding the scenarios, the first scenario (S1) includes all attributes. The second scenario (S2) relies on the nine attributes that have more weight according to the operator *Weight by Chi Squared Statistic*. On the other hand, the third scenario (S3) has the nine attributes that have more weight according to the operator *Weight by Correlation*. The fourth scenario (S4) includes the nine attributes that have more weight according to the operator *W-GainRatioAttributeEval*. The fifth scenario (S5) has the nine attributes that have more weight according to the operator *Weight by Information Gain Ratio*.

- S1 = {All attributes}
- S2   =   {MA_detection_0.5,   MA_detection_0.6, MA_detection_0.7, MA_detection_0.8, MA_detection_0.9, MA_detection_1, exudate_detection_0.3, exudate_detection_0.9, exudate_detection_1,}
- S3   =   {MA_detection_0.5,   MA_detection_0.6, MA_detection_0.7, MA_detection_0.8, MA_detection_0.9, exudate_detection_0.7, exudate_detection_0.8, exudate_detection_0.9, exudate_detection_1}
- S4 = {quality_asses, MA_detection_0.5, MA_detection_0.6, MA_detection_0.8, exudate_detection_0.6,   exudate_detection_0.7,   exudate_detection_0.8, exudate_detection_0.9, exudate_detection_1}
- S5   =   {MA_detection_0.5,   MA_detection_0.6,   MA_detection_0.7, MA_detection_0.8,   exudate_detection_0.6,   exudate_detection_0.7, exudate_detection_0.8, exudate_detection_0.9, exudate_detection_1}

In this study, the models were built according to DMM = {1 Approach, 5 Scenarios, 6 DMT, 4 Sampling Methods, 1 Data Approach, 1 Target}.

3.5 Evaluation

After testing the models, it is necessary to evaluate the results obtained and verify the impact they have on the initially defined objectives.

Since this study is a binary classification problem, the most suitable evaluation criterion to use is the confusion matrix. This predictive classification matrix shows the value of True Negatives (TN), False Negatives (FN), False Positives (FP) and True Positives (TP). Combining these values, it is possible to extract other important metrics to evaluate how well the DMM performed.

The metrics used in this study were the accuracy, precision, and recall (also known as sensitivity or true positive rate). In this case study, accuracy refers to the proportion between the information that was correctly classified (with or without DR) and all the classified cases. Precision checks if the model is capable of detecting only patients with DR and can be defined as the ratio between TP and the instances classified as belonging to the positive class. Finally, sensitivity measures the ratio between the number of positive cases that were correctly classified and all the number of positive cases. This means that sensitivity demonstrates if the model has the ability to correctly identify all the sick patients. Equations Eq. 2, Eq. 3, and Eq. 4 show the metrics used.

$$Accuracy = \frac{TP + TN}{TP + TN + FP + FN} \tag{2}$$

$$Precision = \frac{TP}{TP + FP} \tag{3}$$

$$Sensitivity = \frac{TP}{TP + FN} \tag{4}$$

For a better understanding of the evaluation metrics in the context of the problem at hand, it was decided to present them in the form of questions. Hence, accuracy, precision and sensitivity provide answers, respectively, to the following questions:

- Accuracy: How many people were correctly labeled out of all the people?
- Precision: How many of those labeled has having DR had actually DR?
- Sensitivity: Of all people who have DR, how many of those were correctly predicted?

Table 1 represents the best ten results in terms of accuracy, sensitivity, and precision, according to each S, DMT and SM.

Table 1. Best results for each S, DMT and SM

DMM	S	DMT	SM	Accuracy	Sensitivity	Precision
1	S5	LR	SV (70%)	76.90%	67.40%	85.92%
2	S3	LR	SV (70%)	75.73%	65.19%	85.51%
3	S3	P	SV (70%)	75.73%	74.03%	78.82%
4	S4	LR	SV (70%)	75.15%	67.40%	82.43%
5	S2	LR	SV (70%)	74.85%	63.54%	85.19%
6	S2	LR	CV (7 folds)	74.54%	66.93%	81.61%
7	S5	LR	CV (10 folds)	74.45%	66.78%	81.59%
8	S5	LR	CV (7 folds)	74.37%	67.09%	81.27%
9	S2	LR	CV (10 folds)	74.36%	67.27%	81.05%
10	S3	LR	CV (10 folds)	74.36%	66.45%	81.67%

4 Discussion

As it can be seen from Table 1, the best accuracy obtained was 76.90% and the best precision value obtained was 85.92%, both recorded in the DMM 1, which uses the S5, with the LR algorithm and Split Validation (70%). As for sensitivity, the best value obtained was 74.03%, in DMM 3, represented by the S3, with the Perceptron algorithm and Split Validation (70%). When comparing all these metrics, it was found that the sensitivity presents values much lower in relation to the accuracy and precision, and the accuracy also presents values much lower than the precision. In this work, it was possible to understand the importance of the sensitivity metric in the health area, since the outcomes of classifying a patient with DR as a healthy individual are worse than diagnosing a normal patient with DR. Therefore, having a highly sensitive model is one of the main focuses of this work.

Regarding attributes, scenarios S2, S3 and S5 were the ones that had the greatest influence on the results. These three scenarios are similar, varying in just one or two attributes. It was found that the S4 scenario also showed reasonable results. This scenario has an attribute related to image quality, being the only scenario that includes attributes of this type, with the exception of the first scenario. This may explain why the scenario was not as influential as the others. On the other hand, scenario S1, which contained all the attributes, did not appear in any of the results presented, which means that there are attributes that do not add any predictive value. In summary, the most relevant attributes in the prediction of DR revealed to be the attributes related to the presence of MAs.

As for the DMT, only two of the six made it to the best results: LR and Perceptron. On the contrary, AutoMLP, SVM, JRip, and NN algorithms did not make it to the top results, suggesting that these techniques were not appropriated for this particular data and goal. As for LR and Perceptron, the first proved to be much better, since the Perceptron algorithm only appears in one of the ten best results. But it is important to bear in mind that, even so, it is the one with the highest value of sensitivity.

In terms of sampling methods used, both presented good results in general. It should be noted that the best results were obtained for Split Validation (70%) and that no results appear in the table for Split Validation (80%). Even so, the result was balanced, with five results with Split Validation and five results with Cross Validation as SM.

In short, the DMM1 (marked in bold), the one defined by S5, LR algorithm, and Split Validation (70%) reunited the best metrics combination above all, since it has the best accuracy, the best precision and the second best sensitivity (which is the most relevant metric to have in consideration, as mentioned before).

5 Conclusion

Healthcare organizations have been increasingly using DM for the treatment of the large amounts of complex data that is generated daily in clinical settings. DM assumes great importance in the healthcare domain as it has the power to uncover useful patterns in the data and to offer new knowledge to doctors and nurses, thus improving healthcare and the quality of the services provided to the patients.

In this study, the DM process was applied to real data from patients with DR. The objective was to create accurate DM models to predict whether or not patients have DR, according to the analysis of different factors and, finally, to help in the early detection of DR and, consequently, its treatment, in order to avoid that this disease reaches an irreversible point.

Various scenarios were tested, along with different DMT and SM. All the models achieved a reasonable performance and were quite similar to the values obtained in the scientific literature. The DMM with the best performance was described by a scenario containing the nine most relevant attributes selected by the Information Gain Ratio criterion (S5), which was composed by the attributes associated with the number of MAs and the detection of exudates, the LR algorithm, and the Split Validation method. The model with the best combination of evaluation metrics achieved an accuracy of 76.90%, a precision of 85.92% and a sensitivity of 67.40%. In this case study, although the results shown were similar to those obtained in the related work, they were not ideal since the sensitivity is a critical evaluation metric in the diagnostic context and the values obtained for this metric were lower than those expected for the model to be implemented in healthcare settings and to be able to successfully assist healthcare professionals in their decision-making.

In a future work, the objective would be to integrate new data containing a greater number of instances and also to explore new attributes and how they are correlated, in order to refine the DM model and, thus, achieve better results. Furthermore, although the precision value is relatively high, the values of accuracy and sensitivity are still not very satisfactory, so it would be pertinent to carry out more experiments using different parameters and DM techniques, in order to increase these values so that it is possible to build a high quality model. Thus, this model could later be implemented in clinical settings to support healthcare organizations in predicting patients who have DR.

Acknowledgment. This work has been supported FCT—Fundação para a Ciência e Tecnologia (Portugal) within the Project Scope: UIDB /00319/2020.

References

1. Palaniappan, S., Awang, R.: Intelligent heart disease prediction system using data mining techniques. In: 2008 IEEE/ACS International Conference on Computer Systems and Applications, pp. 108–115. IEEE (2008)
2. Ferreira, D., Silva, S., Abelha, A., Machado, J.: Recommendation system using autoencoders. Appl. Sci. **10**(16), 5510 (2020)
3. Neto, C., Brito, M., Lopes, V., Peixoto, H., Abelha, A., Machado, J.: Application of data mining for the prediction of mortality and occurrence of complications for gastric cancer patients. Entropy **21**(12), 1163 (2019)
4. Milovic, B., Milovic, M.: Prediction and decision making in health care using data mining. Arab. J. Bus. Manag. Rev. **1**(12), 126 (2012)
5. Vislisel, J., Oetting, T.A.: Diabetic retinopathy: from one medical student to another. Eye Rounds (2010)
6. Zhang, X., Saaddine, J.B., Chou, C.F., et al.: Prevalence of diabetic retinopathy in the United States, 2005–2008. JAMA **304**(6), 649–656 (2010)
7. Paetkau, M.E., Boyd, T.A.S., Winship, B., Grace, M.: Cigarette smoking and diabetic retinopathy. Diabetes **26**(1), 46–49 (1977)
8. Pinto, C.C., Silva, K.C., Biswas, S.K., Martins, N., Lopes De Faria, J.B., Lopes De Faria, J.M.: Arterial hypertension exacerbates oxidative stress in early diabetic retinopathy. Free Radical Res. **41**(10), 1151–1158 (2007)
9. Cho, H., Sobrin, L.: Genetics of diabetic retinopathy. Curr. Diab.Rep. **14**(8), 515 (2014)
10. Fong, D.S., Aiello, L., Gardner, T.W., et al.: Retinopathy in diabetes. Diab. Care **27**(Suppl. 1), s84–s87 (2004)
11. Moreno, A., Lozano, M., Salinas, P.: Retinopatia diabetica. Nutr. Hosp. **28**, 53–56 (2013)
12. Chun, M.Y., Hwang, H.S., Cho, H.Y., et al.: Association of vascular endothelial growth factor polymorphisms with nonproliferative and proliferative diabetic retinopathy. J. Clin. Endocrinol. Metab. **95**(7), 3547–3551 (2010)
13. Gadsden Eye Associates, Diabetic Retinopathy. https://gadsdeneye.com/diabetic-retinopathy/. Accessed 30 Oct 2020
14. Ockrim, Z., Yorston, D.: Managing diabetic retinopathy. BMJ **341**, c5400 (2010)
15. Sopharak, A., Dailey, M.N., Uyyanonvara, B., et al.: Machine learning approach to automatic exudate detection in retinal images from diabetic patients. J. Mod. Opt. **57**(2), 124–135 (2010)
16. Ananthapadmanaban, K.R., Parthiban, G.: Prediction of chances-diabetic retinopathy using data mining classification techniques. Indian J. Sci. Technol. **7**(10), 1498–1503 (2014)
17. Maliha, M., Tareque, A., Roy, S.S.: Diabetic retinopathy detection using machine learning (Doctoral dissertation, BRAC University) (2018)

18. Wirth, R., Hipp, J.: CRISP-DM: towards a standard process model for data mining. In: Proceedings of the 4th International Conference on the Practical Applications of Knowledge Discovery and Data Mining, pp. 29–39. Springer, London (2000)

19. Silva, C., Oliveira, D., Peixoto, H., Machado, J., Abelha, A.: Data mining for prediction of length of stay of cardiovascular accident inpatients. In: International Conference on Digital Transformation and Global Society, pp. 516–527. Springer, Cham (2018)

20. UCI Machine Learning Repository: Diabetic Retinopathy Debrecen Data Set Data Set. https://archive.ics.uci.edu/ml/datasets/Diabetic+Retinopathy+Debrecen+Data+Set. Accessed 30 Oct 2020

21. Messidor – ADCIS. https://www.adcis.net/en/third-party/messidor/. Accessed 30 Oct 2020

22. RapidMiner – Operators. https://docs.rapidminer.com/latest/studio/operators/rapidminer-studio-operator-reference.pdf. Accessed 30 Oct 2020

Prediction Models for Polycystic Ovary Syndrome Using Data Mining

Cristiana Neto[1] (ID), Mateus Silva[2] (ID), Mariana Fernandes[2] (ID), Diana Ferreira[1] (ID), and José Machado[1(✉)] (ID)

[1] Algoritmi Research Center, University of Minho, Campus of Gualtar, Braga, Portugal
{cristiana.neto,diana.ferreira}@algoritmi.uminho.pt,
jmac@di.uminho.pt
[2] University of Minho, Campus of Gualtar, Braga, Portugal
{a81952,a81728}@alunos.uminho.pt

Abstract. Polycystic Ovary Syndrome is an endocrine abnormality that occurs in the female reproductive system and is considered a heterogeneous disorder because of the different criteria used for its diagnosis. Early detection and treatment are critical factors to reduce the risk of long-term complications, such as type 2 diabetes and heart disease. With the vast amount of data being collected daily in healthcare environments, it is possible to build Decision Support Systems using Data Mining and Machine Learning. Currently, healthcare systems have advanced skills like Artificial Intelligence, Machine Learning and Data Mining to offer intelligent and expert healthcare services. The use of efficient Data Mining techniques is able to reveal and extract hidden information from clinical and laboratory patient data, which can be helpful to assist doctors in maximizing the accuracy of the diagnosis. In this sense, this paper aims to predict, using the classification techniques and the CRISP-DM methodology, the presence of Polycystic Ovary Syndrome. This paper compares the performance of multiple algorithms, namely, Support Vector Machines, Multilayer Perceptron Neural Network, Random Forest, Logistic Regression and Gaussian Naïve Bayes. In the end, it was found that Random Forest provides the best classification, and the use of data sampling techniques also improves the results, allowing to achieve a sensitivity of 0.94, an accuracy of 0.95, a precision of 0.96 and a specificity of 0.96.

Keywords: Data mining · Classification · CRISP-DM · Decision support systems · Health information systems · Polycystic ovary syndrome

1 Introduction

Polycystic Ovary Syndrome (PCOS) is the most common endocrine disorder in reproductive-age-women and it was first described in 1721 by Antonio Vallisneri [1]. In its classic form, PCOS can be characterized by chronic anovulation, irregular menses, and hyperandrogenism that may be associated with hirsutism, acne, seborrhea and obesity [1]. Due to the endocrine nature of this disease, it can cause long term consequences such as coronary disease, endometrial hyperplasia, and type 2 diabetes mellitus [5]. To

T. Antipova (Ed.): ICADS 2021, AISC 1352, pp. 210–221, 2021.
https://doi.org/10.1007/978-3-030-71782-7_19

further complicate the diagnosis of PCOS, polycystic ovaries have also been described in a high percentage of women who are absolutely normal with no alterations of the ovarian and/or the endocrine phenotype [1, 2]. Thus, this condition is characterized by heterogenous clinical and endocrine features, which has led to a considerable debate about its definition [3].

The criteria for the diagnosis of PCOS have yet to be agreed upon universally [4]. Therefore, International Classification of Diseases 9 (ICD-9) codes may not accurately capture the diagnostic criteria necessary for large scale PCOS identification [6].

Due to the large amount and the complex nature of data generated by transactions in healthcare environments, the interest of healthcare organizations in using Data Mining (DM) has been increasing since it can greatly benefit all parties involved. Namely, the analysis provided by DM allows healthcare institutions to improve operating efficiency while maintaining a high level of quality of care [7]. As health organizations generate and store large volumes of data every day, clinical decisions could be made not only based on doctor's intuition and experience but also based on hidden knowledge stored over time in healthcare databases [9]. Decision Support Systems (DSS) can be defined as the class of computer-based information systems that support decision making activities such as performing a medical diagnosis [8].

Considering the negative impact of PCOS in women's lives and the lack of universally accepted means of diagnosis, this paper has the main purpose of implementing DM techniques (DMT) in order to develop predictive models that are capable of forecast the presence of PCOS. This study used classification techniques with Machine Learning (ML) algorithms in order to predict the target class, which refers to the presence/absence of PCOS, potentially allowing to build a DSS that can help diagnose PCOS.

Regarding the structure of the present document, it is possible to identify in Sect. 2 the methodology, material and methods used during the development of this article, which includes the algorithms used and the steps followed. Then, Sect. 3 presents the results and the discussion. Finally, Sect. 4 outlines the main conclusions achieved with this work and some future work.

2 Methodologies, Material and Methods

The data manipulated in this article belongs to a dataset collected from 10 different hospitals in Kerala, India. The dataset has information relative to 541 patients, including demographic data, clinical history, and clinical analysis distributed in 40 different features [15].

The Cross Industry Standard Process for Data Mining (CRISP-DM) methodology was followed in this study, which is a hierarchical model that divides the process of data mining into six phases: Business Understanding, Data Understanding, Data Preparation, Modeling, Evaluation and Deployment [10].

In this study, Python and Scikit-learn were used. Also, the following classifiers were implemented to create the models: Support Vector Classification (SVM), Random Forest (RF), Multilayer Perceptron (MLP), Logistic Regression (LR) and Gaussian Naïve Bayes (GNB). The SVM algorithm used has a linear kernel and is based on the idea of finding a hyperplane that best divides a dataset into two classes [14]. The RF classifiers, like the name implies, consist of a large number of individual decision trees that operate as an ensemble [13]. MLP is a simple Artificial Neural Network (ANN) where data flows in one direction from input to output [15]. The ANN used has two hidden layers with 8 and 4 nodes, respectively. In addition, the activation function used was the rectified linear unit (relu) and the solver used for weight optimization was the Adam optimization, which is a stochastic gradient descent method that is based on adaptive estimation of first-order and second-order moments. The LR is a statistical model that in its basic form uses a logistic function to model, in this case, a binary dependent variable. The Naïve Bayes (NB) classifiers are a family of simple probabilistic classifiers based on applying Bayes' theorem with strong independence assumptions between the features [14]. In GNB, when dealing with continuous data, the continuous values associated with each class are distributed according to Gaussian distribution.

2.1 Business Understanding

The business goal of the work presented in this paper is to predict if a patient has PCOS or not, given a set of patient's clinical and lifestyle features. The prediction must be highly sensitive and accurate, taking into account its appliance to the healthcare domain. Furthermore, considering the negative impact of this disease in women's lives and the future consequences if it is wrongly diagnosed, we must assure the accuracy of the results.

2.2 Data Understanding

Each data instance of the 541 existing entries consists of a set of 41 attributes: serial number, patient file number, patient's age in years, patient's weight in kg, patient's height in cm, patient's body mass index (BMI), patient's pulse rate in bpm, respiratory rate in breaths/min, hemoglobin in g/dl, the length of fertile period in days, number of years the patient has been married, whether the patient has been pregnant before, number of abortions the patient had, follicle-stimulating hormone (FSH) value in mIU/mL, luteinizing hormone (LH) value in mIU/mL, FSH/LH ratio, patient's hip size in inches, patient's waist size in inches, waist/hip ratio, thyroid stimulating hormone (TSH) value in mIU/L, anti-mullerian hormone value (AMH) in ng/mL, prolactin (PRL) in ng/mL, Vitamin D3 in ng/mL, progesterone (PRG) in ng/mL, random blood sugar test (RBS) in mg/dl, whether the patient has gained weight, whether the patient had hair growth, whether the patient's skin tone has darkened, whether the patient has hair loss, whether the patient has gained pimples, whether the patient eats fast food, whether the patient practices regular exercise, systolic blood pressure in mmHg, diastolic blood pressure in mmHg, number of left follicles, number of right follicles, average size of left follicles in mm, average size of right follicles in mm, insulin levels in µIU/mL, endometrial thickness in mm and PCOS.

The target variable (PCOS) is binary and represents whether the patient has PCOS (PCOS = 1) or not (PCOS = 0). Figure 1 shows the data distribution of this variable on the used dataset and, as it can be observed, 364 of the instances did not had PCOS, which is equivalent to 67% of the dataset.

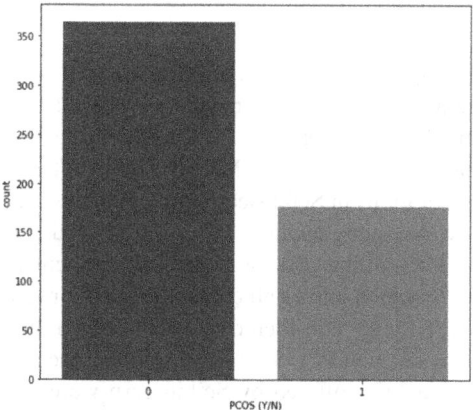

Fig. 1. Data distribution of the target variable 'PCOS'

2.3 Data Preparation

This phase of the CRISP-DM methodology involved the selection and preparation of the data that was used by the DM models. In this stage, Python and set of related libraries were used.

The first step was to remove attributes that were irrelevant in the forecasting process, which were the serial number and the patient's file number. Then, it was necessary to treat all missing or noisy data to ensure that there was no incomplete and inconsistent information. With regards to missing values, there were only 3 instances with missing attributes, so, due to the small number of cases, it was decided to remove those instances instead of using some method of value imputation.

Next, a correlation analysis was carried out between the PCOS target feature and the remaining attributes. It was observed that the insulin level attribute had a correlation with the target feature of 0.9, which was considered a too high value. Despite being an important attribute when it comes to classify the target class, allowing the achievement of accuracy values above 99%, this feature has been removed as it is almost linearly proportional to the target, which leads ML models to give a high importance to this feature and disregard the rest [11].

Once again, a correlation analysis was performed, but this time between all the features, and it was noted that some were highly correlated. First, weight and height were removed due to the great correlation between them and BMI. Then hip and waist due to the high correlation with waist/hip ratio. FHS and LH were also eliminated because they were highly correlated with FSH/LH. Lastly, the years of marriage attribute was also removed because it had a high correlation with age.

Finally, the attributes with continuous values were normalized, with the objective of placing them between the same interval ([0, 1]) and preventing the different intervals from undesirably influencing the results of the ML algorithms applied.

2.4 Modeling

This phase consisted of applying the ML algorithms using the prepared data. As mentioned above, 5 different classification techniques were used: SVM, RF, MLP, LR and GNB. These algorithms were developed in Python relying to Scikit-learn, which is an open-source library that implements a set of ML algorithms. These algorithms were used with the standard configurations in Scikit-learn implementations.

For each DMT e two sampling methods (SM) were tested: Holdout sampling, with 70% of the data used for training and the remaining amount for testing, and Cross Validation using 10 folds, which allows all data to be used for testing.

Different scenarios (S) were considered to assess which attributes were the most relevant to predict the presence of PCOS. In order to select the attributes that were used in each scenario, functionalities offered by Scikit-learn were used to perform feature selection. The functions used were *chi2*, which computes chi-squared stats between each non-negative feature and the class, *f_classif*, which computes the ANOVA F-value for the provided sample and *mutual_info_classif*, which estimates mutual information for a discrete target variable, that in turn measures the dependency between the variables.

The first scenario (S1) includes all the attributes. The scenarios S2, S3 and S4, used the top 10 features selected by *chi2*, *f_classif* and *mutual_info_classif*, respectively. To assess the influence of the number of selected features, two additional scenarios were created, one that selects only 5 features, S5, and another that selects 15 features, S6, both using *f_classif*. This process resulted in the following scenarios:

- S1 = {All Attributes}
- S2 = {'BMI', 'Cycle length(days)', 'AMH(ng/mL)', 'Weight gain(Y/N)', 'hair growth(Y/N)', 'Skin darkening (Y/N)', 'Pimples(Y/N)', 'Fast food (Y/N)', 'Follicle No (L)', 'Follicle No (R)'}
- S3 = {'FSH/LH', 'AMH(ng/mL)', 'Vit D3 (ng/mL)', 'PRG(ng/mL)', 'Weight gain(Y/N)', 'hair growth(Y/N)', 'Skin darkening (Y/N)', 'Fast food (Y/N)', 'Follicle No (L)', 'Follicle No (R)'}
- S4 = {'Cycle length(days)', 'FSH/LH', 'AMH(ng/mL)', 'PRL(ng/mL)', 'Weight gain(Y/N)', 'hair growth(Y/N)', 'Skin darkening (Y/N)', 'Fast food (Y/N)', 'Follicle No (L)', 'Follicle No (R)'}
- S5 = {'Weight gain(Y/N)', 'hair growth(Y/N)', 'Skin darkening (Y/N)', 'Follicle No (L)', 'Follicle No (R)'}
- S6 = {'Age (yrs)', 'BMI', 'Cycle length(days)', 'AMH(ng/mL)', 'Weight gain(Y/N)', 'hair growth(Y/N)', 'Skin darkening (Y/N)', 'Hair loss(Y/N)', 'Pimples(Y/N)', 'Fast food (Y/N)', 'Follicle No (L)', 'Follicle No (R)', 'Avg F size (L) (mm)', 'Avg F size (R) (mm)', 'Endometrium (mm)'}

It its interesting to note that the attributes 'Follicle No (L)', 'Follicle No (R)', 'AMH', 'Weight gain(Y/N)', 'hair growth(Y/N)', 'Skin darkening (Y/N)' and 'Fast food (Y/N)' where considered good predicters (in the top ten) by all three feature selection techniques.

To check if the removal of outlier values would have a positive impact on the result, S2 was also tested applying this condition. The technique used to detect these outliers was Interquartile Range (IQR). In addition, two data approaches were also tested in

S2, namely oversampling, where instances of the minority class are replicated, and undersampling, where random instances are selected to be kept among the class with the most instances.

In the end, 90 different models were compared and analyzed.

2.5 Evaluation

The application of different ML algorithms requires a process that ensures that the results are reliable and statistically significant. Consequently, performance metrics were applied to assure the evaluation of the quality and characteristics of the models, guaranteeing the reliability of the results. These metrics are numerical measures that quantify the performance of a given classifier [12].

The performance of each DMM was assessed through its confusion matrix, obtained with Scikit-learn functionalities, which is composed by the number of True Positives (TP), False Positives (FP), True Negatives (TN) and False Negatives (FN). With these values, it is possible to calculate accuracy, sensitivity, specificity and precision. These numerical metrics were obtained through the following the equations:

$$Accuracy = TP + TN/(TP + TN + FP + FN) \qquad (1)$$

$$Sensitivity = TP/(TP + FN) \qquad (2)$$

$$Specificity = TN/(TN + FP) \qquad (3)$$

$$Precision = TP/(TP + FP) \qquad (4)$$

In more detail, accuracy is the ratio of the correctly labeled instances to the whole pool of instances. Precision is the ratio of the correctly positive labeled by our algorithm to all positive labeled instances. Sensitivity or recall is the ratio of the correctly positive instances labeled by the algorithm to all those that did have PCOS. Specificity is the ratio between the number of correctly negative labeled instances and the all those that are, in fact, healthy. The main objective in the medical context is, in addition to good precision and specificity in order to avoid FP, to also have a good sensitivity to avoid FN, since assuming that the patient does not have the disease when she does have it can result in serious complications.

3 Results and Discussion

Table 1 and Table 2 present the scenarios that achieved the best results, by DMT, and for both SM, holdout sampling and cross validation, respectively.

As it can be observed there are no major differences, in general, between the best results obtained by the different algorithms in the different scenarios, however it is possible to note that better values of accuracy and specificity are achieved with the holdout sampling method, while sensitivity is higher using cross validation.

It is also important to note that only RF algorithm performed better with S1, that included all attributes. Overall, the other algorithms performed better in the scenarios with feature selection. Peculiarly, the GNB algorithm obtained its best results using only 5 features (S5).

The results obtained in these scenarios are satisfactory, since, of all the values, the best precision and sensitivity achieved were 0.906977 and 0.880682, respectively.

Table 1. Best results obtained, by technique, with Holdout SM

DMT	Accuracy	Precision	Sensitivity	Specificity
SVM	S3, S6	S6	S3	S6
	0.870370	0.906977	0.750000	0.962264
RF	S1	S1	S1	S1
	0.864198	0.886364	0.696429	0.952836
MLP	S5	S6	S3, S5	S6
	0.876543	0.904762	0.767856	0.962264
RL	S2	S2	S1, S2, S4, S6	S2
	0.845679	0.897436	0.625000	0.962264
GNB	S5	S5	S5	S5
	0.895062	0.842105	0.857143	0.915094

Table 2. Best results obtained, by technique, with Cross Validation SM

DMT	Accuracy	Precision	Sensitivity	Specificity
SVM	S4	S4	S1, S4, S6	S2, S3, S4
	0.907063	0.888889	0.818182	0.950276
RF	S1	S1	S2	S1
	0.888476	0.876623	0.784091	0.947514
MLP	S2	S5	S6	S5
	0.901487	0.882759	0.812500	0.953039
RL	S6	S6	S1, S6	S6
	0.907063	0.903846	0.801136	0.958564
GNB	S5	S5	S5	S5
	0.897770	0.820106	0.880682	0.906077

Table 3 presents the values obtained for each metric in S2, with and without the outlier removal using the both sampling methods. The best results for each metric observed in this table are highlighted in bold. Looking this table, it is possible to conclude that there are no significant differences that justify the removal of the outliers, because, although there is a slight improvement in the precision values, it is also observed that the sensitivity values worsen.

Table 3. Comparison between S2 and S2 with outlier removal with different DMT and SM

DMT	S	SM	Accuracy	Precision	Sensitivity	Specificity
SVM	S2	Holdout	0.851852	0.823333	0.714286	0.924528
		Cross validation	0.903346	0.887500	0.806818	**0.950276**
	S2 with outlier removal	Holdout	0.851852	0.863636	0.678571	0.943396
		Cross validation	**0.950276**	**0.899628**	**0.896104**	0.784091
RF	S2	Holdout	0.839506	0.826087	0.678571	0.924528
		Cross validation	0.882900	0.846626	0.784091	0.930939
	S2 with outlier removal	Holdout	0.845679	0.860465	0.660714	0.943396
		Cross validation	0.930939	0.886617	0.861635	0.778409
MLP	S2	Holdout	0.833333	0.784314	0.714286	0.896226
		Cross validation	0.901487	0.881988	0.806818	0.947514
	S2 with outlier removal	Holdout	0.827160	0.818182	0.642857	0.924528
		Cross validation	0.886617	0.857143	0.784091	0.936464
RL	S2	Holdout	0.845679	0.897436	0.625000	0.897436
		Cross validation	0.899628	0.891026	0.789773	0.953039
	S2 with outlier removal	Holdout	0.827160	0.850000	0.607143	0.943396
		Cross validation	0.895911	0.884615	0.784091	0.950276
GNB	S2	Holdout	0.858025	0.811321	0.767857	0.905660
		Cross validation	0.888476	0.808511	0.863636	0.900552
	S2 with outlier removal	Holdout	0.876543	0.833333	0.803571	0.915094
		Cross validation	0.899628	0.824468	0.880682	0.908840

Table 4 shows a comparison of the metrics obtained in S2, with and without under-sampling and oversampling, with the different SM. The best results for each metric observed in this table are highlighted in bold.

In these cases, we can notice a significant improvement in the values of the metrics relatively to the previous tests. It should be noted that, in all cases presented by this table, there was an increase in sensitivity values which is a very important metric in the current problem. To be noted that in the cases using oversampling and holdout sampling there was an increase in all the values of the metrics in all algorithms. The main objective of increasing precision and sensitivity was achieved with these methods.

In order to choose the most suitable model, a threshold was established, and the models were ranked according to their results. The defined threshold was sensitivity >92% and specificity and accuracy >90%. Table 5 presents the two models that achieved the threshold by their ranking order. Comparing all the cases, we can conclude that the best results are obtained with S2 when oversampling is applied, and the method of data splitting is holdout sampling.

Table 4. Comparison between S2 and S2 with different DMT, SM and DT

DMT	SM	DA	Accuracy	Precision	Sensitivity	Specificity
SVM	Holdout	No sampling	0.851852	0.823333	0.714286	0.924528
		Undersampling	0.877358	0.923077	0.842105	0.918367
		Oversampling	0.894495	0.950000	0.840708	0.952381
	Cross validation	No sampling	0.903346	0.887500	0.806818	0.950276
		Undersampling	0.866477	0.886228	0.840909	0.892045
		Oversampling	0.892265	0.896648	0.886740	0.897790
RF	Holdout	No sampling	0.839506	0.826087	0.678571	0.924528
		Undersampling	0.849057	0.886792	0.824561	0.877551
		Oversampling	**0.949541**	**0.963636**	0.938053	**0.961905**
	Cross validation	No sampling	0.882900	0.846626	0.784091	0.930939
		Undersampling	0.855114	0.869822	0.835227	0.875000
		Oversampling	0.932520	0.906494	0.964088	0.900552
MLP	Holdout	No sampling	0.833333	0.784314	0.714286	0.896226
		Undersampling	0.849057	0.918367	0.789474	0.918367
		Oversampling	0.885321	0.948980	0.823009	0.952381
	Cross validation	No sampling	0.901487	0.881988	0.806818	0.947514
		Undersampling	0.860795	0.875740	0.840909	0.880682
		Oversampling	0.889503	0.883152	0.897790	0.881215
RL	Holdout	No sampling	0.845679	0.897436	0.625000	0.897436
		Undersampling	0.877358	0.923077	0.842105	0.918367
		Oversampling	0.876147	0.957447	0.796460	0.961905
	Cross validation	No sampling	0.899628	0.891026	0.789773	0.953039
		Undersampling	0.843750	0.866667	0.812500	0.875000
		Oversampling	0.871547	0.874652	**0.967403**	0.875691
GNB	Holdout	No sampling	0.858025	0.811321	0.767857	0.905660
		Undersampling	0.839623	0.857143	0.842105	0.836735
		Oversampling	0.871560	0.920792	0.823009	0.922310
	Cross validation	No sampling	0.888476	0.808511	0.863636	0.900552
		Undersampling	0.868477	0.864407	0.869318	0.863636
		Oversampling	0.864641	0.852941	0.881215	0.848066

It can be observed that these two models used the same ML algorithm but used different sampling methods and oversampling data.

Within the two possibilities, the chosen ML model is the RF in the S2 scenario, with oversampling and with holdout sampling, which despite having less sensitivity, has

Table 5. Best results obtained

DMT	S	SM	DA	Accuracy	Precision	Sensitivity	Specificity
RF	S2	Cross validation	Oversampling	0.932520	0.906494	0.964088	0.900552
RF	S2	Holdout	Oversampling	0.949541	0.963636	0.938053	0.961905

greater accuracy, precision and specificity. This model has an accuracy of 0.949541, a precision of 0.963636, a sensitivity of 0.938053 and a specificity of 0.961905.

In Fig. 2 it is possible to see another metric to evaluate the chosen algorithm that consists of the Receiver Operator Characteristic (ROC) curve, that is created by plotting the true positive rate (sensitivity) against the false positive rate (specificity). The corresponding Area Under the Curve (AUC) has a value of 0.982.

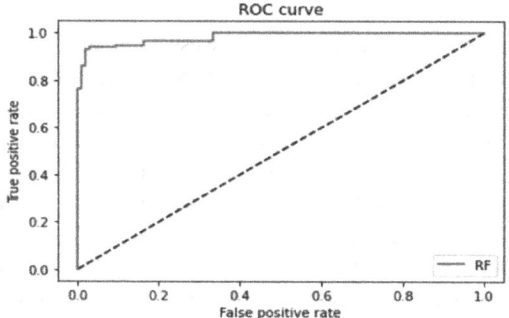

Fig. 2. ROC curve for RF in S2 with oversampling and using holdout sampling

In addition, Fig. 3 shows an additional metric that consists of another curve, the Precision-Recall (PR) curve, that shows how precision varies with sensitivity (recall) for the best model.

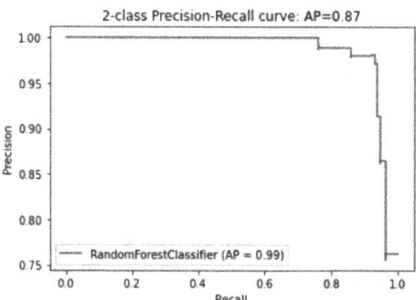

Fig. 3. Precision-Recall curve for RF in S2 with oversampling and using holdout sampling

4   Conclusions and Future Work

DM applied to medicine seems to be one of the technological revolutions of the coming years. It offers the possibility of discovering unknown patterns in the data collected by health organizations and that can be used by DSS for diagnosis, serving as support for health professionals. The focus of this paper was to predict the existence of PCOS, which, as mentioned, is a disease that is very prevalent in reproductive-age-women and that has a great impact in their lives. The best model implemented RF and achieved a sensitivity of 0.94, an accuracy of 0.95, a precision of 0.96 and a specificity of 0.96, which are considered satisfactory results. This model fits into the second scenario, which uses the *f_classif* for feature selection with holdout sampling and oversampling data. The features used by the referred model are: 'BMI', 'Cycle length(days)', 'AMH(ng/mL)', 'Weight gain(Y/N)', 'hair growth(Y/N)', 'Skin darkening (Y/N)', 'Pimples(Y/N)', 'Fast food (Y/N)', 'Follicle No (L)', 'Follicle No (R)'. The use of this model would give healthcare professionals a quality forecast with a low percentage of error in diagnosing PCOS. The forecast offered by this model could be combined with other indicators to offer a more reliable diagnosis.

For future work, it would be important to assess the quality of the prediction using only the patient's lifestyle data and easily obtainable clinical data, such as blood analysis, instead of using features that are obtained by performing more complex medical examinations. This system could serve as a primary indicator to help doctors decide if there is a need to perform complex exams. In addition, more DM models could be tested, changing the parameters used in this study and, as the best model uses oversampling, new instances with PCOS could be added to the dataset.

It is also important to note that the conclusions drawn here should not be generalized, as the data is from a very restricted population whose characteristics may vary in other regions. Hence, it would be valuable to conduct a study with a bigger population, ideally, using data collected across the world, as well as to study the variability of the research results over time.

Funding.   This work has been supported FCT—Fundação para a Ciência e Tecnologia (Portugal) within the Project Scope: UIDB /00319/2020.

References

1. Speca, S., Napolitano, C., Tagliaferri, G.: The pathogenetic enigma of polycystic ovary syndrome. J. Ultrasound **10**(4), 153–160 (2007)
2. Polson, D., Wadsworth, J., Adams, J., Franks, S.: Polycystic ovaries—a common finding in normal women. Lancet **331**(8590), 870–872 (1988)
3. Franks, S.: Polycystic ovary syndrome: a changing perspective. Clin. Endocrinol. **31**(1), 87–120 (1989)
4. Balen, A., Conway, G., Kaltsas, G., Techatraisak, K., Manning, P., West, C., Jacobs, H.: Andrology: polycystic ovary syndrome: the spectrum of the disorder in 1741 patients. Hum. Reprod. **10**(8), 2107–2111 (1995)
5. Soni, P., Vashisht, S.: Exploration on polycystic ovarian syndrome and data mining techniques. In: 3rd International Conference on Communication and Electronics Systems (ICCES), pp. 816–820 (2018)

6. Castro, V., Shen, Y., Yu, S., Finan, S., Pau, C., Gainer, V., Keefe, C., Savova, G., Murphy, S., Cai, T., Welt, C.: Identification of subjects with polycystic ovary syndrome using electronic health records. Reprod. Biol. Endocrinol. **13**(1), 116 (2015)
7. Morais, A., Peixoto, H., Coimbra, C., Abelha, A., Machado, J.: Predicting the need of neonatal resuscitation using data mining. Procedia Comput. Sci. **113**, 571–576 (2017)
8. Power, D.: Decision Support Systems. Quorum Books, Westport (2002)
9. Abirami, N., Kamalakannan, T., Muthukumaravel, A.: A Study on Analysis of Various Datamining Classification Techniques on Healthcare Data (2013)
10. Chapman, P., Clinton, J., Kerber, R., Khabaza, T., Reinartz, T., Shearer, C., Wirth, R.: CRISP-DM 1.0 Step-by-step data mining guide (2001)
11. Sumalatha, M., Ananthi, M., Arvind, A., Navin, N., Siddarth, C.: Highly correlated feature set selection for data clustering. In: 2014 International Conference on Recent Trends in Information Technology, pp.1–4 (2014)
12. Reis, R., Peixoto, H., Machado, J., Abelha, A.: Machine learning in nutritional follow-up research. Open Comput. Sci. **7**(1), 41–45 (2017)
13. Neto, C., Brito, M., Lopes, V., Peixoto, H., Abelha, A., Machado, J.: Application of data mining for the prediction of mortality and occurrence of complications for gastric cancer patients. Entropy **21**(12), 1163 (2019)
14. Silva, E., Cardoso, L., Portela, F., Abelha, A., Santos, M.F., Machado, J.: Predicting nosocomial infection by using data mining technologies. In: New Contributions in Information Systems and Technologies, pp. 189–198. Springer, Cham (2015)
15. Kottarathil, P.: Polycystic ovary syndrome (PCOS), Version 3. https://www.kaggle.com/pra soonkottarathil/polycystic-ovary-syndrome-pcos. Accessed 29 Oct 2020

Latent Failures of the Individual Human Behavior as a Root Cause of Medical Errors

Yuriy Voskanyan[1] , Irina Shikina[3] , Fedor Kidalov[2] , Saida Musaeva[4(⊠)] ,
and David Davidov[2]

[1] Russian Medical Academy of Continuing Professional Education of the Ministry
of Health of Russia, Bld. 1, 2/1 Barrikadnaya St., Moscow 125993, Russia
[2] Moscow State Budgetary Institution "Information and Analytical Center of Healthcare",
Bld. 10, Basmannaya Novaya St., Moscow 107078, Russia
[3] Central Research Institute for Organization and Informatization of Health Care at the Ministry
of Health of Russia, Bld. 11, Dobrolyubova str., Moscow 127254, Russia
shikina@mednet.ru
[4] Kazan Federal University, 18 Kremlyovskaya ulitsa, Kazan 420008, Russia
saida.musaeva.r@gmail.com

Abstract. The article has analyzed the main aspects of individual human behavior
and its regulation in a three-contour context (individual, social and situational).
It was shown that the visible phase of human behavior (the behavioral act) is
a result of preceding complex psychological processes of interaction between
mental thinking, emotions, mindset and motivation. The attention condition and
stress level affect almost every subprocess of human behavior. The authors present
the genesis of the medical errors at each subprocess of human behavior: perception
(illusion and hallucination), information processing (cognitive illusion), decision
making and planning (mistake), implementation (slip, laps and tripping). The work
demonstrates the main types of violations in individual human behavior and also
the fundamental differences between violations and errors. Taking into account the
complexity of individual human behavior and the three contours of its regulation,
the main principles to prevent medical errors and violations were formulated.

Keywords: Medical help safety · Human behavior · Medical errors · Violations

1 Introduction

The rate of adverse events associated with medical errors and violations (preventable
events) is 45.5% [1, 2, 21, 25, 28, 29]. The systemic model for safety evaluates medical
errors as a result of normal, not aberrant psychological processes, leading to behavioral
acts that are not in the context of the situation. This "abnormal" behavioral act is based
on constant replenishment of roots or latent failures that are associated with individual
human behavior, teamwork, struggles within the organization and also with external
factors [3–6, 22, 26].

There are two fundamental and important subprocesses of individual human behavior: invisible (psychological) and visible (physiological phase or behavioral act). Human

© The Author(s), under exclusive license to Springer Nature Switzerland AG 2021
T. Antipova (Ed.): ICADS 2021, AISC 1352, pp. 222–234, 2021.
https://doi.org/10.1007/978-3-030-71782-7_20

behavior is regulated in the three-contour context (individual, collective and situational), not followed by the strict logical reasons but being the result of interaction between mental thinking, emotions, mindset and motivation [7–10].

Human errors, which are unintentional actions and not related to the situation context, are divided into categories: mistakes (as a result of false perception, defective information processing, decision making and planning) and wrong actions with correct decision making and planning (slip, lapse and tripping). Unlike human errors, the violations are the result of the intentional actions of the staff. It is important to know three main types of violations: routine, associated with intentional actions of perpetrators, and violations that are out of the scope of the perpetrator (situation violations and violations forced by the organization) [5, 7, 10, 11, 24].

The aim of the study was to analyse the root causes associated with individual human behavior, as a potential source of human errors in the field of medical care.

2 Materials and Methods

The study represents the literature review on the problem of individual human behavior regulation, as a root cause of medical errors and violations. The information search was conducted by three independent researchers during the period 1980–2020 using the medical databases MEDLINE, Cochrane Collaboration; EMBASE, SCOPUS, ISI Web of Science. The prospective and retrospective observational studies with high methodological quality, analytical digest and original scientific articles were evaluated for the analysis.

The sample included literature with the fundamental issues of human behavior, original articles that studied the root causes, frequency, severity and preventability of adverse events in short-stay hospitals, as well as issues of quality and safety management of medical care in medical organizations. As a result, 179 scientific publications were reviewed, 31 of which were selected for the purposes of this study. The selection criteria for the literature were the presence of the relationship and interdependence of issues related to individual human behavior and the following genesis of errors and violations, as well as the effectiveness of organizational and administrative decisions.

3 Results

3.1 Perception and Information Processing

The visible and invisible subprocesses of human behavior are parts of a continuous cycle [7–10, 12], where the invisible part includes perception, information processing, decision making and planning, and visible is a behavioral action or the action that leads to the specific result, which is perceived by a person as signals (Fig. 1).

The first subprocess of human behavior – perception, which leads to the integration of the received sensory information with previously processed information and other sensory data and aims to form a cognitive model of the world – mental model. Perception helps a person to be environmentally oriented. Perception is always subjective. It aims at the feasibility of intention (meeting the needs to save the viability) and not at the truth.

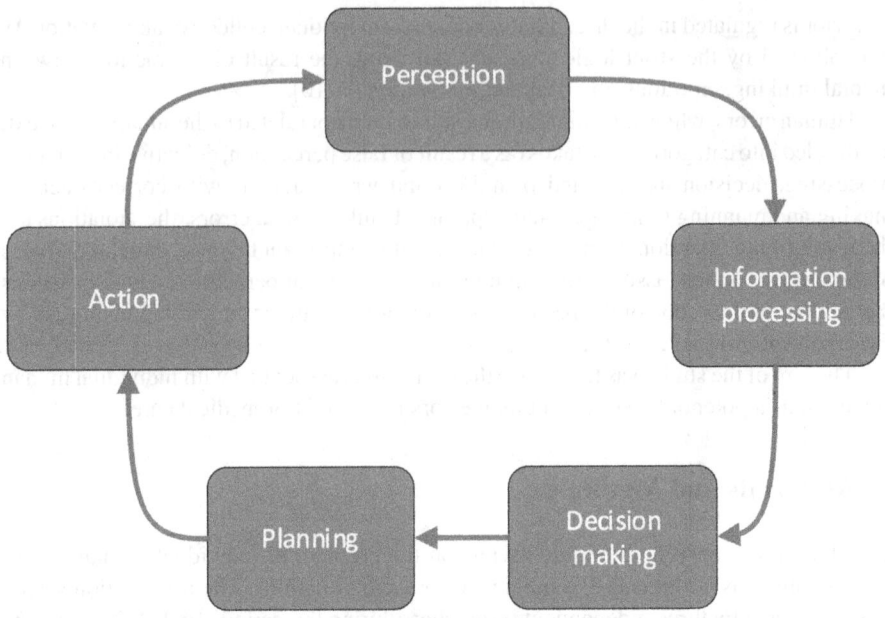

Fig. 1. Illustration of the cycle for subprocesses of human behavior

People build up the relevant sector of reality from not completely perceptive sensory signals, which are necessary for their survival, in the first place [9, 12].

Perception includes three continuous stages: registration [32], identification and interpretation. During the registration, the sensation is forming by sensory organs and subcortical areas. The absolute and relative thresholds impact the registration stage. The first one determines the range of the external stimulus to be detected (depends on the biology and changes during adaptation and fatigue). The second one defines how two stimuli should be different from each other so that they could be distinguished (depends on the level of attention, personal motives and other signals, e.g. noise) [9, 10, 12].

Identification process is characterized by integration of sensation with previously processed information from the memory as mental setups, hypotheses. Identification is under the influence of present personal motives. Mental setups include expectations, emotions and beliefs related to the object (events). They determine the pattern to detect sensation ("we see what we used to see and we hear what we used to hear"). Hypotheses that are based on the knowledge and rules of the memory classify the perceived object (occasion) as the well-known category of events (occasions). Hypotheses are retrieved from the memory under the influence of mental setups, particularly expectations. The result of identification is the complete situation image named "Gestalt". Gestalt is a figure (image) separated from the background (space). Gestalt is not a result of the full imaging of the surrounding reality but an image of united relevant parts in one that was completed in one's mind under the influence of mental sets, hypotheses and personal motives. Competing motives (for example, willingness to sleep) form the Gestalt suitable for the unfulfilled need [7, 8, 13].

At the stage of interpretation, the perception process enters the mind for the first time. The result is a meaningful image of the situation - a concept that answers the question "what is it?". In this case, the figure turns into a conscious object and the background into a semantic context. Interpretation is influenced by mental sets, hypotheses and personal motives, retrieved from the memory [7, 9, 10].

Information processing is the integration of a meaningful image (concept) with knowledge, rules and skills retrieved from the memory. As a result, a mental model of the situation is formed (a cognitive image of the sector of reality in causal relationships) and it answers the question "why is it going this way and what is the forecast?" Information processing is influenced by emotions, mental sets retrieved from memory, existing and emerging motives. Information processing follows certain protocols: skills and rules are retrieved from the memory in the first place, only then - the knowledge (the principle of the economy); selection of the most possible and consistent alternative (the principle of avoiding ambiguity and uncertainty); building a mental model of the situation, which is most related to the existing knowledge, rules and skills of the memory (the principle of protecting one's competence) [8, 14–16, 23].

The processes of perception (identification, interpretation) and information processing involve two essential parts: remembrance and impression.

Remembrance is the retrieval of relevant information of the memory as concepts, diagrams and mental models. Remembrance, as part of identification, interpretation and information processing, is influenced by mental sets, emotions, existing and new motives, and also follows several key principles: remembering the information that requires the least effort (economy); remember the information associated with the dominant motive (importance); an event that happened recently is recalled better (chronological proximity); events that happened at the same time are recalled better (association); frequent and more possible events are recalled better than those that are unlikely to happen and rare; obvious and unique events are recalled faster [7–10].

An impression is the process of subconscious evaluation of incoming information within neural nets. The impression ends up in the emotions as a result of holistic, fast and subconscious information processing. The volume of processed information as emotions is significantly larger than that of consciously processed information, which has great biological importance. In some cases, the emotional and cognitive evaluation of the situation does not match (mind and intuition contradict each other). Emotions in a behavioral act are implemented at five levels: readiness for action (arousal); the level of differentiation of perception and information processing; suppression of other motives; the degree of externalization (openness of intentions for the others); living through the consequences (relief or frustration) [7–10].

3.2 Decision Making, Planning, Behavioral Act

Decision making, planning, behavioral act were studied thoroughly by J. Rasmussen from the Technical University of Eindhoven [17], according to which, at this point, human behavior has three levels of abstraction: based on skills (to solve routine tasks); based on rules (to solve the well-known problems); based on knowledge (to solve unknown problems).

Decision making is a part of associative and analytical thinking, which defines a common goal (goal setting) and chooses an adequate method to achieve the goal (choosing a solution or choosing an alternative). Behaviour-based on skills turns on when problems do not exist. At the same time, there is no "choosing an alternative" stage, as the decision itself is embedded in the sensory-motoric pattern of behavior. In the case of a well-known problem, the solution is based on rules. An unknown problem requires the full involvement of consciousness to identify the problem, to compare and choose alternatives [7, 10, 14–16].

Planning is an approach by imagination to achieve a goal, turning into a plan, which is a virtual model of how to proceed. The plan includes a general goal, intermediate goals (checkpoints), tasks and conditions for completing the tasks (perpetrators, place, time, resources). Behaviour-based on skills has no planning stage since a routine situation generates an automatic sensory-motoric pattern [31] of behavior as a response. Rule-based behavior requires a known plan (a script) retrieved from the memory to solve a known problem. An unknown problem requires the development of a completely new plan, considering the risks, uncertainties and existing restrictions [7, 8, 10].

A behavioral act is a visible part of behavior, which represents consecutive and parallel actions in accordance with the selected plan. In general, a behavioral act is a set of implemented skills in the structure of sensory-motoric patterns, scripts and plans [7, 8, 10, 15].

3.3 Regulation of Human Behavior

Individual human behavior is regulated at the psychological and physiological levels. The latter includes the regulation of the visible part of human behavior (behavioral act) and regulation of the main physiological processes building the functions of the central and peripheral nervous system and also the sense organs [7, 10].

Psychological regulation of behavior exists in a three-contour context: individual, social and situational [7, 10, 14–16].

The individual context of psychological regulation of behavior is provided by the following components:

- Self-perception: self-image (traits and qualities), self-esteem, self-control, focus on control, self-efficacy;
- Psychotype (personality typology based on dispositional characteristics);
- Personal motives: competing motives, problems, illnesses;
- Value system (components of the social system, which particularly important in the individual and collective mind and do not depend on motivation);
- Emotional intelligence: self-awareness, self-regulation, motivation, empathy, social skills;
- Individual competence: skills, rules, knowledge, experience;
- State of attention;
- Stress level.

The social context of individual human behavior is provided by the following components:

– Group perception of a personality: a person's vision of the group and the group's vision of the person
– The individual's perception of the values and standards accepted by the group;
– Interpersonal relationships;
– Commitment to the organization;
– Trust in the directory;

The situational context of individual human behavior is provided by the following components:

– Disposition - a person's preparedness for a certain behavioral act in a certain situation;
– Sets - consistent patterns of behavior in a particular situation, based on the dispositions (see mental sets).

3.4 Classification and Origin of Human Errors and Violations

Human error is a group of unintentional human actions that lead to a result, different from the goal and not correlating with the context of the situation.

Human errors (including medical errors) can be the result of a wrong decision or plan (miscalculation), or an incorrect behavioral act with the right decision and plan [5, 11, 30].

A miscalculation can be the result of a perception error, an information processing error, errors in decision making and building a plan (errors as a result of using rules and knowledge).

There are two groups of errors that could happen during the perception process [5, 7, 10, 11, 27]:

– Illusion (distorted image of the real world): a distorted object, or an existing object that is missing in perception;
– Hallucination (perception of a nonexistent object in a particular place) - a Gestalt that does not exist in reality is retrieved from the memory.

Table 1 summarizes the main root causes that could lead to having illusions and hallucinations.

Table 1. Root causes of illusions and hallucinations

Illusion – distorted object; Hallucination			Illusion – missing object	
Incomplete sensations (defect of registration)	Incorrect Gestalt (defect of identification)	Incorrect conception (defect of interpretation)	Elevation of the absolute threshold	Elevation of the relative threshold
- Attention-deficit - Stress - Competency deficiency (conscious restriction of the information flow) - Elevation of the relative threshold (noise)	- "Dangerous" mindsets[a] - Difficulty in the interface of the working place - Visual resemblance - Similar names of the objects - Stress	- Mindsets (especially expectations) - Emotions - Competency deficiency - Attention-deficit	- Tiredness (fatigue) - Sleepiness - Illnesses and traumas	- Attention-deficit - Personal motives - Other sensory signals (noise)

[a] "Dangerous" mindsets: "macho", "anti-authority", "impulsivity", "invulnerability", "a sense of humility"

At the information processing, the error results in an inadequate mental model (cognitive illusion), which does not correlate to causal linkages in the real world [7, 10]. Table 2 summarizes the main types of inadequate mental models and their main root causes.

Table 2. Root causes of inadequate mental models

Fixed mental model (linked to the existing in the memory mental model)	Simplified mental model (highly simple mental model to explain the complex situation)	Incorrect mental model
- Fixation error[a] - Methodism[b] - Competing motive (for example, protection of a sense of self-competence)	- Principles of the cognitive economy[c] - Methodism	- Competency deficiency - Incorrect evaluation of possibilities

[a] Fixation error – adherence to one mental model, despite the certainty of its failure;
[b] Methodism – "blind" adherence to the rules for determining cause-effect linkages;
[c] Principles of the cognitive economy: simplicity (choosing the most "simple" and well-known mental model); avoiding ambiguity (choosing the most "understandable" mental model) and uncertainty (choosing the most "predictable" mental model).

Miscalculation can form an adequate mental model. In this case, it will be a consequence of defects at the decision making and planning as two abstractions - rules and knowledge. Errors in the application of the rules are always more severe since the confidence to apply a "good" rule strictly limits the possibilities for permanent control of intermediate states. At the same time, an error in the implementation of knowledge has much less severe consequences, since the uncertainty of the situation requires more careful monitoring of the further developments [5, 7, 10, 11, 30]. Table 3 summarizes the main root causes of errors in decision making and planning.

Table 3. Root causes of inadequate decision making and planning

Poor targets	Poor methods	Poor plans
- Lack of goals (ideas instead)	- The chosen method is not	- Methodism instead of
- Lack of priorities for	suitable for achieving goals	planning *(using the proper*
multiple goals	*(unsuitable rule, unsuitable*	*plan in an unsuitable*
- The conflict of contradicting	*implementation of the suitable*	*situation)*
goals are not considered	*rule, knowledge deficit)*	- Fixed plan instead of a
- Unclear goals		flexible one
- The impact of conflicting		- Lack of a backup plan
motives		- Poor plan concretization in
- Goal-setting without		terms of aims, time and
sufficient information		resources
- Fixation on the negative		- Improper risk and ambiguity
goals		evaluation
- Unrealistic goals		

Errors in the implementation phase lead to an incorrect behavioral act when the endpoint is different from the result made by the correct decision and plan. Table 4 summarizes the main types of errors at the implementation phase and their root causes [7, 10, 18, 30].

Attention is the concentration of perception and thinking onto the object. Attention includes two parallel actions (split attention): focalization and background control. Focalization (concentration) fixes relevant information with the full integration of mind and is accompanied by the elevated threshold of perception for other signals and suppression of other motives. Focalization exists as selective attention (selective concentration) or vigilance (close attention over a long period of time). Background control is a constant scanning of the environment for the new relevant information; it is implemented within the neural networks with minimal involvement of mind and is experienced as emotions [7, 10]. The causes of two main forms of attention deficit are presented in Table 5.

Stress is a state of physical and mental activity as a response to the imbalance between the situation demands, personal capabilities and available resources [7, 10]. Stress develops in response to acute and chronic stressors, causing equal responses in the human body (Table 6).

Table 4. Root causes, associated with an incorrect behavioral act

Slip	Lapse	Tripping
- Attention-deficit	- Stress - Forgetting[a] - Lack of situational awareness[b] - Lack of a scope on the consequences[c]	- "Raw" sensory-motoric pattern of behavior[d]

[a] *Forgetting – the loss of linkages between the schemes in the memory and the inability to connect them with the new information perceived;*
[b] *Lack of the situation awareness – lack of monitoring the situation (updating information) during the repeated perception;*
[c] *Lack of a scope on the consequences – the lack of control perception of the object state (the result of a behavioral act);*
[d] *"Raw" sensory-motoric pattern of behavior – the action is processed within the ability level instead of the skill level when it requires mental control to complete it.*

Table 5. Attention-deficit

Attention-deficit	Full focus *(excessive concentration on the object, leading to the loss of background control)*
- Tiredness (fatigue) - Sleepiness - Monotony	Stress

Table 6. Stress response

Adequate	Inadequate	
Acute stress	*Acute stress*	*Chronic stress*
Coping-behavior (overcoming): - *Active (fighting)* - *Passive (avoiding the situation)*	- Full focusing of attention onto the stressor - Choosing simple alternatives[a] - Methodism - Excessive externalization[b] - Refusing the self-reflection	Professional burnout: - Emotional breakdown - Depersonalization - Loss of self-esteem

[a] *Choosing simple alternatives – using what the person knows best, instead of what would be best;*
[b] *Excessive externalization – the main emphasis is on the person's behavior and not on goals.*

It is possible to assume, while considering the complexity of the three-contour context of psychological regulation of behavior, that the miscalculations portion leading to incorrect decisions and plans will prevail in the structure of medical errors [7, 10].

Violations, unlike mistakes, are always intentional, meaning that an adequate mental model of the situation is accompanied by decision making and planning contradicting the existing and known to the performer rules of behavior in a particular situation [7, 10, 19]. Table 7 summarizes the main types of violations and their root causes.

Table 7. Types of violation

Routine (defined by the perpetrator's behavior)	Forced by the organization	Situational
- Personal motives (Competing motives, problems, illnesses) - Value system - Psychotype - Excessive qualification *(the current job position does not allow existing competencies to develop)*	- Defects in the management *(poor rule, poor management decision)* - *Defects in the* support system *(lack or deficit of the resources)*	- A deficit in time *(time-pressing)* - Sudden rise in resource needs

3.5 Modern Strategy for Preventing Active Failures Associated with Staff Performance (Errors and Violations)

Considering the very limited possibilities to influence the psychological regulation of behavior in the majority of cases, the prevention of medical errors and their consequences

Table 8. The strategy for preventing medical errors

Reducing the probability	Early detection	Preventing the incident
- Team-building - Additional functions[a] - Duplication of functions[b] - Formalization of routine operations (rules) - Automatization of routine operations (in the information systems) - Eliminating organizational failures that lead to attention deficit and stress	- Constant preparedness of a failure - Team-building - Additional functions - Duplication of functions	- Error acknowledgement - Team-building and using the team resources - Response algorithm for the error (rules) - Continuous repetition of the response algorithms

[a]*Additional (backup) functions: checklists, technical and behavioral insurance, decision support system;*
[b]*Duplication of functions – cross control in the teamwork (increasing the frequency of modality), multimodal control (increasing the number of modalities).*

is achieved by team building, proper management amendments and redundancy in the medical care system, which requires additional time and resources [2, 7, 10, 18, 31, 32]. Table 8 summarizes the main directions for preventing medical errors.

Preventing the violations is achieved in most cases with the elimination of the administrative failures at the level of directory and support systems, and also by making correct management decisions [7, 10, 19, 20]. This strategy allows to avoid revealing violations by government auditors [30]. Table 9 summarizes the main approaches to prevent various types of violations.

Table 9. The strategy for preventing the violations

Routine violations	Violations forces by the organization	Situational violations
- Negative support - The authority based on force - Eliminating organizational failures that lead to attention deficit and stress	- Optimization of the support system - Optimization of the management system - Negative support for the leader	- Algorithm for an urgent increase in productivity (rules) - Constant repetition of the algorithm for an urgent increase in productivity

4   Conclusion

Individual human behavior is a very complex management object. A fair desire to guide human behavior into the right context is associated with minimal opportunities to influence any of its stages, especially the perception and information processing, decision making and planning. With regard to the above, preventing active failures associated with the individual human behavior (errors and violations) should be based on team building, proper management amendments and redundancy in the medical care system. Modern strategy for preventing active failures associated with the staff performance (errors and violations) Modern strategies for managing medical errors are primarily aimed at reducing the likelihood of their occurrence, early detection and prevention of an incident.

These fundamental principles of managing errors and violations that are based on understanding their root causes could form the educational basis for the staff of medical organizations on quality and safety management because this process is rather declarative or normative as for now. Understanding the root causes and the modern strategies for active failures management that emerge from this knowledge will significantly help to increase the commitment of healthcare professionals to fair actions for safety, modern methods of managing failures and vulnerabilities in healthcare, and to participate in the development of new effective solutions.

The main limitation of the study is the lack of experimental confirmation of the conclusion on improving the effectiveness of quality and safety management after changing the way of teaching this section by the staff of medical organizations based on understanding the root causes of errors and violations. Conducting prospective clinical and psychological research in this direction represents a great prospect in replicating the modern strategy for quality and safety management of medical care.

References

1. Zegers, M., Bruijne, M.C., Wagner, C., et al.: Adverse events and potentially preventable deaths in Dutch hospitals: results of a retrospective patient record review study. Qual. Saf. Health Care **18**(4), 297–302 (2009). https://doi.org/10.1136/qshc.2007.025924
2. Voskanyan, Y., Shikina, I., Kidalov, F., Davidov, D.: Medical care safety - problems and perspectives. In: Integrated Science in Digital Age, Batumi, pp. 291–304 (2019). https://doi.org/10.1007/978-3-030-22493-6_26
3. Macchi, L., Pietikäinen, E., Reiman, T., Heikkilä, J., Ruuhilehto, K.: Patient safety management: available models and systems. VTT Working Papers, no. 169. VTT Technical Research Centre of Finland, Espoo (2011)
4. Lawton, R., Carruthers, S., Gardner, P., Wright, J., McEachan, R.R.: Identifying the latent failures underpinning medication administration errors: an exploratory study. Health Serv. Res. **47**(4), 1437–1459 (2012). https://doi.org/10.1111/j.1475-6773.2012.01390.x
5. Reason, J.: Human error: models and management. Bmj **320**(7237), 768–770 (2000). https://doi.org/10.1136/bmj.320.7237.768
6. Hoffmann, B., Rohe, J.: Patient safety and error management: what causes adverse events and how can they be prevented? Deutsches Arzteblatt Int. **107**(6), 92 (2010). https://doi.org/10.3238/arztebl.2010.0092
7. St Pierre, M., Hofinger, G., Simon, R.: Crisis Management in Acute Care Settings: Human Factors and Team Psychology in a High-Stakes Environment. Springer, Heidelberg (2016)
8. Mattsson, S., Fast-Berglund, Å., Stahre, J.: Managing production complexity by supporting cognitive processes in final assembly. In: Swedish Production Symposium 2014, SPS14, 16–18 September 2014, Gothenburg, pp. 16–18 (2014)
9. Akintunde, E.A.: Theories and concepts for human behavior in environmental preservation. J. Environ. Public Health **1**(2), 120–133 (2017)
10. Mitchell, L.: Safer Surgery: Analysing Behavior in the Operating Theatre. CRC Press (2017)
11. Scherer, D., Maria de Fátima, Q.V., Neto, J.A.D.N.: Human error categorization: an extension to classical proposals applied to electrical systems operations. In: IFIP Human-Computer Interaction Symposium, pp. 234–245. Springer, Heidelberg (2010). https://doi.org/10.1007/978-3-642-15231-3_23
12. Dörner, D.: Blueprint for a Soul. Rowohlt, Reinbek (2001)
13. Hartmann, G.W., Poffenberger, A.T.: Gestalt Psychology: A Survey of Facts and Principles. Kessinger, Whitefish (2006)
14. Johnson-Laird, P.N.: Mental Models: Towards a Cognitive Science of Language, Inference, and Consciousness. No. 6. Harvard University Press (1983)
15. Kahneman, D., Slovic, S.P., Slovic, P., Tversky, A. (eds.): Judgment Under Uncertainty: Heuristics and Biases. Cambridge University Press (1982)
16. Klein, G.A.: A recognition-primed decision (RPD) model of rapid decision making. Decis. Making Action: Models Methods **5**(4), 138–147 (1993)
17. Rasmussen, J.: Skills, rules, and knowledge; signals, signs, and symbols, and other distinctions in human performance models. IEEE Trans. Syst. Man Cybern. **3**, 257–266 (1983). https://doi.org/10.1109/TSMC.1983.6313160

18. Sirriyeh, R., Lawton, R., Gardner, P., Armitage, G.: Coping with medical error: a systematic review of papers to assess the effects of involvement in medical errors on healthcare professionals' psychological well-being. Qual. Saf. Health Care **19**(6), e43–e43 (2010). https://doi.org/10.1136/qshc.2009.035253

19. Cuschieri, A.: Medical errors, incidents, accidents and violations. Minim. Invasive Ther. Allied Technol. **12**(3–4), 111–120 (2003). https://doi.org/10.1080/13645700310007698

20. Voskanyan, Y., Shikina, I., Andreeva, O., Kidalov, F., Davidov, D.: Multifactorial model of adverse events and medical safety management. J. Digit. Sci. **2**(1), 29–39 (2020). https://doi.org/10.33847/2686-8296.2.1_3

21. Anderson, G., Abrahamson, K.: Our health care may kill you: medical errors. Build. Capacity Health Inf. Future **234**, 13–17 (2017). https://doi.org/10.3233/978-1-61499-742-9-13

22. Boulanger, J., Keohane, C., Yeats, A.: Role of patient safety organizations in improving patient safety. Obstet Gynecol. Clin. N Am. **46**, 257–267 (2019). https://doi.org/10.1016/j.ogc.2019.02.001

23. Chapuis, C., Chanoin, S., Colombet, L., Calvino-Gunter, S., Tournegros, C., Terzi, N., Bedouch, P., Schwebel, C.: Interprofessional safety reporting and review of adverse events and medication errors in critical care. Therap. Clin. Risk Manage. 549–556 (2019). https://doi.org/10.2147/TCRM.S188185

24. Grossmann, N., Grautwohl, F., Musy, S., Nielen, N., Donze, J., Simon, M.: Describing adverse events in medical inpatients using the global trigger tool. Swiss Med. Wkly. **149**, w20149 (2019). https://doi.org/10.4414/smw.2019.20149

25. Halfon, P., Staines, A., Burnard, B.: Adverse events related to hospital care: a retrospective medical records review in a Swiss hospital. Int. J. Qual. Health Care **29**(4), 527–533 (2017). https://doi.org/10.1093/intqhc/mzx061

26. Hooker, A., Etman, A., Westra, M., Van der Kam, W.: Aggregate analysis of sentinel events as a strategic tool in safety management can contribute to the improvement of healthcare safety. Int. J. Qual. Health Care **31**(2), 110–116 (2019). https://doi.org/10.1093/intqhc/mzy116

27. Parand, A., Dopson, S., Renz, A., Vincent, C.: The role of hospital managers in quality and patient safety: a systematic review. BMJ Open **9**, e005055 (2014). https://doi.org/10.1136/bmjopen-2014-005055

28. Rafter, N., Hickey, A., Conroy, R., Condell, S., O'Connor, P., Vaughan, D., Walsh, G., Williams, D.: The Irish National Adverse Events Study (INAES): the frequency and nature of adverse events in Irish hospitals – a retrospective record review study. BMJ Qual. Saf. **26**, 111–119 (2017). https://doi.org/10.1136/bmjqs-2015-004828

29. Schwendimann, R., Blatter, C., Dhaini, S., Simon, M., Ausserhofer, D.: The occurrence, types, consequences and preventability of in-hospital adverse events – a scoping review. BMC Health Serv. Res. **18**, 521 (2018). https://doi.org/10.1186/s12913-018-3335-z

30. Antipova, T.: Fraud prevention by government auditors. In: 12th Iberian Conference on Information Systems and Technologies (CISTI), p. 1–6 (2017). https://doi.org/10.23919/CISTI.2017.7976024

31. Riurean, S., Antipova, T., Rocha, Á., et al.: VLC, OCC, IR and LiFi reliable optical wireless technologies to be embedded in medical facilities and medical devices. J. Med. Syst. **43**, 308 (2019). https://doi.org/10.1007/s10916-019-1434-y

32. Rozhkova, N., Blinova, U., Rozhkova, D.: The concept of management accounting based on the information technologies application. Inf. Technol. Sci. 89–95 (2018). https://doi.org/10.1007/978-3-319-74980-8_8

Clinical Features of Peripheral Vestibulopathies

Irina Borodulina$^{(\boxtimes)}$ ⓘ, Nataliia Voronchikhina ⓘ, Yulia Karakulova ⓘ,
Aleksei Elovikov ⓘ, and Olga Khlynova ⓘ

Perm State Medical University, Perm, Russia
borodulina35@mail.ru

Abstract. We examined 25 patients with dizziness who were treated in the otorhinolaryngology department of the hospital. Dizziness was shown to be systemic peripheral; its leading causes were: benign paroxysmal position vertigo, Meniere's disease, vestibular neuritis. Physical examination, complaint collection, anamnesis of the disease remained a priority in the diagnosis. Establishing the cause of dizziness is especially important, as it allows choosing an adequate treatment tactics. About half of patients experienced symptoms of anxiety and depression, which is important to pay attention to in order to achieve the best effect of therapy. There was also a significant decline in the quality of life in this group.

Keywords: Benign paroxysmal position vertigo (BPPV) · Meniere's disease · Vestibular neuritis

1 Introduction

Dizziness is the illusory movement of a fixed environment in any plane, as well as a sense of rotation or movement of your own body [1]. Dizziness is one of the most common complaints among patients of any age (about 5% of the world's population suffer from dizziness of various genesis) [3]. To date, there are no unified approaches in terminology, classification, study of pathogenetic factors of dizziness formation. Diagnostics and differential diagnosis of dizziness are complex and debatable [4]. In 30–40% of cases, the etiology of dizziness remains unclear. Treatment issues remain underdeveloped. In addition, the number of patients with dizziness has recently increased [6, 7].

Dizziness often leads to a significant deterioration of the patient's quality of life, changes in the familiar lifestyle, social disadaptation, loss of ability to work, depriving you of the opportunity to perform your professional duties of quality, may be a possible cause of falls and injuries [4]. Over the last decade, much attention has been paid to studying the quality of life and satisfaction in various diseases worldwide [3, 5, 13]. In addition, the interaction between vestibular and psychological mechanisms, which have a multidirectional effect in patients with different types of dizziness and contribute to the formation of neurotic reactions with extremely individual features, is also discussed [2].

Dizziness can be vestibular (systemic, or true) or non-systemic. Vestibular dizziness may be central (due to lesions of the vestibular nuclei of the brain stem, vestibular

T. Antipova (Ed.): ICADS 2021, AISC 1352, pp. 235–241, 2021.
https://doi.org/10.1007/978-3-030-71782-7_21

pathways in the brain or lesions of the cerebellum) or peripheral, associated with lesions of the vestibular nerve and maze [1]. The non-systemic dizziness usually hides three groups of states: lipotimic (pre-fainting) states, balance and gait disorders (instability), and mental disorders [1, 8]. Patients describe systemic dizziness as a sense of rotation of surrounding objects in one plane or another, or a sense of rotation of one's own body around objects [1].

This work presents empirically results and their discussions, give practical examples and describe existing experience.

2 Methodology

The aim of the study is to analyze the clinical features of dizziness in the practice of an otorhinolaryngologist. We examined 25 patients who were inpatient in the otorhino-laryngological department of the Perm regional clinical hospital for systemic peripheral dizziness from January to December 2018. Women made up 80% (20 people), men - 20% (5 people). All patients underwent clinical otoneurological examination with evaluation of static and dynamic coordination of movements (Romberg's pose, finger-nose sample, adiadochokinesis sample), gait (straight, flank), nystagmus (spontaneous/positional), its direction, plane-bone, amplitude, degree, frequency), vestibular-ocular reflex (its safety or prolapse), functions of the auditory analyzer (acupuncture, audiometry) and cranial nerves, the results of positional samples (Dick-sa-Hollpike, Roll-test). During research we used digital equipments as computed or magnetic resonance imaging of the brain made, ultrasound examination of the neck vessels, and radiography of the cervical spine.

The study of the vestibular-ocular reflex (Impulse Test, Halmagi Test) is essential to determine the function of the peripheral vestibular system. The loss of the vestibular-ocular reflex (a positive symptom of the "doll") is a diagnostic sign of lesion on the peripheral part of the vestibular analyzer, the side of the loss of the reflex corresponds to the affected side (Fig. 1).

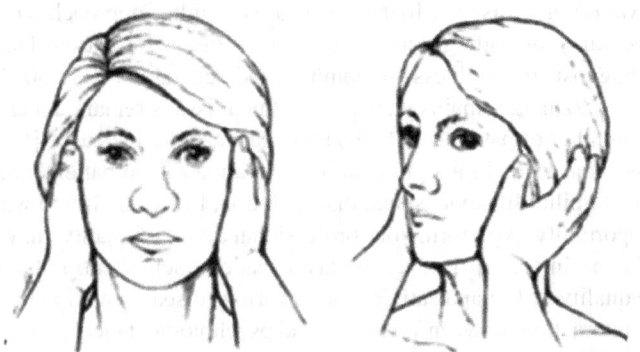

Fig. 1. Vestibular-ocular reflex study (Source: [9, 10]).

As shown on Fig. 1, in case of a sharp turn of the head towards the patient's eye, the patient should remain fixed (on the nose of the doctor), the appearance of a corrective saccade indicates the loss of this reflex.

Positioning tests are the leading method for diagnosing benign paroxysmal position vertigo (BPPV). The Dix-Hallpike test is used to diagnose canalolithiasis of the posterior semicircular canal, the McClure-Pagnini test is used to diagnose cupulolithiasis and canalolithiasis of the horizontal semicircular canal (Fig. 2, Fig. 3).

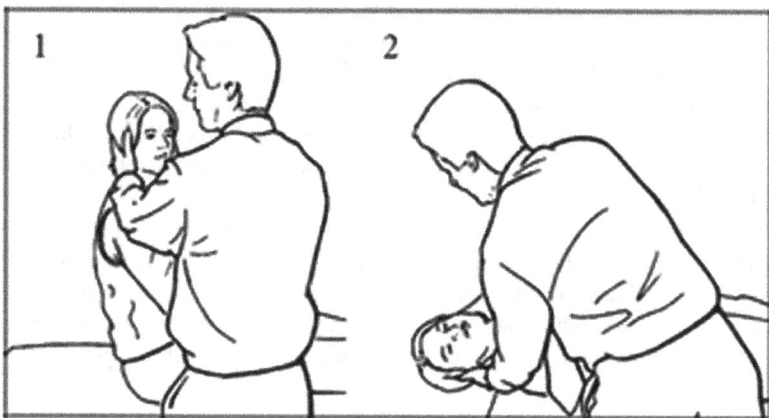

Fig. 2. Dix-Hallpike test. Turning the head 45° towards the researcher, laying on the couch with the head thrown back. Assessment of the appearance of horizontal nystagmus with a rotary component (Source: Zamergrad, M.V. Benign Paroxysmal Positional Vertigo: Current Approaches to Diagnosis and Treatment (in Russian). *Journal of Effective pharmacotherapy. And Psychiatry.* 2017. №4 (33), pp 52–55).

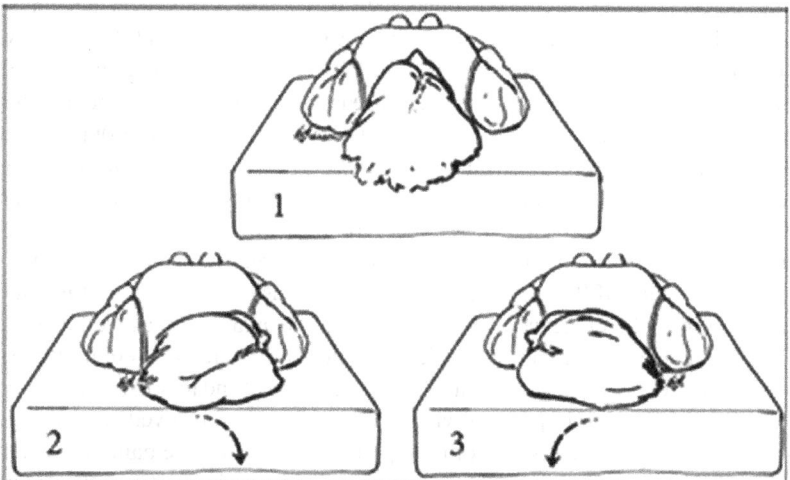

Fig. 3. McClure-Pagnini test (Roll-test). Turn the head 90° in both directions. Assessment of the presence, direction (geotropic, apogeotropic) and severity of nystagmus (Source: [11]).

To exclude central and other causes of dizziness, all patients underwent computed or magnetic resonance imaging of the brain made on medical digital equipment, ultrasound examination of the neck vessels, and radiography of the cervical spine. All patients were consulted by a neurologist.

In collecting complaints and anamnesis of the disease, it was important to determine the duration of an attack of dizziness, the frequency of attacks, preceding factors, as well as the presence of other symptoms (hearing loss, tinnitus, nausea, vomiting). In addition to the otoneurological study, to determine the state of the psycho-emotional sphere, a screening psychological study was carried out, which included a test for the presence and severity of depression using the CES-D questionnaire scale and the use of the hospital anxiety and depression scale (HADS). Further, the level of quality of life with dizziness was assessed by using the Vestibular Rehabilitation Benefit Questionnaire (VRBQ), which shows the percentage deficit of the current state in comparison with the state before the onset of dizziness.

Statistical data processing was carried out using digital technologies, such as Microsoft Office Excel and Word 2010 computer programs. To measure the static significance of the values, the Student's t-test was used, to confirm the dependence of the parameters on each other, the Pearson's correlation coefficient was calculated. For qualitative indicators, the frequency of occurrence was estimated as a percentage.

3 Results

Dizziness in all cases had the character of a systemic peripheral, associated with the pathology of the vestibular nerve and the labyrinth. Among nosological forms, benign paroxysmal position vertigo (BPPV), prevailed in 14 cases (56%). Vestibular neuritis was diagnosed in 5 patients (20%), in 6 patients - Meniere's disease (24%). The average duration of stay in the hospital for patients with BPPV was 6.14 ± 1.17 bed days, with Meniere's disease - 10.2 ± 1.6 bed days, with vestibular neuritis - 11.4 ± 3.4. Average age of patients differed: among patients with BPPV, elderly people prevailed (62.8 ± 6.7 years), vestibular neuritis and Meniere's disease were more often found in young age groups (45.0 ± 14.9 and 43.5 ± 12.0 years respectively), and these groups of patients had significant age differences (from 24 to 52 years). The diagnosis was made mainly on the basis of complaints data, disease history and objective status. It was important to determine the duration of dizziness attack, frequency of attacks, previous factors, as well as the presence of other symptoms (hearing loss, tinnitus, nausea, vomiting).

For patients with BPPV in order to find out the cause of the disease it is important to pay attention to the presence of head injuries in the anamnesis, as well as long stay in one position. Thus, 1 patient had a clear indication of a craniocerebral injury in the anamnesis, 1 patient noted the fact of a long stay of the head and torso in a stationary state (in a dental chair). 5 out of 14 patients with BPPV (36%) were observed in neurologist for cerebrovascular disease, and symptoms of peripheral dizziness remained undiagnosed for a long time. BPPV of the posterior semicircular canal was the most frequently observed - 8 cases (57%), the posterior and horizontal canals were involved in 4 people (29%), 2 patients had an isolated lesion of the horizontal semicircular canal. All patients were tested for Dix-Halpyke (posterior semicircular canal) and Roll-test (horizontal canal) as a diagnostic technique to make this diagnosis. Repositioning maneuvers (Epley, Lemperta, Brand-Daroff) were used in all cases as the main treatment method. 100% of patients with this nosology had no objective signs of dizziness at the time of discharge (absence of nystagmus when repeating the samples), and no patient had any complaints about systemic peripheral dizziness.

All patients with Meniere's disease took caloric samples to determine the functional state of the labyrinth for 6–8 days from the moment of hospitalization. Hyper-reflexion of the affected labyrinth (initial stage of the disease) was detected in 3 out of 6 patients, normofunction (debut of the disease) in 2 patients, hyporeflexion (degenerative stage of the disease) in 1 patient. Neurosensory hearing loss of the 1st century was noted in half of cases, in other 3 patients - neurosensory hearing loss corresponded to the 2nd degree. As a treatment, all patients received a course of systemic hormonal therapy (Dexamethasone 16 mg intravenously drops once a day every 10 days), Betahistine in a dose of 24 mg orally 2 times a day, in the first 1–3 days after admission - diuretics (Diacarb 125 mg orally once a day). All patients were discharged with improvement, against the background of the treatment it was noted the cessation of attacks of systemic dizziness with the preservation of mild instability, precariousness during rapid walking and sharp changes in body position, head and trunk turns.

To diagnose vestibular neuritis, in all cases, we used a study of the vestibular-ocular reflex, the loss of which from the affected side was the main criterion in making the diagnosis. According to the literature, to date, the leading role in the development of vestibular neuritis is attributed to the infection with Herpes simplex virus type 1, but only 1 of the studied patients had information about the appearance of herpes rashes on the eve of a long and pronounced attack of systemic dizziness. In order to detect the viral etiology of the disease, all patients with vestibular neuritis were examined for the herpes virus group (blood antibodies Ig M and Ig G to Herpes simplex virus types 1 and 2, Cytomegalovirus, Epstein-Barr virus, determination of the avidity of antibodies), but in no case were received any signs of fresh infection or data for reactivation of the Herpes virus. To obtain more reliable results, the study should be continued, as in this case the sample of patients is not sufficient. All patients with vestibular neuritis received hormonal infusion therapy (Dexamethasone at high doses of 16–24 mg with a course of up to 8–10 days), empirically all patients were prescribed antiviral therapy (Aciclovir 400 mg inwards 5 times a day).

On the background of treatment in all cases there was a positive dynamics, all patients were discharged with improvement, there was a partial recovery of vestibular-ocular reflex.

Analysis of the results of testing of patients with vestibulopathy using the CES-D questionnaire revealed the presence of mild depression in 23% of cases, medium depression - in 12% of cases, severe depression was noted in 8% of patients. The average emotional state of patients with dizziness on this scale corresponded to mild depression (18.0 ± 10.5 points). A total of 42% of patients reported having depression on this scale.

Subclinical anxiety on the HADS scale was found in 27% of patients with vertigo, subclinical depression - in 8% of patients. Also, one third of the patients (27%) showed signs of clinically expressed anxiety, and 15% of the patients showed signs of clinically expressed depression. Average test scores on the scale were 8.2 ± 4.2 for anxiety level (subclinical anxiety) and 6.4 ± 4.3 for depression level. Total depression on this scale was recorded in 23% of patients, and anxiety in 54% of patients.

The quality of life in patients with systemic peripheral vestibulopathy was reduced by $56.4 \pm 9.5\%$. Moreover, the contribution of the dizziness symptom itself to the quality of life deficit was $62.7 \pm 6.8\%$, the percentage indicator of anxiety was $35.3 \pm 7.5\%$.

4 Conclusion

Dizziness is a significant medical and social problem, Dizziness is a significant medical and social problem, occupying one of the leading positions in neurological and otorhinolaryngological practice. Among patients in the otorhinolaryngological hospital, dizziness is of a systemic peripheral nature, which is due to the lesion of the vestibular nerve or maze. Among the leading causes of systemic peripheral dizziness are BPPV, Meniere's disease, vestibular neuritis. The basis for the diagnosis of these diseases is still a physical examination (otorhinolaryngological, neurological examination, performance of vestibular tests and samples), careful collection of complaints and anamnesis of the disease.

Carrying out differential diagnostics and determining the cause of dizziness is of particular importance, as it allows choosing adequate treatment tactics, ensuring the implementation of etiopathogenic treatment, achieving good results that improve the quality of life of this group of patients. Psychological testing to assess the presence and severity of the signs of depression and anxiety can also contribute to the right choice of therapy for this group of patients, attracting an interdisciplinary approach to the treatment of vestibulopathy, improving the doctor-patient balance, and achieving favorable outcomes of treatment.

In the future we plan to use trained neuronet models to diagnose and choose optimal way of treatment Peripheral Vestibulopathies [14].

References

1. Bronstein, A., Lempert, T.: Dizziness, Transl. Parfenova, V.A. (ed.), 2nd ed. GOTAR-Media, Moscow (2019)
2. Veltishchev, D.Yu, et al.: Psychopathological aspects of dizziness (in Russian). J. Neurol. Psychiatry, **7**(8), 69–72 (2010)
3. Illarionova, E.M., et al.: J. Neurol. Psychiatry. (8), 50–52 (2011). (in Russian)
4. Illarionova, E.M., et al.: Modern aspects of dizziness diagnostics. Russian otorhinolaryngol. (8), 70–73 (2011)
5. Samartsev, I.N., Zhivolupov, S.A.: Vertigo. The Newest Interpretation in Neurology. MEDpress-Inform, Moscow (2019)
6. Bittar, R.S.M., Oiticica, J., Bottino, M.A., Gananca, F.F., Dimitrov, R.: Population epidemiological study on the prevalence of dizziness in the city of Sao Paulo. Braz. J. Otorhinolaryngol. **79**(6), 688–698 (2013)
7. Caldas, M.A., Gananca, C.F., Gananca, F.F., Gananca, M.M., Caovilla, H.H.: Clinical features of benign paroxysmal positional vertigo. Braz. J. Otorhinolaryngol. **75**(4), 502–506 (2009)
8. Mangabeira Albernaz, P.L.: Vertigo in elderly patients: a review of 164 cases in Brazil. Ear Nose Throat J. **93**(8), 322–330 (2014)
9. Brandt, T., Dieterich, M., Shtrupp, M.: Vertigo: trans. from English. Practice (2009)
10. Parfenov, V.A., Zamergrad, M.V., Melnikov, O.A.: Dizziness: Diagnosis and Treatment, Common Diagnostic Errors: A Study Guide. MIA (2009)

11. Zamergrad, M.V.: Benign paroxysmal positional vertigo: current approaches to diagnosis and treatment. J. Effect. Pharmacother. Psychiatry. (4), (33), 52–55 (2017). (in Russian)

12. Mirskikh, I., Mingaleva, Z., Kuranov, V., Matseeva, S.: Digitization of medicine in Russia: mainstream development and potential. Lecture Notes in Networks and Systems, vol. 136, pp. 337–345 (2021)

13. Buzin, V.N., Mikhailova, Y.V., Chukhriyenko, I.Y., Buzina, T.S., Shikina, I.B., Mikhailov, A.Y.: Russian healthcare through the eyes of the population: dynamics of satisfaction over the past 14 years (2006–2019): Review of sociological studies. Prev. Med. **23**(3)

14. Khlynova, O.V., Yasnitsky, L.N., Skachkova, I.V.: Neural network system for medical diagnostic of gastrointestinal diseases. In: Antipova, T., Rocha, A. (eds.) Digital Science. DSIC18 2018. Advances in Intelligent Systems and Computing, vol. 850. Springer, Cham (2019). https://doi.org/10.1007/978-3-030-02351-5_41

Analysis of Sustainable Health Development Goals in Improving Public Health

Eko Priyo Purnomo[1]([✉]), Mochammad Iqbal Fadhlurrohman[1], Lubna Salsabila[1], Aqil Teguh Fathani[1], Sujud Sujud[2], and Yeni Widowaty[2]

[1] Department of Government Affairs and Administration, Jusuf Kalla School of Government, Universitas Muhammadiyah Yogyakarta, Yogyakarta, Indonesia
eko@umy.ac.id, lubna.salsa@gmail.com, aqil.teguh.psc19@mail.umy.ac.id
[2] Magister of Law, Universitas Muhammadiyah Yogyakarta, Yogyakarta, Indonesia
sujud.pasca18@mail.umy.ac.id, yeniwidowaty@umy.ac.id

Abstract. This research aims to assess public health sustainability priorities and explore multiple factors that affect sustainable development goals. The government is obliged to carry out activities to optimize development, such as health services that must be provided adequately to the community to achieve quality development. This study used qualitative descriptions with data from various newspaper articles, online media, and other research-related data. Data collected were reduced and used to determine variables and indicators. The analysis shows that three health resource development goals play an important role in implementing performance to deliver public health. There are three health resource development goals. Various factors, such as human settlement, water infrastructure, economy, and health, cannot be distinguished from program implementation success.

Keywords: Public health · Sustainable health development · SDGs

1 Introduction

This study aims to analyze the sustainability development goals in public health and look at several factors that influence sustainable development goals. The Sustainable Development Goals are a set of global priorities for all social elements to achieve prosperity [1]. The Sustainable Development Goal is a global policy framework to answer the countries' fundamental challenges in implementing state development [2]. Sustainable health development aims to end hunger, protect the world, and ensure that all people, now and in the future, achieve stability and prosperity [3]. Thus, the SDG framework in enhancing development optimizes the government's performance by looking at the problems that occur by improving sustainable development indicators to improve their quality in solving problems in the relevance of indicators [3].

In the sustainable development goal, there are 17 goals, 169 targets, and 230 indicators covering several fields, one of which is in the health sector [4]. In health development is not new; many studies talk about sustainable development in the health sector,

T. Antipova (Ed.): ICADS 2021, AISC 1352, pp. 242–246, 2021.
https://doi.org/10.1007/978-3-030-71782-7_22

as stated [5] that health is an integral part of national development goals to encourage more sustainable development. Meanwhile, according to [6], health can develop human resources to carry out activities to meet the community's needs. The goal of sustainable health development is health development related to the quality of public health by meeting requirements such as nutrition for the community [7], health insurance for mothers and children, and others related to improving public health [8]. The crucial mission of a country for the future created by environmental politics is to achieve sustainable human development by optimizing the government's role in transforming socio-economy and ensuring society's security [9]. So the government is obliged to carry out activities to optimize development that must be carried out, such as in health services that must be adequately provided to the community to carry out quality development [9].

In implementing the Sustainable Development Goals, every country must improve in implementing public health development; in implementing public health improvement, it must be seriously taken because health is a human right that must be fulfilled to create a prosperous society [10]. This study used data analysis with qualitative descriptions with data from a range of newspaper articles, online media and other research-related data. The first research was conducted to reduce secondary data. The analysis was performed by assessing the variables and indicators chosen in the study came from several previous studies and literature reviews.

2 Discussion

Evaluation is a processor that investigates systematically to determine the value and benefits of the object being evaluated [11]. The definition of evaluation described by William Dunn (2003) in [12] evaluation is the most crucial policy process because with evaluation we can assess how far needs, values, and opportunities go through public action, where the objectives are specific can be achieved, so that new policy alternatives or revised policies can ascertain the appropriateness of the policies.

Sustainable development aims to eliminate gaps in society to make society prosperous and targets creating a policy or implementing development [13]. The Sustainable Development Goals (SDGs) are a set of global goals, from the ecological biosphere to local populations, for equitable and balanced health at all levels [1]. Health is the goal of achievement in guaranteeing human rights that must be provided in a sustainable development carried out to fulfill social justice for society; that is the crucial role of SDG's [14]. The implementation of SDGs cannot be separated from several factors supporting its success, such as human settlement, water infrastructure, economy, and health [15]. Also, sustainable development goals focus on environmental, economic, and social aspects [16, 17].

Health development is one of the development challenges that cannot be ignored; all countries always consider and improve the health sector. Health development is not new; many studies talk about sustainable development in the health sector, as stated [5] that health is an integral part of national development goals. One of the efforts to improve health in the sustainable development goals (TPB) of the government must make a significant investment to eliminate gaps in society [18]. According to Walpole & Mortimer (2017), increasing awareness about health can be done by managing complex systems ideally and considering environmental aspects and understanding from an early age [19–21]. One of the complex system management is regulations in optimizing health funds better [13] due to a very complex problem [22] in improving health, namely the community economy [23].

In the study [24], four principles can support the care and development of sustainable health practices, including the health education system, sustainable health human resources, and medical infrastructure/equipment. [25] states that in carrying out sustainable health development, it must also be supported by seeing the proportion of good quality medical personnel. In [6], it is also stated that the direction of sustainable health development is determined by one of them a quality medical team [26] who provides the best service for the community. To improve quality in sustainable health development, nurses play a significant role or offer useful services to the organization [27]. Health human resources have a significant role in implementing performance to provide public health. There are three goals of health human resource development, according to [26]:

1. Able to develop and update science and technology in health promotion by mastering and understanding scientific approaches, methods, and principles accompanied by application skills in developing and managing human health resources.
2. Able to identify and formulate solutions to problems in developing and managing human health resources through research activities.
3. Develop/improve professional performance as shown by a sharp analysis of health problems, formulate, and advocate for health programs and policies to develop and manage human health resources.

In Utami (2019), a complex health problem is a lack of nutrition for the community. The higher the Public Health Development Index (IPKM), the lower the effect of malnutrition. As in research conducted by [7], health is essential for a company, so providing a child, the government must do nutrition. To determine the level of success of sustainable health development [25], one of them is by looking at the increase in the Community Development Index (HDI) and the public health index [28]. As mentioned by research [29], nutrition services for the community are essential to provide the community's right to life and improve newborn children's nutrition [30]. So, health insurance must be fulfilled by prioritizing mothers and children [8]. There is one program from the community, such as Posyandu or integrated service posts [31]. The city's help, such as PKK mothers accompanied by the local health office or Puskesmas [32].

3 Conclusion

The Sustainable Development Goals (SDGs) are a series of global goals for inclusive and balanced health at all levels, from the natural biosphere to local communities. Sustainability Health development goals are the purpose of achieving human rights that must be ensured in sustainable development to deliver social justice. One of the sustainability issues that should not be overlooked is healthy growth, and all countries always recognize and make changes in the health sector. Nurses play a vital role in improving sustainable health growth or providing valuable resources to the organization. Human capital in the area of health has a crucial role to play in implementing public health efficiency. A variety of factors promoting its success, such as human settlement, water infrastructure, economy, and health, cannot be distinguished from SDGs' introduction. Regulations to help optimize health funds are one of the dynamic system management regulations.

References

1. Morton, S., Pencheon, D., Squires, N.: Sustainable Development Goals (SDGs), and their implementation. Br. Med. Bull. **124**(1), 81–90 (2017)
2. Deacon, B.: Assessing the SDGs from the point of view of global social governance. J. Int. Comp. Soc. Policy **32**(2), 116–130 (2016)
3. Hák, T., Janoušková, S., Moldan, B.: Sustainable development goals: a need for relevant indicators. Ecol. Indic. **60**, 565–573 (2016)
4. Panuluh, S., Fitri, M.R.: Perkembangan pelaksanaan sustainable development goals (SDGs) di Indonesia. Biefing Pap. 02, vol. infid, no. Sustainable Development Goals (SDGs), pp. 1–25 (2016)
5. Berman, P.A.: Health sector reform: making health development sustainable. Heal. Sect. Reform Dev. Ctries. **32**, 13–33 (1996)
6. Walpole, S., Barna, S., Richardson, J., Rother, Hanna-Andrea.: Sustainable healthcare education: integrating planetary health into clinical education. Lancet Planet. Health **3**(1), e6–e7 (2019)
7. Hendrawati, M.A.: Pemberdayaan kader kesehatan dalam pencegahan dan penatalaksanaan stunting pada anak di wilayah kerja puskesmas jatinangor. Dharmakarya **7**(4), 274–279 (2018)
8. Lestary, H., Sugiharti, S., Suparmi, S.: Pemanfaatan jaminan kesehatan dalam pelayanan kesehatan ibu di tujuh kabupaten/kota di Indonesia. J. Ekol. Kesehat. **18**(2), 111–121 (2019)
9. Nasirin, C.: Penguatan peran petugas kesehatan: implementasi kelembagaan daerah dalam meningkatkan kesehatan keluarga prasejahtera di mataram, vol. 3, no. 1, pp. 87–99 (2017)
10. Dewi, A., et al.: Global policy responses to the COVID-19 pandemic: proportionate adaptation and policy experimentation: a study of country policy response variation to the COVID-19 pandemic. Heal. Promot. Perspect. **10**(4), 359–365 (2020)
11. Jaya, P., Ndeot, F.: Penerapan model evaluasi cipp dalam mengevaluasi program layanan paud holistik integratif. PERNIK: J. Pendidikan Anak Usia Dini **1**(01), 10–25 (2019)
12. Nur, A., Utami, F., Mutiarin, D.: Evaluasi program jaminan kesehatan nasional pada fasilitas kesehatan tingkat i kabupaten sleman tahun 2016 (2016)
13. Suarsih, S.: Analisis Kebijakan dana desa untuk pembangunan kesehatan di kabupaten malinau dengan pendekatan segitiga kebijakan. J. Sist. Kesehat. **2**(4), 211–217 (2017)
14. Garay, J.E., Chiriboga, D.E.: A paradigm shift for socioeconomic justice and health: from focusing on inequalities to aiming at sustainable equity. Public Health **149**, 149–158 (2017)
15. Vörösmarty, C.J., et al.: Ecosystem-based water security and the sustainable development goals (SDGs). Ecohydrol. Hydrobiol. **18**(4), 317–333 (2018)
16. Habibie, W.L., Hardjosoekarto, S., Kasim, A.: Health reform in Indonesia towards sustainable development growth (Case study on BPJS Kesehatan, health insurance in Indonesia). Rev. Integr. Bus. Econ. Res. **6**(3), 375–383 (2017)
17. Purnomo, E.P., Ramdani, R., Salsabila, L., Choi, J.-W.: Challenges of community-based forest management with local institutional differences between South Korea and Indonesia. Dev. Pract. **30**(8), 1082–1093 (2020)
18. Pablos-Mendez, A., Cavanaugh, K., Ly, C.: The new era of health goals: universal health coverage as a pathway to the sustainable development goals. Heal. Syst. Reform. **2**(1), 15–17 (2016)

19. Kosasih, C.E., Isabella, C., Sriati, A.: Upaya peningkatan gizi balita melalui pelatihan kader kesehatan. Media Karya Kesehatan **1**, 90–100 (2018)
20. Nurmandi, A., Purnomo, E.P.: Making the strategic plan work in local government: a case study of strategic plan implementation in yogyakarta special province (YSP). Int. Rev. Public Adm. **16**(2), 143–164 (2011)
21. Purnomo, E.P., Anand, P.B., Choi, J.W.: The complexity and consequences of the policy implementation dealing with sustainable ideas. J. Sustain. For. **37**(3), 270–285 (2018)
22. Gumilar, S.: Tanggung jawab sosial perusahaan dan kesehatan anak balita (Kasus Pada Csr Pt. Pertamina Tbbm Bandung Group). Share Soc. Work J. **8**(2), 225 (2019)
23. Pinem, M.: Pengaruh pendidikan dan status sosial ekonomi kepala keluarga bagi kesehatan lingkungan masyarakat. J. Ilmu Pemerintah. dan Sos. Polit. UMA **4**(1), 97–106 (2016)
24. Mortimer, F.: The sustainable physician. Clin. Med. J. R. Coll. Phys. London **10**(2), 110–111 (2010)
25. Indrawati, T.: Peran indikator pelayanan kesehatan untuk meningkatkan nilai sub indeks kesehatan reproduksi dalam indeks pembangunan kesehatan masyarakat (IPKM). Penelit. dan Pengemb. Upaya Kesehat. Masyarakat, Badan Penelit. dan Pengemb. Kesehat. Kementeri. Kesehat. RI, vol. 28, no. 2, pp. 95–102 (2018)
26. Putri, A.: Kesiapan sumber daya manusia kesehatan dalam menghadapi masyarakat ekonomi asean (MEA). J. Medicoeticol. dan Manaj. Rumah Sakit. **6**(1), 55–60 (2017). https://doi.org/10.18196/jmmr.2016
27. Carryer, A.: Nurse practitioners as a solution to transformative and sustainable health services in primary health care: a qualitative exploratory study. Collegian **24**(6), 525–531 (2017)
28. Tjandrarini, D.: Mubasyiroh, "Pencapaian Indonesia Sehat Melalui Pendekatan Indeks Pembangunan Kesehatan Masyarakat Dan Indeks Keluarga Sehat." Bul. Penelit. Sist. Kesehat. **21**(2), 90–96 (2018)
29. Hadi, H.: Gizi lebih sebagai tantangan baru dan implikasinya terhadap kebijakan pembangunan kesehatan nasional. J. Gizi Klin. Indones. **1**(2), 47 (2004)
30. Setyowati, A.: Pengelolaan data pelayanan kesehatan ibu dan anak bagian gizi balita di peskesmas berbasis android mobile untuk mendukung pencapaian sustainable develoment goal's (SDG's). In: Prosding-Seminar Call Paper, pp. 22–26 (2012)
31. Tse, A.D.P., Suprojo, A.: Peran kader posyandu terhadap pembangunan kesehatan masyarakat. J. Ilmu Sos. dan Ilmu Polit. **6**(1), 60–62 (2017)
32. Shalfiah, R.: Peran pemberdayaan dan kesejahteraan kkeluarga (PKK) dalam mendukung program-program pemerintah kota bontang. J. Ilmu Pemerintah. Unmul **1**(3), 975–984 (2013)

Advances in Digital Social Media

Social Media Engagement Through Video Advertising: Informativeness and Self Brand Connection as Predictors

Sónia Ferreira[1]([⊠]) [iD], Sara Santos[2] [iD], and Pedro Espírito Santo[3] [iD]

[1] Centre for Studies in Education and Innovation (CI&DEI), Instituto Politécnico de Viseu, Viseu, Portugal
sonia.ferreira@esev.ipv.pt
[2] Instituto Politécnico de Viseu, Viseu, Portugal
[3] ESTGOH - Instituto Politécnico de Coimbra, Coimbra, Portugal

Abstract. The evolution of social media has been great and people have been looking for more information on these networks. Globally, there are more than 4 billion active users on social networks who brands seek to attract their products and services. In this context, our investigation found factors that influence social media engagement: informativeness, self-brand connection and advertising stimulation.

Through the literature review, a conceptual model was proposed that was later tested using PLS-SEM. Data were collected from 237 consumers and it was revealed that engagement in social media is explained by the variables identified by our model. Important contributions to the theory and management of the brand will be found in this investigation.

Keywords: Social media engagement · Video advertising · Self brand connection

1 Introduction

In recent years, the rapid expansion of the use of social networks by increasingly heterogeneous audiences has brought new challenges and opportunities in the area of advertising. The use of advertising videos for social networks in marketing strategies increasingly create engagement with the brand.

Compared to other advertising formats, videos have the possibility of becoming more personal, using creativity to optimize interaction with the public and achieve commercial goals. For that and according to Escadas [1] ads should not be only persuasive, but they must have an affective component that directs to the consumer unconsciously. Structurally, narratives must be based on a plot with incidents and surprises, on characters to whom conflicts, and events take place and at a climax moment where the resolution of the story is presented [2] and the greater conflict finally resolved. In addition, the symbiosis between the structure of the narrative and the elements of the story creates a powerful tool for transmitting information [3] and, when created effectively, allows for

© The Author(s), under exclusive license to Springer Nature Switzerland AG 2021
T. Antipova (Ed.): ICADS 2021, AISC 1352, pp. 249–260, 2021.
https://doi.org/10.1007/978-3-030-71782-7_23

more favorable cognitive responses, warm feelings and positive attitudes of the narrative ad, when compared to argument advertising [4].

On the other side, consumers tend to purchase brands and products similar to them, that coincide with their ego and they could build a brand-consumer emotional relationship [5]. Consumer engagement plays a central role in social media marketing strategies, however further research is needed [6] because studies in this topic are limited [7]. In this study we analyze self-brand connection, informativeness and advertising stimulation as predictors of social media engagement on video advertising.

2   Theoretical Background

2.1   Advertising Informativeness and Stimulation

It is undeniable that technological development has brought with it new social, cultural and economic contexts. With it, advertising also gained power and advertising communication became part of a context never before achieved. It is, in fact, a reality that reaches consumers of the most varied classes and profiles, informing about products and services, persuading for consumption and conquering space for brands. In the commercial complex, advertising is a communication force between the product and the public. The direct and immediate function is not to make the purchase just happen. Of course, it contributes to this, but its specific objective is to act on the public's state of mind and lead them to purchase [8]. Establish favorable conditions for consumption, create attributes so that, even after the purchase, the consumer feels privileged by the choice. Advertising objectives, dependent on those of marketing, respond to three established basic needs: to inform, persuade and remember [9].

According to Lee and Hong [10], the informativeness of ads is as important as creativity, often emphasized in public literature, they reveal themselves to be the main drivers of behavior favorable to advertisements on social networking services, promoting purchase intentions. On the other hand, advertising campaigns also have more specific objectives, such as helping to introduce a product to the market, maintaining the market, informing about new products and their use, selling the brand image, among others.

To achieve these goals, and in digital contexts, advertisers are integrating social media in planning advertising strategies in order to boost digital engagement as well. For Hide, Noort, Muntinga and Bronner [11], one of the important variables in the analysis of this involvement is the stimulation that advertising can offer. Here, the analysis of the enthusiasm of the recipients and the seduction and originality of advertising is valued through informativeness. Therefore, a clear awareness of these variables and the needs of the public is fundamental to the success of an advertising campaign. Thus, we proposed the following:

H1: Informativeness has positive effects on advertising stimulation.

2.2   Self-brand Connection

Brands represents who consumers are or want to be, as their life projects, goals or personal concerns [12], are perceived as brands with personality and congruent to their own personality [13].

The concept of brand-self connection derived from 'Self-Expansion Theory' and it is formed by consumers when conceptualize their self into brands to others [14]. In this way, self–brand connection is a crucial dimension of the consumer–brand relationship [15] and it is defined by Escalas and Bettman [16] as "the degree to which consumers have incorporated the brand into their self-concept".

Park et al. [13] conceptualize brand attachment in two dimensions: self-connection and prominence. While self-connection is the expression of individuals their selves similar to brand personality, brand prominence is the easiness and frequency that brand brought into consumers' mind [13].

Brands are used by consumers to express their self-concept and identity, allowing differentiation from others and express of individuality [17]. When consumers have a high self–brand connection also have a higher tendency to engage, refuse negative information about brand and show positive WOM [15]. As brand image and consumer's ego image coincide it increases the intention to purchase brand and positive evaluations about it [18].

Thomson et al. [19] demonstrates that brand attachment has consequences for brands such as brand loyalty [20] and the willingness to pay a price premium. It is also reflected in emotional responses, brand responsiveness [22], commitment [21] and minimization of the effects of negative information or unethical firm behavior [20]. Therefore, we analyze advertising stimulation as an emotional response. Thus, we propose the following:

H2: Self-brand connection has positive effects on advertising stimulation.

2.3 Social Media Engagement

In recent years, brands have seen changes in the way consumers interact and get engaged especially through social media [23]. Calder, Isaac and Malthouse [24] consider consumer engagement as "a multilevel, multidimensional construct that emerges from the thoughts and feelings about one or more rich experiences involved in reaching a personal goal". This engagement varies depending on type of social media such as Facebook, Instagram, Twitter or YouTube [25] and is expressed through likes, comments or shares in brand's posts that reflect cognitive, affective and behavioral aspects [26]. According to the Consumer Online Brand-Related Activity (COBRA) model proposed by Muntinga et al. [26] engagement can be: low (users just consume content, viewing videos and pictures, have a passive posture - lurkers), medium (users contribute, comment on posts, etc.) or high (users share and create content about the brand).

In a study about advertising and engagement with social media platform, Hilde et al. [11] demonstrated that engagement is highly context specific, it depends on the experience in each platform, i.e., people evaluate advertising differently depending on the platform used and not just the ad content. Engagement with the medium or platform influences responses and engagement with advertising [27].

Engagement has several consequences for brands, and numerous studies have pointed out this [28–30]. It has impacts on brand loyalty [29, 30], brand awareness [28], commitment [31] and perceived quality [28].

In social media, some studies demonstrated that engagement also increase brand performance [32] and influence purchase decisions and sales [33, 34].

Several studies also confirm that people high engaged can be more responsive to brand advertisements [35]. Consequently, we propose to test the following research hypotheses:

H3: Self-brand connection has positive effects on social media engagement.
H4: Advertising stimulation has positive effects on social media engagement.

Based on the previous information, our study aims to test the following conceptual model (see Fig. 1).

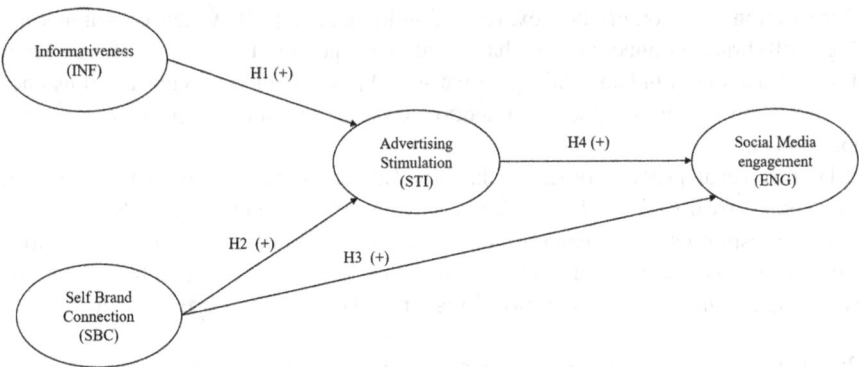

Fig. 1. Conceptual model

3   Method

To validate the proposed model, a survey was conducted among Staples brand consumers. We collected data through a self-administered survey. This survey had two phases. First, we presented an advertising video to the participants, and after, we ask about the video and about the brand.

To measure the constructs, we used items adapted from previous studies. Hence, to measure the constructs' items, for informativeness we used items from Jeon, Lee and Hong [36]. Self-brand connection was measured through items used by Jeon, Lee and Jeong [36]. The scale adopted for Advertising Stimulation were adapted from Voorveld, van Noort, Muntinga and Bronner [3] and we measured social media engagement following Schivinski, Christodoulides and Dabrowski [28]. All the items used in this study were measured using a 5-point Likert scales, ranging from strongly disagree (1) to strongly agree (5).

3.1   Sample

The sample has 237 valid questionnaires. Table 2 show that 59.5% are female and most of the observations (N = 185; 78.1%) originate from individuals below 40 years old. Many of them (N = 134; 56.5%) have primary or secondary school qualifications (Table 1).

Table 1. Demographic profiles (N = 237)

Variables	Category	N	%
Gender	Male	96	40.5
	Female	141	59.5%
Age	≤ 20	69	29.1%
	20–29	97	40.9%
	30–39	19	8.0%
	40–49	27	11.4%
	≥50	25	10.5%
Education	Primary school studies	18	7.6%
	Secondary school studies	116	48.9%
	High school	103	43.5%

4   Results

In order to estimate the model, we choose Partial Least Squares – Structural Equation Modelling (PLS-SEM) because it enables the researchers to assess causal relationships among items and causal relationships between latent variables. The PLS-SEM is appropriate for exploratory research and does not require normality of data [37]. PLS-SEM is executed in two steps. First, we analysed the reliability and validity of measurement model and secondly we analysed the relations between constructs as suggested by [37]. The PLS-Algorithm was executed on SMART PLS v3.3.2.

4.1   Common Method Bias

Responses were collected from the same respondents and, there was a possibility of common method bias. First, we performed Harman's one-factor test [38] with factor one representing 30.52% of the variance. Furthermore, we carry out a preliminary data analysis to validate VIF – Variance Inflactor Factor. VIF values are ranged between 1.393 and 3.286 which is below the threshold value (VIF < 5). Therefore, there is no multicollinearity. Thus, common method bias would not be a concern. Moreover, we analysed the Skewness (Sk) and Kurtosis (Ku), which reveal that the items do not diverge from normality (Sk < 3; Ku < 7) [39].

4.2   Measurement Model

To achieve construct validity and reliability, standardized item loadings (λ) were analysed for all reflexive constructs. Table 2 shows that standardized item loadings are above the minimum threshold value of 0.7 [40], which were acceptable for further analysis.

Table 2 also shows that Average Variance extracted (AVE) (ranging from 0.583 to 0.793) and composite reliability (CR) (ranging from 0.874 to 0.939) are above the

Table 2. Items, Descriptive Statistics, and coefficient loadings (N = 237)

	Mean	SD	λ	t values	p values
Informativeness (α = 0.905; ρA = 0.905; CR = 0.934; AVE = 0.778)					
INF01	3.36	0.910	0.859	39.000	0.000
INF02	2.82	1.160	0.819	33.462	0.000
INF03	3.34	0.980	0.904	57.152	0.000
Self Brand Connection (α = 0.826; ρA = 0.826; CR = 0.896; AVE = 0.742)					
SBC01	2.22	1.075	0.890	46.568	0.000
SBC02	2.55	1.209	0.887	42.610	0.000
SBC03	2.53	1.174	0.900	46.472	0.000
SBC04	2.29	1.162	0.885	42.637	0.000
Advertising Stimulation (α = 0.913; ρA = 0.922; CR = 0.939; AVE = 0.793)					
STI01	3.13	1.063	0.786	27.536	0.000
STI02	2.77	1.113	0.773	21.754	0.000
STI03	3.50	1.060	0.701	12.686	0.000
STI04	3.01	1.169	0.726	16.921	0.000
STI05	2.67	1.086	0.874	60.179	0.000
Social Media Engagement (α = 0.820; ρA = 0.845; CR = 0.874; AVE = 0.583)					
ENG01	2.50	1.174	0.857	37.282	0.000
ENG02	2.06	1.106	0.908	58.753	0.000
ENG03	2.31	1.170	0.880	39.287	0.000
ENG04	2.22	1.156	0.883	36.732	0.000

Note: All items are measured with a 5 point Likert Scale ranging between (1) strongly disagree and (5) strongly agree

SD = Standard Deviation; α = Cronbach's Alpha; λ = Standardized Loadings; AVE = Average Variance extracted; CR = Composite Reliability; INF = Informativeness; SBC = self-brand connection; STI = Advertising Stimulation; ENG = Social Media Engagement.

threshold values (AVE > 0.5; CR > 0.7) [41]. These values showed convergent validity and reliability for all constructs.

Discriminant validity has been confirmed in three steps. First, through the Fornell and Larcker criteria [42] it was verified that the correlations between latent constructions are below than values of the square root diagonals of the AVE (Table 3). In the second step, the discriminant validity was verified through the cross-loadings criterion [43]. Table 4 shows a comparison of the column loadings and each indicator exhibits that indicator's loadings on its construct is higher in all cases compared to all its cross-loadings with other constructs. Finally, in the third step, the discriminant validity was examined through the Heterotrait-Monotrait (HTMT) ratio of correlations and, as can be seen in Table 5 the HTMT values are below 0.90 [43]. Therefore, discriminant has been established between variables.

Table 3. Discriminant validity: Fornell and Larcker criterion [42]

	ENG	INF	SBC	STI
ENG	0.882			
INF	0.548	0.862		
SBC	0.544	0.442	0.890	
STI	0.674	0.679	0.584	0.764

INF = Informativeness; SBC = self-brand connection; STI = Advertising Stimulation; ENG = Social Media Engagement.

Table 4. Cross-loadings

	ENG	INF	SBC	STI
INF01	0.436	0.859	0.353	0.553
INF02	0.550	0.819	0.428	0.619
INF03	0.422	0.904	0.355	0.575
SBC01	0.576	0.436	0.890	0.557
SBC02	0.397	0.352	0.887	0.456
SBC03	0.445	0.403	0.900	0.477
SBC04	0.492	0.374	0.885	0.570
STI01	0.522	0.446	0.457	0.786
STI02	0.509	0.470	0.491	0.773
STI03	0.319	0.353	0.303	0.641
STI04	0.476	0.624	0.411	0.726
STI05	0.673	0.638	0.529	0.874
ENG01	0.857	0.524	0.497	0.605
ENG02	0.908	0.467	0.504	0.572
ENG03	0.880	0.460	0.449	0.622
ENG04	0.883	0.481	0.468	0.576

INF = Informativeness; SBC = self-brand connection; STI = Advertising Stimulation; ENG = Social Media Engagement.

Table 5. Discriminant validity HTMT ratio [43]

	ENG	INF	SBC	STI
ENG				
INF	0.630			
SBC	0.589	0.503		
STI	0.760	0.804	0.657	

INF = Informativeness; SBC = self-brand connection; STI = Advertising Stimulation; ENG = Social Media Engagement.

4.3 Structural Model

The study began with the assessment of structural equation model by globally evaluating fit through the standardizing root mean residual (SRMR). We obtained SRMR = 0.076, which is below than the threshold value [44]. Next, we analysed adjusted R2 of the endogenous variables in our model, which are 0.557 for Advertising Stimulation and 0.484 for Social Media Engagement.

Next, we analyzed the results of the hypotheses tests by evaluating the significance of the path coefficients bootstrapping procedure with 10000 subsamples. We display on Table 6 the results of hypotheses testing.

Table 6. Hypotheses testing

Path	β	t values	p values	95% confidence interval	Effect size (f^2)	Hypothesis
INF → STI	0.523	10.777	0.000	[0.427 .. 0.617]	0.500	H1: Supported
SBC → STI	0.353	7.013	0.000	[0.254 .. 0.452]	0.228	H2: Supported
SBC → ENG	0.228	3.620	0.000	[0.102 .. 0.352]	0.067	H3: Supported
STI → ENG	0.540	10.281	0.000	[0.434 .. 0.641]	0.376	H4: Supported

β = Standardized path coefficients; INF = Informativeness; SBC = self-brand connection; STI = Advertising Stimulation; ENG = Social Media Engagement.

H1 checks if informativeness has positive effects on advertising stimulation. Our results indicate that higher value of informativeness are perceived as higher stimulation ($\beta_{INF \to STI}$ = 0.523; p < 0.05). Additionally, H1 has an effect size (f^2) equal to 0.500. According to Cohen [45] f^2 values of 0.02, 0.15, and 0.35 for the significant independent variables represent weak, moderate and substantial effects, respectively. Therefore, informativeness has substantial positive effects on advertising stimulation.

Our investigation tested the positive effects of self brand connection on advertising stimulation (H2). Our results support H3 ($\beta_{SBC \to STI}$ = 0.353; p < 0.05) and suggests that self-brand connection has considerable effects on advertising stimulation (f^2 = 0.228).

H3 proposes that self-brand connection has positive effects on social media engagement. We find support for such effect ($\beta_{SBC \rightarrow ENG} = 0.228$; $p < 0.05$). Accordingly, the effects ($f^2 = 0.067$) from self-brand connection on social media engagement is moderate.

Finally, H4 proposes that advertising stimulation positively influences social media engagement. Our results show that this path is significant ($\beta_{STI \rightarrow ENG} = 0.353$; $p < 0.05$) with positive large effects ($f^2 = 0.376$).

5 Discussion and Implications

The recent evolution of social media has directed brand communication actions towards these media. Naturally, new challenges have arisen for brands, who want to know how their consumers move on the digital channel. This investigation aimed to identify social media engagement through the presentation of an advertising video, because this way of communicating optimizes the interaction between consumers and brands through creativity.

5.1 Theoretical Implications

This investigation tested the proposed conceptual model and found relevant effects among the variables studied. Following Hide, Noort, Muntinga and Bronner [11], we found significant effects so that informativeness influences enthusiasm and emotional stimulation of an advertising video.

The study also found that the self-brand connection has a determining role in advertising stimulation. This conclusion is reflected in the literature, since higher values of self-brand connection translate into emotional responses to brand advertisements [22].

In the study on advertising, Hilde et al. [11] found that social media engagement depends on previous experiences. This study found that self-brand connection and advertising stimulation influence social media engagement.

5.2 Implications for Management

According to our findings, managers must understand that social media engagement is determined by informativeness, self-brand connection and advertising stimulation.

Therefore, they must identify the utility that consumers attribute to the information presented in their ads. In addition, social media ads must contain characters that consumers can identify with. It is also relevant that the video ad scenes are congruent with consumers in order to create a self-connection with each brand.

This study also suggests that advertising videos on social networks must be original and unique in order to create something new in consumers that generates advertising stimulation and, consequently, higher values of social media engagement.

6 Conclusions, Limitations and Future Research

This research has shown that social media engagement is determined by informativeness, self-brand-connection and advertising stimulation. In general, the variables identified explain the social media engagement. Therefore, brands must recognize the factors identified in this investigation as determinants for social media engagement.

It is important to point out the limitations of our research. First, the composition of the sample in terms of age and education may have influenced the results obtained. Second, the presentation of a specific announcement to the respondents may have caused specific results. Third, our conclusions must be analysed from a transversal perspective.

In this way, it will be relevant to develop future research on social media engagement. Therefore, we suggest that new investigations should be carried out, with samples different from ours. Also, it will be relevant to carry out studies that analyse advertising videos from other brands. In addition, we indicate that longitudinal studies can be carried out in order to understand the evolution of social media engagement over time.

Acknowledgment. This work is funded by National Funds through the FCT - Foundation for Science and Technology, I.P., within the scope of the project Refª UIDB/05507/2020. Furthermore, we would like to thank the Centre for Studies in Education and Innovation (CI&DEI) and the Polytechnic of Viseu for their support.

References

1. Escalas, J.E.: Self-referencing and persuasion: narrative transportation versus analytical elaboration. J. Consum. Res. **33**, 421–429 (2007)
2. van Laer, T., de Ruyter, K., Visconti, L.M., Wetzels, M.: The extended transportation-imagery model: a meta-analysis of the antecedents and consequences of consumers' narrative transportation. J. Consum. Res. **40**(5), 797–817 (2014)
3. Brechman, J.M., Purvis, S.C.: Narrative, transportation and advertising. Int. J. Advert. **34**(2), 366–381 (2015)
4. Chang, C.: Being hooked by editorial content: the implications for processing narrative advertising. J. Advert. **38**(1), 21–34 (2009)
5. Kleine, S.S., Baker, S.M.: An integrative review of material possession attachment. Acad. Mark. Sci. Rev. **1**(1), 124 (2004)
6. Ge, J., Gretzel, U.: Emoji rhetoric: a social media influencer perspective. J. Mark. Manag. **34**(15–16), 1272–1295 (2018)
7. Sobhanifard, Y., Sadatfarizani, S.: Triplex modeling of the political messages consumer behavior in social networks. J. Consum. Behav. **17**(3), 187–196 (2018)
8. Ferreira, B., Marques, H., Caetano, J., Rasquilha, L., Rodrigues, M.: Fundamentos de Marketing. Lisboa, Portugal: Edições Sílabo (2012)
9. Kotler, P., Keller, K.L.: Marketing Management. Prentice Hall, Boston (2012)
10. Lee, J., Hong, I.B.: Predicting positive user responses to social media advertising: the roles of emotional appeal, informativeness, and creativity. Int. J. Inf. Manag. **36**, 360–373 (2016)
11. Hilde, A.M.V., van Noort, G., Muntinga, D.G., Bronner, F.: Engagement with social media and social media advertising: the differentiating role of platform type. J. Advert. **47**(1), 38–54 (2018)
12. Mittal, B.: I, me and mine: how products become consumers' extended selves. J. Consum. Behav. **5**(6), 550–562 (2006)
13. Park, C.W., MacInnis, D.J., Priester, J., Eisingerich, A.B., Iacobucci, D.: Brand attachment and brand attitude strength: conceptual and empirical differentiation of two critical brand equity drivers. J. Mark. **74**, 1–7 (2010)

14. Sirgy, J.M.: Self-concept in consumer behaviour: a critical review. J. Consum. Res. **9**, 287–300 (1982)
15. Swaminathan, V., Page, K.L., Gürhan-Canli, Z.: My' brand or 'our' brand: the effects of brand relationship dimensions and self-construal on brand evaluations. J. Consum. Res. **34**(2), 248–259 (2007)
16. Escalas, J.E., Bettman, R.J.: You are what they eat: the influence of reference groups on consumers' connections to brands. J. Consum. Psychol. **13**(3), 339–348 (2003)
17. Richins, M.L.: Valuing things: the public and private meanings of possessions. J. Consum. Res. **21**(3), 504–521 (1994)
18. Hogg, M.A.: Subjective uncertainty reduction through self-categorization: a motivational theory of social identity processes. Eur. Rev. Soc. Psychol. **11**(1), 223–255 (2000)
19. Thomson, M., MacInnis, D.J., Park, C.W.: The ties that bind: measuring the strength of consumers' emotional attachments to brands. J. Consum. Psychol. **15**, 77–91 (2005)
20. Schmalz, S., Orth, U.R.: Brand attachment and consumer emotional response to unethical firm behavior. Psychol. Mark. **29**, 869–884 (2012)
21. Thomson, M.: Human brands: investigating antecedents to consumers' strong attachments to celebrities. J. Mark. **70**, 104–111 (2006)
22. Park, C.W., MacInnis, D.J., Priester, J.: Beyond attitudes: attachment and consumer behavior. Seoul J. Bus. **12**, 3–5 (2006)
23. Kabadayi, S., Price, K.: Consumer – brand engagement on facebook: liking and commenting behaviors. J. Res. Interact. Mark. **8**, 203–223 (2014)
24. Calder, B.J., Malthouse, E.C., Schaedel, U.: An experimental study of the relationship between online engagement and advertising effectiveness. J. Interact. Mark. **23**(4), 321–331 (2009)
25. Khan, M.: Social media engagement: what motivates user participation and consumption on YouTube? Comput. Hum. Behav. **66**, 236–247 (2017)
26. Muntinga, D.G., Moorman, M., Smit, E.G.: Introducing COBRAs: exploring motivations for brand-related social media use. Interact. J. Advert. **30**(1), 13–46 (2011)
27. Calder, B.J., Isaac, M., Malthouse, E.C.: Capturing consumer experiences: a context-specific approach to measuring engagement. J. Advert. Res. **57**(1), 39–52 (2016)
28. Schivinski, B., Dabrowski, D.: The impact of brand communication on brand equity through Facebook. J. Res. Interact. Mark. **9**(1), 31–53 (2015)
29. Leckie, C., Nyadzayo, M.W., Johnson, L.W.: Antecedents of consumer brand engagement and brand loyalty. J. Mark. Manag. **32**(5–6), 558–578 (2016)
30. Kim, J., Park, J., Glovinsky, L.: Customer involvement, fashion consciousness, and loyalty for fast-fashion retailers. J. Fash. Mark. Manag. **22**(3), 301–316 (2018)
31. Hutter, K., Hautz, J., Dennhardt, S., Füller, J.: The impact of user interactions in social media on brand awareness and purchase intention: the case of mini on Facebook. J. Prod. Brand Manag. **22**(5), 342–351 (2013)
32. Dessart, L.: Social media engagement: a model of antecedents and relational outcomes. J. Mark. Manag. **33**(5–6), 375–399 (2017)
33. Pentina, I., Guilloux, V., Micu, A.: Exploring social media engagement behaviors in the context of luxury brands. J. Advert. **47**(1), 55–69 (2018)
34. Guercini, S., Mir, P., Prentice, C.: New marketing in fashion e-commerce. J. Global Fash. Mark. **9**(1), 1–8 (2018)
35. Wang, A.: Advertising engagement: a driver of message involvement on message effects. J. Advert. Res. **46**(4), 355–368 (2006)
36. Jeon, M.M., Lee, S., Jeong, M.: Perceived corporate social responsibility and customers' behaviors in the ridesharing service industry. Int. J. Hosp. Manag. **84**, 102341 (2020)
37. Hair, J.F., Hult, G.T.M., Ringle, C., Sarstedt, M.: A Primer on Partial Least Squares Structural Equation Modeling (PLS-SEM). Sage Publications, Thousand Oaks (2016)

38. Podsakoff, P.M., MacKenzie, S.B., Lee, J.-Y., Podsakoff, N.P.: Common method biases in behavioral research: a critical review of the literature and recommended remedies. J. Appl. Psychol. **88**, 879–903 (2003)
39. Hair, J.F., Black, W.C., Babin, B.J., Anderson, R.E.: Multivariate Data Analysis. Pearson Education Limited, London (2013)
40. Chin, W.W.: The partial least squares approach for structural equation modeling. In: Modern Methods for Business Research, pp. 295–336. Lawrence Erlbaum Associates Publishers, Mahwah (1998)
41. Bagozzi, R.P., Yi, Y., Phillips, L.W.: Assessing construct validity in organizational research. Adm. Sci. Q. **36**, 421–458 (1991)
42. Fornell, C., Larcker, D.F.: Evaluating structural equation models with unobservable variables and measurement error. J. Mark. Res. **18**, 39–50 (1981)
43. Henseler, J., Ringle, C.M., Sarstedt, M.: A new criterion for assessing discriminant validity in variance-based structural equation modeling. J. Acad. Mark. Sci. **43**, 115–135 (2015)
44. Hu, L.-T., Bentler, P.M.: Fit indices in covariance structure modeling: sensitivity to under-parameterized model misspecification. Psychol. Methods **3**, 424–453 (1998)
45. Cohen, J.: Statistical Power Analysis for the Behavioral Sciences. Taylor & Francis, London (2013)

Using Social Media as Tools of Social Movement and Social Protest in Omnibus Law of Job Creation Bill Policy-Making Process in Indonesia

Arissy Jorgi Sutan[1]([⊠]) [iD], Achmad Nurmandi[1] [iD], Dyah Mutiarin[1] [iD], and Salahudin Salahudin[2] [iD]

[1] Department of Government Affairs and Administration, Universitas Muhammadiyah Yogyakarta, Bantul, Indonesia
`arissy.jorgi.psc20@mail.umy.ac.id`
[2] Department of Government Studies, Universitas Muhammadiyah Malang, Malang, Indonesia

Abstract. Social media has become a current important media of a social movement. The paper investigates the relationship between social media and social protest in the case of the Omnibus Law Protest in Indonesia and to understand the narrative content that spread in social media on the Omnibus Law. This study used social media data that is Twitter hashtags related to Omnibus Law issues. This study has been using Q-DAS (Qualitative Data Analysis Software) (Nvivo 12 plus) to collect data and analyze data with Chart analysis, Cluster analysis, and Word cloud Analysis. These research findings reveal that the social movement on Omnibus Law protest is namely through 1) Social media has been used to mobilize the activities of the action; 2) Social media has also been used to spread the voice and spread the crucial issues on the contra of Omnibus Law; 3) Social media have a strong relation to circulating the urgent issue and public policy debate. According to the Indonesian public demands, this movement's implication is the change necessary items on Omnibus Law. The limitation of this research is; 1) Only use social media data, 2) Data focused on hashtags, 3) This research focused on social media and the social movement relation and the narrative on social media. The recommendation to another article; 1) Using digital or mass source as an option, 2) If using social media data combined the account data and hashtags data, 3) Research focus on the process of a social movement.

Keywords: Social media · Omnibus Law · Protest · Social movement · Government

1   Introduction

On October 5, 2020, Indonesia Government has certified the Omnibus Law Job Creation Bill as a law and legal product. With the government making the Omnibus Law Job Creation Bill, there is the voice to reject the Omnibus Law. On October 8, 2020, they demonstrate to leave the Omnibus Law in Jakarta and another region in Indonesia. Most

© The Author(s), under exclusive license to Springer Nature Switzerland AG 2021
T. Antipova (Ed.): ICADS 2021, AISC 1352, pp. 261–274, 2021.
https://doi.org/10.1007/978-3-030-71782-7_24

of the protest participants are the college students, the laborers' alliance, and the civil society. Besides the protest, the voice of rejection about the Omnibus Law also trends on Twitter. They are using hashtags as the rising voice in the political aspect. Chang, Chu, and Welsh argue that "The advent of social media provides triple major providences for civil society's development: 1. providing alternative sources of information; 2. Reducing the cost of political participation, 3. Increasing the ability of opposition forces to mobilize" [1]. From that, we can know that in this era, social media has a massive impact on everyone to get information and also make another perspective about a political issue. There are two primary schools of thought about the effect of the development of the internet on politics. The first one claims that the different technologies provided by the internet would be at best unrelated to politics within the grand scheme of things. At worst, they have numerous adverse effects such as hemophilia, decreased institution-building, and authoritarian regimes' empowerment. The second model claims that the internet fundamentally shifts political equilibrium by dramatically lowering transaction costs, making it impossible to coordinate collective action [2].

In the case of social media and the protest of Omnibus Law, we can see hashtags to raise of voice about the controversial issue on the Omnibus Law Job Creation Bill such as the labor rights, the obligation of investors to workforce health insurance, a controversial issue about the biological environment, and press freedom. Social media also arise to coordinate the mass to get the protest action about the Omnibus Law Issue. Not just coordinate, but the social media used to get the attention and information about this issue. To get the attention and make a trending on social media about a political issue, isn't the first time, in the middle east, about the Arab springs using social media ass tool. Social media is an alternative platform to raise the voice and freedom to speak. This paradigm is relevant in this era situation, who using social media is more comfortable and more accessible than a rising voice in digital media and mass media.

There is a lot of the previous studies only focused on Social media and the political phenomenon. This study focused on two sides; the first one talks about the relationship between Social media and the Social movement [3]. The second one talks about the narrative topic or content spread in the social media about the social movement or protest. This study talks about the relation between social media and social movement and the narrative spared in social media about Omnibus Law Job Creation Bill In Indonesia. This study using a Qualitative approach. This study has been using Q-DAS (Qualitative Data Analysis Software) (Nvivo 12 plus) to collect data and analyze data with Chart analysis, Cluster analysis, and Word cloud Analysis. This paper explores the social media and social movement connection in the Omnibus Law Job Creation Bill protest. The social movement can be initiation from the social media platform. For example, the umbrella movement in 2014 happened in Hongkong. The social movement, usually using songs, posters, rituals, artwork, symbols, slogans, lectures, and social media, like the Umbrella Movement, created a lively, fluid, and participatory public political space and culture. Both the assembly of bodies in public space and the political performance expressed and enacted the people's political aspirations and visions for a different political future [4].

This study focuses on the function and relevance of social media and social movement on the Omnibus Law Protest. This paper seeks to answer the following Questions: Q1 What kind the function and the connection between social media and social movement in the Omnibus Law protest. Q2 What type of narrative that spreads in social media about the Omnibus Law Protest.

2 Literature Review

2.1 Social Media to Rising Voice

Social media and the political aspect are related to the social movement. Social media consists of a collection of Internet-based applications that enable people to build and maintain online profiles and share and receive data on their online networks. Social media offers the opportunity to understand better how communities shape and retain collective identities around political issues [5, 6]. In the United States, we can see social movements like Black Lives Matter. A social movement rooted in Black people's collective and individual experience in this country promotes constructive resistance to their lives being continuously dehumanized and devalued [7].

"Twitter user profiles who used the hashtag # BlackLivesMatter and found blacks to be more involved with the hashtag" [8]. There is a key aspect of framing made by the users. Platforms such as Twitter allow people to create their own "content" that other users can discuss or connect to, unlike print or broadcast media. ICT in this specific social media was implemented in many protests in the 21 century [9]. From the Arab spring until the umbrella movement in 2014, all use social media to uprising the voice and protest. In the fact that social media accommodate the instrument like spread video, status even until the emojis [10, 11].

Social media shows that even small groups of activists who facilitated social media could spread the information on any topic and linked collective action. There are two primary schools about the internet effect on politics; the first argues that with the "big scheme", the development of the internet brings the political situations unstable, such as homophobia, decreased institution-building, and authoritarian empowerment regimes[12]. The second told the internet development to change the political equilibrium for a short example to make the political campaign cheaper [13, 14]. Social media also bring the netizen to more accessible political knowledge. In the other side, social networking changes the logic of collective action to the logic of connective action [15, 16].

The quality and quantity of individual people about social, political talks are increased because of the variety of sources on social media and the internet. Twitter, for example, has proved a more big impact on the social-political topic in social media than the other social media platform [17]. Social media, also the open, are to mobilize and encourage change at the local, national, or international level [18, 19]. Usage of social media like Facebook, Twitter, and Blogs are real-world protest engagement indicators like in Singapore, the United States, Egypt, Chile, and then transform become a social movement [20, 21]. Personalized frames also appear in the interaction of netizens on social media [22]. Social movement strategically controls the distribution or repression of such feelings to organize and sustain the movement, and it can coordinate using social media. Without the sustainable movement, the possibility of 'slacktivism' or 'clicktivism' probability appears because they lull people believe the false causing progress [18, 23].

2.2 Social Movement in Practice

Slogans are important in the social movement and protest like the umbrella movement using slogans like we need real universal suffrage! The slogans appear because the police attack the protest using tear gas. On the other side, frontline, many various actions using in the protest like the blockade of the street, oration, even until confronting the police [24, 25]. The civil rights movement, like the traditional way in this era irrelevant. New Movement movements have this, in general, is that they've been immediately formed to fight for a particular case but with an overarching resistance to domination undertone. This situation is also what the food movement observes. What unites the food movement is not a single alternative philosophy, but a single adversary, the industrial food system [26, 27].

Civil society in the field of social interaction in various aspects forms political until the economy. Civil society can be transformed like what happened in China using economic market growth. The social movement like happened in the Hongkong shows the protection of local identity motive [28]. Activists and ordinary people use advantages from the government's relative openness to the internet and social media to post the story online through Facebook and Twitter, turning fear into anger and outrage into hope for a better humanity, coming from the back of the Arab Spring. The umbrella movement shows the desire of the people of Hong Kong to resist Beijing's to take their destiny and regain unique local cultural identities as distinct from China. Activist academics and learners dominated the Umbrella Movement scene [29]. Demonstrations that made it so efficient in mobilizing massive protests rapid might not be as suitable for maintaining longer-term activism. Instead of fostering well-developed political visions and reform strategies, such demonstrations tend to arise as spontaneous reactions to perceived wrongs. Moreover, the ten-year-old power structure of networked movements will make it difficult for them to develop the collective and institutional leadership that is usually required to maintain a campaign and make substantial progress over time [30, 31].

This study focuses on the function and relevance of social media and social movement on the Omnibus Law Protest. From all of those literature reviews, we can see that social media and political issues in this era are connected. The difference between this paper with another paper is to see the trend and connection of social media and social movement about the Omnibus Law Job Creation Bill's protest.

3 Method

This paper using a qualitative approach. The qualitative approach provides the exploration and description of the political issue, the descriptive use to give the information, and the data explanation. This study has been using the Q-DAS (Qualitative Data Analysis Software), which the toll using was Nvivo 12 plus to analytical data. This research used three different analyses; there are Chart analysis, Cluster analysis, and Word cloud Analysis. Nvivo 12 Plus is used as a tool to analyze data and collect data from social media. All data used in this research is social media data. The social data media is relevant to users to see the trending political issue in social media. This study used 14 hashtags that are as Table 1 follows.

Table 1. Hashtags and description

Hashtags	Description
#CabutOmnibusLaw	The hashtags are used to promote the rejection of Omnibus Law
#GagalkanRUUCiptaker	The hashtags were used to defeat the Omnibus Law narration
#JogjaMemanggil	The social movement in the Yogyakarta region to reject the Omnibus Law
#MosiTidakPercaya	The hashtag that used refuse to believe DPR in Omnibus Law issue
#OmnibusLawBermanfaat	The aim to make netizen think that Omnibus Law make a benefit
#omnibuslawtaataturan	The drive to image enhancement Omnibus Law as a legal product
#RakyatBukanMusuhNegara	To show that society is not the enemy of the nation, about the chaos protest
#RakyatButuhUUCiptaKerja	The aim of image enhancement Omnibus Law shows that society needs the Omnibus Law
#RuuCiptaker	Hashtags aim to protest the Omnibus Law
#SaatnyaJokowiTurun	Hashtags show some parts of society want the president to step out of office
#tenggelamkandemokrat	To shows that the democrat party must be dissolved because they reject the Omnibus Law
#MahasiswaBergerak	The slogans to raise the voice of college student
#RakyatInginJokowiTurun	Hashtags show the president must step out of the office

N-capture application of Nvivo 12 plus has been used to download the content of the hashtags. Analyze data using three different analyses: Chart analysis, Cluster analysis, and Word Cloud analysis. The chart analysis of Nvivo 12 plus has been used to analyze three words that are: "Pemerintah", "Masyarakat" and "Komunitas Aksi". The three words are related to public protest to the omnibus law issues. The cluster analysis is used to understand the content similarity of hashtags. At the same time, the word cloud has been used to know topics of the public protest on omnibus law.

4   Result and Discussion

4.1   Function and the Connection Between Social Media and Social Movement in Omnibus Law Protest

In this case, social media has an important impact on spreading the information, narrative, and coordinating to take action. In this part, we use the Nvivo 12 Pro as a tool of analysis. In the analysis, nvivo 12 plus use the social media hashtags about the social-political phenomenon of Omnibus Law Job Creation Bill In Indonesia. 14 Hashtags

are using as the main source of this article. The 14 Hashtags are #CabutOmnibus-
Law, #GagalkanRUUCiptaker, #JogjaMemanggil, #MosiTidakPercaya, #OmnibusLaw-
Bermanfaat, #omnibuslawtaataturan, #RakyatBukanMusuhNegara, #RakyatButuhUU-
CiptaKerja, #RuuCiptaker, #SaatnyaJokowiTurun, #tenggelamkandemokrat, #Maha-
siswaBergerak, #RakyatInginJokowiTurun. All those Hashtags include trending Twitter
as massive social media. The narration in social media, especially on Twitter, is split
into two perspectives. The first one is protest and rejects the OmnibusLaw Job Creation
Bill. Another one is acceptable and supports the OmnibusLaw Job Creation Bill. In
this case, we can see the trend in social media about OmnibusLaw Job Creation Bill.
The coding process using three nodes there are Community or "Pemerintah" Society or
"Masyarakat", and Community Protest or "Komunitas Aksi".

Collect data in this article using three analyses. The first one is using a chart to know
the presentation of a tweet in the hashtag. The second one is using the cluster analysis
to see the relation between the hashtags. And the last is using the word cloud analysis to
understand the popular word that appears on all of the hashtags. The result of the coding
by nvivo 12 plus about perception and the narrative content social media (Twitter) with
the chart analysis in Fig. 1 and Table 2 follows:

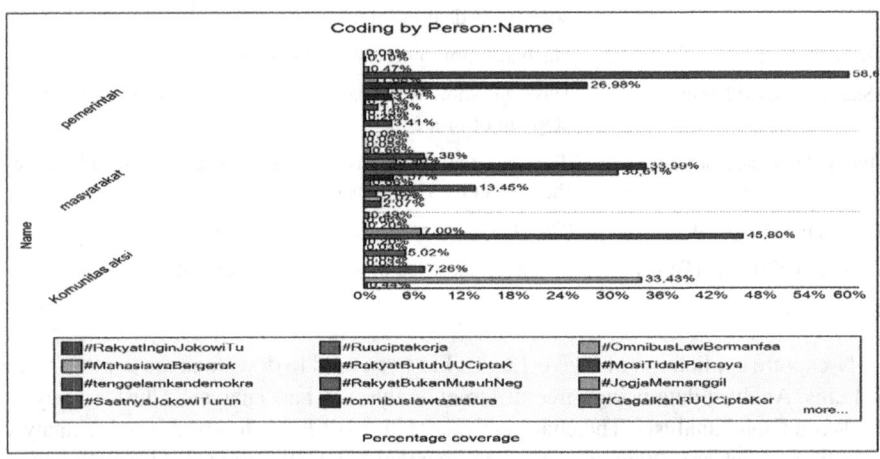

Fig. 1. The chart analysis of the hashtags

The data shows that Nodes of Komunitass aksi or Community protest the highest
result is #MosiTidakPercaya with 45,80% with rounding to 46%. The Lowest on nodes
of Community protest or komunitas aksi is ##tenggelamkandemokrat with 0%. The
Nodes Society or rakyat with the highest result is #omnibuslawtaataturan, with 33,9%
rounding to 34%. The lowest on nodes of society or rakyat is DPRRIKhianatiRakyat
and #GagalkanRUUCiptaker with 0,05% rounding to 0%. Nodes of Pemerintah or Government with the highest result is #MosiTidakPercaya with 58,6% or rounding to 59%.
The lowest result is #CabutOmnibusLaw with 0,03% or rounding to 0%. The nodes of
Masyarakat, the highest score, get the #omnibuslawtaataturan with 33,99% Coverage.

Table 2. Result chart analysis of hashtags

Hashtags	Komunitas Aksi (Community Protest)	Rakyat (Society)	Pemerintah (Government)
#RakyatInginJokowiTumbang	0%	2%	3%
#MahasiswaBergerak	33%	2%	0%
#tenggelamkandemokrat	0%	1%	0%
#SaatnyaJokowiTurun	7%	13%	2%
#Ruuciptakerja	0%	1%	0%
#RakyatButuhUUCiptaKerja	0%	4%	3%
#RakyatBukanMusuhNegara	5%	31%	3%
#omnibuslawtaataturan	0%	34%	27%
#OmnibusLawBermanfaaat	0%	4%	2%
#MosiTidakPercaya	46%	7%	59%
#JogjaMemanggil	7%	1%	0%
#GagalkanRUUCiptaKerja	0%	0%	0%
#DPRRIKhianatiRakyat	0%	0%	0%
#CabutOmnibusLaw	0%	0%	0%

The nodes Komunitas Aksi gets the highest reach the #MosiTidakPercaya with 45,80% Coverage.

These data show that the social media users to spread the narrative content and its split become two sides the acceptance of Omnibus Law and the contra with Omnibus Law. That data shows that social media is a right place to share the content of criticizing the government and also the right platform to share the doctrine by the buzzer. In this point, we can see the argument by Castells that "activists and ordinary people took advantage of the government's relative openness to the internet and social media to post the story online through Facebook and Twitter, turning fear into anger and outrage into hope for a better humanity, coming from the back of the Arab Spring" [32]. This happens in social media, especially Twitter, with all of those hashtags showing the netizen's anger and disappointment as represent of Indonesian people. The connection between all of those hashtags as Fig. 2 and the explanation in Table 3 follows.

The data shows the ten strongest relation scores between those hashtags. The highest score is relation hashtags of #MosiTidakPercaya and #JogjaMemanggil with 0,68899. In this case, we use a score between the zero lower limit and one upper limit similarity. All those hashtags are connected, but the value is weak because it's only seen using the lower limit 0 similarities and upper limit 1. Using cluster analysis is essential to know the relevant topic in social media, in this case, Twitter and the social movement. Besides that, we can see that the social movement's counter topic still appears but irrelevant to the narration and social movement topic. The explanation of those data shows that

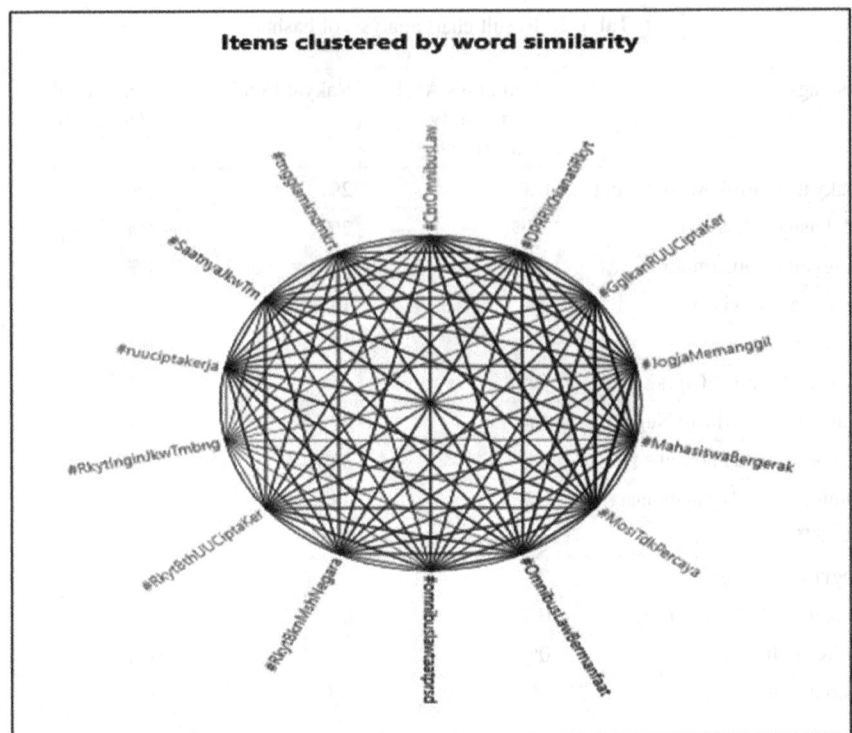

Fig. 2. The cluster analysis of the hashtags

Table 3. Ten highest score of cluster analysis

Hashtags 1	Hashtags 2	Relation Score
#MosiTidakPercaya	#JogjaMemanggil	0,68899
#MosiTidakPercaya	#CabutOmnibusLaw	0,680913
#MosiTidakPercaya	#GgalkanRUUCiptaKerja	0,652245
#MosiTidakPercaya	#DPRRIKhianatiRakyat	0,628695
#JogjaMemanggil	#CabutOmnibusLaw	0,588144
#OmnibusLawBermanfaat	#MosiTidakPercaya	0,587053
#MahasiswaBergerak	#DPRRIKhianatiRakyat	0,582743
#GagalkanRUUCiptaKerja	#DPRRIKhianatiRakyat	0,565542
#MosiTidakPercaya	#MahasiswaBergerak	0,562233
#omnibuslawtaatprosedural	#OmnibusLawBermanfaat	0,546612

the connection of all the hashtags is not strong enough. The relationship can see in the zero lower limits and one upper limit. All are connected. With this analysis, we can see that the more vital link or connection has the same content and goals. Like the #MosiTidakPercaya and #JogjaMemanggil with 0,682368 points, this connection has the same range and goals to reject the Omnibus Law.

The social movement sees as a collective challenge of several groups interest who has the same goals, and solidarity in interaction with elite, opponents, and government [33]. The social movement has characteristics like informal interaction network, feeling, and mutual solidarity, conflict as the focus in collective action, priority on a form of protest. If this social media phenomenon and social movement about the Omnibus Law protest, we tendency of the social movement as a collective behavior perspective. Social movement as a collective behavior is a perspective to see the social movement that appears from injustice, disappointment, and dissatisfaction. And become one in action, and none single government can't respond to the group's disappointment [33]. In this case, the disappointment and injustice felling when the Omnibus Law only make it profitable to investors and authorities. And the social movement spread the narrative and content using social media. Besides that, social media is also used to coordinate mass to take action protests.

The main point in these parts is the function and relation of social media and social movement in the case of the protest Omnibus Law Job Creation Bill is:

1. Social media is used to coordinate the act of movement. Social media is the right platform to coordinate mass, and almost the action community using the social media ass tool to coordinate mass.
2. Social media is also used to spread the voice and spread the information about the rejection of Omnibus Law. Using hashtags is an efficient step to make the topic is trending on social media. But on the other side, social media is also used to spread the doctrine by the buzzer.
3. Social movements and social media have a strong relation to spread the urgent issue. The type of social movement can see using the social movement as a collective behavior perspective.

For those explanations, we can see that there are connections between social media and social movement, but it's not strong enough; we can see only using lower limit 0 point and upper limit 1 point using Nvivo 12 plus. We can see the type of social movement from the Collective Behavior perspective that shows the social movement appears because it has the same feeling about injustice, disappointment, and anger about the Omnibus Law. Social media has the function to Spread the narrative content and voice about the social, political theme, in this case, Omnibus Law. On the other side, social media also used to coordinate the mass using social media we can see with the rejection of Omnibus Law using the hashtags to protest and make a social movement like happened on October 8, 2020.

4.2 The Narrative that Spreads in Social Media About the Omnibus Law Protest

To know the narrative that spread in social media about the Omnibus Law protest, we can see using the word cloud analysis using a word cloud analysis to understand what kind of discussion in public space like social media. In a word, the cloud shows 100 words most appear in social media, all from the hashtags; the results in Fig. 3 follow.

Fig. 3. The word cloud analysis

The most important word in the word cloud is "Masyarakat and Pemerintah", which is from the hashtags most appear is the word of "Masyarakat" and "pemerintah". On the other side, a phrase like Aksi, Serikat, hoax, and disinformation, and "melakukan" is the other word with a big presentation to more of this topic. Not just a single word, but there are also appear hashtags too; the first is #mositidakpercaya,#omnibuslawtaatprosedural, later #jogjamemanggil, and then #saatnyajokowiturun.

In either then there is a word of the profession like polisi, buruh, pengamat, LSM, mahasiswa, etc. from the word cloud, we can see that the most appear word is a universal and neutral word to use by the hashtags of omnibus law, like the accept the omnibus law or also the opposite who rejected the omnibus law. Why the biggest word like Masyarakat and pemerintah using by the two sides, its because the topics of pro and contra the omnibus law brings effect to masyarakat and pemerintah, its mean that masyarakat and pemerinath in this topic of omnibus law are like the subject and object the subject is pemerintah, and the object is masyarakat. So those two word shows as a position of the narration of topics.

Form the word analysis; we can know that the content narrative to the social media protest of omnibus law is using general and also specific word. The general word such as "Masyarakat, Pemerintah, melakukan" in the other side there is the specific word like "aksi, Serikat, hoax, disinformation' and many other. Framing also appears in social media like happened in Argentina using hashtags # Tarifazo that become two sides of the corruption topic and wasteful protest [34]. In the Omnibus Law Protest, the farming flew it by the buzzer. The buzzer using unmatchable hashtags like #omibuslawtaataturan, #OmnibusLawBermanfaat, and #TenggelamkanDemokrat. They are using the unmatchable hashtags, its' to make a contra the major of hashtags that from narrative spread contra about Omnibus Law. Why the content, social media can affect the social movement, so lets us think about this argument. Protests, many have seen ICT as indispensable: "For example, social media use was so prevalent during the Arab Spring that numerous commentators dubbed the events as some permutation of" the (Twitter/Facebook) and (Uprisings protest)" [35]. The uniqueness of Social media is the social media can deliver all variant information and all connected, and the netizen can interact and make a Discussion Forum without face to face. Still, the essential value can be interpreted directly or indirectly. The main point in these questions is:

1. The narrative of hashtags is split into two perceptions. The first become to accept the omnibus law. And the second one is the contra about omnibus law. Both of those Hashtags have the most similar word using. And split into the general like "Pemerintah, Masyrakat" and the specific like "Aksi, Serikat, Hoax, Disinformasi" and many others. The difference is the general word using by both perceptions. And the specific use in the specific perspective.
2. The most word shows that what the public thinks about the Omnibus law Issue. And there are also using framing to counter the issue by the buzzer. A comment like #omnibuslawtaataturan, flew up in social media to counter the Omnibuslaw Issue that trend in social media.

At this point, we can see that the narrative that spread in social media defied into two sides the accept and the contra. Both of them using the universal word like "Masyarakat" and "Pemerintah" that using the accept and contra side as major word content. But they are also using the specific word like "aksi", "serikat", "hoax", and many others. The most significant word means the most used word by the two sides in social media. Besides that, social media also shows the like hashtags #omnibuslawtaataturan to farming the content. All of those words and hashtags using to counter the argument and counter content in social media about the Omnibus Law to affect the netizen.

5 Conclusions

Social media has been part of the social movement. Social media can deliver information and issue and coordinate them into an act of social movement. Social media and social movement in the Omnibus Law Protest, using social media, can coordinate using social media and spread the issue real fast on social media. Social media also the rights way to raise the voice of netizens to criticize the Omnibus Law issue. On the other side, social media content can even be split into different perspectives; both use hashtags to elaborate on the opponent. Social movement about the Omnibus Law Issue is the collective behavior perspective.

There is the limitation of this research that is *first,* research data uses social media data. The social media data choose to get the perception of netizens about the Omnibus Law Issue. Besides that, Social media data use has developments in existing topics. *Second*, data Focused only on hashtags, which have spread in social media with various main points. *Third,* this research focused on social media and the social movement relation, and the narrative on social media. It does not include why the protest happened, what kind the protest happened, and the protest's perspective or the buzzer.

Recommendation to another article who wants to take Omnibus Law Job Creation Bill and social media as the theme there some suggest: *first*, the data also using digital data, like the online newspaper, the protest video is using to bring the new perspective in the research. *Second*, using social media data can combine hashtags data and account data. It probably helps check the Shifting content in social media. *Third*, research focus on the process of a social movement.

References

1. Bui, T.H.: The influence of social media in Vietnam's elite politics. J. Curr. Southeast Asian Aff. **35**(2), 89–111 (2016). https://doi.org/10.1177/186810341603500204
2. Wilson, S.L.: Detecting mass protest through social media. J. Soc. Media Soc. **6**(2), 5–25 (2017). https://www.thejsms.org/index.php/TSMRI/article/view/239/123
3. Kholid, A., Husein, R., Mutiarin, D., Listiya E.R.S.: The influence of social media towards student political participation during the 2014 indonesian presidential election. J. Gov. Polit. **6**(2) (2015). https://doi.org/10.18196/jgp.2015.0019
4. Kim, G.J.: Theological reflections on the Hong Kong umbrella movement (2016)
5. Ozturkcan, S., Kasap, N., Cevik, M., Zaman, T.: An analysis of the Gezi Park social movement tweets. Aslib J. Inf. Manag. **69**(4), 426–440 (2017). https://doi.org/10.1108/AJIM-03-2017-0064
6. Brown, M., Ray, R., Summers, E., Fraistat, N.: #SayHerName: a case study of intersectional social media activism. Ethn. Racial Stud. **40**(11), 1831–1846 (2017). https://doi.org/10.1080/01419870.2017.1334934
7. Ince, J., Rojas, F., Davis, C.A.: The social media response to black lives matter: how Twitter users interact with black lives matter through hashtag use. Ethn. Racial Stud. **40**(11), 1814–1830 (2017). https://doi.org/10.1080/01419870.2017.1334931
8. Gready, P., Robins, S.: Rethinking civil society and transitional justice: lessons from social movements and 'new' civil society. Int. J. Hum. Rights **21**(7), 956–975 (2017). https://doi.org/10.1080/13642987.2017.1313237
9. Munmun, D.C., et al.: Social media participation in an activist movement for racial equality. Physiol. Behav. **176**(3), 139–148 (2017)
10. Aitchison, B., Soundy, A., Martin, P., Rushton, A., Heneghan, N.R.: Lived experiences of social support in Paralympic swimmers: a protocol for a qualitative study. BMJ Open **10**(9), e039953 (2020). https://doi.org/10.1136/bmjopen-2020-039953
11. Brunner, E.: Wild public networks and affective movements in China: environmental activism, social media, and protest in Maoming. J. Commun. **67**(5), 665–677 (2017). https://doi.org/10.1111/jcom.12323
12. Burgess, J., Marwick, A., Poell, T., Poell, T., van Dijck, J.: Social media and new protest movements. In: SAGE Handbook Social Media, pp. 546–561 (2017). https://doi.org/10.4135/9781473984066.n31
13. Setiawan, D., Nurmandi, A.: Sandiaga Uno: personal branding di Twitter. J. Public Policy **6**(1), 19 (2020). https://doi.org/10.35308/jpp.v6i1.1657
14. Rahmat, A.F., Purnomo, E.P.: Twitter media platform to set-up political branding: analyzing @Kiyai_Marufamin in 2019 presidential election campaign. Nyimak J. Commun. **4**(1), 73 (2020). https://doi.org/10.31000/nyimak.v4i1.2268
15. Chu, D.S.C.: Media use and protest mobilization: a case study of umbrella movement within Hong Kong schools. Soc. Media Soc. **4**(1) (2018). https://doi.org/10.1177/2056305118763350

16. Lee, S.: The role of social media in protest participation: the case of candlelight vigils in South Korea. Int. J. Commun. **12**, 18 (2018)
17. Jost, J.T., et al.: How social media facilitates political protest: information, motivation, and social networks. Polit. Psychol. **39**(3), 85–118 (2018). https://doi.org/10.1111/pops.12478
18. Poell, T., Abdulla, R., Rieder, B., Woltering, R., Zack, L.: Protest leadership in the age of social media. Inf. Commun. Soc. **19**(7), 994–1014 (2016). https://doi.org/10.1080/1369118X. 2015.1088049
19. Ouassini, A.: We are all amina filali: social media, civil society, and rape legislation reform in Morocco. Women Crim. Just. **31**(1), 1–6 (2019). https://doi.org/10.1080/08974454.2019. 1698488
20. Zhu, Q., Skoric, M., Shen, F.: I shield myself from thee: selective avoidance on social media during political protests. Polit. Commun. **34**(1), 112–131 (2017). https://doi.org/10.1080/105 84609.2016.1222471
21. Kharroub, T., Bas, O.: Social media and protests: an examination of twitter images of the 2011 Egyptian revolution. New Media Soc. **18**(9), 1973–1992 (2016). https://doi.org/10.1177/146 1444815571914
22. Clayton, D.M.: Black lives matter and the civil rights movement: a comparative analysis of two social movements in the United States. J. Black Stud. **49**(5), 448–480 (2018). https://doi. org/10.1177/0021934718764099
23. Specht, D., Ros-Tonen, M.A.F.: Gold, power, protest: digital and social media and protests against large-scale mining projects in Colombia. New Media Soc. **19**(12), 1907–1926 (2017). https://doi.org/10.1177/1461444816644567
24. Stevens, T.M., Aarts, N., Termeer, C.J.A.M., Dewulf, A.: Social media as a new playing field for the governance of agro-food sustainability. Curr. Opin. Environ. Sustain. **18**, 99–106 (2016). https://doi.org/10.1016/j.cosust.2015.11.010
25. Kou, Y., Kow, Y., Gui, X., Cheng, W.: One social movement, two social media sites: a comparative study of public discourses. Comput. Supp. Coop. Work (CSCW) **26**(4–6), 807–836 (2017). https://doi.org/10.1007/s10606-017-9284-y
26. Lee, F.L.F., Chan, J.M.: Digital media activities and mode of participation in a protest campaign: a study of the Umbrella Movement. Inf. Commun. Soc. **19**(1), 4–22 (2016). https:// doi.org/10.1080/1369118X.2015.1093530
27. Lee, F.L.F.: Internet alternative media, movement experience, and radicalism: the case of post-Umbrella movement Hong Kong. Soc. Mov. Stud. **17**(2), 219–233 (2018). https://doi. org/10.1080/14742837.2017.1404448
28. Dong, T., Liang, C., He, X.: Social media and internet public events. Telemat. Inform. **34**(3), 726–739 (2017). https://doi.org/10.1016/j.tele.2016.05.024
29. Wang, J., et al.: Effect of digitalized rumor clarification on stock markets. Emerg. Mark. Financ. Trade **55**(2), 450–474 (2019). https://doi.org/10.1080/1540496X.2018.1534683
30. Wang, Y.: Local identity in a global city: Hong Kong localist movement on social media. Crit. Stud. Media Commun. **36**(5), 419–433 (2019). https://doi.org/10.1080/15295036.2019.165 2837
31. Ley, B.L., Brewer, P.R.: Social media, networked protest, and the march for science. Soc. Media Soc. **4**(3), (2018). https://doi.org/10.1177/2056305118793407
32. Ouassini, A.: The Ummah racial project: Arab satellite television, Islamic movements, and the construction of Spanish Moroccan identity. Ethn. Racial Stud. **43**(4), 751–767 (2020). https://doi.org/10.1080/01419870.2019.1587174
33. Manulu, D.: Gerakan Sosial Dan Perubahan Kebijakan Publik Kasus Perlawanan Masyarakat Batak vs PT. Inti Indorayon Utama, di Porsea, Sumatera Utara. Populasi **18**(1), 27–50 (2016). https://doi.org/10.22146/jp.12066

34. Aruguete, N., Calvo, E.: Time to #protest: selective exposure, cascading activation, and framing in social media. J. Commun. **68**(3), 480–502 (2018). https://doi.org/10.1093/joc/jqy007
35. Little, A.T.: Communication technology and protest. J. Polit. **78**(1), 152–166 (2016). https://doi.org/10.1086/683187

Promoting Cities as Cultural Destinations Through Events. Case Study: Aarhus European Capital of Culture

Claudiu Coman[1](✉) (iD), Maria Cristina Bularca[1] (iD), and Adrian Otovescu[2]

[1] Transilvania University of Brasov, Brasov, Romania
claudiu.coman@unitbv.ro,
maria-cristina.bularca@student.unitbv.ro
[2] University of Craiova, Craiova, Romania
adiotovescu@yahoo.com

Abstract. In the context of today's fast developing society in which peoples' preferences are constantly changing, in order to gain competitive advantages, cities must develop powerful brands and adopt effective strategies to promote themselves. Thus, having the title of European Capital of Culture can be a major opportunity for cities to enhance their cultural and social life and thus to promote themselves as cultural and tourist destinations. Even more, events play an important part in the life of a city by providing new possibilities for entertainment, by improving the quality of life of the citizens, by favoring intercultural communication. In this regard, the purpose of the paper was to assess the way the city of Aarhus promoted itself through one of its events, during the time it was European Capital of Culture order to extract information and strategies that can be used as a frame of reference by other cities. While analyzing a representative event for the overall strategy of the program - the only opera festival for children in Denmark, our findings reveal that Aarhus used culture as a method to innovate and renew its cultural and social dimensions. The event was aligned with the concept of the program, it had a well consolidated strategic plan and managed to determine new audiences to take part in the cultural activities of the city.

Keywords: Cultural destination · City promotion · Capital of culture

1 Introduction

In today's society people's preferences in terms of cultural, touristic and social activities are constantly changing. With people being always on the move and having the liberty to travel or live in any place they desire, cities are in the need to effectively build their brands and to promote themselves in order not to just attract tourists, but also to offer them experiences that live up to their expectations.

In this context, a growing competition between cities can be observed, and in order to become more visible and attractive, cities develop strategies in which process like marketing and branding have an essential role. [1] In this regard, obtaining the title of

© The Author(s), under exclusive license to Springer Nature Switzerland AG 2021
T. Antipova (Ed.): ICADS 2021, AISC 1352, pp. 275–286, 2021.
https://doi.org/10.1007/978-3-030-71782-7_25

European Capital of Culture can be a major opportunity for cities to grow, build strong identities, gain popularity and create positive associations among people. Even more, organizing events that appeal to both to citizens and tourists is also important for the development of the city and of people's feeling of belonging to that city.

Taking into account the aspects previously mentioned and the idea that having the title of European Capital of Culture can provide cities with the opportunity to develop alternative methods of promotion, the paper focuses on analyzing the activities carried out by the city of Aarhus in the year it held the above mentioned title. Thus, the purpose of the paper is to assess the way the city of Aarhus promoted itself through the events it organized, during the period in which it was European Capital of Culture, in order to extract information and strategies that can be used as a frame of reference by other cities in their promotion strategies.

2   Literature Review

2.1   Place Branding

Brand and branding are two concepts which have been defined in multiple ways over the years, and applied to products, services, cities, nations or institutions. While a brand can be considered a name, a logo, image, symbol or a combination of these elements that aims to differentiate a certain product or service [2], branding can be seen as the process of building and managing identity, the process that adds value and vitality to products or services [3]. When the international commerce started to develop and flourish in the 14th century, many forms of branding appeared, such as: services branding, corporate and place branding [4].

In a broad and simple way, place branding refers to strategies and activities that are meant to improve the image and the brand of a place, thus being "a way to make places famous" [5]. However, in comparison with product or service branding, there are some different aspects that must be taken into account in place branding. In this regard, Simon Anholt states that when branding places, it is very important for public institutions, the government and members of the society to clearly and effectively communicate with each other if they want to influence the opinion of the public [6].

Place branding is also seen as the activities developed by local governments in order to attract people and increase the economic and social capital of a place: city, country. In this regard many governments try to promote a positive image regarding the quality of life of people living in that certain place, by adopting specific slogans, or by organizing specific events such as festivals, or sports events [7]. In other words, place branding refers to attracting people who can then "choose to live, work, study or just visit a certain place" [8].

Place branding is closely related to place marketing, and in order to successfully brand a place, attention should be payed to both marketing and branding approaches [9]. In place marketing, the place must be developed in a way that is adapted to the needs of the consumers [10], and in place branding, the concepts of image and identity are also relevant. Brand identity is the element that helps creating and establishing the essence of the brand [11]. Identity is managed by the brand, or in this case, by the city, and it can also be considered an asset, that helps the brand differentiate itself, express its unique characteristics, build awareness and make the brand more visible [12]. Visual identity also plays an important part in branding a place. Thus, in the process of differentiation, brands could gain competitive advantage through the logo, slogan and symbols that they use, and symbols that are easy to remember can help the place (city, region) position itself and be more visible [13]. The image of a brand represents the way consumers perceive the brand and it consists of people's beliefs and opinions about the brand [14]. Thus, a place brand is also based on people's associations about the place and about what that place offers [15].

Destination and City Branding

Destination branding and city branding are concepts that are often used interchangeably. Strategies used in order to brand destinations and places emerged around the 90's when cities like New York developed and adopted strategies meant to help them build the image of the city through logos and slogans such as "I love New York" [16], that can be found in Fig. 1. Other clever examples of city branding strategies are those of Amsterdam, which adopted the slogan "I amsterdam", with emphasis on "I am" [17], which is represented in Fig. 2, Lyon, with "OnlyLyon" slogan [18], Berlin, with "Be-Berlin", Seoul with "I Seoul U" [19].

Fig. 1. New York logo and slogan. Source: BestDesigns, https://www.bestdesigns.co/best-design/the-i-love-new-york-logo-is-an-iconic-widely-imitated-tourism-symbol

Fig. 2. Amsterdam logo and slogan in 3D letters. Source: The conversation, https://theconversation.com/rescaling-through-city-branding-the-case-of-amsterdam-71956

The process of destination branding is complex, it involves cooperation and revolves around the brand's image and identity. Thus, Cai Liping's model of destination branding places identity at the center of the process, and states that this type of branding involves relations between brand image building and brand element mix [20].

City branding can be considered an important process that helps a city gain competitive advantage, develop economic capital, tourism, but also a process that determines citizens and the local community to identify with the city [21]. In this regard, the objective of city branding does not only resume to the development and promotion of the elements that make the city special and attractive to outsiders, but also to residents, thus increasing their sense of belonging [22]. Thus, city brands cannot be built over night: the process is complex and involves leadership, collaboration between authorities, government and members of the community [23].

The city brand has become today an essential tool for promoting tourism and is mainly used to create the image of the city [24], as well as to build its reputation. While city branding encompasses elements such as: uniqueness, authenticity and image [25], the image created and projected must correspond to the reality. In this context city branding requires concrete actions, such as urban development, building the community, or organizing events [26].

City branding must also be approached in relation to the concept of culture. Thus, city branding can be the process that helps cities progress and thrive and a main preoccupation of cities should be to find ways to combine "culture, history, social development, infrastructure, architecture and landscapes in order to create an identity that is accepted and shared by most people" [27]. Thus, the importance of culture for a city brand is highlighted by the fact that it is often used as a technique to renew and improve the economic, social or environmental dimensions of the city [24].

2.2 Role of Events in Promoting Cities as Tourist Destinations

More and more people today desire to live in cities with rich cultural life, that offer them many opportunities and that are open and tolerant with different cultures [28].

The cultural resources of a city are represented by many elements such as: local history, festivals and events, art activities, places of entertainment, universities and architectural buildings [29]. Thus, it is important for a city to have a rich and vibrant cultural life [30], and events are a major part of the cultural life of a city.

Events and cultural activities have an essential role in developing and promoting cities. Events help the city gain competitive advantage, they can offer new meanings to a city, they provide new opportunities for entertainment, they improve the quality of life of the city, they can help the city develop partnerships, they promote intercultural communication and bring together people from different cultures [31]. Even more, events create dynamic mediums of communication for stakeholders, and many cities base their promotion strategies on events: Barcelona [32], Edinburgh who developed its cultural image through international festivals [33], or Budapest, which is focused on mega-events like Sziget Festival [34].

2.3 European Capital of Culture Program

The European Capital of Culture program offers many opportunities for cities to develop and promote themselves as cultural and tourist destinations. The program was developed in 1985 as a way of "celebrating European cultural diversity" [35]. Every year, European cities compete in order to gain the title of European Capital of Culture, (more than 40 cities already had this title), and events organized on this occasion generally positively impact the city, and provide opportunities for its cultural, social and economic development [36]. Thus ECOC events, can represent and alternative method for promotion and development, for they facilitate the projection of the desirable image to tourists and investors [37]. In this context, the cities that implement successful programs and events are the ones that focus on creating a strong identity that citizens can relate to, as well as on projecting an image that can enhance the city's cultural identity [38]. The first city who was awarded this title was Athens in 1985 [39], and over the years many other cities managed to implement successful programs among which we mention: Cork, in 2005 with "The city of making" program [35], and Aarhus in 2017 [39].

In 2017 Aarhus won the title of European Capital of Culture and thus implemented a very complex program. The entire program revolved around the "Let's Rethink" concept, according to which the city carried out activities throughout the year. The concept and the logo for Aarhus 2017 are represented in Fig. 3 With an ambitious and creative program, Aarhus desired to "transform the Central Denmark Region into a cultural laboratory" [40]. "Let's Rethink" was more than just a theme, it was "a mindset for change, innovation and courage", meant to consolidate the identity of the city [40]. While having the vision to use art and culture as a means for rethinking challenges, and the mission to create sustainable development in terms of cultural, social and economic growth [41], the city also wanted to redefine itself as an international city whose main key elements are culture and creativity [42]. The city wanted to rethink and improve every aspect of its cultural and social life, and it was focused on organizing creative events and mega events that would bring together people with different ages, cultures or interests. Thus, there were organized many street performances, (Off road festival), visual festivals (Visual Art 2017), music and theater festivals, as well as children events (GrowOP festival), [43]. The overall impact of the program was a positive one: 60% of citizens attended the events, more than 1,950 new full time jobs were created and the program received 92% positive feedback from the public [41].

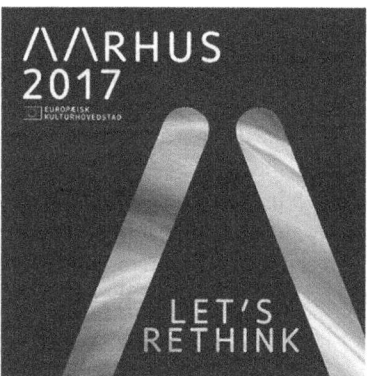

Fig. 3. Logo and concept of Aarhus 2017. Source: https://via2017.via.dk/kulturbyaebning-fra-vias-campus-c

3 Present Study

The present study focuses on analyzing the way the city of Aarhus promoted itself through one of its main events during the time it was European Capital of Culture.

Situated in the Central Denmark Region, Aarhus is the second largest city in Denmark. It has a rich cultural life that encompasses many museums, festivals, it is an educational center, Aarhus University having more than 40000 students, and it is a dynamic city that promotes progress and healthy life style: 20% of trips in Aarhus are being made by bike. [44] Aarhus tried to use culture in a creative way and innovate its cultural life on every level, and the program it developed had positive outcomes. In this context, we considered relevant and useful an analysis of one of the events it organized in order to discover strategies that brought Aarhus success. In this regard we focused our research of one of the city's representative and strategic events, GrowOP! Festival, an opera festival for children, whose interesting logo is represented in Fig. 4.

Fig. 4. GrowOP! Festival logo. Source: International community, https://internationalco mmunity.dk/en-US/News/Nyhedsvisning?Action=1&NewsId=4015&currentPage=69&M=New sV2&PID=261

4 Methodology

4.1 Purpose and Objectives

The purpose of the research was to assess the way the city of Aarhus promoted itself through its events, during the time it was European Capital of Culture, in order to extract information and strategies that can be used as a frame of reference by other cities in their development and promotion strategies.

The objectives of the research include: identifying the main strategies that the event used in order to promote the city and attract the target audience, analyzing how the event was in line with the overall concept of the program, identifying the messages sent to the public through the event, and how the event contributed to the development of the city's cultural life.

4.2 Method

The method used in order to conduct the research was content analysis. Although the city organized many thematic events, the event we analyzed, GrowOP! Festival, is representative for the strategy adopted by Aarhus in its program, due to the fact that it was a new event that targeted a different kind of audience.

Thus, in order to conduct the research, firstly the official website of the Aarhus 2017 program was analyzed, in order to see how the event is presented within the program. Next, the official website of the event was analyzed, in order to discover the vision of the organizers, their objectives, the way the event was carried out, what type of techniques were used to promote the event. Moreover, other websites that mentioned the event and its outcomes were analyzed too.

5 Results and Discussion

5.1 The Concept of the Event

Our research revealed that GrowOP! Festival is an innovative and creative event focused on increasing the interest of children and young people for a type of cultural activity not so popular among them, opera shows. In this regard, the event is in line with the overall concept of the city, because it tries to think of ways to develop and innovate musical theatre. GrowOP! Festival revolves around the idea that even from a young age, children should be exposed to cultural activities such as opera shows. The festival took place between 11 and 26 November 2017, and is the only opera festival in Denmark that is dedicated entirely to children and teenagers [45]. The event desires to rethink and change the way opera is perceived, thus transforming it into a modern art for children. The organizers of the event wanted to create a link between opera and children and determine this new type of audience to participate more in opera events [46]. The 2017 edition included, in addition to opera performances in which music was combined with theater and fairy tales, the newest version of the series of opera for children aged 0–3 years, but also interactive workshops, an international seminar on art and infancy that emphasized how important it is for babies to come in contact with music, and a seminar on youth and talent.

GrowOP! Festival also tries to optimize the collaboration between European institutions and to give young people and children the chance to share ideas and experiences, having as a central point opera presented in new forms: with stories and characters from fairy tales. The performances included "Fogonogo" a show presenting children adventures while doing various experiments, "Little red biting hood", and also "Snow white", operas that had a language appropriate for the age of the children but also a message that was easy to understand. However, although it also had many international guests, one disadvantage of the event was the fact that some of the performances were only presented in Danish, thus making the people who did not speak Danish unable to attend those shows.

5.2 The Objectives of the Event

Through its new approach on opera and musical theatre, the event desired to help the city become more visible. Even more, the objectives of the event refer to involving more children and teenagers in cultural activities, developing collaborations with other opera institutions, and creating positive associations among people regarding the initiatives of the city of Aarhus and this type of music.

5.3 Target Audience

The unique character of the event is given mainly by the audience it targeted: children, teenagers, and families with children. The event was especially designed for children but the opera shows were also promoted as a cultural event that parents could attend together with their children.

5.4 Messages Sent to the Public

While communicating with the target audience of the event, the organizers used messages that highlighted the distinctive character of the event, the concern for the participation of children to cultural activities even from an early age. The organizers emphasized that GrowOP! is the only opera event that is entirely created and adapted for children, that it does not only offer musical shows but also workshops and seminars were children could learn about music and bond with their parents.

5.5 Strategies and Tactics

The research revealed that, in order to develop the event, promote and communicate information about it, several strategies were used. The first strategy, defined as the "Artscapes remix" [47] is focused on rethinking or creating an interesting theme for the event. While implementing this strategy, GrowOP! managed to stand out from the crowd by promoting opera shows as suitable for a new audience - children, thus bringing this type of cultural events closer to a category of people that did not used to or did not had the intention to participate in opera shows.

In order to promote and create positive associations among people, the organizers also adopted the strategy of involving the community in the event, strategy that generates commitment and brand attachment. Citizens were offered the possibility to volunteer at the event and have the chance to also attend the shows.

Another strategy that the event adopted was the strategy of increasing cultural mobility in the region, by attracting and developing new audiences. People were attracted to this event because the shows were presented in a new and interesting manner, they were focused on heroes and fairytales, and the language was adapted so that children would easily understand it. Thus, because the event was designed for children and families with children that usually do not attend this type of events, through its creative approach, GrowOP! Festival increased cultural mobility in the city of Aarhus.

The strategy adopted for communicating with the public, the media mix strategy is well consolidated. The event promoted itself both in the online and the offline environment. Online, it integrated social media instruments and the information was communicated mainly through the official website of the event, through its Facebook page, and Youtube page where shows were briefly presented, but also through websites of the partners of the event such as, Den jyske opera. Offline, information about the festival was given on the radio station Da Radio Denmark, but also on TV on the TV channel Danmark 1 and on channels for children. The event was also promoted through brochures containing the program of the event.

5.6 Outcomes

Being part of the Aarhus 2017 program, the development of GrowOP! Festival had positive outcomes. Tickets for most activities and shows were sold out, the performances of the national and international guests (UK, Portugal, Italy, Hong Kong) was appreciated and the festival was considered an event that provided high quality artistic acts. Even more, the positive impact on the cultural life of the city was proven by the Prize for innovation of the year that it received in 2018.

6 Conclusions

In the context of globalization, cities compete not only in order to gain popularity and attract tourists, but also to be recognized as powerful, developed cultural centers, that people would also like to live in. When trying to define themselves and create their brand, cities must focus on building their identity and must take into account the opinions and belief of the citizens.

The purpose of our paper was to assess the way the city of Aarhus promoted itself through events it organized during the time it was European Capital of Culture. Thus our paper shows that through its overall concept, "Let's rethink", that is very well embedded in the GrowOP! Festival, Aarhus managed to improve the cultural life of the city, by using culture as an element to innovate itself.

Thus, we analyzed a representative event for the strategic approach of the entire program, the GrowOP! Festival. Our findings are in line with other studies regarding the role of events in the life of a city [31], and reveal that the festival analyzed provided new opportunities for entertainment, it improved the cultural life of the city by attracting younger audiences, it helped the city to differentiate itself and also be recognized for hosting the only opera event designed for children.

The concept of the festival was innovative, creative and aligned with the overall concept of the program. The festival had clear objectives, it sent messages that were meant to highlight the opportunities people could have by attending the events and it emphasized how important it was for children to participate from a young age, to cultural activities. However, one aspect that represented a disadvantage for the festival was that some shows were only presented in Danish and only people who spoke Danish were able to attend them. The strategic plan of the event was well consolidated: the festival used methods to differentiate itself, it offered community the chance to be involved and contribute to developing the cultural life of the city, and it oriented itself towards an audience that was not usually fond of the shows it provided managing to make the public participate, thus creating positive associations among them.

As other studies suggest [37], our paper supports the idea that events are an essential part of the cultural life of the city, and that the title of European Capital of Culture can be used as an alternative method of promoting and developing cities. Even more, since Aarhus organized events in which it used culture as a way to renew its cultural and social dimensions, and succeeded, our paper is in line with another paper that states that culture favors innovation, cultural, economic and social growth [24]. Thus, the research provides information regarding the way the city of Aarhus managed to thrive while holding the title of European Capital of Culture, through the help of one of the events it organized.

Therefore, while the event we analyzed and the strategies adopted by it can be used as a frame of reference for other cities that desire to improve and promote themselves as cultural destinations, the research also has some limitations. One limitation is represented by the fact that only one event and one city was analyzed. In this regard, a future research should expand and should take into consideration the analysis of other cities that were European Capitals of Culture in order to make a comparative analysis.

References

1. Petrea, R., Petrea, D., Olău, P.E., Filimon, L.: Place branding as efficient management tool for local government. Transylv. Rev. Adm. Sci. **9**(SI), 124–140 (2013)
2. Kotler, P., Keller, K.L., Brady, M., Goodman, M., Hansen, T.: Marketing Management. Pearson Education Limited, London (2009)
3. Bastos, W., Levy, S.J.: A history of the concept of branding: practice and theory. J. Histor. Res. Mark. **4**(3), 347–368 (2012)
4. Briciu, V.A., Briciu, A.: A brief history of brands and the evolution of place branding. Bull. Transilv. Univ. Brasov **9**(58), 137–142 (2016)
5. Anholt, S.: Definitions of place branding – working towards a resolution. Place Brand. Public Dipl. **6**, 1–10 (2010). https://doi.org/10.1057/pb.2010.3
6. Anholt, S.: Place branding: is it marketing, or isn't it? Place Brand. Public Dipl. **4**(1), 1–6 (2008). https://doi.org/10.1057/palgrave.pb.6000088
7. Andersson, I.: Geographies of place branding: researching through small and medium sized cities. Doctoral dissertation, Stockholm University (2015). https://www.diva-portal.org/smash/get/diva2:798832/FULLTEXT01.pdf. Accessed 22 Nov 2020
8. Briciu, V.A.: Differences between place branding and destination branding for local brand strategy development. Bull. Transilv. Univ. Braşov Ser. VII Soc. Sci. Law **1**, 9–14 (2013)
9. Govers, R.: From place marketing to place branding and back. Place Brand. Public Dipl. **7**(4), 227–231 (2011). https://doi.org/10.1057/pb.2011.28
10. Lazăr, D.T., Chirca, A.: Place marketing. Cluj-Napoca City case study. Transylv. Rev. Adm. Sci. **3**(20), 30–42 (2007)
11. Hanna, S., Rowley, J.: Towards a strategic place brand-management model. J. Mark. Manag. **27**(5–6), 458–476 (2011)
12. Wheeler, A.: Designing Brand Identity: An Essential Guide for the Whole Branding Team, 3rd edn. Wiley, New York (2009)
13. Ruzinskaite, J.: Place branding: the need for an evaluative framework. Doctoral dissertation, University of Huddersfield (2015)
14. Jain, R.: Basic branding concepts: brand identity, brand image and brand equity. Int. J. Sales Mark. Manag. Res. Dev. (IJSMMRD) **7**(4), 1–8 (2017)
15. Kavaratzis, M., Kalandides, A.: Rethinking the place brand: the interactive formation of place brands and the role of participatory place branding. Environ. Plan. A **47**(6), 1368–1382 (2015). https://doi.org/10.1177/0308518X15594918
16. Almeyda-Ibáñez, M., George, B.P.: Place branding in tourism: a review of theoretical approaches and management practices. Tour. Manag. Stud. **13**(4), 10–19 (2017)
17. Govers, R.: Why place branding is not about logos and slogans. Place Brand. Public Dipl. **9**(2), 71–75 (2013). https://doi.org/10.1057/pb.2013.11
18. Heeley, J.: City Branding in Western Europe. Goodfellow Publishers Ltd, Oxford (2011)
19. Galí, N., Camprubí, R., Donaire, J.A.: Analysing tourism slogans in top tourism destinations. J. Destin. Mark. Manag. **6**(3), 243–251 (2017)
20. Cai, L.A.: Cooperative branding for rural destinations. Ann. Tour. Res. **29**(3), 720–742 (2002)

21. Kavaratzis, M.: From city marketing to city branding: towards a theoretical framework for developing city brands. Place Brand. **1**(1), 58–73 (2004)
22. Ferreira, P., Dionísio, A.: City brand: what are the main conditions for territorial performance? Sustainability **11**(14), 3959 (2019)
23. Cozmiuc, C.: City branding-just a compilation of marketable assets? Econ. Transdiscipl. Cogn. **14**(1), 428–436 (2011)
24. Ruiz, E.C., De la Cruz, E.R.R., Calderón Vázquez, F.J.: Sustainable tourism and residents' perception towards the brand: the case of Malaga (Spain). Sustainability **11**(1), 292 (2019). https://doi.org/10.3390/su11010292
25. Riza, M., Doratli, N., Fasli, M.: City branding and identity. Procedia-Soc. Behav. Sci. **35**, 293–300 (2012)
26. Prilenska, V.: City branding as a tool for urban regeneration: towards a theoretical framework. Archit. Urban Plann. **6**, 12–16 (2012)
27. Bidgoli, S.J., Arani, A.A., Bidgoli, F.O.: City branding position and challenges. Indian J. Fundam. Appl. Life Sci. **4**(1), 166–173 (2014)
28. Zenker, S., Petersen, S., Aholt, A.: Development and implementation of the citizen satisfaction index (CSI): four basic factors of citizens' satisfaction. Res. Pap. Mark. Retail. **39**(1), 1–19 (2009)
29. Papazoglou, G.E.: Society and culture: cultural policies driven by local authorities as a factor in local development—the example of the municipality of Xanthi-Greece. Heritage **2**(3), 2625–2639 (2019). https://doi.org/10.3390/heritage2030161
30. Rosenstein, C.: Cultural development and city neighborhoods. City Cult. Soc. **2**(1), 9–15 (2011)
31. Richards, G., Palmer, R.: Why cities need to be eventful. In: Richards, G., Palmer, R. (eds.) Eventful Cities: Cultural Management and Urban Revitalization, pp. 1–37. Elsevier (2010). https://doi.org/10.1016/b978-0-7506-6987-0.10001-0
32. Karabağ, S.F., Yavuz, M.C., Berggren, C.: The impact of festivals on city promotion: a comparative study of Turkish and Swedish festivals. Tourism (13327461) **59**(4), 447–464 (2011)
33. Van der Borg, J., Russo, A.P.: The impacts of culture on the economic development of cities. In: European Institute for Comparative Urban Research (EURICUR) Erasmus University Rotterdam (2005)
34. Popescu, G.V.: From local to global with city branding. EcoForum **6**(1), 1–27 (2017)
35. Quinn, B.: The European capital culture initiative and cultural legacy: an analysis of the cultural sector in the aftermath of Cork 2005. Event Manag. **13**(4), 249–264 (2009)
36. Liu, Y.D.: Event and sustainable culture-led regeneration: lessons from the 2008 European Capital of Culture, Liverpool. Sustainability **11**(7), 1869 (2019)
37. Binns, L.: Capitalising on culture: an evaluation of culture-led urban regeneration policy. Futures Academy, Technological University Dublin (2005). https://arrow.tudublin.ie/future sacart. Accessed 25 Nov 2020
38. Palmer, R., Richards, G.: European cultural capital report. ATLAS, Arnhem (2009)
39. Green, S.: Capitals of Culture. An introductory survey of a worldwide activity [report] (2017). https://prasino.eu/wp-content/uploads/2017/10/Capitals-of-Culture-An-int roductory-survey-Steve-Green-October-2017.pdf. Accessed 25 Nov 2020
40. About Aarhus 2017. https://www.aarhus2017.dk/en/about-us/what-is-aarhus-2017/about-aar hus-2017/index.html. Accessed 25 Nov 2020
41. Aarhus 2017: Welcome future. https://www.aarhus2017.dk/media/1297/welcome_future_ eng_online.pdf. Accessed 25 Nov 2020
42. Cultural policy, City of Aarhus. https://www.aarhus.dk/media/6713/cultural-policy-2017-2020.pdf. Accessed 25 Nov 2020

43. Aarhus 2017, Programme. https://www.aarhus2017.dk/en/programme/index.html. Accessed 25 Nov 2020
44. CityLogo - Urbact, Innovative place brand management [Report] (2012). https://urbact.eu/sites/default/files/import/Projects/CityLogo/documents_media/baseline_citylogo_2012.pdf. Accessed 25 Nov 2020
45. Aarhus 2017: GrowOP! Festival. https://www.aarhus2017.dk/en/programme/children-and-young-people/growop-festival/index.html. Accessed 25 Nov 2020
46. Behind GrowOP! https://growopfestival.com/bag-om-growop. Accessed 25 Nov 2020
47. Aarhus 2017: European Capital of Culture, EcoC. https://issuu.com/aarhus2017/docs/application_for_ecoc_2017_aarhus. Accessed 26 Nov 2020

The Impact of Using Social Media Twitter to Promote Tourism in Indonesia

Dicky Izmi Syahputra[1]([✉]) [iD], Achmad Nurmandi[2] [iD], Salahudin[3] [iD], Dyah Mutiarin[2] [iD], and Suswanta[2] [iD]

[1] Departement of Government Affairs and Administration,
Universitas Muhammadiyah Yogyakarta, Yogyakarta, Indonesia
`dicky.izmi.psc20@mail.umy.ac.id`
[2] Department of Government Affairs and Administration, Jusuf Kalla School of Government,
Universitas Muhammadiyah Yogyakarta, Yogyakarta, Indonesia
[3] Department of Goverment Science, Social and Political Science Faculty,
Universitas Muhammadiyah Malang, Malang, Indonesia

Abstract. Considering its broad and active users in Indonesia, Twitter is one of the media chosen by the government and the public to promote Indonesia's tourism. This article addresses the impact of government programs via social media on Twitter in promoting tourism in Indonesia. The accounts used were the official accounts used to promote Indonesian tourism by the Ministry of Tourism and Creative Economic, namely @indtravel. The second account is @genpi_id, an account collectively generated by Indonesian netizens with the same purpose. The data used in this paper came from Twitter and focused on the Department of Tourism and Creative Economic successes. The data were analyzed using one of the tools for data analysis, NVivo. The data showed that the number of foreign and local tourists increased every year from 2015 to 2019, and revenue from the tourism sector increased significantly. However, this rise is not only due to promotions on social media Twitter. The government has also introduced several other programs. The researchers did not compare Twitter with other social media in this article because of limited data, so that whether it is effective or not can still be discussed because it depends on how the public perceives the content of the promotion. Thus, more research should be done comparing social media to promote tourism in Indonesia to know how social media promotes tourism.

Keywords: Tourism · Kemenparekraf · Social media · Twitter · Government · Indonesia

1 Introduction

Indonesia is an archipelagic country. On each island, it has diversity in terms of geography, society, culture, etc. This diversity is an advantage possessed by this country because it will be a wealth that will not run out and be profitable if managed properly. For example, cultural diversity has given birth to beautiful temples, beautiful mosques, and beautiful churches to different human characters. Geographic diversity has given

T. Antipova (Ed.): ICADS 2021, AISC 1352, pp. 287–297, 2021.
https://doi.org/10.1007/978-3-030-71782-7_26

birth to beautiful beaches, beautiful mountains, and forests with various beautiful flora and fauna.

Technology is increasingly advanced, accessibility, effectivity, productivity, and government integrity can be better with IT systems in government processes [1]. Twitter is one of the social media that has many users worldwide, so information from all over the world is easy to get. Many individuals use this social media for advertising their products by sharing them to build awareness, recognition, and loyalty to the brand [2]. This social media is not only used for promotion but also used to Transform interaction into interactive dialogue [3].

The Indonesian government loses this opportunity. The government encourages tourism in Indonesia through the Ministry of Tourism and Creative Economic since it can provide significant state revenue [4]. The government supports tourism in different ways through the Ministry of Tourism and Creative Economic, holding festivals, branding, engaging in international tourism forums, and social media branding, including Twitter [5]. In this article, the researchers will present some of the information gathered to promote Indonesia's tourism through social media.

Since 2015, the government has been promoting tourism through the @indtravel account (an account created by the Ministry of Tourism and Creative Economy to promote tourism in Indonesia). On the other hand, people independently also make efforts that they can do to promote tourism. Their @genpi_id account posts tourist attractions in their area, attracting foreign tourists' attention. In this essay, the researchers will use these two accounts as subjects of study and see from each post of these two accounts how they affect tourists' interest in visiting Indonesia. Compared to previous research, the research focuses on the national branding logo or hashtag (#Wonderfulindonesia), technology-based tourism development. The data management were carried out independently and obtained data using interviews. While data management using the NVivo software, the research would concentrate on social media Twitter and the @indtravel and @genpi_id accounts instead of the branding logo. The researchers can determine whether the posts' intensity will affect tourists' willingness to vacation in Indonesia.

2 Literature Review

2.1 Social Media Technology

Social media technology is used to accommodate a quick and open information distribution system that could be structured to provide a service involving public engagement and potentially to improve the government's positive picture. (Budiana, H. R., Sjoraida, D. F., Mariana, D., & Priyatna, 2016). In the public sector context, social media can be described as a group of technologies that enable public agencies to generate greater citizen engagement. There are possibilities for public engagement, co-production, and crowdsourcing solutions and developments when the government uses social media channels [6].

The latest move in using the government's Internet or ICT-enabled services is social media, which is part of Web 2.0 applications. Social media provides government agencies an opportunity to be more open, sensitive and successful than ever, thus enabling the public to share their opinions on government policy and government services quality [7].

The government's involvement on social media sites is no longer considered an option but a new rule. Therefore, the government's social media use is becoming a significant subject in worldwide e-government research and practice [8].

The use of social media in government has a great prospect in raising the perceptions of people about government, such as government images and benefiting the livelihood of people if the government can develop governance capabilities and carry citizens pleased social media activity in terms of information quantity and quality, the right mindset, and use of social media [9].

2.2 Social Media Usage by Government

The presence of social media, particularly about the dissemination of information and awareness and the enhancement of global connectivity, is beneficial [10]. In the general classification, social media is used to 1) define the appropriate target audience corresponding to technographic segmentation; 2) choose and build the target audience's social media account; 3) establish and upload tagging messages; 4) track conversation; 5) respond to the audience's statement, feedback, or question; 6) analyze and extract entire input from tagging audience as policy reviews development/improvement; 7) recommend that activity, program, or policy be implemented according to the feedback and aspiration of the audience; and 8) disseminate policy and follow up on the implementation of the program [11]. Social networks are also an important information channel for the Federal Executive Board's non-formal contact with all people, not limited to users of official Internet services [12].

Social media has been described primarily as "the many fairly affordable and widely available online technologies that make it easier for everyone to publish and access data, collaborate on a common effort, or develop relationships" [13]. These platforms can differ significantly from government agencies: while most still use them merely as an additional medium of communication and representation (one-way information transmission), more advanced applications include public participation (bidirectional communication) and networking (co-production of knowledge) [14].

2.3 The Impact of Social Media Use by Government

Social media use has several benefits in government activities, such as easier public participation, social empowerment, shorter duration to delivering information, and easier government collaboration [15].

A new form of public participation, known as e-participation, has emerged in ICT, which aims to encourage active citizenship with the latest technology advancement by increasing access and participation to promote a fair and effective partnership between society and government [16]. A better-informed public and improved decision-making can lead to public involvement. Participation is imperative for legitimacy since participation is central to the idea of democracy. Participation is likely to increase the public's awareness of problems through attention and participation [17].

Using social media to deliver information can lead to an open government that can increase awareness and decision-making transparency and is a 'multilateral, political and social mechanism' that transforms democracies and strengthens the relationship between government and the public [18]. Of the many benefits, social media also has several problems; for example, it is a hoax. Hoax is fake data ready to be released that is hidden by a mask of fact. Fake news is a fact, but the substance of the news is deceptive. It means that an article may be aimed at cornering or defaming a person or group of individuals or organizations from the beginning to create hate among community members [19].

Another problem is about social media user's data privacy. Users of social media have all agreed to a set of terms and conditions for each social media site they use, and there are also provisions within these terms and conditions about how third parties, the data can be considered in the public domain if users have agreed to these terms [20], which mean, these data can be used to harmful activities.

Nowadays, it is not known who the "strangers" are, particularly in the social media sector, or we usually call it "anonymity" in social media [13]. With this anonymity, the researchers do not know who is commenting on or posting information on social media, leading to information uncertainty. The risk of harm is most likely where the privacy and anonymity of a social media user have been breached and is also more significant when dealing with more sensitive data that can expose a social media user to the risk of embarrassment, reputational damage, or prosecution when revealed to new audiences [20].

3   Method

A qualitative descriptive approach is used in this article, explaining how the government has used social media to promote Indonesia's tourism. The social media that this research selected was Twitter since the government uses Twitter as one of its promotional media through the Ministry of Tourism and Creative Economy. The Ministry of Tourism and Creative Economic's official account aims to promote tourism in Indonesia, that is @indtravel. @genpi account is a public account that netizens used the Twitter account to promote tourism in Indonesia. Both @indtravel and @genpi account were used as the analysis object of this study. The data analyzed used a data processing program, Nvivo 12 Plus. This study's period was from 2015 to 2019, since the Ministry of Tourism's Twitter account has been involved since 2015.

4   Result and Discussion

The government used social media such as Twitter to advertise tourist attractions in Indonesia to encourage tourism in Indonesia. The Ministry of Tourism and Creative Economic (Kemenparekraf) is responsible for growing the tourism economy and the creative economy. The Ministry of Tourism and Creative Economic has a Twitter account devoted to promoting Indonesia's tourism (@indtravel).

Besides the Twitter account of the Ministry of Tourism, there is a group that also promotes Indonesian tourism, namely Genpi (@genpi_id). Genpi is a group made up of netizens involved in tourism. Via their Twitter account, they engage in the promotion of tourism in Indonesia and also conduct offline events to grow regional tourism (Fig. 1).

Fig. 1. @genpi_id word cloud

It can be seen from the word cloud above, the word that most often appears on Genpi's Twitter account is @Genpi, which means that many netizens mention genpi in their photos with the theme of tourism in Indonesia. Other words that appear the most in Indonesian tourism mean this Twitter account focuses on promoting Indonesia's tourism. In other words, such as "#pesonaindonesia" and "#wonderful Indonesia," which are hashtags often used by netizens for their photos on the beauty of tourism in Indonesia (Fig. 2).

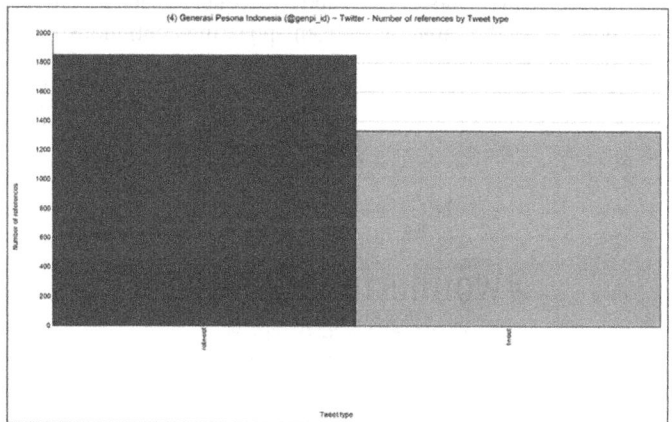

Fig. 2. @genpi_id tweet type references

From the table above, it can be seen how the @genpi id account differentiates between tweets and retweets. The number of retweets indicates that the @genpi id account's followers contributed to posting tourism stuff in their respective areas by retweeted it from them. This situation is performed as a mutual form of society to support tourism in the area where they live and individually help improve their respective regions' strengths. The number of tweets, however, is not that far from their retweets (Fig. 3).

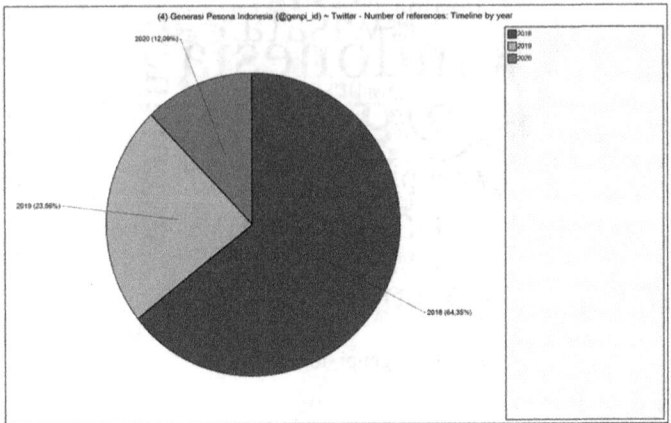

Fig. 3. @genpi_id timeline by year

This account has been actively tweeting since 2018, not only on Twitter, this community also does it on other social media such as Instagram and Facebook. The posting trend has continued to decline this year and can be caused by several factors such as the posting of tourist attractions in the area, as the growth of new tourist attractions does not occur in a matter of months, requires preparation and processes to build tourist attractions. It can be seen from the intensity posts in 2018 and subsequent years are reduced, which is likely to prevent posts from the same place from appearing continuously (Fig. 4).

Fig. 4. @indtravel wordcloud

It can be seen from the word cloud above on the @indtravel Twitter account, the word that appears most often is #Wonderfulindonesia, and netizens use this hashtag to promote tourism in Indonesia. Not only netizens, but this account is also a promotional account from the Ministry of Tourism and Creative Economy to promote activities or events created by the Ministry of Tourism and Creative Economy and collaborative activities (Fig. 5).

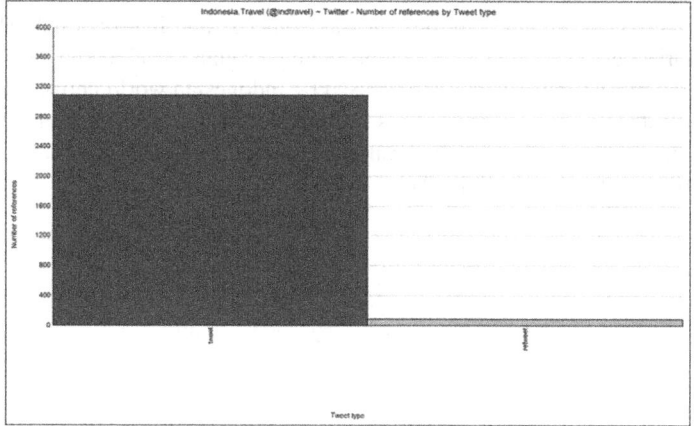

Fig. 5. @indtravel tweet type references

The difference between the number of tweets and the number of retweets from the @indtravel account can be seen in the graph above. There is a substantial distinction between the two. The Ministry of Tourism and Creative Economic's objective in this report is to promote tourism in Indonesia based on government-scheduled programs

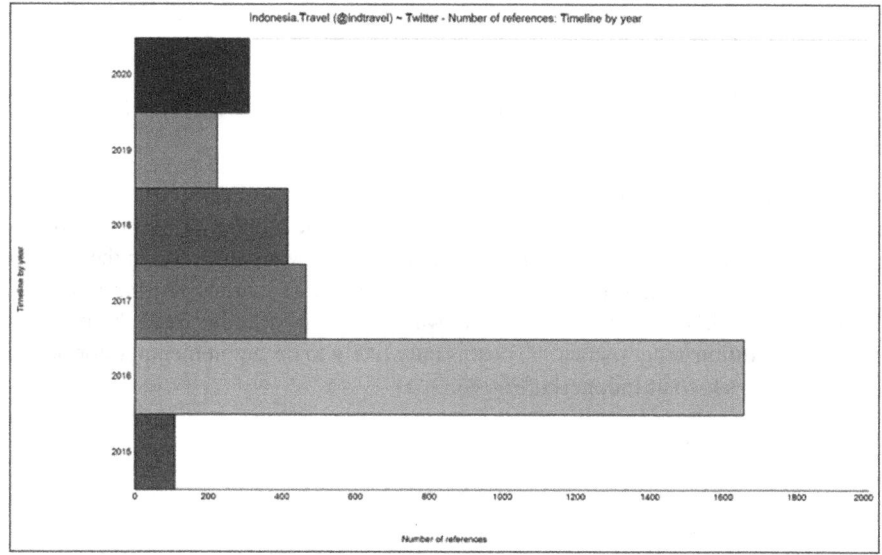

Fig. 6. @indtravel timeline by year

or activities, so it can be seen that the number of tweets is higher than the number of retweets. Retweeted items are only linked to the promotion of the activities and programs being carried out (Fig. 6).

From the figure above, it can be seen how intensive the tweets are from 2015 to 2020. The value increases, and sometimes it goes down. In 2016, there was an intense increase due to the many government activities such as domestic and international exhibitions, festivals, tourist destination development promoted through social media, and Twitter is still one of the most popular social media used by Indonesian world netizens 2016. If it does not indicate that the government has declined in terms of services in the following years, then that is one reason, but the branding is no longer based on social networking on Twitter but by other media (Fig. 7).

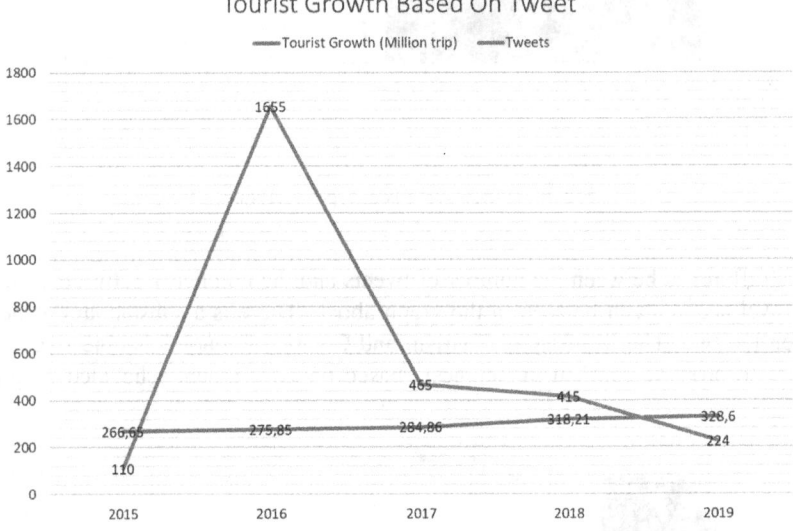

Fig. 7. Tourist growth based on tweet (Kemenparekraf, 2019 & NVivo)

The number of tweets shared by @indtravel impacts the number of visitors each year is based on the graph above. It can be seen that, even as the number of tweets continues to decline, the number of visitors continues to rise. There was a high rise in the number of tweets posted by @indtravel in 2016, but the number of tourists visiting Indonesia was not increased. The number of tourists was not much different from the previous year. This condition suggests that the post is only likely to be supplemental information to attract visitors to visit Indonesia (Fig. 8).

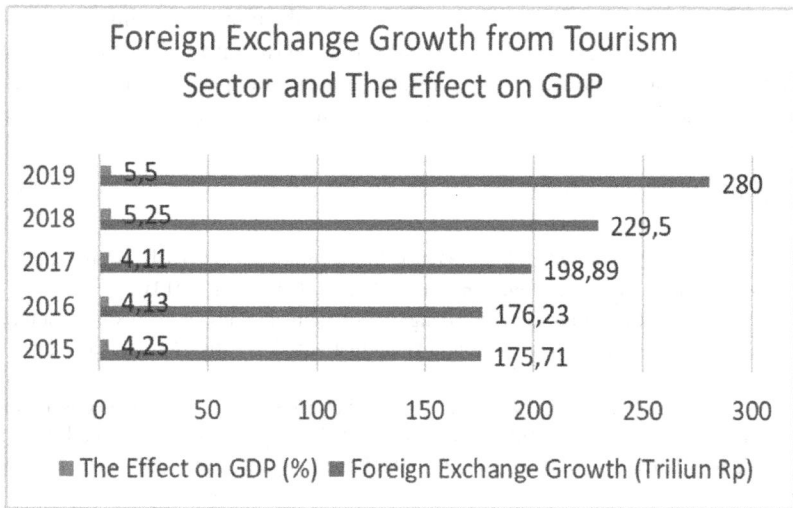

Fig. 8. Foreign exchange growth from the tourism sector and the effect on GDP (Kemenparekraf, 2019)

Of course, with so many tourism sectors in Indonesia, whether they have been revealed or not, it is expected that the tourism sector will become one of the critical sectors in rising the foreign exchange rate of the country. Income from the tourism sector has continued to increase since 2015, also increased sharply in 2019. This condition also means that the government's attempts to encourage tourism to Indonesia's international community have achieved positive results. The government's ongoing promotion has opened up space for international tourists to consider Indonesia one of their priority tourist destinations. Although its contribution to domestic GDP is only 5.5%, it is better than in the previous year.

5 Conclusion

Twitter was selected as one of the platforms used in government tourism marketing based on the discussion. It can be inferred that the government is active in posting on the Twitter account @indtravel, and netizens are also helping to support it independently via the account @genpi id. Every year, while the intensity of the posts made by both accounts decreases, the growth of tourists coming to Indonesia continues to increase last year. The researchers hope that there will be more studies about how social media impacts people's willingness to travel in Indonesia by comparing other social media such as Instagram, Facebook, etc. and with more detailed branding details on this social media.

It is not easy to decide how the effects of posts on the @indtravel and @genpi_id accounts will undoubtedly affect tourists' interest to go on vacation to Indonesia because of the data available on visitor growth and total rise in revenue from the tourism sector in general. As it is known, many things have been done to increase, including one of which is not only Twitter branding on social media, and also the trends regarding social media are changing, so we conclude that because social media is an informative media, so every post made by both accounts have an informative impact, to provide information related tourism in Indonesia and also have some impact to the growth of traveler in Indonesia.

References

1. Setiawan, H., Mustofa, K.: Metode audit tata kelola teknologi informasi di instansi pemerintah Indonesia. J. IPTEKKOM J. Ilmu Pengetah. Teknol. Inf. **15**(1), 1 (2013)
2. Isman, I., Patalo, R.G., Pratama, D.E.: Pengaruh sosial media marketing, ekuitas merek, dan citra destinasi terhadap minat berkunjung ke tempat wisata. J. Stud. Manaj. dan Bisnis **7**(1), 30–36 (2020)
3. Suryani, I.: Pemanfaatan media sosial sebagai media pemasaran produk dan potensi Indonesia dalam upaya mendukung ASEAN community 2015. (Studi Social media marketing pada twitter kemenparekraf ri dan facebook disparbud provinsi jawa barat). J. Komun. **8**, 123–138 (2014)
4. Mudrikah, A., Sartika, D., Yuniarti, R., Satia, A.B.: Kontribusi Sektor Pariwisata Terhadap GDP Indonesia Tahun 2004–2009. Econ. Dev. Anal. J. **3**(2), 362–371 (2014)
5. Kemenparekraf: Laporan Kinerja Kementerian Pariwisata Tahun 2019 (2019)
6. Reddick, C.G., Chatfield, A.T., Ojo, A.: A social media text analytics framework for double-loop learning for citizen-centric public services: a case study of a local government Facebook use. Gov. Inf. Q. **34**(1), 110–125 (2017)
7. Khan, S., Zairah, A., Rahim, N., Maarop, N.: A systematic literature review and a proposed model on antecedents of trust to use social media for e-government services. Int. J. Adv. Appl. Sci. **7**(2), 44–56 (2020)
8. Gintova, M.: Social media use by government in Canada. In: SMSociety17: Proceedings of the 8th International Conference on Social Media & Society, pp. 1–5 (2017)
9. Lu, B., Zhang, S., Fan, W.: Social representations of social media use in government: an analysis of Chinese government microblogging from citizens' perspective. Soc. Sci. Comput. Rev. **34**(4), 416–436 (2016)
10. Erawati, L.D.: Islamic Ethics in Social Media. Epistemé J. Pengemb. Ilmu Keislam.**14**(1), 17–34 (2019)
11. Rahmanto, A.N., Dirgatama, C.H.A.: The implementation of e-government through social media use in the local government of Solo Raya. In: 2018 International Conference on Information and Communications Technology, ICOIACT 2018, vol. 2018, no. 83, pp. 765–768 (2018)
12. Bundin, M., Martynov, A.: Use of social media by the federal government in Russia. In: ACM International Conference Proceeding Series, vol. 01–03, no. November, pp. 394–395 (2016)
13. Siddiqui, S., Singh, T.: Social media its impact with positive and negative aspects. In: DS 38: Proceedings of E and DPE 2006, the 8th International Conference on Engineering and Product Design Education, vol. 5, no. 2, pp. 207–212 (2006)
14. Tagliacozzo, S., Magni, M.: Government to Citizens (G2C) communication and use of social media in the post-disaster reconstruction phase. Environ. Hazards **17**(1), 1–20 (2018)
15. Alotaibi, R., Ramachandran, M., Kor, A.-L., Hosseinian-Far, A.: Factors affecting citizens' use of social media to communicate with the government: a proposed model. Electron. J. e-Gov. **14**(1), 60–72 (2016)
16. Khadzali, N.R., Utara, U.: E-participation: a systematic understanding on public participation in the government in 21st century. J. Intelek **13**(2), 29–46 (2018)

17. Manaf, H.A., Mohamed, A.M., Lawton, A.: Assessing public participation initiatives in local government decision-making in Malaysia. Int. J. Public Adm. **39**(11), 812–820 (2016)
18. Schmidthuber, L., Piller, F., Bogers, M., Hilgers, D.: Citizen participation in public administration: investigating open government for social innovation. R D Manag. **49**(3), 343–355 (2019)
19. Suyanto, T., Zen, I.M., Prasetyo, K., Isbandono, P., Gamaputra, G., Purba, I.P.: The study perception of social sciences and law faculty students for a hoax in social media. In: Journal of Physics Conference Series, vol. 953, no. 1 (2018)
20. Townsend, L., Wallace, C.: Social media research: a guide to ethics. Univ. Aberdeen, **1**, 13–14, 16, 18 (2016)

Analysis of the Anti-corruption Movement Through Twitter Social Media: A Case Study of Indonesia

Danang Kurniawan[1]($\boxtimes$) [ID], Achmad Nurmandi[1] [ID], Salahudin[2] [ID], Dyah Mutiarin[1] [ID], and Suswanta[1] [ID]

[1] Department of Government Affairs and Administration, Jusuf Kalla School of Government, Universitas Muhammadiyah Yogyakarta, Yogyakarta, Indonesia
kurniawand949@gmail.com

[2] Department of Goverment Science, Social and Political Science Faculty, Universitas Muhammadiyah Malang, Malang, Indonesia

Abstract. Social media, as the delivery of information to the public, has high effectiveness. This research looks at the function of social media accounts of official government agencies of the Corruption Eradication Commission (KPK) as Integration Marketing Communication (IMC) of the Anti-Corruption Movement on social media. The method in this study used Q-DAS (Qualitative Data Analysis Software) Nvivo 12 plus. Stages of data analysis with Nvivo 12 plus captured data, data import, data coding, data classification, and data display. This study's findings revealed that the Anti-Corruption Eradication Commission (KPK) conducted anti-corruption movements by delivering substance information such as anti-corruption education, law enforcement, bureaucrat's reform government official integrity. The implication of this movement was to prevent and eradicate corruption in Indonesia. The research's limitations were that this research data used only social media data to submit to the Corruption Eradication Commission (KPK RI). Recommendations for future research is to explore the impact of the anti-corruption movement on social media on the suppression and eradication of corruption in Indonesia.

Keywords: Social media · Social movement · Anti-corruption · Government · Indonesia

1 Introduction

The Government of Indonesia prevents and eradicates corruption crimes through Law No. 30 of 2002 on Corruption Eradication Commission, Law No. 19 of 2019 on The Second Amendment of Law No. 30 of 2002 on The Eradication Commission of Corruption Crimes. Corruption is the biggest problem facing Indonesia today, its eradication has increased in the last three years, and there have been no convincing signs that this problem can be solved immediately. Based on [1] Public Partition survey measuring people's perceptions, attitudes, and behavior about corruption and KPK, corruption is common /often occurring in Indonesia (78%). Indonesia is at the level of tending to be

T. Antipova (Ed.): ICADS 2021, AISC 1352, pp. 298–308, 2021.
https://doi.org/10.1007/978-3-030-71782-7_27

corrupt (98%). Corruption perpetrators, according to the community, have a high percentage of Members of the House of Representatives (>77%), and the Head of (>97%). The results show the survey findings of the Corruption Eradication Commission (KPK). The cause of corruption by the public is weak law enforcement (>72%), and KPK Law Enforcement Index in 2016 was 62.27%, 71.03% % of 2018 low corruption sentencing (>63%).

The Indonesian Corruption Eradication Commission or Komisi Pemberantasan Korupsi (KPK), in 2018, was appointed as the National Team's coordinator for The Prevention of Corruption. Four ministries were appointed as part of the team, namely the Minister of National Development Planning/Head of the National Development Planning Agency, the Minister of Home Affairs, the Minister of Utilization of State Apparatus, and Bureaucratic Reform, and the Chief of The Presidential Staff. Over the past two decades, the role of internet politics and digital social media has evolved into an established topic of political communication and political participation [2]. Media, education systems, politicians, and civil society organizations need to contribute to eradicating this social disease, as proposed by this work [3]—corruption Eradication Commission (KPK). There is a significant multiplier effect when users share information across their networks, expanding the reach and impact of information [2].

This study's urgency is that social networks and society are continually interacting in modern society. Social media function to broker connections between previously disconnected groups, spread shared grievances beyond the small community activist leaders, and globalize the domestic movement's reach and appeal for democratic change [4]. Social media can deliver a message of substance to the Corruption Eradication Commission (KPK) [5].

Previous movement research focused only on social media and politics. This archived social media conversation reveals a public sphere that is heavily skewed towards the side of the majority opinion, often destroying the voices of lone minorities [6]. This study, the latest look at the Corruption Eradication Commission (KPK) strategic role in delivering anti-corruption movements through social media. This paper looked at the anti-corruption movement The Corruption Eradication Commission (KPK) to the public [7].

The study used Q-DAS (Qualitative Data Analysis Software) Nvivo 12 plus to collect data and analyze data with Chart analysis, Cluster analysis, and Word cloud analysis. The purpose of this research is to look at the effectiveness of the anti-corruption movement information delivery strategy conducted by the Corruption Eradication Commission (KPK) through social media Twitter accounts [8]. This research focuses on the relevance of information delivery through social media in mobilizing the anti-corruption movement. This research seeks to answer the question of substance: Q1 What is the function and association between social media and the anti-corruption movement?; Q2 What kind of narrative to create the anti-corruption movement in the public that the Corruption Eradication Commission (KPK) uses?.

2 Literature Review

2.1 Social Media as the Mobilization of the Anti-corruption Movement

There is a positive dynamics of growth of the number of communities on corruption in social networks, which indicates the actuality of the problem in civil society. The article aims to highlight the effectiveness of the fight against corruption or question how to counter it and anti-corruption activities in social networks [9]. The research urgency is caused by the fact that social networks and society are continually interacting in modern society. Social media functioned to broker connections between previously disconnected groups, spread shared grievances beyond the small community activist leaders, and globalize the domestic movement's reach and appeal for democratic change [4]. Social media refers to several online tools that facilitate creating and sharing highly interactive and user-generated content. These include networking sites, bulletin board systems, chat rooms, micro-blogging, instant message services, and photo and video-sharing sites [10].

Social Network Analysis (SNA) identifies multiple roles of Twitter users in disseminating information about corruption [11]. The anti-corruption movement's Twitter feeds suggest. The movement had a strong presence on social media, which was ostensibly translated into online activism in the form of demonstrations and big turnouts in public meetings [12]. This study's results suggest that social media can discipline corruption even in a country with limited political competition and heavily censored traditional media [13]. Emerging new media has recently become an alternative source of independent information for citizens and, potentially, an agent of political change in nondemocratic regimes. Especially the young generation is actively participating in these Internet discussions, thereby forming their civic position. Therefore, the quality of the content has a strong influence on the attitudes of young people. State authorities can use thematic groups' emergence on anti-corruption for rapid response to facts of corruption phenomena and reactions to them [9].

Social media have achieved a mobilizing impact on policy formulation and public opinion. These developments demonstrate that the definition of 'sensitivity' to various problems is heavily conditioned on the ability of pacing, framing, and actors to push the limits of agreed expression [14]. Words such as 'transparency', 'accountability' and 'advocacy' are now in common usage, previously deemed off-limits, and even 'civil society' is used more frequently than in the past. In Vietnam, it has become increasingly complex and sophisticated to use social media as interactive forms of information and communication technologies (ICT) involving people in the generation and sharing content [14]. Although the consumption of information is an obvious way to elicit feelings and attitudes about collective action's psychological context, interpersonal communication is another essential collection of variables that can be considered in future models [15]. Becker and Salton argue that "As internet connectivity becomes more prevalent, scholars from the future of teledemocracy addressed the relationship between digital media and civil politic [16, 17].

Social media must be understood as the result of organic interactions between technology and structures and social, political and cultural relationships [18]. The possibility factor should be used to motivate engagement in collective action to study the role of visual social media content in political mobilization. In this case, particularly in the

2011 Egyptian revolution, the researchers recognize visual content known to inspire political participation [19]. Numerous legislators and researchers have argued that, in particular, social media platforms empower organizations involved in protests and abuse to identify and hire potential supporters and activists more easily. Via persuasion and mobilization, progressives can use social media networks to increase the possibility of political engagement [20].

2.2 Movement Anti-corruption

Government, business, and civil society fight against 'evil corruption' to free the world from corruption. Corruption has been one of the worst problems, hindering socio-economic and political advancement. The successive regime's efforts to fight the scourge since the country got her flag independence did not yield the desired result [21]. An increasing number of countries are joining anti-corruption movements specified as e-government strategies, open government initiatives (e.g., US Open Government Initiative and the Open Government Partnership), and transparency efforts in data and information and policy processes [22]. Corruption is "an insidious plague undermines democracy and the rule of law, leads to violations of human rights, distorts markets, erodes the quality of life and allows organized crime, terrorism and other threats to human security to flourish [23].

Anti-corruption policies should create conditions favorable to the development of trust and reciprocity among citizens, public agents, and clients who can associate and mobilize in collective action to encourage whistle-blowing, exposing, and enforcing anti-corruption rules [24]. Political corruption manifests two main areas: activities connected with election and succession and manipulating people and rules/tradition to retain power and office [21]. The media, education system, politicians, and civil society organizations need to contribute to eradicating this social disease, as this work proposes [3]. The internet's prominent role in the diffusion of popular protest across the Arab world and the ouster of authoritarian regimes in Tunisia and Egypt has re-energized the debate on the implications of social media networks for political mobilization and patterns of protest diffusion, as well as the impact of social [2].

Such accounts extol the role of social media and the internet as the loci for the mobilization of popular protest, so much so that news narratives and scholarly commentary both see these technologies as shaping these revolutions, as enabling such upheavals in civil society. Using a recent famous mobilization case in India, the anti-corruption movement was inspired in 2011 by Anna Hazare [12]. Rampant corruption has been denounced by social movements, which have developed specific diagnostic and prognostic frames and knowledge and practices for the social accountability of political and economic powers [24]. In democracies, mass media is an essential instrument for monitoring public officials' behavior, limiting corruption, and reducing political rents [13].

3   Research Methods

The analysis software of this study used Nvivo 12 plus. Nvivo 12 plus as a qualitative analysis tool displays data in quantitative form, referred to as qualitative to quantitative analysis (Bazeley, 2013). The qualitative approach provides exploration and description of anti-corruption problems, descriptive use to provide information, and an explanation of data efficiently. This study's data is the content of the official Twitter of the Corruption Eradication Commission (KPK) (@KPK_ RI). Nvivo 12 plus in this research was Chart analysis, cluster analysis, and Word cloud analysis. The use of Nvivo as an analysis tool had five stages: (1) capturing data, (2) import data, (3) data coding, (4) data classification, and (5) displaying data. The data that has been displayed is committed to the data by the qualitative data analysis method.

4   Results and Discussion

4.1   Marketing Objective

Anti-corruption policies must create conditions that support the development of trust and reciprocity among citizens, public agents, and clients who can unite and engage in collective action to encourage secret disco-surfing, uncovering, and enforcing anti-corruption rules [24]. The submission of information on the Corruption Eradication Commission (KPK) shows that the last three years' comparison has increased significantly. In 2018 only (100) activities existed on social media @KPK_RI. Different conditions in 2019 saw a significant increase in (1000) information delivery activities to the public. The proposed formation in 2020 also has the same percentage of activity as 2019 (1000).

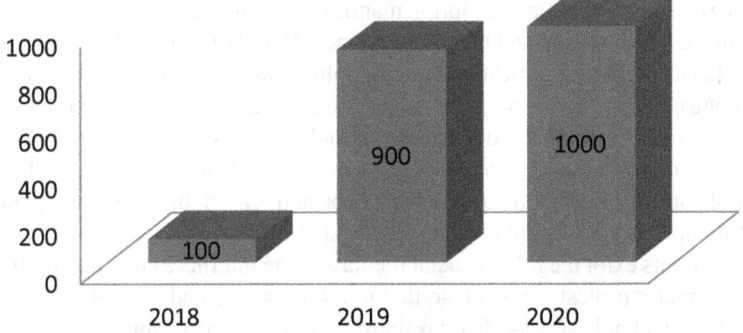

Fig. 1. Social Media Activity @KPK_RI Corruption Eradication Commission (KPK)

Information delivery activities @KPK_RI shows an increasing trend each year. This condition became the Corruption Eradication Commission (KPK) strategy in mobilizing the period as an educational effort of the anti-corruption movement. Over the past five years, professional medical conferences have seen a massive increase in real-time Twitter use [26]. The anti-corruption submission content as a whole and dominates each charge is seen in Word frequency in Fig. 1, showing the Corruption Eradication Commission (KPK).

Eradication of corruption must be followed up with a comprehensive, integral, and holistic strategy to achieve the expected results truly. The anti-corruption movement's

Fig. 2. Word Frequency Corruption Eradication Commission (KPK)

information content as a whole that the Corruption Eradication Commission (KPK) has the anti-corruption movement's message content. Word Frequency data show the words that dominate the delivery of @KPK_RI information in Fig. 2 are classified according to the content and purpose of the delivery of information into variable anti-corruption movements. Content that dominates @KPK_RI #kawanaksi, suspects, and conferences (Table 1).

Table. 1. Anti-corruption movement content classification

Indicator	Reference	Percentage
Strengthening and Accelerating Bureaucratic Reform	55	16%
Community Anti-Corruption Cultural Development	143	41%
The firm, Consistent and Integrated Law Enforcement	97	28%
Improving the Integrity and Ethics of State Organizers	54	15%

The submission of information on the anti-corruption movement Commission Eradication Corruption (KPK) has the substance of an even proportion of content from the four indicators. The anti-corruption movement related to the Cultural Development Anti-corruption Society's colonial narrative has the highest of 143 (41%). The percentage is close to the amount of content submitted, while the submission of Strict, Consistent, and Integrated Law Enforcement information has a total of 97 with a percentage of (28%). Anti-corruption education to the public is often produced by the Corruption Eradication Commission (KPK) to prevent and eradicate corruption through social media. Although tweets are in the form of a conversation, mostly not a conversation function, which shows Twitter is used as a broadcasting tool in the context of providing an understanding related to anything [27]. Strengthening and Accelerating Bureaucratic Reform in delivering information on anti-corruption movements has a relatively equal percentage to the Improvement of Integrity and Ethics of State Organizers amounting to 55/54 with a percentage of less than (20%).

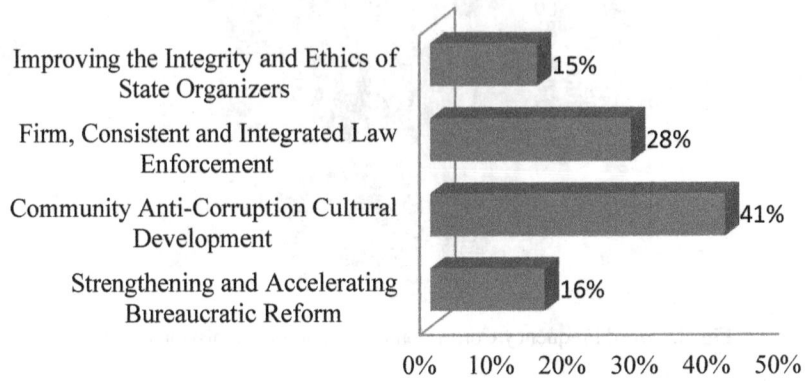

Fig. 3. Anti-Corruption Movement Content Classification

Figure 3 shows the instrument aspects of the anti-corruption movement conducted by the KPK (Corruption Eradication Commission), dominated by the Anticorruption Cultural Development movement. Information is conveyed by the content of increasing the community's role in controlling markup-prone central government policies, such as the implementation of "Social Assistance". The development of anti-corruption culture is also followed up through virtual learning facilitated by KPK (Corruption Eradication Commission), aired LIVE on KPK_RI's Youtube channel. Virtual learning informs the account of @KPK republic of Indonesia in 126 learning activities of Anti-Corruption Cultural Development in the community. The latest activities in the virtual learning of the anti-corruption movement are:

Commemorating the 92nd Youth Oath Day, #KawanAksi can watch the 'Anti-corruption Youth Dialogue', live on Youtube @SpakIndonesia, October 28, 2020, 09.45 to 12.30 WIB. #KawanAksi can also follow via Zoom by registering via the link: https://t.co/9FfcQQX0IW https://t.co/wCn2h7GnUu

The tweet from @KPK_RI above is part of the anti-corruption movement's strategy through education and developing an anti-corruption culture in the community. The anti-corruption movement strategy in Strict, Consistent, and Integrated Law Enforcement is the second most crucial dominance in delivering anti-corruption information @KPK Republic of Indonesia. The issue of hokum issues and the success of hokum enforcement into content in information delivery production. Overall hokum enforcement information, accounts of @KPK_RI are active in controlling the punishment of suspected corruption cases amounting to (100) press activities.

4.2 Integrated Marketing

The submission of information on the anti-corruption movement through the posting of @KPK_RI conducts government agencies, media, and civil society. The strategy of delivering mentions is the forwarding of the delivery of information to the intended party. Official @KPK_RI:

Table 2. Mentions Akun Media Sosial Twitter

Mentions	Reference	Percentage
kemendagri	10	11,76%
kongresdesa	9	10,59%
tiiindonesia	9	10,59%
bappenasri	9	10,59%
provjateng	7	8,24%
kemenpanrb	6	7,06%
kemenkeuri	5	5,88%
dprri	4	4,71%
ganjarpranowo	4	4,71%
jokowi	4	4,71%
kemendikbudri	3	4,71%
hariankompas	3	3,53%
kejaksaanri	3	3,53%
kemenagri	3	3,53%
kumparan	3	3,53%
detikcom	2	2,35%

Although some institutional steps have been taken to reduce corruption, there is still room for improvement [27]. Collaboration with the Corruption Eradication Commission (KPK) in carrying out effectiveness in delivering information on the anti-corruption movement, through mentions to the intended party's official account. Based on Table 2, the Corruption Eradication Commission (KPK) has intense collaboration with official government agencies at the ministerial level, including the Ministry of Home Affairs @kemendagri, National Planning and Development Agency/PNPP @bappenasri, Ministry of Finance @kemenkeuri, Ministry of Culture and Education @kemendikbudri. The activity is seen as the largest percentage of the Corruption Eradication Commission (KPK) conducting the Ministry of Home Affairs @kemendagri (11.76%). Regional Government of Central Java Province includes @provjateng and Regional Head @ganjarpranowo.

4.3 Type of Information (Product)

The social movement, in this case, anti-corruption, is an integral part of social media by using the hashtag section of communication strategy encoding information content. The Corruption Eradication Commission (KPK) aims to convey the anti-corruption movement to the public. The movement's substance is Community Anti-Corruption Cultural Development, Law Enforcement, Bureaucratic Reform, Enhanced Integrity of State Organizers. The strategy of delivering the anti-corruption movement message through several hashtags that massif carried out 12. Table 3 is carried out to explain the hashtag type of Corruption Eradication Commission (KPK) in the anti-corruption movement campaign:

Table 3. Hashtag Twitter Social Media Accounts

Hashtag	Reference	Percentage
kawanaksi	326	70,3%
perscope	40	8,6%
ajlk2020	28	6,0%
jagabansos	16	3,4%
acffest2020	13	2,8%
pilihyangjujur	13	2,8%
gratifikasi	7	1,5%
duatahunnovel	6	1,3%
jaga	6	1,3%
cegahkorupsidariujungjari	5	1,1%
generasianti korupsi	4	0,9%

The submission of information on the anti-corruption movement submitted by the Corruption Eradication Commission (KPK) using hashtag type classifies each content's substance. The #kawanaksi has a total of 326 and a percentage of (70.3%). Hashtag @KPK_RI is highly variable from content analysis, and the overall content hashtag content has the content of anti-corruption movements (prevention, suppression, and eradication). The anti-corruption movement's overall information substance is focused on the prevention stage of corruption through the hashtag #kawanaksi. When evaluated collectively, Twitter posts are part of a much larger and more sustainable narrative in a social setting than to make a difference [28].

The information submitted by the @KPK_RI account uses several hashtag messages to conduct different anti-corruption campaigns, such as #kawanaksi has a percentage (70.3%) the highest of the whole. Focus #kawanaksi socialize and provide anti-corruption culture education to the broader community. Content messages contain mobilizing netizen #kawanaksi from official accounts @KPK_RI. Substantial information messages #perscop (8.6%) related to the enforcement of corruption crimes handled by the KPK and designated as suspects. The Corruption Eradication Commission (KPK) has an application (JAGA) to oversee public services, #Jaga (1.3%) branding application. The Novel case has a fairly dominant information #duatahunnovel percentage (1.3%). The use of hashtags in the submission of information on the anti-corruption movement Corruption Eradication Commission (KPK) seen from its content has the message of the anti-corruption movement's substance.

5 Conclusion

Information on Twitter anti-corruption movement, the Corruption Eradication Commission (KPK) has an integrated communication strategy. Integrate information delivery on aspects of Marketing Objectives, Integrated Marketing, and Type In Formation. The substance of the Corruption Eradication Commission's anti-corruption movement creates goals, anti-corruption education, creating an active society, law enforcement, bureaucratic reform, and improved integration of State organizers. Targets of the Corruption Eradication Commission's anti-corruption movement from all circles, (Society, Civil Society, Academia, Minis-tries/Institutions). Limitations for this research; first, this research data only used social media data. Second, the study did not look at the process of reciprocal interaction in the anti-corruption movement in the social environment. Future research recommendations are: first, to display data by comparing the Corruption Eradication Commission (KPK) in the Medium and Long Term Work Plan; second, to support social media user interaction data with the social media accounts of the Corruption Eradication Commission (KPK).

References

1. Pelayanan, L., Publik, I.: Laporan pelayanan informasi publik, pp. 1–35 (2019)
2. Breuer, A., Landman, T., Farquhar, D.: Social media and protest mobilization: evidence from the Tunisian revolution. Democratization **22**(4), 764–792 (2015). https://doi.org/10.1080/135 10347.2014.885505
3. Hope, K.R.: The corruption problem in swaziland: consequences and some aspects of policy to combat it. J. Dev. Soc. **32**(2), 130–158 (2016). https://doi.org/10.1177/0169796X15609715
4. Lim, M.: Clicks, cabs, and coffee houses: social media and oppositional movements in Egypt, 2004–2011. J. Commun. **62**(2), 231–248 (2012). https://doi.org/10.1111/j.1460-2466.2012. 01628.x
5. Rahmat, A.F., Purnomo, E.P.: Twitter media platform to set-up political branding: analyzing @Kiyai_Marufamin in 2019 presidential election campaign. Nyimak J. Commun. **4**(1), 73 (2020). https://doi.org/10.31000/nyimak.v4i1.2268
6. Eipe, J., Varghese, T.: 'India Against Corruption' Movement: An Online Version of a Non-Violent Mass Movement, no. December 2012 (2014)
7. Setiawan, D., Nurmandi, A.: Sandiaga Uno: Personal Branding di Twitter. J. Public Policy **6**(1), 19 (2020). https://doi.org/10.35308/jpp.v6i1.1657
8. Kholid, A., Husein, R., Mutiarin, D., Listiya E. R, S.: The influence of social media towards student political participation during the 2014 Indonesian presidential election. J. Gov. Polit. **6**(2) (2015). https://doi.org/10.18196/jgp.2015.0019
9. Frolova, I., Antonova, I., Khamitova, L., Zakirova, L., Chukmarova, L.: Studies the social media on corruption issues through content analysis. Espacios **38**(49), 167–173 (2017)

10. Gunitsky, S.: Corrupting the cyber-commons: social media as a tool of autocratic stability. Perspect. Polit. **13**(1), 42–54 (2015). https://doi.org/10.1017/S1537592714003120

11. Prabowo, H.Y., Hamdani, R., Sanusi, Z.M.: The new face of people power: an exploratory study on the potential of social media for combating corruption in Indonesia. Australas. Acc. Bus. Finance J. **12**(3), 19–20 (2018). https://doi.org/10.14453/aabfj.v12i3.3

12. Harindranath, R., Khorana, S.: Civil society movements and the twittering classes in the postcolony: an indian case study. S. Asia J. S. Asia Stud. **37**(1), 60–71 (2014). https://doi.org/10.1080/00856401.2012.744285

13. Enikolopov, B.R., et al.: Social Media and Corruption, vol. 10, no. 1, pp. 150–174 (2018)

14. Bui, T.H.: The influence of social media in Vietnam's elite politics. J. Curr. SE Asian Aff. **35**(2), 89–111 (2016). https://doi.org/10.1177/186810341603500204

15. Jha, C.: Can social media and the internet reduce corruption?, pp. 4–7 (2017)

16. Liu, Y., Rui, J.R., Cui, X.: Are people willing to share their political opinions on Facebook? Exploring roles of self-presentational concern in spiral of silence. Comput. Hum. Behav. **76**, 294–302 (2017). https://doi.org/10.1016/j.chb.2017.07.029

17. Choi, D.H., Shin, D.H.: A dialectic perspective on the interactive relationship between social media and civic participation: the moderating role of social capital. Inf. Commun. Soc. **20**(2), 151–166 (2017). https://doi.org/10.1080/1369118X.2016.1154586

18. Lim, M.: Freedom to hate: social media, algorithmic enclaves, and the rise of tribal nationalism in Indonesia. Crit. Asian Stud. **49**(3), 411–427 (2017). https://doi.org/10.1080/14672715.2017.1341188

19. Kharroub, T., Bas, O.: Social media and protests: an examination of Twitter images of the 2011 Egyptian revolution. New Media Soc. **18**(9), 1973–1992 (2016). https://doi.org/10.1177/1461444815571914

20. Zeitzoff, T.: How social media is changing conflict. J. Conflict Resolut. **61**(9), 1970–1991 (2017). https://doi.org/10.1177/0022002717721392

21. Aderonmu, J.A.: Civil society and anti-corruption crusade in Nigeria'S Fourth Republic. J. Sustain. Dev. Afr. **13**(1), 75–86 (2011)

22. Nam, T.: Examining the anti-corruption effect of e-government and the moderating effect of national culture: A cross-country study. Gov. Inf. Q. **35**(2), 273–282 (2018). https://doi.org/10.1016/j.giq.2018.01.005

23. Goswami, D., Bandyopadhyay, K.K.: The Anti-Corruption Movement in India, pp. 1–22 (2011)

24. Della Porta, D.: Anti-corruption from below: social movements against corruption in late neoliberalism. Partecip. e Conflitto **10**(3), 661–692 (2017). https://doi.org/10.1285/i20356609v10i3p661

25. Giroux, H.A.: The iranian uprisings and the challenge of the the new media: rethinking the politics of representation. Fast Capital. **5**(2), 87–92 (2009). https://doi.org/10.32855/fcapital.200902.009

26. Cohen, D., et al.: #InSituPathologists: How the #USCAP2015 meeting went viral on Twitter and founded the social media movement for the United States and Canadian Academy of Pathology. Mod. Pathol. **30**(2), 160–168 (2017). https://doi.org/10.1038/modpathol.2016.223

27. Jacobson, J., Mascaro, C.: Movember: Twitter conversations of a hairy social movement. Soc. Media Soc. **2**(2) (2016). https://doi.org/10.1177/2056305116637103

28. Hosterman, A.R., Johnson, N.R., Stouffer, R., Herring, S.: Twitter, social support messages and the #MeToo movement. J. Soc. Media Soc. Fall **7**(2), 69–91 (2018)

Analysis of Twitter's Election Official as Tools for Communication and Interaction with Indonesian Public During the 2019 Presidential Election in Indonesia

Dimas Subekti[1]([✉]) [iD], Achmad Nurmandi[2] [iD], Dyah Mutiarin[2] [iD], Suswanta[2] [iD], and Salahudin[3] [iD]

[1] Department of Government Affairs and Administration,
Universitas Muhammadiyah Yogyakarta, Yogyakarta, Indonesia
dsubekti05@gmail.com
[2] Department of Government Affairs and Administration, Jusuf Kalla School of Government,
Universitas Muhammadiyah Yogyakarta, Yogyakarta, Indonesia
dyahmutiarin@umy.ac.id
[3] Department of Goverment Science, Social and Political Science Faculty,
Universitas Muhammadiyah Malang, Malang, Indonesia

Abstract. The General Election Commission (KPU) and the Election Supervisory Agency (Bawaslu) are public institutions that must show their performance transparent and accountability by using several social media, including Twitter. Therefore, this study seeks to describe the communication and interactions between the KPU RI and Bawaslu RI through social media Twitter during the 2019 Indonesian Pre elections. This study used Q-DAS (Qualitative Data Analysis Software) (Nvivo 12 plus) to collect data and analyze data with Chart analysis, Cluster analysis, and Word cloud Analysis. This study showed that the *first* scene from the role of election officials on Twitter was that Bawaslu RI had more dominant communication and interactions with the public than the KPU RI. *Second,* the parts of official election roles were interrelated. However, the strong correlation was the socialization of election stages with campaign socialization. *Third,* KPU RI and Bawaslu RI's communication and interactions on social media Twitter discussed election content. *Fourth,* KPU RI and Bawaslu RI on social media Twitter had high communication intensity with the public in 2018 and 2019. It is caused by in 2018 and 2019. There were simultaneous regional head elections and simultaneous general elections.

Keywords: General Election Commission (KPU) · Election Supervisory Agency (Bawaslu) · Social media Twitter · Communication · Interaction

1 Introduction

Law number 7 of 2017 concerning general elections states that an election organizer is an institution that organizes elections consisting of the General Election Commission,

© The Author(s), under exclusive license to Springer Nature Switzerland AG 2021
T. Antipova (Ed.): ICADS 2021, AISC 1352, pp. 309–323, 2021.
https://doi.org/10.1007/978-3-030-71782-7_28

the Election Supervisory Body, and the Honorary Council of Election Administrators as an integral function of the Election Administration to elect members of the House of Representatives, members of the Council. Regional Representatives, President and Vice President, elect members of the House of Representatives directly by the people. From now on, referred to as KPU, the General Election Commission is an Election Management institution that is national, permanent, and independent in implementing Elections. The Election Supervisory Agency, or Bawaslu, is an Election Management institution that oversees the Republic of Indonesia election.

The General Election Commission (KPU) and the Election Supervisory Agency (Bawaslu) are public institutions that must show their performance transparent and accountability by using several social media, including Twitter. The reason for using social media is because social media exists and has interactions between accounts and followers. China's social media appears vibrant and extensive as in any western country, with more than 1,300 social media companies and websites. Every day, millions of posts were written by people all over the country [1]. As articulated by Bertot et al. (2010), social media within governments is a way of growing transparency and openness [2]. Accountability is ensured through citizens' participation in governance through the broad space of social media. Social media is a group of Internet-based applications built on Web 2.0's ideological and technological foundations and allows user-generated content to be created and shared [3]—the use of web 2.0 tools and their interactions with citizens within government organizations [4]. Social media enhances communication between citizens and government better than e-government sites because of personalization or community sense [5]. Social media platforms facilitate the exchange of emotional and motivational content, including messages emphasizing anger, social identification, group effectiveness, and concerns about justice and deprivation, as well as explicitly ideological themes, to support and oppose protest activity [6].

The problem is related to the communication and interaction between the General Election Commission of the Republic of Indonesia (KPU RI) and the Election Supervisory Body of the Republic of Indonesia (Bawaslu RI) and the public using social media, especially Twitter. Moreover, in 2018 there were simultaneous regional head elections, and in 2019 there were simultaneous general elections. The KPU RI and Bawaslu RI played an essential role in organizing the general election. Therefore, so that the implementation runs with community participation, it is necessary to interact with the community using social media. Because social media has a significant impact on political and social life, it is an academically worthwhile endeavor to focus on and examine social media [7]. With the advent of social media and its real achievement on how people do things and how they can do something, it is interesting to find out if they have made inroads into local governance [3]. Social media facilitates citizen partnership in a structured way to strengthen partnerships, build new connections, and facilitate better decision making [8]. With little censorship, social media interactions are mostly open, thus increasing democracy and participation [9].

Based on the explanation above, the focus of previous research studies explains the role of both central and local governments use social media as tools for public services, policymaking, and interaction with the community. Therefore, this study's novelty is that this research focuses on social media analysis to explore independent state institutions' communication and interaction responsible for election administration and implementation. This study's question is: Does this article answer how communication and interactions the KPU RI and Bawaslu RI with the public through social media Twitter during the 2019 Indonesian Pre elections?

2 Literature Review

2.1 Social Media Platform

In 2017, Facebook was reported to be the Research Center's most used central social media platform. Its user base is mostly reflective of the population as a whole. Facebook users accounted for 68% of U.S. adults in January 2016. Two social media networks, the usefulness of various social media platforms, including Twitter, Facebook, Google+, Youtube, and Instagram, have been used by the city of Baton Rouge in disaster responses, such as their feasibility and reliability as an emergency information disseminator [10]. Worldwide, governments are beginning to harness the internet and related Information and Communication Technologies (ICTs) to address citizens' desire for greater access to information, institutional transparency, participatory decision-making, and public service access. Social media, such as off-the-shelf networking sites such as Facebook, microblogging services such as Twitter, and information dissemination platforms such as Youtube, is one channel through which these objectives are pursued [11].

To increase individual capacity in the economic, social, political, and cultural sectors, the social media phenomenon, particularly in urban areas, can not be separated from its core purpose of communicating and collecting information. It is the function of the third wave of information culture, coined by Alvin Toffler (1980) Social media in the governmental context [12]. Social media platforms enable emergency managers to increase traditional crisis communication approaches [13]. Social media is considered a platform on which human opinions, comments, thoughts, and attitudes are expressed, shared, exchanged, or even influenced, with both popularity and prevalence. For example, by sharing brief messages of interests and activities, Twitter users build social relationships with friends and strangers. This user-generated social media content has become a valuable asset for businesses and organizations, as it often contains important information for better strategies and decision-making [14].

2.2 Function Social Media in Government

Definition Social media use by governments refers to web-based tools to provide government information and services, like Facebook and Twitter, and promote social involvement with stakeholders [15]. According to Margo, social media apps have become fundamentally important communication, leisure, and change tools that will increasingly impact the networked society's future life. While according to Criado, the use of government social media is becoming one of the main trends in the study and practice of e-government. According to Song and Lee, for years, social media has transformed how people communicate; thus, governments embrace social media as participation channels that allow people to access government services [16]. 'Push' refers to the simple provision to citizens of public and accurate information, which is related to transparency;

'pull' refers to the agency's interaction with citizens to acquire citizen information and feedback that can be accomplished by observing user actions or explicitly asking for feedback, and networking or collaboration refers to activities in which agencies and their activities [17].

Governments' use of social media is seen as a new wave of technology adoption. Governments on social media platforms are no longer seen as a choice but as a new standard. Therefore, in e-government research and practice worldwide, government use of social media is becoming a big topic. A primary objective for social media's benefit is to establish a new way of engaging with the public and improving participation and collaboration among citizens [18]. Understanding user views in social media communications has gained greater attention as social media has become an essential platform for organizations to interact with users [14]. Definition the use of social media by governments refers to web-based tools to provide government information and services, including Facebook and Twitter, and encourage social participation with stakeholders [15]. The Facebook accounts of Jordanian government bodies have produced a low number of followers and low numbers of tweets and likes, as have their Twitter accounts. In reality, their Youtube channels have a minimal number of subscribers, indicating a low public participation level. Thus, reflecting that dialogue is a one-way dialogue in Jordan by social media platforms, not government-to-user interactivity [16]. Data on 75 local governments in 15 countries were collected and evaluated for both governments using and civic engagement. The data show that the usage of Facebook by the local governments of Western Europe has become commonplace. The audiences on the official Facebook pages are relatively widespread, but citizen participation is soft [19].

To connect with those they represent, government agencies are regularly using social media. Such ties have the power to broaden government services, demand new ideas, and facilitate decision-making and problem-solving [20]. Social media analysis can shed light on Islamic groups' polarisation in the 2019 election by illustrating Islamic groups' interconnections on Twitter [21]. Public sector agencies use social media to connect with people and promote more robust relationships with local governments across Europe and East Asia [22]. Using online commentators to form conversations on the internet and social media media, government or political party organizations using three ways: first pushing pro-government or pre-party propaganda; second attacking the opposition or mounting smear campaigns; or third neutral tactics involve diverting conversations or criticism from essential issues or checking information [23]. Via social media, government initiatives have taken some governments a step further by using social media as a platform for political campaigns. Although this issue is beyond this paper's scope, it is useful to recognize the value that governments put on social media to understand the broader context of the govt-social media relationship [24].

2.3 Using a Social Media Platform for Government

This study found that social media has not yet affected its internal processes in the three cities. The use of social media is also not appropriate as a space for interaction among people and governments. It is used only to disseminate information to the public; social media seems to have been used only to collect citizens' data but not involve them in the decision-making process [3]. The results obtained show that Twitter is preferred to

Facebook to engage in local government issues. The level of online transparency, mood, the activity level in social media, and the local government website's interactivity are other factors relevant to citizen engagement [25]. Research has found that Bandung City has achieved significant achievements at the engagement level by engaging in daily city affairs with the city community. Iligan City and Phuket City, meanwhile, are both in the process of transforming their resilient internal organizations. Top and middle leadership, legal standing for social media use, policy framework, and internal change management are the critical variables of social media use in three cities [7].

The results indicate a wide variety of social media approaches, from a closed and outsourced system where posted materials go through layers of high-level approval to open and internal structures of greater autonomy in developing content. Social media are now considered essential components of communication strategies, but how they are implemented by each agency differs. There is a high engagement with audiences for some agencies, considering that these are official government social media accounts [26]. The empirical analysis demonstrates that through decentralized channels, most Twitter communication takes place. While on Twitter, a minority of police officers use personal names, most use their formal identity. Twitter is mostly used for external correspondence, but the shared interest in other police officers' Twitter communications is essential. The research affirms the principle of transformative change: in the hybrid organization of social media communications, the old bureaucratic and new models manifest themselves [27]. The findings suggest that social media's assimilation in local government agencies is significantly influenced by technology competence, top management support, perceived benefits, and citizen readiness. The strongest predictor of social media assimilation in a government agency is maximum management support [28].

This study finds that social media can increase public trust as long as it perceives transparency and interactivity in social media communication [29]. This case study revealed the type of information sent and received by either a particular local government through its Facebook page and what this type of data shows about government-citizen communication. In similar studies of municipalities' Facebook pages, the set of categories elaborated here, possibly refined, can be used to identify and characterize the exchange of information between local governments and citizens and organizations [2]. The spread of social media has significantly expanded the possibilities to communicate with their stakeholders, the local government units. Indeed, in general profiles and specific profiles dedicated to individual projects and events, local government units are becoming increasingly active. However, it should be stressed that social media are used to a varying extent [30].

3   Research Method

This study used Q-DAS (Qualitative Data Analysis Software) (Nvivo 12 plus) to collect data and analyze data with Chart analysis, Cluster analysis, and Word cloud Analysis. This study chose the Twitter KPU RI and Bawaslu RI social media accounts because the KPU RI and Bawaslu RI were election officials in Indonesia responsible for organizing the elections in Indonesia. The use of Twitter as a social media is one way to make expectations reliable with the community. The data collection period on the Twitter social media accounts of the KPU RI and Bawaslu RI ranged from 2018–2019. The reason is that in 2018 there was a simultaneous regional head election in Indonesia. Likewise, in 2019, Indonesia held general elections simultaneously electing the President and Vice President, DPR RI, DPD, Provincial DPRD, Regency/City DPRD. Data taken on the social media accounts of the Indonesian KPU and Bawaslu RI included necessary information such as followers, following, tweets, retweets, communication content, and communication intensity.

The number of tweets on the KPU RI social media Twitter account reached 1683. Simultaneously, the number of retweets from the KPU RI Twitter account reached 46 (see Fig. 1).

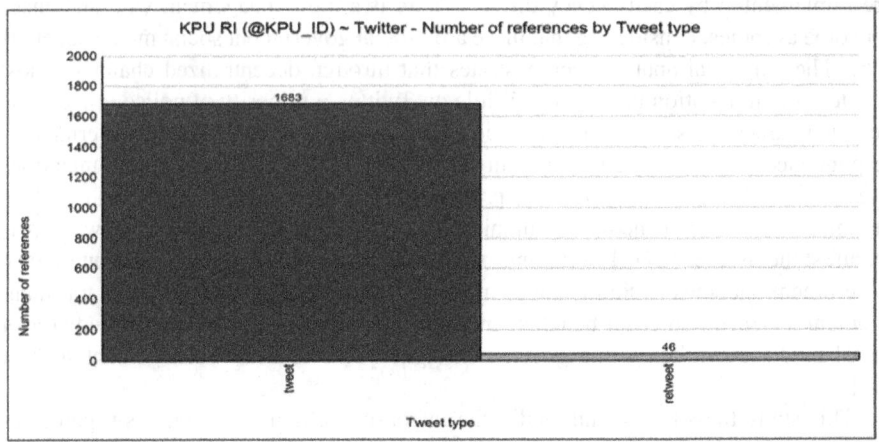

Fig. 1. Tweet and retweet the KPU RI Twitter account

While the number of tweets on the Bawaslu RI social media Twitter account reached 2397, and the number of retweets from the Bawaslu RI social media Twitter account reached 375 (see Fig. 2).

The KPU RI Twitter account has 235.3 thousand followers and 203 following numbers. Meanwhile, the Twitter social media account from Bawaslu RI has 87.6 thousand followers, and the following Twitter social media account is 144. The number of tweets, retweets, followers, and the KPU RI and Bawaslu RI Twitter social media accounts above shows that the Twitter account is an active social media account.

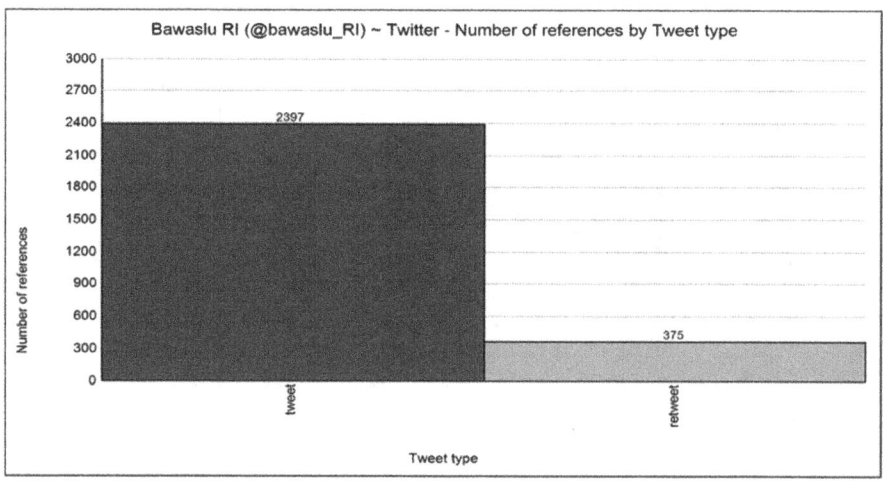

Fig. 2. Tweet dan retweet the Bawaslu RI social media Twitter account

4   Research Findings

4.1   The Role of Election Administrators Based on Twitter Social Media

The role of election administrators in carrying out their duties and functions is visible in technical activities and can also be seen from social media, especially Twitter. The coding chart wizard in Fig. 3 shows the KPU RI and Bawaslu RI election officials' role on Twitter. The coding chart wizard in Fig. 3 helps understand how significant the role is between the KPU RI and Bawaslu RI on Twitter.

Based on the results of the coding wizard chart above, Bawaslu RI's role in member activities is 66.42% greater than KPU RI members' actions, which are only 33.58%. Bawaslu's part related to agency activities shows a relatively large percentage of 96.43%, much higher than the Indonesian KPU, which is only 3.57%. In Bawaslu RI's role in election administration, the rate is 59.31%, which is also more significant than the KPU RI, which shows a percentage of 40.69%. Bawaslu RI's role in campaign socialization based on twitter was 56.92%, while the RI KPU was 43.08%. Bawaslu RI and KPU RI's role in the electoral stages' socialization, the percentage results are not too far away, Bawaslu RI is 52.43% and KPU RI 47.57%. Looking at data based on twitter's social media accounts related to election officials' role shows that Bawaslu RI looks more active than the KPU RI in each of its parts.

The KPU RI and Bawaslu RI are closely related to their roles. The relationship between election administrators' position based on Twitter from member activities, agency activities, election administration, campaign socialization, and The socialization of election stages. Cluster analysis Fig. 4 shows the interrelation of the roles between the KPU RI and Bawaslu RI. The cluster analysis results below help determine the connectivity, the relationship between the parts of the KPU RI and Bawaslu RI officials.

Based on the cluster analysis results above, the campaign's election socialization stage has the highest relationship. Following the campaign for socialization with the

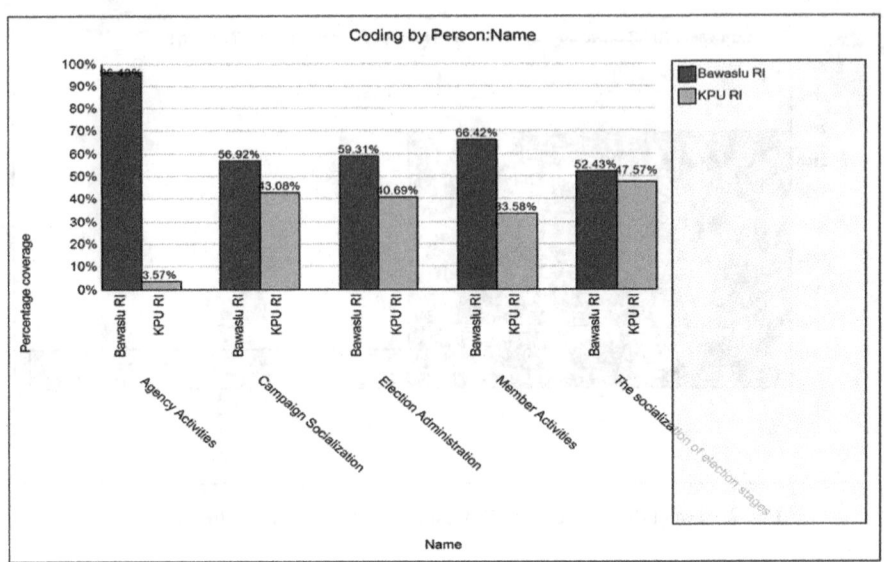

Fig. 3. The role of KPU RI and Bawaslu RI based on Twitter

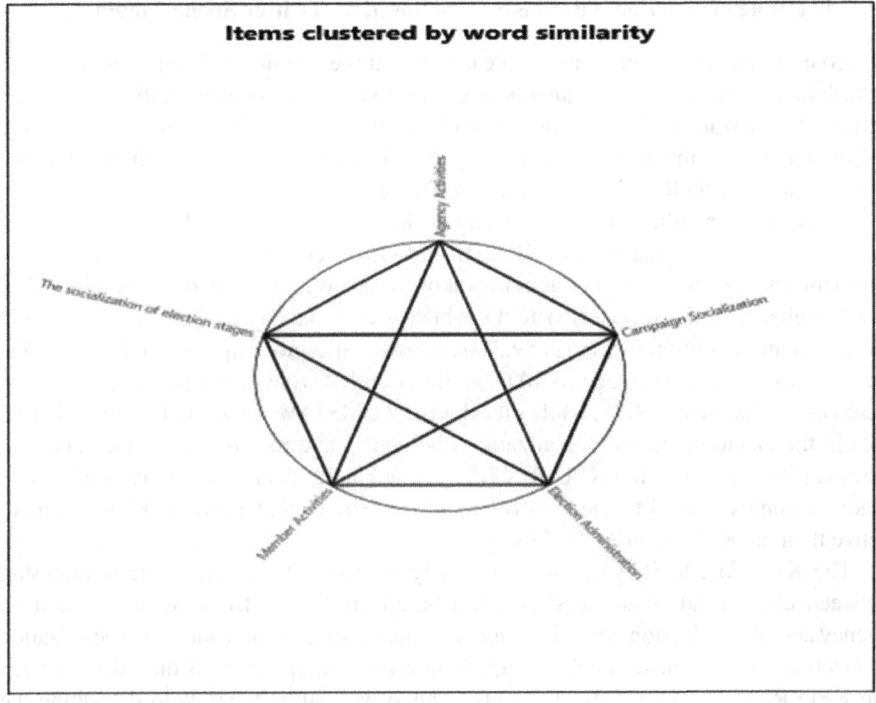

Fig. 4. The relationship between the roles of KPU RI and Bawaslu RI based on Twitter

holding of elections, cluster analysis shows the Indonesian KPU and Bawaslu RI intensively discuss the relationship between these roles. Furthermore, the socialization of elections with the implementation of elections, the fourth is campaign socialization with member activities. The fifth is the relationship between campaign socialization and agency activities. The sixth stage is socialization in the election of member activities. The seventh is the relationship between agency activities and member activities. The eight roles are socialization of the election stages with agency activities, then election administration with agency activities, and finally, the lowest correlation between election administration and member activities. Below is a cluster analysis of the coefficient numbers of the relationship between the roles of KPU RI and Bawaslu RI based on Twitter (Table 1).

Table 1. The coefficient figure of the relationship between the roles of KPU RI and Bawaslu RI based on Twitter

Role A	Role B	Pearson correlation coefficient
The socialization of election stages	Campaign Socialization	0.955435
Election Administration	Campaign Socialization	0.876671
The socialization of election stages	Election Administration	0.845917
Member Activities	Campaign Socialization	0.83973
Campaign Socialization	Agency Activities	0.837769
The socialization of election stages	Member Activities	0.799109
Member Activities	Agency Activities	0.77897
The socialization of election stages	Agency Activities	0.755943
Election Administration	Agency Activities	0.695311
Member Activities	Election Administration	0.69234

Based on the table above are coefficient figures. The socialization of the election stages with the Socialization campaign has the most vital relationship with the coefficient of 0.955435, followed by election administration with a campaign socialization coefficient of 0.876671. While the weakest relationship between member activities and election administration is with a coefficient of 0.69234.

4.2 Election Official Communication and Interaction Content on Twitter Social Media

Based on the word cloud analysis of the two Twitter accounts of the election officials of the KPU RI and Bawaslu RI, the election officials discussed the election intensively and consistently; the dense use of the word illustrates that the KPU RI and Bawaslu RI have a high level of concern with elected officials' responsibilities. The density of the use of "Pemilu" was followed by the thickness of the use of the words #KPUmelayani, 2019, #sahabatbawaslu, #bawaslumengawasi, 2018, #sukseskan-pemilu2019, pilkada, #bawaslujagahakpilihkabamatan, #temanpemilih, #election2019, #pemilihberdaulat, and several other words that describe the seriousness of the RI KPU and Bawaslu RI in holding elections.

Fig. 5. KPU RI and Bawaslu RI communication content based on Twitter

The word cloud analysis results from the election officials' Twitter account, such as (see Fig. 5), confirm that the Indonesian KPU and Bawaslu RI have excellent attention to their elected officials' responsibilities. KPU RI and Bawaslu RI officials consistently discuss election issues in twitter content, emphasizing their respective duties. The KPU RI and Bawaslu RI emphasize service to voters with #KPUmelayani and #sahabatbawaslu. Bawaslu RI is aware of the risk of fraud in the #bawaslumengawasi and #bawaslujaga-hakpilihkabamatan elections. However, the Indonesian KPU and Bawaslu RI are trying to invite the public to succeed in the elections held in Indonesia #sukseskanpemilu2019, #pemilu2019, and Regional head elections "Pilkada". The KPU RI and Bawaslu RI also know that voters are the most crucial element in a #temanpemilih and #pemilihberdaulat election. The consistency of conversation and communication of election management officials on the official Twitter accounts of the KPU RI and Bawaslu RI shows that the Republic of Indonesia's election officials responded to the implementation of elections at national and regional levels. By implementing policies to serve voters, supervise the implementation of elections, and call for the election administration's success.

Overall, the KPU RI and Bawaslu RI Twitter account discuss elections, including implementing the 2018 regional elections and the 2019 election. Discussing parties participating in the election starting from candidates and election participants, socializing election stages, and campaign violations. Among the discussion topics, organizing the

2019 election was most frequently discussed by RI KPU and Bawaslu RI officials. This condition is inseparable from holding simultaneous elections in 2019 to be the first and largest in history. In the 2019 election, the Indonesian people simultaneously elect the president and vice president, DPR RI, DPD, Provincial DPRD, and Regency/City DPRD. Also, the high intensity of discussions about the 2019 simultaneous elections is due to the many problems related to the distribution of election logistics and the complexity of the implementation. The parties involved are so enormous and related to the occurrence of election violations. Election officials are trying to overcome these problems through intensive conversations and communication on the official Twitter social media of the KPU RI and Bawaslu RI.

4.3 Election Official Communication and Interaction Intensity in Twitter Social Media

The Bawaslu RI officials responded to the elections in January–March, increased from April to June, and increased again from July to September, and the highest was in October–December 2018 (see Fig. 7). The intensity of Bawaslu RI communication in 2018 cannot be separated from the simultaneous regional elections in 171 regions and the start of the campaign period at the end of the year for the 2019 simultaneous polls. One of the elected officials is tasked with overseeing the campaign period to take action against violations. In the 2018 simultaneous regional elections, Bawaslu RI found 1,792 violations, such as being late at the TPS opening, unavailability of tools for blind voters, damaged ballots, and assistants did not sign a statement letter until the DPT was posted on the announcement board [31]. From September 2018 to early April 2019, the campaign period for 2019 was a simultaneous general election. The campaign period was so long, and the election participants were so many, that Bawaslu RI massively supervised the campaign nets in collaboration with the regional Bawaslu (Fig. 6).

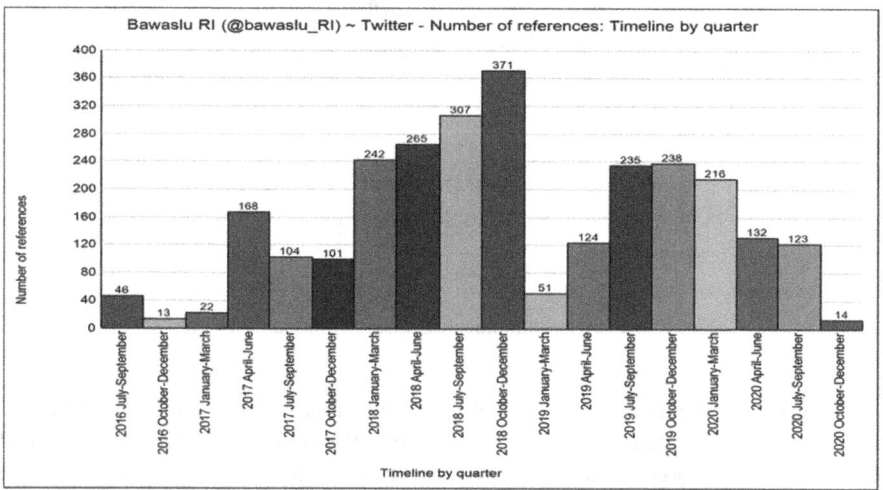

Fig. 6. Bawaslu RI communication intensity based on Twitter

Furthermore, in mid-April, the elections for president and vice president, DPR RI, DPD, Provincial DPRD, Regency/City DPRD, will be held simultaneously. Bawaslu RI paid serious attention to the election's implementation because of the political turmoil that began to be dynamic between election participants, both candidates, and political parties. July–September and the end of the year are busy months for Bawaslu RI because they have to process and follow up on violations in the 2019 simultaneous elections. Based on the RI Bawaslu report on the results of handling 16,134 administrative violations, 373 breaches of the code of ethics, 585 criminal violations, 1,475 other violations of the law [32].

Likewise, the KPU RI responded to its responsibilities regarding elections starting in April–June. Its communication intensity increased dramatically in July–September, stable the following six months, October–December and early 2019 January–March (see Fig. 7). The power of communication in 2018 is also inseparable from the simultaneous regional elections in 2018 in several Indonesian regions. The Indonesian KPU has the responsibility to coordinate with the local KPU to ensure that the elections run smoothly. The researchers are preparing to implement the elections starting from the budget, regulation, or legal framework, forming the elements needed for the performance of the regional head elections, and so on. Moreover, at the end of 2018, the KPU RI was busy carrying out the 2019 simultaneous election stages. Starting from the voter data stage, determining election participants, the campaign period, and others.

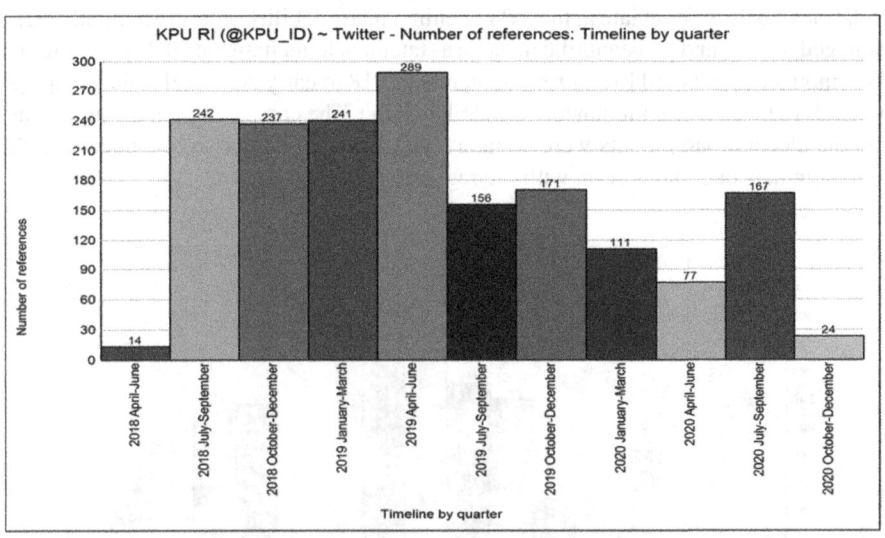

Fig. 7. KPU RI communication intensity based on Twitter

Furthermore, the highest intensity of communication between the KPU RI Twitter account occurred in April–June 2019 because the 2019 simultaneous elections were held in that month, the largest Indonesia's largest election. The 2019 elections indicated a direct democracy system where people can participate in political choices. The intensity of the KPU RI communication in these months was also due to the smooth 2019 simultaneous elections' preparation, which was so complicated. There were several problems in the implementation of the first 2019 simultaneous elections, handling election logistics. Nationally, 10,520 polling stations experience a shortage of election logistics related to voter data handling. The second is related to the KPPS workload. May 16, 2019, data from the health ministry showed that 527 KPPS officers died, and 11,239 fell ill. The third is an error in the recapitulation of the 2019 concurrent election vote count.

5 Conclusions

The conclusion of this study is; *first,* election officials' role on social media Twitter, Bawaslu RI, has more dominant communication and interactions with the public than KPU RI seen from members' activities, agencies, election administration, campaign socialization, and election stages. *Second*, the parts of official election roles are interrelated. However, the strong correlation is the socialization of election stages with campaign socialization. *Third*, KPU RI and Bawaslu RI Twitter's communication and interactions with the public talk a lot about election content. *Fourth*, KPU RI and Bawaslu RI on social media Twitter had high communication intensity with the public in 2018 and 2019. It is caused by in 2018 and 2019. There were simultaneous regional head elections and simultaneous general.

This research's limitation is that the source of data taken only comes from the official social media accounts of the KPU RI and Bawaslu RI Twitter. Therefore, recommendations for further research can be taken from social media accounts, Twitter and Facebook so that the data obtained is complete.

References

1. King, G., Pan, J., Roberts, M.E.: How the Chinese government fabricates social media posts for strategic distraction, not engaged argument. Am. Polit. Sci. Rev. **111**, 484–501 (2017)
2. F. AggelikiTsohou Habin Lee Zahir Irani Vishanth Weerakkody Ibrahim H. Osman Abdel L. Anouze Tunc Medeni: Facebook usage in a local government - a content analysis of page owner posts and user posts. Emerald Insight **7**(2), 240–255 (2013)
3. Roengtam, S., Nurmandi, A., Almarez, D.N., Kholid, A.: Does social media transform city government ? A case study of three ASEAN cities. Emerald Insight **11**(3), 343–376 (2017)
4. Alotaibi, R.M., Ramachandran, M., Kor, A., Hosseinian-far, A.: Factors affecting citizens' use of social media to communicate with the government: a proposed model. Electron. J. e-Gov. **14**(1), 60–72 (2016)
5. Khan, G.F., Yoon, H.Y., Kim, J., Park, H.W.: From e-government to social government: Twitter use by Korea's central government. Online Inf. Rev. **38**(1), 95–113 (2014)
6. Jost, J.T., et al.: How social media facilitates political protest: information, motivation, and social networks. Polit. Psychol. **39**(3), 85–118 (2018)
7. Nurmandi, A., et al.: To what extent is social media used in city government policymaking? Case studies in three asean cities. Public Policy Adm. **17**(4), 600–618 (2018)
8. Vakeel, K.A.: Social media usage in e-government: mediating role of government participation. J. Glob. Inf. Manag. **26**(1), 1–19 (2018)
9. Azad, B., Faraj, S.: What shapes global diffusion of e-government: comparing the influence of national governance institutions. J. Glob. Inf. Manag. no. May 2014

10. Kim, J., Hastak, M.: Social network analysis: characteristics of online social networks after a disaster. Int. J. Inf. Manag. **38**(1), 86–96 (2018)
11. Tursunbayeva, A., Franco, M., Pagliari, C.: Use of social media for e-Government in the public health sector: a systematic review of published studies. Gov. Inf. Q. (October 2016), 0–1 (2017)
12. Budiana, H.R., Sjoraida, D.F., Mariana, D., Priyatna, C.C.: The use of social media by Bandung city government in increasing public participation. In: International Conference on Communication and Media Studies, CCCMS, no. 1, pp. 63–70 (2016)
13. Wukich, C.: Government social media messages across disaster phases. J. Contingencies Cris. Manag. (2016)
14. Chen, H.M., Franks, P.C., Evans, L.: Exploring government uses of social media through Twitter sentiment analysis. J. Digit. Inf. Manag. **14**(5) (2016)
15. El-taliawi, O.G.: Global encyclopedia of public administration, public policy, and governance. Glob. Encycl. Public Adm. Public Policy, Gov., no. July 2020
16. Al-Masaeed, S.: Social media use by the government: adoption and efficiency. Int. J. Electron. Gov. **11**(2), 205–216 (2019)
17. DePaula, N., Dincelli, E., Harrison, T.M.: Toward a typology of government social media communication: democratic goals, symbolic acts and self-presentation. Gov. Inf. Q. **35**(1), 98–108 (2018)
18. Gintova, M.: Understanding government social media users: an analysis of interactions on Immigration, refugees and citizenship Canada Twitter and Facebook. Gov. Inf. Q. (June), 101388 (2019)
19. Bonsón, E., Royo, S., Ratkai, M.: Facebook practices in western european municipalities: an empirical analysis of activity and citizens' engagement. Adm. Soc. **49**(3), 320–347 (2017)
20. Bertot, J.C., Jaeger, P.T., Hansen, D.: The impact of policies on government social media usage: issues, challenges, and recommendations. Gov. Inf. Q. **29**(1), 30–40 (2012)
21. Salahudin, Nurmandi, A., Jubba, H., Qodir, Z., Jainuri, Paryanto: Islamic political polarisation on social media during the 2019 presidential election in Indonesia. Asian Aff. (Lond). **51**(3), 656–671 (2020)
22. Porumbescu, G.A.: Linking public sector social media and e-government website use to trust in government. Gov. Inf. Q. (2016)
23. Bradshaw, S., Howard, P.N.: Challenging truth and trust: a global inventory of organized social media manipulation. Comprop.Oii.Ox.Ac.Uk, p. 26 (2018)
24. Dwivedi, Y.K., Rana, N.P., Tajvidi, M., Lal, B., Sahu, G.P., Gupta, A.: Exploring the role of social media in e-government: an analysis of emerging literature. In: ACM International Conference Proceeding Seriers, vol. Part F1280, pp. 97–106 (2017)
25. Haro-de-Rosario, A., Sáez-martín, A., Caba-pérez, M.C.: Using social media to enhance citizen engagement with local government: Twitter or Facebook ? New Media Soc. (2018)
26. David, C.C.: Social media use by frontline government agencies: review and recommendations. Public Policy 50–72 (2017)
27. Meijer, A.J., Torenvlied, R.: Social media and the new organization of government communications: an empirical analysis of twitter usage by the dutch police. Am. Rev. Public Adm. (2014)
28. Zhang, H., Xiao, J.: Assimilation of social media in local government: an examination of key drivers. Emerald Insight (2017)
29. Marpianta, D.A.: Influence of use of social media of government agencies on trust to the government: study on social media owned by Dinas Penanaman modal dan Pelayanan Terpadu Satu Pintu Provinsi DKI Jakarta. J. Komun. Indones. **VIII**(2P) (2019)
30. Kuźniar, W., Szopiński, W.: The use of social media by local government units to communicate with stakeholders. Nierówności społeczne a wzrost Gospod **45**(1), 247–254 (2016)

31. Andayani, D.: Bawaslu Temukan 1.792 Pelanggaran di Pilkada Serentak 2018. Detiknews.com (2018). https://news.detik.com/berita/d-4086830/bawaslu-temukan-1792-pelanggaran-di-pilkada-serentak-2018. Accessed 07 Nov 2020
32. Data Pelanggaran Pemilu Tahun 2019 4 November 2019. Bawaslu.co.id (2019). https://www.bawaslu.go.id/id/hasil-pengawasan-pemilu/update-data-pelanggaran-pemilu-tahun-2019-4-november-2019. Accessed 09 Nov 2020

Advances in Digital Technology
and Applied Sciences

The Practice of Creating Intelligent Manufacture Management Systems Based on a ERP

Artem Vozhakov$^{(\boxtimes)}$ (iD)

Perm National Research Polytechnic University, Perm 614990, Russia
vozhakov@ya.ru

Abstract. The practice of building an intelligent enterprise management system based on the existing ERP is considered. The approach is based on intellectual elements that allow reducing the degree of uncertainty and making informed management decisions in an ever-changing environment. Intelligent elements are developed in accordance with the principles of optimal production management set out in the most popular concepts today: "Lean Manufacturing", "Theory of Constraints", "Quick Responsible Manufacturing" applied to a wide range of automated systems: ERP, MES, APS. The main task was the development of ERP from a push-system to a pull-system. The tools for building pulling systems, such as a supermarket, FIFO, drum-buffer-rope, WIP limit, were considered and summarized. A mathematical model and tools used to build intelligent systems in conditions of fuzzy initial information, formulation and solution of production management problems are considered. The principles of the practical implementation of intelligent elements in the form of loosely coupled elements integrated into the main information system are stated.

Keywords: ERP · Intelligent enterprise · Enterprise resource planning · Intelligent production management system · Mathematical modeling · APS · MES · Lean manufacturing · Theory of constraints · Quick response manufacturing · Fuzzy sets · Knowledge bases · Expert systems · Artificial intelligence · Production planning · Synchronization of small-scale production · Intelligence actual element

1 Introduction

The current level of technology evolution has significantly influenced the manufacturing processes in all industries. The use of numerically controlled machining centers, industrial robots, high-precision measuring instruments, automated warehousing complexes and automatic production lines has transformed manufacturing halls beyond recognition, reducing human participation in the production process to a minimum. The use of automatic equipment has increased the speed of production processes by an order of magnitude. At the same time, the flow of information has radically changed, influencing the decisions made. The rate of change in the external environment has increased, and the

© The Author(s), under exclusive license to Springer Nature Switzerland AG 2021
T. Antipova (Ed.): ICADS 2021, AISC 1352, pp. 327–339, 2021.
https://doi.org/10.1007/978-3-030-71782-7_29

required response time to external changes has decreased. To maintain its sustainability, an enterprise must take into account rapidly changing demand. Large-scale production is found less and less in the range of products. Production is more and more focused on products, each instance of which can have individual characteristics. The company must be ready to quickly change partners without losing productivity and quality of products.

The process of managing a manufacturing system, as a particular case of socio-technical systems [1], is associated with significant difficulties caused by incomplete information, conflicts of interests and goals, rapid and numerous changes in the environment of an industrial enterprise. This necessitates intellectualization and informatization of management processes, which ultimately leads to the concept of building an intelligent control system as the main management mechanism of a production enterprise.

2 Concept of Intelligent Manufacture Management System

We will call an intelligent manufacture management system an automated manufacture management system with a wide use of artificial intelligence technologies to solve management problems in an automatic mode (without involving a person) or in an automated mode (with the involvement of decision-makers). These properties of an intelligent system distinguish it from classical automated management systems that specialize at various levels of enterprise management [2]. Consider the main types of automated manufacture management systems that are relevant today:

1. Enterprise resource planning (ERP) is the integrated management of main business processes, ERP is a category of business management software – typically a suite of integrated applications – that an organization can use to collect, store, manage, and interpret data from many activities. ERP provides an integrated and continuously updated view of core business processes using common databases maintained by a database management system. ERP systems track business resources – cash, raw materials, production capacity and the status of business commitments: orders, purchase orders, and payroll. The applications that make up the system share data across various departments (manufacturing, purchasing, sales, accounting, etc.) that provide the data [3] ERP facilitates information flow between all business functions and manages connections to outside stakeholders [2].
2. Advanced planning and scheduling (APS, also known as advanced manufacturing) refers to a manufacturing management process by which raw materials and production capacity are optimally allocated to meet demand [3]. APS is especially well-suited to environments where simpler planning methods cannot adequately address complex trade-offs between competing priorities. Production scheduling is intrinsically very difficult due to the (approximately) factorial dependence of the size of the solution space on the number of items/products to be manufactured.
3. MES may operate across multiple function areas, for example: management of product definitions across the product life-cycle, resource scheduling, order execution and dispatch, production analysis and downtime management for overall equipment effectiveness (OEE), product quality, or materials track and trace. MES creates the "as-built" record, capturing the data, processes and outcomes of the manufacturing

process. This can be especially important in regulated industries, such as food and beverage or pharmaceutical, where documentation and proof of processes, events and actions may be required [4].

It is necessary to highlight ERP systems, which consist of a large number of software modules that can be purchased separately and help manage many activities in various functional areas of the business, including modules for sales and distribution, financial accounting, financial controlling, production planning, asset management, personnel management, etc.

The evolution of ERP systems led to the fact that most of the developers of ERP systems began to include their own MES modules in their systems, or to include tools for integrating with existing MES into ERP. A similar situation happened with APS systems, many ERP developers have created their own modules and included them in the delivered ERP solutions. This approach has many benefits, including:

1. the data in the system is guaranteed to be up-to-date, without the possibility of discrepancies;
2. a single tool allows you to build end-to-end enterprise management.

ERP, APS, MES − although they are completely different systems with different functionalities, designed for different purposes, but at the same time they can not only perfectly coexist, but also complement each other in terms of creating a powerful planning system at the enterprise, covering all existing tasks [5].

After the completion of the ERP system implementation stage (in the extended sense, including the functionality of APS and MES), the management of the enterprise receives at its disposal a huge amount of primary data about its business, which contains structured information necessary for decision-making [6].

The first step in the intellectualization of systems can be considered the construction of advanced reporting and a business intelligence subsystem based on primary ERP data. A high-quality business analysis system forms almost ready-made decisions: which orders should be canceled, what additional products should be produced to make up for defects, which orders should be given higher priority, which clients should be suspended due to arrears, etc. At the same time, in real conditions, the timeliness of decision-making is extremely important (the earlier the problem is identified and measures are taken to solve it, the less negative consequences for the enterprise). Therefore, the next step in the development of automated systems is the transfer of decision-making powers directly to the automated control system: automatically open orders for production, suspend production of unclaimed goods, send an email to the client with notification of changes in delivery conditions, etc. This moment can be considered the moment of the emergence of an intelligent information system.

3 Manufacturing Systems Management Models

In parallel with the development of automated management systems, approaches to the organization and management of production also developed. There are three most well-developed approaches to the organization and management of production:

1. Lean production [7] is an approach to management that focuses on cutting out waste, whilst ensuring quality. This approach can be applied to all aspects of a business – from design, through production to distribution. Lean production aims to cut costs by making the business more efficient and responsive to market needs. This approach sets out to cut out or minimise activities that do not add value to the production process, such as holding of stock, repairing faulty product and unnecessary movement of people and product around the business. Lean production originated in the manufacturing plants of Japan, but has now been adopted well beyond large and sophisticated manufacturing activities. The key aspects of lean production that you should be aware of are [8]:

   a. 5S Visual Workplace;
   b. Standardized Work Instructions;
   c. VSM Value Stream Mapping;
   d. TPM Total Productive Maintenance;
   e. Kaizen Blitz Events;
   f. Error & Mistake-Proofing;
   g. Self-Directed Work Teams;
   h. Mixed/Level-Loaded Production;
   i. SMED Setup Reduction;
   j. Inventory & Lead Time Reduction;
   k. Lean Visioning;
   l. TOC Constraint Management;
   m. 2-Bin Auto-Replenishment System;
   n. Quality System Certification;
   o. KanBan Implementation;
   p. Lean Six Sigma.

2. The theory of constraints (TOC) is a management paradigm that views any manageable system as being limited in achieving more of its goals by a very small number of constraints. There is always at least one constraint, and TOC uses a focusing process to identify the constraint and restructure the rest of the organization around it [9]:

   a. method "buffer - drum - rope";
   b. critical chain method.

3. Quick response manufacturing (QRM) is an approach to manufacturing which emphasizes the beneficial effect of reducing internal and external lead times. Shorter lead times improve quality, reduce cost and eliminate non-value-added waste within the organization while simultaneously increasing the organization's competitiveness and market share by serving customers better and faster. The time-based framework of QRM accommodates strategic variability such as offering custom-engineered products while eliminating dysfunctional variability such as rework and changing due dates [10]. For this reason, companies making products in low or varying volumes have used QRM as an alternative or to complement other strategies such as Lean Production, Total quality management, Six Sigma or Kaizen. However, the

benefits of QRM are still mooted and contested by experts around. Many opposers of QRM criticize its approach being very "marketing-style" rather than academic or statistical. The main tool in QRM is the POLCA. POLCA is a card-based production control system for high mix/low volume, and customer-specific manufacturing. It helps to decrease the throughput time on a shop floor with production cells, and products that have to follow their own route during their assembly or manufacturing.

It should be noted that these practices fully work only for mass production lines and are often inapplicable for small-scale custom-made production of complex science-intensive products. As a rule, an ERP system is introduced in such industries, the 5S principle is introduced, and this is where the use of modern tools ends. At the same time, the efficiency of these industries remains low, there are wastes, long production cycles and a high level of stocks at all stages of production remain.

The evolution of automated management systems went hand in hand with the evolution of approaches to the organization and management of manufacture. Using manufacture optimization tools in intelligent manufacture management systems seems to be relevant. The intelligent management system independently will make and implement decisions formed on the best practices and models of production management. The development of mathematical models of production processes based on information in ERP for solving management problems using the best practices is proposed.

4 Extension of the Manufacturing Management Concept in ERP

The ERP-system is a push system because releases are made according to a master production schedule without regard to system status. Hence, no prior work in process (WIP) limit exists [11]. It means the organization of the material flows through the production system, in which material resources are supplied from the previous operation to the next in accordance with a pre-formed static schedule. Material resources are "pushed" from one point to another. The main schedule sets the time for each operation by which it must be completed. The resulting product is "pushed" further and becomes a stock of work in progress at the input of the next operation. The calculated plan is mandatory for production. If at the stage of assembling products a serious problem arises that does not allow for a long time to produce a certain type of product, the ERP system will not stop blank production for this order, which will further lead to production unrequired work-in-progress parts, capacity utilization with unclaimed products, etc.

Various sources devoted to the practice of introducing lean production, the theory of constraints and QRM provide detailed justifications for the fact that push systems, faced with constant deviations and lacking internal mechanisms to control deviations, rapidly increase the level of work in progress. Produce unclaimed products and incur other losses.

In contrast to the "push-out" system, the concept of a "pull-out" logistic system is introduced, i.e. such an organization of material flows, in which material resources are supplied to the next technological operation from the previous one as needed, therefore there is no static schedule for material flows. There is also a simpler definition, a pull system is a production management system in which the volume of work in progress is limited in a certain way.

The lean production literature describes the following types of pull production that have proven themselves well in high volume production:

- Supermarket;
- FIFO limited queues;
- Drum-Buffer-Rope;
- WIP limit.

In QRM, the POLCA control system is proposed (paired-cell over-lapping loops of cards with authorization), designed for use in factories organized according to QRM principles.

Below is a generalized model for constructing a pulling system for discrete production, which does not impose additional restrictions on the method of organizing production (production cells, etc.).

As the model object, consider small-scale discrete-stop manufacture with a wide range of highly-diverse product nomenclature.

Manufacture is organized in several shops $n \in \overline{1, N}$ where N - total number of plants, which in turn are divided into manufacturing areas $k \in \overline{1, K}$ where K - number of stations in all the shops in which concentration a certain amount of process equipment, united, or by usage of technological operations, or on the principle of formation of QRM cells. At the same time, to ensure manageability, the plot size is not too large in scale of manufacture. We introduce the portion belonging shop via n_k, $k \in \overline{1, K}$ - shop number, in which the portion.

Figure 1 shows an example of manufacture and the division into sections as well as the arrows indicate the main direction of motion of the material flow forming process in the manufacture chain. Analysis of processing chains is crucial for the construction of efficient manufacture process and for synchronizing manufacture in particular.

It is believed that the company has implemented an automated enterprise management system based on the standard for the MRP II data model:

- Given resource specifications for products and semi-finished products.
- Given technological fabrication routes that contain at least the processing route of parts and assemblies and expandable list of technological operations and indicating the complexity of the equipment used.
- Enter the calendar specifications duration route transitions.
- There is main schedule, which determines the need for the release of products by date needs, divided by customer orders and output parties inside the order.
- Manufacture of parts and assemblies is carried batches whose size is determined in the ERP-system batch numbered index p, $p \in \overline{1, P}$.
- For each batch of known processing route. We know the number of block transitions for each party - m_p for each route section of the transition defined by the Executive Room: M_{mp}, $m \in \overline{1, m_p}$, $p \in \overline{1, P}$
- At any given time we know the current status of work in progress. For each batch of parts, defined the current routing transition $C_p \in \overline{0, m_p}$, where the value 0 corresponds to the state of completion of the processing of the party details.

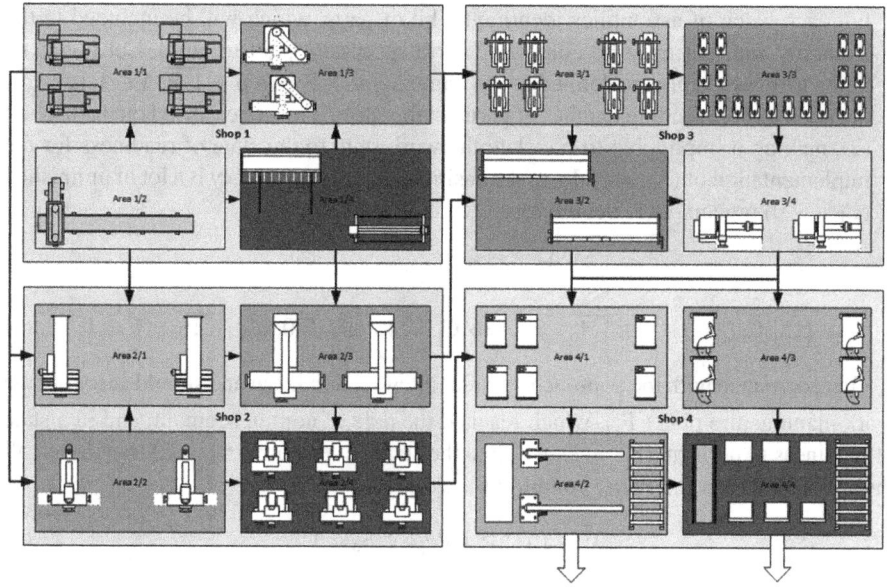

Fig. 1. An example of manufacture and dividing into sections

– Part details proceeds to the next step immediately after the completion of processing at the previous stage, after the completion of the last stage of the current state automatically switches to state 0, the transition current value change according to formula (1):

$$C_p^* = \begin{cases} M_{C_p+1p}, C_p < m_p \\ 0, C_p = m_p \end{cases} \qquad (1)$$

– For the current state of work in progress more discrete state determined, where 0 - batch transport step is in or acquisition; 1 - the party is on the workshop section, resourced and ready for processing; 2 - the party determined to fulfill; 3 - launched batch processing; 4 - completed batch processing on the workshop section: $S_p \in \{0, 1, 2, 3, 4\}$
– It should be noted that the party status of translation in the next state is irreversible, the translation of the party in the previous state is impossible.
– For a more detailed account of the additional current status states can be allocated in ERP-system, such as: Incoming expectation, waiting for the start of processing, handling, etc. launched [12].
– For each party defined the order priority, containing aggregated information on the importance of a sales order, profitability, value and other parameters of the order which affect the importance of the order for the enterprise $W_p \in [0, 1]$, where one corresponding to the maximum value of the order of importance.
– Intended manufacture schedule is determined for each point of the route for each part details the planned date of commencement of treatment DP_{mp}, duration of treatment DL_{mp} and the planned effective date DF_{mp} of the start of processing batches of parts in areas where: $m \in \overline{1, m_p}$, $p \in \overline{1, P}$.

– For each batch of assemblies identified a kit of parts, which will be included in the assembly and that must be completed for a successful start to build. Lot numbers, going to an assembly are defined in the array: $L_{pl} \in \overline{1,P}$, где $p \in \overline{1,P}$, $l \in \overline{1,l_p}$. l_p is the number of preceding batches of parts. In this case, it must be carried out restriction (2), making it impossible to translate the work state to the state of readiness for the implementation of ($S_p = 1$) the first transition ($C_p = 1$) assembly if a lot of unfinished ($C_{L_{pl}} > 0$) precursors is not empty:

$$\{L_{pl}|C_{L_{pl}} > 0, \; l \in \overline{1,l_p}\} \neq \varnothing \Rightarrow C_p = 1 \wedge S_p = 0, p \in \overline{1,P} \qquad 1 = 1 \qquad (2)$$

In accordance with the approach «just in time» section k can and should carry out the work manufacture plan ($\vec{F}$), which reached the date of commencement, and in a state of readiness to perform ($S_p = 1$), while in the status to perform ($S_p = 2$) transferred all the work for which following conditions are satisfied:

$$\vec{F} = \{p \,|\, t \geq DP_{C_{pp}} \wedge S_p = 1\} \qquad (3)$$

However, manufacture management experience shows that blind adherence calculated in ERP-system of manufacture as a consequence of the plan are inevitable deviations lead to skew areas of activity.

We introduce the concept of synchronization cards, designed to synchronize the activity areas in order to minimize the manufacture of unclaimed goods, excess inventory, increase responsiveness and reduce manufacture cycle times. The essence of the concept is similar cards POLCA, used in QRM [10], but the logic will be different, which requires separate isolation determination. A synchronization card has two mandatory attribute - portion sender and the recipient workshop section. Each portion (receiver) supplies at least one card for each portion, which is the supplier of the technological chains elements (assemblies) to the sector. The number of cards for a recipient provider may be increased if the material flow between the portions particularly intense.

The number available for use cards determine the synchronization of the array $K_{k_1 k_2}$, where k_1 in - the receiver portion, and k_2 - a workshop section provider. Number of cards k_2 used supplier k_1 portion is determined by counting the number of lots of parts which are in the shop k_1, which is the portion of the previous processing k_2. At every point in time, the number of cards currently in use $K^*_{k_1 k_2}$ shall not exceed the number of available cards synchronization:

$$K^*_{k_1 k_2} = \sum_{p=1}^{P} q(p, k_1 k_2) \leq K_{k_1 k_2}, \text{где } q(p, k_1 k_2) =$$
$$\begin{cases} 1, & M_{C_p p} = k_2 \wedge M_{C_p+1\,p} = k_1 \wedge S_p \in [2,3] \\ 0, & M_{C_p p} \neq k_2 \vee M_{C_p+1\,p} \neq k_1 \vee S_p \in [0,1,4] \end{cases} \qquad (4)$$

If a certain operation is performed p ($S_p \in [2, 3]$) in the area k_2 ($M_{C_p p} = k_2$) with subsequent transfer to the portion k_1 ($M_{C_p+1p} = k_1$) (it is considered that the synchronization uses one card k_1 k_2).

Need to develop a system of synchronization of manufacture, which is based on the information in the ERP-system in a real-time and sends to the implementation of

$(S_p = 2)$ the optimal set of parties $\overrightarrow{FA}$, the implementation of which is most expedient at the current time t based on drawing principles, restrictions on work in progress and prioritization orders.

In other words, you want to define, and translate to the implementation of such parties $p^* \in \overrightarrow{FA}$, for which the following conditions are met: $p^* \in \overrightarrow{FA}$

$$p^* \in \vec{F} = \{p \mid t \ge DP_{C_p p} \wedge S_p = 1\} \tag{5}$$

$$\sum\nolimits_{p^* \in \overrightarrow{FA}} \omega(p^*, k_1, k_2) + K^*_{k_1 k_2} \le K_{k_1 k_2} .$$

$$\text{где } \omega(p, k_1, k_2) = \begin{cases} 1, & M_{C_p p} = k_2 \wedge M_{C_{p+1} p} = k_1 \wedge C_p > 1 \\ 0, & \text{else} \end{cases} \tag{6}$$

$$J_1 = \sum\nolimits_{p^* \in \overrightarrow{FA}} W_{p^*} \to \max \tag{7}$$

$$J_2 = \sum\nolimits_{p^* \in \overrightarrow{FA}} DP_{C_{p^*} p^*} \to \min \tag{8}$$

The above statement of the problem is two-criteria optimization problem with two independent criteria. For the problem to be solvable, it is required to introduce a generalized criterion optimality of the plan. In practice, usually one of the criteria is selected as a "master", that is, having the highest importance. However, all experts agree, what about other criteria and can not be forgotten - they are all important. However, the ratio of importance of these criteria is very unclear. To solve this problem generalized optimality criterion may be used with a special fuzzy set extended over private optimality criteria [13].

On the basis of the proposed partial criteria may be introduced generalized optimality criterion with extended special fuzzy set over particular optimality criteria:

$J^r = \{\mu_1/J_1; \mu_2/J_2; \}$ where μ_1 and μ_2 - the importance of particular criteria J_1 and J_2 respectively. Note that $\mu_i \in [0, 1]$, $i = 1, 2$, r - number of this variant of the manufacture plan. For definiteness, we assume that both private is necessary to minimize the optimality criterion. In this case it is sufficient to criteria (7) to change the sign to "-".

Now, using the clear function of the fuzzy argument (special index ranking), comparison of variants of the manufacture plan r1 and r2 by the generalized criterion Jr can be carried out according to the formula:

$$H(J^{r1}, J^{r2}) = \text{sign } R_i, \text{где } R_i = \frac{\mu_i^{r1} \cdot J_i^{r1} - \mu_i^{r2} \cdot J_i^{r2}}{\max(J_i^{r1}, J_i^{r2})} \tag{9}$$

Of all the values Ri is selected modulo a maximum value, the sign of which determines the ranking of the index (9).

Now, if the value $H(J^{r1}, J^{r2}) = +$ then $J^{r1} > J^{r2}$ else $J^{r1} < J^{r2}$.

This the problem posed above two-criteria discrete optimization can be reduced to the following single-criterion problem with the generalized fuzzy optimality criterion:

Party find such optimal $p^* \in \overrightarrow{FA}$, at which the minimum of the generalized criterion

$$J^r \to \min \tag{10}$$

and executed limit (5) and (6).

5 Practical Implementation of an Intelligent Manufacture Management System

Let's call it an intellectual element − a mechanism that allows, in certain situations, to partially or completely replace a person making a decision. To implement an intelligent system, an approach was chosen to create an intelligent management system based on an existing ERP system at the enterprise by embedding intelligent elements into the ERP, or by closely integrating intelligent modules (subsystems) into the ERP. For these purposes, any ERP is suitable, which develops in accordance with general trends in the ERP systems market:

1. Modularity of solutions, allowing enterprises to choose the functionality that they need, and to collect those configurations of ERP solutions that meet the needs of the enterprise.
2. The transition to a microservice architecture of software, aimed at the interaction of, as far as possible, small, loosely coupled and easily modified modules. This architecture allows enterprises to create a single information system from modules from different manufacturers, based on their unique needs.
3. Development of cloud solutions, which allows using the products both on our own server capacities and in the form of cloud solutions, which increases the availability of solutions for remote work, and reduces infrastructure costs.
4. Support for mobile clients. Already today, up to half of all information is generated and consumed on mobile devices. This trend will undoubtedly continue, including for business applications. Mobile clients for all platforms will soon become an industry standard.
5. Expansion of the functionality of solutions without withdrawing from support. To date, mechanisms for expanding functionality have been developed without removing the solution from the manufacturer's support and the ability to quickly disable such extensions, these mechanisms make customization of the ERP system fast and safe.
6. Development of means of online integration with other systems and the Internet of things. This direction of development is key for the development of ERP systems. More and more data is generated automatically in related systems. The use of duplicate data entry in ERP by a person is already considered bad form today.
7. Development of B2B, B2C directions. This direction will lead to the fact that ERP systems of various enterprises will exchange data on orders, status of execution, delivery and payment terms automatically, without the need for operators. Using online stores and other related systems, interaction with private buyers can also be fully automated.

It should also be noted that most of the competing systems on the market have a similar data model that complies with MRP II, MES, etc. Thus, the software implementation of the mathematical model can be used to extend the functionality of various systems through additional development of integration modules. The developed mathematical models of smart elements must comply with the standards in terms of the data model as much as possible, especially with regard to the input data for tasks. The use of additional information that is not provided for in the standards can be performed only when determining the method of obtaining this information (automatic data collection, manual input, etc.).

As the main intellectual element (not the only one), a production synchronization agent was implemented, which allows organizing pull production in real time by monitoring the state of production as a whole, identifying deviations and automatically suspending the execution of unclaimed works and, on the contrary, increasing the priority of works, the implementation of which is currently the most relevant.

The intelligent production synchronization system was implemented as a separate service that exchanges data with the ERP system online. To solve the optimization problem in the developed System, an empirical algorithm is implemented that is capable of solving the optimization problem with the required speed and accuracy.

In Fig. 2 shows the automation scheme of the System. This scheme contains the minimum number of connections, and, therefore, minimizes the complexity of the task of integrating information systems. The data necessary for the operation of the System is received from the ERP system online. Then the optimization problem is solved. Information about batches of parts that are now available for processing is displayed on the information panels of the areas and is also transferred to the ERP. The foreman can take only those orders that the System has activated [12].

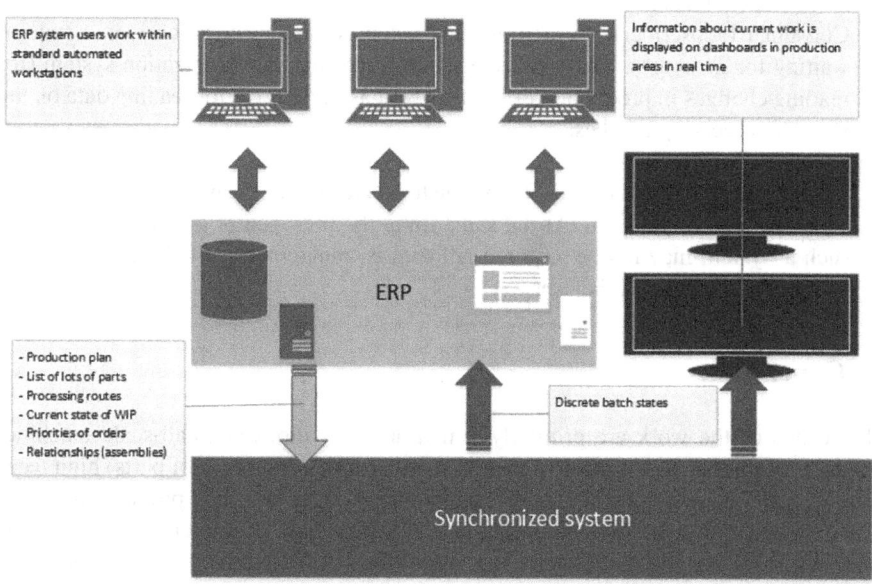

Fig. 2. System automation scheme

For the stable operation of the system, it was necessary to build a reliable integration system that allows real-time exchange of messages about all changes in the production system. One of the most interesting options for the implementation of integration mechanisms in multi-platform environments was chosen based on the use of message-broker software (in particular RabbitMQ), that originally implemented the Advanced Message Queuing Protocol (AMQP) and has since been extended with a plug-in architecture to support Streaming Text Oriented Messaging Protocol (STOMP), MQ Telemetry Transport (MQTT), and other protocols [14].

RabbitMQ allows interoperability of various programs, is a well-proven solution for building SOA (service-oriented architecture) and the distribution of deferred resource-intensive tasks. RabbitMQ is a message broker used to send and receive messages with the principle of the post office, when one system drops a letter in a mailbox, the post office guarantees that sooner or later the postman will deliver it to the addressee. RabbitMQ has flexible routing capabilities that make the system unique.The following components are involved in the integration:

1. Producer – the program that sends messages. This system component is built into ERP for sending production data. A similar component is built into the synchronization system to transmit the data stream on the status of processing batches.
2. Queue – the name of the "mailbox". It exists inside RabbitMQ. Although messages go through RabbitMQ and applications, they are only stored in queues. The queue has no limit on the number of messages. For each data item (production order, production stage, etc.), its own queue is created, in which messages about all new and changed records are written.
3. Consumer (subscriber) – a program that receives messages, which is in a state of waiting for messages. The subscriber is built into the synchronization system (for reading changes in production orders), as well as in the ERP for reading data on the state of processing batches.

The use of this approach made it possible to create an intelligent system independent of the technological platform. At the same time, the presence of a graphical interface for such a System may not be required, all data is transferred to the ERP and there is brought to the attention of the user.

6   Conclusion

The results of the work are primarily aimed at applications at small-scale machine-building enterprises that produce complex (consisting of thousands of parts) high technology products, with long chains of technological stages and long production cycles. It is expected that it is for such enterprises that the effect of implementation will be maximal. However, it is assumed that the use of the system in other types of production will also improve the efficiency of production, or at least reduce the operating load to maintain the control system.

Despite the fact that the early stages of intellectualization of the management system took place back in the 60s of the twentieth century, it can be stated with confidence that the main achievements and discoveries in this area of knowledge are still ahead. Today there are a lot of ideas and approaches that have not yet been implemented in practice due to any objective limitations. Technologies will become more accessible every year, more and more data will accumulate in a digital and structured form, and all this will lead to new and new breakthroughs in the field of enterprise intellectualization.

The proposed approach will allow, over time, to make a smooth transition to a fully intelligent enterprise management system. Synchronizing the production process across different shops is only the first step. In the future, it is possible to implement additional intelligent elements, such as an intelligent element for manufacture planning [15], real time intelligent element for distributing work between work centers, an intelligent element for selective quality control based on defect statistics, and others.

The creation of an intelligent manufacture management system will become the next stage in the development of manufacture management systems, and the combination of best practices and approaches to manufacture management will create a synergistic positive effect in the practical implementation of the system.

References

1. Gitman, M.B., Stolbov, V.Y., Gilyasov, R.L.: Management of socio-technical systems, taking into account fuzzy preferences. In: LENAND 2011, p. 272 (2011)
2. Browne, J., Harhen, J., Shivnan, J.: Production Management Systems: An Integrated Perspective, 2nd edn, p. 284. Addison-Wesley Publishing Company (1996)
3. Cox, J.F.: APICS Dictionary: APICS – The Educational Society for Resource Management, p. 104 (2006)
4. Meyer, H., Fuchs, F., Thiel, K.: Manufacturing Execution Systems: Optimal Design, Planning, and Deployment, p. 274. McGraw Hill, New York (2009). ISBN 9780071623834
5. Fogarty, D.W., Blackstone Jr., J.H., Hoffmann, T.R.: Production & Inventory Management, 2nd edn, p. 870 South-Western Publishing Co., Cincinnati (1993)
6. Gaither, N., Frazier, G.V.: Production and Operations Management, 8th edn, p. 874. Southwestern College Publishing, Cincinnati (1999)
7. Ohno, T.: Toyota Production System: Beyond Large-Scale Production, p. 176. CRC Press (1988). ISBN 0915299143, 9780915299140
8. Womack, J.P., Jones, D.T.: Lean Thinking, 1st edn, p. 350. Taylor & Francis (1996). ISBN 0684810352, 9780684810355
9. Goldratt, E.M.: Theory of Constraints, 1st edn, p. 160. North River Press (1999). ISBN ISBN-10: 9780884271666
10. Suri, R.: It's About Time: The Competitive Advantage of Quick Response Manufacturing, p. 228. CRC Press (2010). ISBN 1439805962, 9781439805961
11. Takeda, H.: The Synchronized Production System, p. 263. Kogan Page Publishers (2006). ISBN 0749447656, 9780749447656
12. Sumner, M.: Enterprise Resource Planning, 1st edn, p. 208. Pearson (2004). ISBN 0131403435, 978–0131403437
13. Vozakov, A., Gitman, M., Stolbov, V.: Synchronization and management of material flows in small-scale production. In: Advances in Engineering Research, vol. 157, pp. 622–626 (2018)
14. Barthel, J.: Getting Started with AMQP and RabbitMQ. InfoQ, 13 September 2009
15. Yevstratov, S.N., Vozhakov, A.V., Stolbov, V.Yu.: Automation of production planning within an integrated information system of a multi-field enterprise. Autom. Remote Control 75(7), 1323–1329 (2014)

Evaluation of Factors Contributing to the Repurchase Intention of the Automobile Industry Using Sparse Modeling

Takumi Kato[✉] [iD]

Saitama University, 255 Shimo-Okubo, Sakura-ku, Saitama 338-8570, Saitama, Japan
takumikato@mail.saitama-u.ac.jp

Abstract. The best competitive advantage for a company is building long-term relationships with highly satisfied customers, by understanding the factors that contribute to their repurchase intentions. Since large amount of point of sales (POS) data can be obtained in the retail industry, past research has focused on the retail industry. Recently, many studies targeting online shopping sites were reported. On the other hand, there are not many studies of durable consumer goods with a long span of replacement. This research evaluates the factors that contribute to the repurchase intention for the Japanese automobile industry. The evaluation should not depend on the parameters defined by the researcher, but on the factors from the consumer's point of view. Therefore, in the online survey, the reason for the repurchase intention was heard by free answer and the factors were extracted by natural language processing. The most important issue in statistics is said to be variable selection in regression models because constructing statistical models using many variables may cause over-fitting of acquired data. Evaluation was performed by sparse modeling which efficiently extracts the maximum feature quantity from data by utilizing the universally inherent sparsity of high-dimensional data. The results show that the most effective factors were trust and design with the odds ratios of 1.339 and 1.163, respectively. This research is expected to be useful for improving the products and services by quantitatively clarifying the factors of repurchase intention in the automobile industry, that has been insufficient till date.

Keywords: Customer relationship management · Loyalty · Emotional value · Functional value · Sparse modeling

1 Introduction

The customer base is said to be the "ultimate management asset". This is because highly satisfied customers show repeat purchase behavior and thus make a significant contribution to profits [1]. Since it is an important management issue to prevent customers from moving to competitors, many companies have long aimed to increase the repurchase intention by introducing a loyalty program [2]. However, while cases of discounting and one-time benefits that are ineffective stand out, there are a few successful cases [3].

© The Author(s), under exclusive license to Springer Nature Switzerland AG 2021
T. Antipova (Ed.): ICADS 2021, AISC 1352, pp. 340–350, 2021.
https://doi.org/10.1007/978-3-030-71782-7_30

It is important to identify the factors contributing to the repurchase intention and offer incentives based on them to customers. It is also necessary to make a distinction between simple repeat purchases and repeat purchases that result from high loyalty.

Research on the factors contributing to repurchase and repurchase intentions has been focused mainly on the retail industry. Point of sales (POS) data can be obtained in the retail industry, which is one of the reasons that it is easy to evaluate in combination with a questionnaire survey. Recently, many studies targeting online shopping sites were reported. On the other hand, there are not many studies of durable consumer goods with a long span of replacement. In particular, in the automobile industry where the average replacement period is 7.5 years and the ownership rate is 25% or more for an ownership period of 10 years or more [4], factor evaluation of the repurchase intention is insufficient.

Therefore, in this research, the factors contributing to the repurchase intention were evaluated for the Japanese automobile industry. Data was obtained by using an online survey asking for the repurchase intentions and the reasons. This research is expected to be useful for improving the products and services including the loyalty program by quantitatively clarifying the factors of repurchase intention in the automobile industry that has been insufficient until now.

2   Purpose of CRM and Factors of Repurchase Intention

2.1   Purpose of CRM

The importance of Customer Relationship Management (CRM) has long been appealed. The aim is to build loyalty over the long term and to increase profits efficiently. In other words, raising loyalty has a large contribution towards profits. One of the factors is the increase in repurchase rates. Furthermore, customers with high loyalty show their intention to continue purchasing products and services consistently in the future even if there is a possibility of changes in future market environment and corporate marketing activities [5].

Improvement of repurchase rate is not the only effect of customer loyalty. Customers with high loyalty have a passion for the brand, understand the product well, and act as brand evangelists [6]. Even if loyal customer feel dissatisfaction, not only it is rare to occur dissident behavior, but suppress even negative word of mouth [7]. Customer loyalty also contributes to raising product prices [8]. As a result, brands with high loyalty increase new customers while reducing marketing costs and gain high market share. These are the reasons why it is said that loyalty is a source of high profitability.

Therefore, managing loyalty is extremely important for a company. In the past, behavioral indicators such as repeat purchase behavior and percentage of purchases were used as measures of loyalty. However, it was pointed out that the attitude index should be considered because observation of simple repetitive behavior includes unconscious continuation and situations where it is necessary to continue due to constraints [9]. One of the representative examples of attitude indicators is the recommendation intention. Customers with high loyalty and commitment to the brand, confidently recommend it to others. Reichheld [10] indexed this recommendation intention as Net Promoter Score (NPS) and showed that there is a correlation between NPS and revenue growth rate. As another indicator, there are many cases where repurchase intention was adopted.

This indicator is more straightforward for the purpose of raising loyalty. Moreover, in recent years many indicators encompassing psychological aspects (relative attitudes) and behavioral aspects (repetitive actions) have been proposed [11].

Repurchase intention was the focus of this research and was used as the objective variable. However, due to the long replacement period in the automobile industry, it is concerned that there are consumers who cannot answer with a real intention to repurchase [12]. This point was taken into consideration in this study. Respondents were also asked to indicate recommendations as control variables in order to rule out repurchase intentions arising from constraints such as proximity from home. Noted that although bias due to giving financial incentives to survey respondents is also a concern [13], it was not considered in this study.

2.2   Factors Influencing Repurchase Intention

The retails industry has a rich accumulation of POS data and is often considered to study factors affecting repurchase intention. The result of examining the factors of repurchase intention for that industry show two main influencing factors, satisfaction and trust [14, 15] and perceived quality and trust [16]. Recently, online shopping sites have attracted attention. Like the retail industry, trust is an important influencing factor for online shopping sites. For example, trust and satisfaction [17], trust and commitment [18] have been reported. There are many reports that emotional value such as perceived usefulness is a more effective factor than functional value [19–21]. It was also confirmed in the automobile industry that the influence of functional value is decreasing [22]. In addition, factors that do not always influence but change with conditions are also discussed. The impact of compensation after service failure on repurchase intention has been shown to vary with the location of responsibility and the frequency of failure [23]. Switching barriers did not affect the repurchase intention when satisfaction was high but indicated positive influence when satisfaction was low [24].

In the automotive industry, large-scale studies of 100,040 automotive customers have shown that satisfaction has some relevance to repurchase intentions [25]. In addition, an online survey of 573 users who had experienced car recalls showed that voluntary recalls or improvement campaigns had a significant positive impact on consumer loyalty and purchase intentions [26]. Specific brand verification has also been reported. The study focused on testing 335 Toyota owners compensated with a cash settlement, who participated in a class action lawsuit against Toyota and 89 Toyota owners who did not participate in a class action. Respondents were interviewed one to two years after participants were compensated in a cash settlement for illegal overcharging. As a result, owners who participants and were satisfied with the settlement were shown to have a higher repurchase intention than owners who did not participate [27]. In the Japanese automobile industry, it was verified the difference in the causal effects of delivery times on purchase intentions between the first and second candidates. As a result, no adverse effect on purchase intentions was observed for the length of delivery time for the first candidate car [28]. However, unlike online shopping site validation, there are few studies that deal with various factors widely for repurchase intention in the automotive industry.

3 Evaluation Methodology

3.1 Online Survey

The data used is the result of an online survey conducted in Japan in August 2020. The subjects are 900 men and women between 20 years to 60 years and own a car. The target brands are Toyota, Nissan, Honda, Suzuki, Mazda, Subaru, Lexus, Mercedes-Benz, and BMW. The number of owners for each brand was 100, for a total of 900 respondents. As shown in the summary of the respondent attributes in Table 1, there are more men than women because of the characteristics of the car product. In this survey, 764 participants owned one car, 115 participants owned two cars, and 21 participants owned three cars. When respondents owned multiple cars, they answered questions regarding the car they use mainly. Six car body types were covered in this survey, Kei (Japanese unique standard smaller than Compact), Compact, Minivan, SUV, Sedan, and Coupe.

The survey consisted of two parts, the screening survey and the main survey. First, the purpose of the screening survey was to acquire the attributes to be used as a control variable, in addition to extracting the respondents who satisfied the above conditions. Respondents answered questions regarding eight parameters, gender, age, household income, number of cars owned, frequency of driving, body type of cars owned, purchase emphasis point (multiple answers allowed), and owned brands (used mainly). At the end of the screening survey, respondents indicated repurchase intention. In the automobile industry, which has a long replacement period, consumers may not be aware of the next purchase. Therefore, as the repurchase intention, "Q1. Do you feel that you want to purchase again from the brand (manufacturer) you currently own when you purchase a car next time? (0: I do not know at this moment, 1: I do not want to purchase, 2: I do not want to purchase probably, 3: I want to purchase probably, 4: I want to purchase)". Respondents who selected "4: I want to purchase" was defined as those who have a repurchase intention. Here, those who chose "0: I do not know at the moment" were excluded. From the point of view of the respondent, the boundary between the screening survey and the main survey cannot be identified on the survey screen.

The main survey has two questions. The first question was "Q 2. Please answer the reasons for the repurchase intention you answered in Q1. (Free answer)". Here, the respondents gave a free answer in order to exclude any bias. That is, the evaluation should not depend on the parameters defined by the researcher, but on the factors from the consumer's point of view. Another reason was listening to the recommendation intention as a variable to control the repurchase intention arising from the unconsciousness/constraint, not the loyalty. The next question, Q 3 states, how much would you recommend the brand (manufacturer) you own to a friend or acquaintance? (0: not at all likely to 10; extremely likely). Based on NPS [10] where 0–6 is judged as "Detractors", 7–8 as "Passives", 9–10 as "Promoter", it was defined that the subject who selected 9–10 has a recommendation intention in this research.

The repurchase intention and the recommendation intention are converted to dummy variables. There were 317 (35.2%) respondents with the repurchase intention and 244 (27.1%) with the recommendation intention. Since the two indicators were asked questions on different scales, there should be no comparison. It should be considered simply

Table 1. Summary of respondent attributes.

Item	Number of respondents
Gender	Male(644), Female(256)
Age	20s(32), 30s(154), 40s(228), 50s(296), 60s(190)
Household income (million yen)	-3.99(85), 4.00-5.99(155), 6.00-7.99(191), 8.00-9.99(183), 10.00-14.99(201), 15.00-(85)
Number of owned cars	1(764), 2(115), 3(21)
Body type	Kei(124), Compact(220), Minivan(152), SUV(171), Sedan(199), Coupe(34)
Number of owned brands	Toyota(100), Nissan(100), Honda(100), Suzuki(100), Mazda(100), Subaru(100), Lexus(100), Mercedes-Benz(100), BMW(100)

as the objective variable and the control variable in the definition of this research. The correlation coefficient between both variables is 0.393.

3.2 Text Analysis and Evaluation

The purpose of this research is to evaluate the factors contributing to the repurchase intention for the Japanese automobile industry. As a reason for the repurchase intention, the factors were extracted from the free answers given by respondents. The language is Japanese because the survey was conducted in Japan. As a result of extracting nouns and adjectives using MeCab, which is the engine of Morphological analysis, 1,411 words which occurred 7,085 times in total were obtained. Among them, given that it is difficult to quantitatively evaluate words that appear infrequently, only words that appeared three or more times were focused on. As a result, 434 words which occurred 5,900 times in total were obtained. The word with the highest frequency of occurrence was "car" which appeared 312 times and the next word was "good" which appeared 290 times. However, words with poor meaning for this research purpose were excluded. Therefore, 14 categories shown in Table 2 were constructed as factors of the repurchase intention. For example, F_01_Driving is composed of words such as "driving performance", "engine", and "turbo". A total of 105 words were registered in 14 categories.

Next, when each response regarding the reason for the repurchase intention included the registered word, the corresponding category was marked as a tag. The most frequent factor was F_01_Driving which appeared 305 times. On the other hand, the smallest factor was F_14_Recall which appeared 30 times. In addition, although words such as "motor sports" and "closeness (of distance)" were also found, those words were excluded because they did not appear more than 30 times. Note that the above count was redundant when the respondent referred to the same tag multiple times. However, dummy variables are more interpretable in a quantitative analysis. As a result of converting count data of each tag to dummy variables, the registered word has appeared 1,251 times. 235 respondents (26.1%) did not receive tags. As a result of converting into dummy variables, one tag is 297 (33.0%), and a plurality of tags (person who mentions a plurality of different tags) is 368 (40.9%).

The largest emerging factor on the whole is F_01_Driving, but as shown in Table 3, when viewed by brand, the characteristics of each brand can be understood. Considering the brands Suzuki, Subaru and BMW, the most frequently mentioned factor was F_01_Driving. The most frequent factor for Nissan, Honda, and Lexus was F_04_SalesStaff. The most frequent factors were F_05_Trust for Toyota, F_02_Design for Mazda, and F_06_Safety for Mercedes Benz. Hence, the most frequently mentioned companies and their features emerge in this study.

The most important issue in statistics is said to be variable selection in regression models because constructing statistical models using many variables may cause overfitting of acquired data. Especially, in recent years with the availability of a large amount of data, attention has been focused on methods for avoiding that risk. The Sparse modeling method is mentioned as a typical example. This method efficiently extracts the maximum feature quantity from data by utilizing the universally inherent sparsity of high-dimensional data. Therefore, it is used to obtain a robust estimate when there are many variables and variables with high correlation. As shown in Table 4, 50 dummy variables were used for modeling. The objective variable was repurchase intention (No. 1), the explanatory variables were factor dummies of repurchase intention (No. 36–49), and the control variables were the respondent attribute dummy (No. 2–35) and the recommendation (No. 50). Because there are many variables, it was judged that variable selection by sparse modeling was suitable.

In this study, three models were constructed, including the above-mentioned sparse models. Model 1 was the logistic regression model in which variables are selected by the stepwise method. Model 2 was the lasso logistic regression, and Model 3 was the elastic net logistic regression. The analysis environment was statistical analysis software R, and Model 1 used the stepAIC function in the MASS package. Model 2–3 calculated λ that minimizes the mean square error with the cv.glmnet function of the glmnet package and built the model with the glmnet function. The accuracy verification of each model was conducted as a 10-fold cross validation with 50% as training data and 50% as validation data. Hence, the factors of the repurchase intention were evaluated from the parameter estimates of the most appropriate model.

Table 2. Dictionary structure of factors of repurchase intention and registered word example.

No	Tag	Example word 1	Example word 2	Example word 3
1	F_01_Driving	driving performance	engine	turbo
2	F_02_Design	design	exterior	stylish
3	F_03_Brand	brand	concept	reputation
4	F_04_SalesStaff	sales staff	kindness	hospitality
5	F_05_Trust	trust	credit	relief
6	F_06_Safety	safety performance	collision avoidance	automatic braking
7	F_07_Comfortable	comfortable	usability	convenient
8	F_08_CostPerformance	cost performance	affordable	inexpensive
9	F_09_Fuel	fuel efficiency	economical	eco-friendly
10	F_10_Service	after service	customer service	maintenance
11	F_11_Technology	technology	cutting edge	R&D
12	F_12_Luxury	luxury	status	high class
13	F_13_Quality	quality	durability	tough
14	F_14_Recall	recall	failure	defect

Table 3. Tag composition ratio of each brand (before dummy treatment).

Tag	Toyota	Nissan	Honda	Suzuki	Mazda	Subaru	Lexus	Mercedes Benz	BMW	Total
F_01_Driving	10.2%	14.0%	13.1%	17.0%	21.5%	36.1%	8.8%	13.6%	31.8%	18.4%
F_02_Design	4.8%	11.8%	9.2%	14.0%	28.0%	7.2%	12.4%	12.3%	22.7%	14.1%
F_03_Brand	19.3%	7.4%	11.8%	14.6%	9.3%	9.0%	12.4%	11.0%	8.1%	11.5%
F_04_SalesStaff	7.5%	25.0%	19.6%	9.9%	15.4%	4.8%	13.5%	5.9%	2.5%	10.9%
F_05_Trust	30.5%	3.7%	11.8%	2.9%	1.4%	4.8%	7.8%	9.7%	12.6%	9.6%
F_06_Safety	4.8%	8.8%	5.9%	4.1%	1.9%	18.7%	5.7%	21.2%	1.5%	8.2%
F_07_Comfortable	2.1%	8.8%	7.8%	2.3%	1.4%	1.8%	8.8%	12.7%	4.5%	5.7%
F_08_CostPerformance	3.7%	4.4%	4.6%	14.6%	6.5%	4.2%	4.7%	3.4%	3.0%	5.4%
F_09_Fuel	4.8%	3.7%	6.5%	14.0%	4.2%	1.2%	1.0%	0.4%	3.5%	4.2%
F_10_Service	4.3%	2.2%	3.9%	3.5%	3.7%	2.4%	9.8%	1.3%	3.0%	3.8%
F_11_Technology	0.5%	5.9%	2.6%	1.2%	4.2%	2.4%	2.1%	0.4%	1.5%	2.2%
F_12_Luxury	0.5%	0.0%	0.7%	0.0%	0.0%	0.0%	7.8%	5.9%	2.0%	2.1%
F_13_Quality	3.7%	0.7%	0.7%	0.0%	0.9%	3.0%	4.1%	1.7%	3.0%	2.1%
F_14_Recall	3.2%	3.7%	2.0%	1.8%	1.4%	4.2%	1.0%	0.4%	0.0%	1.8%
Total	100.0%	100.0%	100.0%	100.0%	100.0%	100.0%	100.0%	100.0%	100.0%	100.0%
Frequency	187	136	153	171	214	166	193	236	198	1,654

Table 4. Variable list.

No	Variable	Description
1	RepurchaseIntention	dummy of recommendation (Top one in five scale)
2	Female	dummy of female
3	Age_30s	dummy of 30s (The standard is 20s.)
4	Age_40s	dummy of 40s (The standard is 20s.)
5	Age_50s	dummy of 50s (The standard is 20s.)
6	Age_60s	dummy of 60s (The standard is 20s.)
7	HouseholdIncome_400_599	dummy of household income 4.00-5.99 million yen (The standard is -3.99.)
8	HouseholdIncome_600_799	dummy of household income 6.00-7.99 million yen (The standard is -3.99.)
9	HouseholdIncome_800_999	dummy of household income 8.00-9.99 million yen (The standard is -3.99.)
10	HouseholdIncome_1000_1499	dummy of household income 10.00-14.99 million yen (The standard is -3.99.)
11	HouseholdIncome_1500	dummy of household income 15.00- million yen (The standard is -3.99.)
12	MultipleOwned	dummy of more than 2 cars owned
13	DailyDriving	dummy of driving everyday
14	BodyType_Kei	dummy of body type of Kei (The standard is Coupe.)
15	BodyType_Compact	dummy of body type of Compact (The standard is Coupe.)
16	BodyType_Minivan	dummy of body type of Minivan (The standard is Coupe.)
17	BodyType_SUV	dummy of body type of SUV (The standard is Coupe.)
18	BodyType_Sedan	dummy of body type of Sedan (The standard is Coupe.)
19	EmphasisPoint_Brand	dummy of purchasing emphasis point for brand
20	EmphasisPoint_Design	dummy of purchasing emphasis point for design
21	EmphasisPoint_UX	dummy of purchasing emphasis point for ux
22	EmphasisPoint_Driving	dummy of purchasing emphasis point for driving
23	EmphasisPoint_Fuel	dummy of purchasing emphasis point for fuel
24	EmphasisPoint_Safety	dummy of purchasing emphasis point for safety
25	EmphasisPoint_WoM	dummy of purchasing emphasis point for WoM
26	EmphasisPoint_Price	dummy of purchasing emphasis point for price
27	EmphasisPoint_DealerAccess	dummy of purchasing emphasis point for dealer access
28	Nissan	dummy of currently owned brand of Nissan (The standard is Toyota.)
29	Honda	dummy of currently owned brand of Honda (The standard is Toyota.)
30	Suzuki	dummy of currently owned brand of Suzuki (The standard is Toyota.)
31	Mazda	dummy of currently owned brand of Mazda (The standard is Toyota.)
32	Subaru	dummy of currently owned brand of Subaru (The standard is Toyota.)
33	Lexus	dummy of currently owned brand of Lexus (The standard is Toyota.)
34	MercedesBenz	dummy of currently owned brand of Mercedes-Benz (The standard is Toyota.)
35	BMW	dummy of currently owned brand of BMW (The standard is Toyota.)
36	F_01_Driving	dummy of tag of F_01_Driving
37	F_02_Design	dummy of tag of F_02_Design
38	F_03_Brand	dummy of tag of F_03_Brand
39	F_04_SalesStaff	dummy of tag of F_04_SalesStaff
40	F_05_Trust	dummy of tag of F_05_Trust
41	F_06_Safety	dummy of tag of F_06_Safety
42	F_07_Comfortable	dummy of tag of F_07_Comfortable
43	F_08_CostPerformance	dummy of tag of F_08_CostPerformance
44	F_09_Fuel	dummy of tag of F_09_Fuel
45	F_10_Service	dummy of tag of F_10_Service
46	F_11_Technology	dummy of tag of F_11_Technology
47	F_12_Luxury	dummy of tag of F_12_Luxury
48	F_13_Quality	dummy of tag of F_13_Quality
49	F_14_Recall	dummy of tag of F_14_Recall
50	Recommendation	dummy of recommendation (Top two in eleven scale)

4 Results and Discussion

As shown in Table 5, looking at Model 2 in which variables were selected by stepwise method, F_05_Trust and F_02_Design were adopted as positive effects and F_09_Fuel as negative effect. Further, the results of sparse modeling of Model 2–3 were confirmed. The lasso regression of Model 3 has the smallest mean square error when $\lambda = 0.024$ ($log_e\ 0.024 = -3.730$). The estimation results of the model use the same variables as

Model 1 among the reasons for the intention to repurchase. Finally, the elastic net of Model 3 has the smallest mean square error when $\lambda = 0.040$ (log_e $0.040 = -3.219$). In addition to the Model 3 variables, F_10_Service was adopted as the reason for the repurchase intention.

It was necessary to select the most appropriate model in the factor evaluation of the repurchase intention. Therefore, 10-fold cross validations were conducted to determine a highly accurate model for estimating repurchase intention. The prediction accuracy of Models 1–3 is 0.727, 0.745, and 0.756, respectively. Hence, elastic net of Model 3 was adopted. As shown in Table 5, the variables adopted for the repurchase intention of Model 3 have a positive effect on F_05_Trust which has the maximum odds ratio of 1.339, followed by 1.163 for F_02_Design, and 1.009 for F_10_Service. As a negative influence, F_09_Fuel has shown an odds ratio of 0.906.

The above discussion shows that elastic net was the most appropriate model and as factors for the repurchase intention, trust and design have a large positive impact and fuel economy has a negative impact. In addition, emphasizing the brand at the time of purchase mostly leads to a repurchase intention. Therefore, in product planning, promotion, and sales sites, it is effective to put emphasis on trust and design, not on driving performance, fuel efficiency, cost performance, etc. In the previous research in Chapter 2, trust and perceived quality/value were influencing factors extracted in many industries and it was found that the same results were generally obtained in the automotive industry. In addition to that, new findings like negative effects such as fuel efficiency were also extracted.

The limitation of this research is that unless a factor can be perceived by more than a certain number of consumers it cannot be evaluated as a factor of repurchase intention. There is a limit to the quantitative analysis that is conducted based on survey data where consumers give a free answer. Therefore, it is difficult to cover the factors by the method of this research, and the purpose of this study is to understand important factors from the consumer's point of view.

Table 5. Odds ratio estimated by each model.

Variable	Model 1	Model 2	Model 3
(Intercept)	0.307	0.309	0.324
Female	0.652	0.980	0.940
HouseholdIncome_1500		1.058	1.128
EmphasisPoint_Brand	1.617	1.281	1.312
EmphasisPoint_Fuel			0.998
Suzuki		0.800	0.772
Mazda			0.978
F_02_Design	1.273	1.157	1.163
F_05_Trust	1.698	1.301	1.339
F_09_Fuel	0.544	0.982	0.906
F_10_Service			1.009
Recommendation	5.572	4.442	4.025

5 Conclusion

Customers who intend to repurchase are important management bases that boost profits. However, research on the factors behind the repurchase intention are often done in the data-rich retail industry, and recently, online shopping sites. On the other hand, there is not much research for repurchase intention of durable consumer goods with a long span of replacement. In particular, in the automobile industry with an average replacement period of 7.5 years and ownership rate of 25% or more for an ownership period of 10 years or more, factor evaluation has been insufficient.

Therefore, in this research, the factors contributing to the repurchase intention were evaluated for the Japanese automobile industry. Data was obtained from an online survey which asked for repurchase intentions and reasons. The factors were extracted from the text of the free answer for reason of the repurchase intention and the contribution was evaluated by a statistical model. The model with the highest prediction accuracy for the repurchase intention was elastic net logistic model. The results show that the factor most contributing to the repurchase intention was trust and the odds ratio was 1.339. The second factor was design with an odds ratio of 1.163. Furthermore, it was estimated that the odds ratio for the consumers who cited fuel economy as the reason for the repurchase intention was 0.906 and was considered a negative influence. In other words, it is effective to put emphasis on emotional values such as trust and design, not on functional values such as driving performance, fuel consumption performance and cost performance.

The best competitive advantage is building long-term relationships with highly satisfied customers. For that purpose, it is important to grasp effective factors quantitatively and to appeal effective factors consistently. This research is expected to be useful for improving the products and services and designing the loyalty program by quantitatively clarifying the factors of the automobile industry's repurchase intention that has been insufficient until now.

References

1. Fryer, B.: Tom siebel of siebel systems: high tech the old-fashioned way. Harv. Bus. Rev. **79**(3), 119–125 (2001)
2. Nunes, J.C., Drèze, X.: Your loyalty program is betraying you. Harv. Bus. Rev. **84**(4), 124–131 (2006)
3. O'Brien, L., Jones, C.: Do rewards really create loyalty? Harvard Bus. Rev. **73**(3), 75–82 (1995)
4. Japan Automobile Manufacturers Association: Passenger car market trends in Japan: Summary of results of JAMA's Fiscal 2015 Survey. Press Release (2016). https://www.jama-english.jp/release/release/2016/160426-1.html. Accessed 1 Nov 2020
5. Oliver, R.L.: Whence consumer loyalty? J. Mark. **63**(4), 33–44 (1999)
6. Aaker, D.A., Joachimsthaler, E.: Brand leadership. The Free Press, New York (2000)
7. Priluck, R.: Relationship marketing can mitigate product and service failures. J. Serv. Mark. **17**(1), 37–52 (2003)
8. Chaudhuri, A., Holbrook, M.B.: The chain of effects from brand trust and brand affect to brand performance: the role of brand loyalty. J. Mark. **65**(2), 81–93 (2001)
9. Jacoby, J., Kyner, D.B.: Brand loyalty vs. repeat purchasing behavior. J. Mark. Res. **10**(1), 1–9 (1973)

10. Reichheld, F.F.: The Ultimate Question. Harvard Business School Press, Boston (2006)
11. Dick, A.S., Basu, K.: Customer loyalty: toward an integrated conceptual framework. J. Acad. Mark. Sci. **22**(2), 99–113 (1994)
12. Kato, T.: Loyalty management in durable consumer goods: trends in the influence of recommendation intention on repurchase intention by time after purchase. J. Mark. Anal. **7**(2), 76–83 (2019)
13. Kato, T., Kishida, N., Umeyama, T., Jin, Y., Tsuda, K.: A random extraction method with high market representation for online surveys. Int. J. Bus. Innov. Res. **22**(4), 569–584 (2020)
14. Zboja, J.J., Voorhees, C.M.: The impact of brand trust and satisfaction on retailer repurchase intentions. J. Serv. Mark. **20**(6), 381–390 (2006)
15. Chinomona, R., Sandada, M.: Customer satisfaction, trust and loyalty as predictors of customer intention to repurchase South African retailing industry. Mediterr. J. Soc. Sci. **4**(14), 437–446 (2013)
16. Noyan, F., Simsek, G.G.: A partial least squares path model of repurchase intention of supermarket customers. Proc. Soc. Behav. Sci. **62**, 921–926 (2012)
17. Fang, Y.H., Chiu, C.M., Wang, E.T.: Understanding customers' satisfaction and repurchase intentions: an integration of IS success model, trust, and justice. Internet Res. **21**(4), 479–503 (2011)
18. Elbeltagi, I., Agag, G.: E-retailing ethics and its impact on customer satisfaction and repurchase intention: a cultural and commitment-trust theory perspective. Internet Res. **26**(1), 288–310 (2016)
19. Aren, S., Güzel, M., Kabadayı, E., Alpkan, L.: Factors affecting repurchase intention to shop at the same website. Proc. Soc. Behav. Sci. **99**, 536–544 (2013)
20. Zhang, Y., Fang, Y., Wei, K.K., Ramsey, E., McCole, P., Chen, H.: Repurchase intention in B2C e-commerce—a relationship quality perspective. Inf. Manage. **48**(6), 192–200 (2011)
21. Chiu, C.M., Chang, C.C., Cheng, H.L., Fang, Y.H.: Determinants of customer repurchase intention in online shopping. Online Inf. Rev. **33**(4), 761–784 (2009)
22. Kato, T., Tsuda, K.: The effect of the number of additional options for vehicles on consumers' willingness to pay. Proc. Comput. Sci. **176**, 1540–1547 (2020)
23. Grewal, D., Roggeveen, A.L., Tsiros, M.: The effect of compensation on repurchase intentions in service recovery. J. Retail. **84**(4), 424–434 (2008)
24. Jones, M.A., Mothersbaugh, D.L., Beatty, S.E.: Switching barriers and repurchase intentions in services. J. Retail. **76**(2), 259–274 (2000)
25. Mittal, V., Kamakura, W.A.: Satisfaction, repurchase intent, and repurchase behavior: Investigating the moderating effect of customer characteristics. J. Mark. Res. **38**(1), 131–142 (2001)
26. Souiden, N., Pons, F.: Product recall crisis management: the impact on manufacturer's image, consumer loyalty and purchase intention. J. Prod. Brand Manage. **18**(2), 106–114 (2009)
27. Peyrot, M., Van Doren, D.: Effect of a class action suit on consumer repurchase intentions. J. Consum. Aff. **28**(2), 361–379 (1994)
28. Kato, T.: Differences in delivery times' effects on purchase intentions by the purchase candidates' sequencing in the Japanese automotive industry. J. Mark. Anal. **8**, 1–11 (2020)

The Old Question: Which Programming Language Should We Choose to Teach to Program?

Sónia Rolland Sobral[(⊠)]

REMIT, Universidade Portucalense, Porto, Portugal
soniarollandsobral@gmail.com

Abstract. When students enter higher education in computer courses, students have, in the first semester, to perceive and streamline computer (or computational) thinking . The names for the unit are diverse, usually algorithm and programming, introduction to programming or even Programming I or algorithm. In research we call CS1, or computer science 1, since that designation was used in an important document with curriculum recommendations prepared by the Association for Computing Machinery (ACM) in 1978. Despite different names, it is intended with these units that students start in the world of programming. However, if the learning objectives do not differ much, the same cannot be said of the adapted programming language. There is not, nor is it likely to be, a consensus regarding the programming language used by the student's startup in the programming world. The various documents with curriculum recommendations for computer science from ACM and the Institute of Electrical and Electronic Engineers (IEEE) do not clearly define which programming language to adopt. Those responsible for the courses have to make this choice, consciously and not just following industry trends. The purpose of this article is to draw a picture of the adoption of the different programming languages that have been used over time, as well as to help in choosing the most accurate and conscious option possible.

Keywords: Programming languages · Undergraduate studies · Introduction to programming

1 Introduction

Computer thinking is addressed in the first semester of courses to students who are new to major computer courses. These courses have different names, such as algorithm and programming, programming I or introduction to programming - generally called CS1 (computer science I), a name adopted by a 1978 document with curriculum recommendations made by Association for Computing Machinery (ACM) [1]. These curricular units (or courses) have different names, but the goal is to start students in computational thinking. It often seems that this course is "just" programming, but what precedes the use of a programming language is very important before the student can program. A student can only try to solve a problem with a computer programming language after

T. Antipova (Ed.): ICADS 2021, AISC 1352, pp. 351–364, 2021.
https://doi.org/10.1007/978-3-030-71782-7_31

knowing how to think and use tools that help him transform "normal" thinking into "computational" thinking and translate it into a programming language. There is the use of tools such as top-down, an algorithm (with syntax and predefined rules for others to understand). Often the technique is to break the problem down into several parts (often called "thinking small to program large"). It's time to use variables, to think about the best data structure. Only after all these tasks, the student (the computer) can "translate" a solution into any programming language.

Programming is objectively a method for the programmer to communicate instructions to the computer [2]. Some call it an art [3], a science [4], a discipline [5] or even the science of abstraction [6]. Often students devalue this previous learning: students want to quickly go to the computer and to the editor of any programming language. Students often have the perception that the focus is on learning the syntax of the programming language, leading them to focus on implementation activities rather than activities like planning, design or testing [7]. Programming is considered by some to be a very difficult task [8, 9] while for others it may be easy [10]. The success of the programming world is achieved through a lot of work, a lot of study, a lot of research, asking for help and resilience, planning, persistence and preferably passion for the activity [2]. The first programming language that someone uses is truly important, as Dijkstra [11] wrote "the tools we are trying to use and the language or notation we are using to express or record our thoughts are the major factors determining what we can think or express at all!". If we look at the following table (Table 1), we see that programming languages are apparently not very different. This table shows how each of the ten most popular programming languages writes the famous "Hello, World!". You can see then that there are no big differences: the different notation, some programming languages use semicolons and the way of writing varies a little, but it is to write "Hello World!" in these languages it looks quite similar.

Table 1. "Hello World!" ten different programming languages.

Prog. language	Write Hello World
C	printf ("Hello World!");
C#	Console.WriteLine ("Hello World!");
C++	cout< <"Hello World"
COBOL	display "Hello world!"
Fortran	print *,"Hello world!"

The notation of programming languages to write "hello world!" is very similar. Therefore, one can ask why there is not a single programming language, even if it is improved over time (evolution) and, therefore, understood by everyone. Or at least, why is there not only one programming language considered the right one for the introduction to programming? There is no consensus, and there never will be, to consider a more correct programming language to start a student. There are several arguments that appear to be religious, political, or futuristic. It seems that there are fans of C, Java or even Pascal.

What is the reason? In this article, we will show the evolution of the first programming language options over time. as well as what arguments can be considered when making the choice.

2 Programming Languages

There are many definitions for the programming language and a computer program. For example: a program is a set of instructions that makes up a solution after being coded in a programming language [13]. In reality, and to put it simply, a programming language is a system that allows for interaction between man and machine, which both perceive to be able to communicate with each other. Like a natural language, a programming language has a set of instructions and rules that are perceived by the computer and the programmer (s). When writing a program, a programmer must do it in a way that other people will understand what is written. It is important to do so because a program may have to be changed by someone (or a group of people) other than the original programmer. Therefore, the writing must be succinct and clear, so that several people can understand it and even modify it [12]. And the obvious question is why are there multiple languages? Why are there so many programming languages? wouldn't it be simpler to have just one programming language? There are, at least, 8945 programming languages [14] according to Online Historical Encyclopedia of Programming Languages. Scott [15] presents three main reason: Evolution, Special Purposes and Personal preference. The revolution in "structured programming" at the end of the 1960s, made the flow control with GoTo of programming languages such as FORTRAN, COBOL and Basic to cycle with loops while and case statements (switch or select case). If for many years there was an almost consensus with the use of Pascal as an initial programming language, such as Algol and Ada on a smaller scale of consensus, object-oriented languages such as Smalltalk, C++ and Eiffel became more relevant. What happened to GoTo, Pascal and object-oriented languages is just one example of what has happened over the years. We know that there are programming languages that are better for one purpose than another: if C is good for low-level system programming, Prolog is good for reasoning about logical relationships between data. Each programming language can be used successfully for a wide range of tasks, but the emphasis is clearly on the specialty. Finally: different people like different things: there are those who like yellow or those who hate it. C is an example as it has many devotees and others who just do not really like the C programming language. According to Stack Overflow Annual Developer Survey [16], with over 65,000 answers, by 2020 and for the eighth year, the most widely used programming language is JavaScript (see next table) (Table 2) .

But, if the previous table shows a continuous first place for JavaScript, the TIOBE Programming Community index [17] shows that there are changes from the year 2019 to the year 2020 (both September): C moves to the first place passing Java to the second position (see next table) (Table 3).

Table 2. Top10, Programming languages most used in 2020 [16].

Programming language	%
JavaScript	67.7%
HTML/CSS	63.1%
SQL	54.7%
Python	44.1%
Java	40.2%
Bash/Shell/PowerShell	33.1%
C#	31.4%
PHP	26.2%
TypeScript	25.4%
C++	23.9%
C	21.8%

Table 3. Top7, Indicator of popularity of programming languages, TIOBE [17].

Sep 2020	Sep 2019	PL	Ratings	Change
1	2	C	15,95%	+0.74%
2	1	Java	13,48%	−3.18%
3	3	Python	10.47%	+0.59%
4	4	C++	7.11%%	+1.48%
5	5	C#	4.58%	+1.18%
6	6	Visual Basic	4,12%	+0.83%
7	7	JavaScript	2.54%	+0.41%

In this table we see that C is in September of 2020 the most popular programming language. However, a year earlier, Java was the most popular programming language. In other words, it is seen that a generation of programming languages is dynamic and evolutionary. In the case of JavaScript, the previous table is in the first position of the most used programming languages, in this table appears in seventh place in the popularity index. Thus, we realize that there is no consensus.

3   Programming Languages for CS1

There was always a lot of discussion when choosing the first programming language. It is a concern of researchers and teachers, and there are different perspectives on the subject [18–22], and there have been a lot of changes over time. The first document with strict curricular recommendations is dated 1968 and was prepared by a task force organized by the Association for Computing Machinery (ACM), Curriculum 68: Recommendations for academic programs in computer science [23]. This document is very important for the area as it presents the first indications of curricular models for programs in computer science and computer engineering. In this document, a course, Introduction to Computing (or B1) is presented as the starting one. For B1, an algorithmic language is recommended and eventually one or two languages would be used so that students know the diversity of programming languages, and SNOBOL for "having elegance" and "being new" can be one of those programming languages for novice students.

Curriculum'78: recommendations for the undergraduate program in computer science [1], the updating of Curriculum 68, presented for the first time the name CS1 (Computer Programming I) for the initial unit. The recommendations of this second document, and for the CS1 course, are to emphasize algorithm development techniques and to be concerned with "programming with style". It also recommends that these concerns be independent of those of the "esoteric resources of a programming language". Recommended curriculum for CS1, 1984 [24] is the next document with updates to curriculum recommendations. This report detailed a first computer science course that emphasizes programming methodology and problem solving. There are three programming languages that are referred to because they fit the objectives: these programming languages are Ada, PL1 and Pascal. The reason these programming languages are used as a good example is because they allow the student to use procedural and data abstraction. And at the same time avoiding languages that are more complicated and that allow many resources that can be used posteriori, such as structured control and data structures. In this document, it is stated that BASIC and FORTRAN would not be a good programming language alternative for freshman students. The same is written in relation to ALGOL, which, despite satisfying the requirements, is not so widely used in the other programming languages mentioned. IEEE (Institute of Electrical and Electronics Engineers) and ACM joined for a new document in 1991, Computing curricula 1991: report of the ACM/IEEE-CS Joint Curriculum Task Force [25]. There are structural changes in the curricular recommendations of this document, where there is a posture of greater flexibility with the institutions. This document presents a set of individual units of knowledge corresponding to a theme that should be addressed at some point during graduation. Thus, institutions create courses according to their needs and assets, considering that these themes should appear at some point in the courses, without the ACM and IEEE imposing any rigid (blind) structure for the institutions. The CC2001 Computer Science document [26], the new document from task force ACM-IEEE, questioned the way in which the beginning of the programming described in previous documents is proposed. The problem is to program before knowing how to program: that is, immediately using a programming language makes students believe that the important thing is to write a program in a programming language, devaluing all the initial work of computational thinking and use of tools that the student must know well (not "computer science = write a program in a programming language"). In 2001, the problem was accentuated with object-oriented programming languages: many of the languages used for object-oriented programming in the industry, like C++, and some Java as well, are significantly more complex than classical languages. The document emphasizes that the use of object models can only be used as a first approach if the complexity of these languages is not used so that the freshman student is not more focused on the details of these object-oriented languages than on the introductory points that are truly important. An eventual form proposed is to limit this complexity so that the details do not overwhelm the introductory students. The 2008 document, called Computer Science Curriculum 2008: An Interim Revision of CS 2001 [27] presents the idea that IT professionals should use multiple languages throughout their professional lives. And they do this because a programming language fits a purpose better than another programming language, depending on the different purposes. The document argues that future computer professionals must be able to learn new languages throughout their careers as the field evolves. In other words, it shows that it is important that students realize the benefits of learning and applying new programming languages and paradigms. The document states that the choice of the programming paradigm can significantly influence the way you think about problems and express solutions to those problems. For all these reasons, this document argues that students should program in several programming languages and in several paradigms. Computer Science Curricula 2013: Curriculum Guidelines for Undergraduate Degree Programs in Computer Science [28] says that the choice of programming languages seems to depend on the chosen paradigm. The document refers (at that time) to a clear trend towards safer and more managed programming languages. One example is to evolve from C to Java, or to more dynamic languages (like Python and JavaScript). In this document it is stated that certain visual programming languages and with a lighter (or apparently easier) syntax, despite being very popular and used at the time (such as Alice and Scratch), can be used for not graduated courses. There is no clear indication of which programming languages to use. The document also states that several intuitions were, at the time, concerned with presenting alternative programming paradigms, such as script vs. Programming. Procedural programming or functional programming vs. Object-oriented programming, to give students a greater appreciation of the diverse perspectives in programming, to avoid the use of language fixation, and to disappoint them with the notion that there is a single "correct" or "better" programming language.

These are the recommendations of the documents, but what has really happened throughout the history of computer education? When the first computer courses started, there was no talk of methodology, everything was assumed to be programming. FORTRAN was selected as a high-level programming language for the first introductory courses, especially those linked to engineering departments. COBOL was also adopted by departments that were closer to information systems [29]. The emergence of BASIC, in the 1960s, led many departments to adopt this language for introductory students [30]. In the early 1970s, almost all universities used LISP, ALGOL and/or FORTRAN, but most data processing programs used COBOL. It depended a lot on schools and even countries. For example, in Britain, BASIC was the most used, although in the late 1960s some departments tried PLI and other programming languages [31]. Dijkstra's manifest [32] caused structured programming to be discussed [33, 34]. With the emergence of the Pascal language [35] seems to become almost consensual [29]. Pascal is a written language for the purpose of programming learning, using a very friendly development environment [36], and obviously because of the proliferation of personal computers and the availability of Pascal compilers [37]. Pascal's decline began in the late 1980s, early

1990s, with object-oriented programming. And also because Pascal has a difficult document reuse, and also because Pascal is not a "real world" language [37]. In 1995–1996, the programming language Pascal was used by 36% and C++ by 32% but 22% intended to make a switch to C++, C, Ada or Java [38]. There are several studies that present the evolution of the languages adopted in initial programming curricular units [39, 40] and even lists of programming languages taught in various courses [41]. It is impossible to speak about programming languages and CS1 without mentioning the Reid list (Reid First Course Language List): a list of programming languages used in CS1 courses from the early 1990s until Reid's retirement in 1999, continued until 2006 by Frances Van Scoy [42] and updated by Robert M. Siegfried (2011 and 2015) [43]: "88% of the Reid List schools use one of only four programming languages: Java, Python, C++ and C".

In Portugal, in the 2016–2017 school year, the most common first-year programming language sequence in 46 courses analyzed was C (48%), followed by Java (22%), C and Haskell (9%), C and Java (4%), Scheme and Java (4%) [44]. Regarding the ten Portuguese first cycle (or with integrated master's degree) courses in Computer Engineering considered most significant [45], it was found that the most common sequences were only Java or Python and C (both with 30%), C (20%), Python and Java or Haskell and C (both with 10%). Concerning to Portuguese courses, 25 out of 33 courses in computer engineering from Portuguese public institutions published information on program content on the internet in 2019–2020 school year. It appears that 44% of these courses use C as the first programming language, with Python being used by 28%, java by 24% and Haskell by one of the courses (4%). For the 2020–2021 academic year, C is adopted in the first semester by 14 of the 29 computer engineering courses that indicate which programming language adopted. Then Java and Python appear in seven courses and one course adopts Haskell as the first programming language. It means that in 2020, the most used programming language (in 48%) for curricular units in Portugal and computer engineering, and, in equal measure, Java and Python appear in close to 14%, as in Portugal, we can see other examples in the world. We found that currently in teaching the initial programming language is somewhere between C, Java and Python, with rare exceptions. Looking at the rest of the world and through studies published in renowned international journal. For example, in the United Kingdom. 73.8% use only one programming language; 21% reported using two. The most widely used language is Java (46%), followed by the "C family" (C, C++ and C #) (23.6%) and Python (13.2%) [46]. 48 courses from Universities of Australia and New Zealand date from 2016: 15 used Java or Python, 8 used C and 5 C #. In other words, it is easy to verify that 62.5% used Java or Python - which is an expressive number. C and C3 (what is called the C family) appears in 27% of these universities. The programming languages most used in Irish introductory programming courses 39 introductory programming courses at 25 third-level institutions from Ireland [47] are Java (49%), Python (28%), JavaScript (18%), C and C# (10% each). A 2018 study with 53 respondents from 13 Ecuadorian Institutions revels that Java (20.7%), JavaScript (13%), C++ (12.4%), PHP (10.1%), C# (9.5%), C (7.1%) and Python (4.1%) are the most used programing languages for them [48]. In 2016, a study ask 218 colleges and 143 universities in 35 European countries for the most used programming language, and the answers was C (30.6%), following C++ (21.9%) and Java (20.7%) [49]. Another study with 496 four-year courses in the United States, refers that Java is used by 41.94%, Python 26.45%, C++ 19.35%, C 4.52%, C # 0.65% [50]. A document for 152 CS1 units from several different countries concludes that Java is by far the most common CS1 language, used in 74 (49%) of the 152 programs. The second most frequent is Python, with 36 (24%). C++ comes in 30 (20%) followed by C in 8 (5%), with the most obvious change being the rise of Python which "probably occurred at the expense of Java and C++" [9]. We can continue to cite an example of studies, such as these in Portugal, United Kingdom, United States, Ecuador, Australia, or New Zealand. It does not seem worth it: what we do think is that the results have been very similar - at least today. Today, with few exceptions, the academy, across and across the world, adopts the "C family" (C, C++, C #), Python, Java, and JavaScript are undoubtedly the programming languages adopted in introductory programming units.

4 How to Choose the Initial Programming Language

Why does an institution or department choose a specific programming language to introduce students to the world of programming? In the previous point, we mentioned several studies. We focus only on the programming languages used, but some of these studies seek to assess the reasons why a programming language is adopted and not another. Eric Roberts talks about the complexity problem and how complexity has increased, referring to teaching tools, languages and programming paradigms [51]. Due to the complexity, it takes more time to know how to handle the material. He also says that programming languages and tools change rapidly, leading to profound instability in the way computer science is taught. Roberts is of the opinion that one should adopt a programming language (In 2004, suggests Java) and not change every semester. When selecting the first programming language for introductory programming courses, it is important to consider whether it is suitable for teaching and learning. Over time various pseudo-code languages have been created in search of the perfect teaching language but no definitive solution has been found [52]. Some discusses the need to teach introductory programming using educational programming languages [49], but in the past these kind of programming languages have been discontinued, Pascal language being the most visible. Chose a programming language for introductory programming courses often seems like a religious or football issue. In reflection-teaser. The strongest is a document called "The wars of the programming language" where (exaggerating) it says that the wars of the programming language are a big social problem, causing serious problems in the discipline [53]. The reason is the duplication of efforts and the desire to reinvent a wheel that was long invented. Choosing the best programming language is often an emotional issue, leading to major debates [54], Teachers must demonstrate to the world (and especially to their students) that the question of war of languages is a non-issue and that students have to program in many different programming languages throughout their lives [55]. In fact, two of the most important points are pedagogical issues and student preparation for the world of work. One way of cataloging programming languages is to define them as pragmatic and pedagogical: acceptance by the industry, market penetration, as well as the employability of graduates [31]. Thinking about small programming needs to be mastered before large programming [56] since traditionally students only "in the third or fourth year are faced with the problems that arise in the design of large programs."

The task of choosing the initial language is not an easy task. Trends change dramatically over time so the programming languages Some factors such as simplicity, suitability for tasks, expressiveness, reliable compilers and availability of accessible resources. Programming languages are the fundamental basis of programming, but. Professionals will not use the same programming language, or even the same programming model, for their entire professional career [56]. It is crucial that students master the essential concepts of programming languages, so that they can choose and use languages based on a deep understanding of the abstractions they express and their ability to solve programming problems [57]. A programming language should be "in the happy medium that allows for beginners to grasp the basics but still be able to eventually use the more advanced concepts, and also "be intuitive enough that beginners will not give up in frustration" [36].

A problem is that Coordinators and instructors often do not know the differences between programming languages. A study compares 8 widely used languages representing the major programming paradigms (procedural: C and Go; object-oriented: C# and Java; functional: F# and Haskell; scripting: Python and Ruby). This is the Comparative Study of Programming Languages in Rosetta Code, and it is a very interesting tool for those who need to make decisions, or at least think about the direction of their institutions [58]. There are numerous comparisons between the most commonly used languages: like Python vs C++ [59], Python vs C [60], Java vs Python [61], C++ vs. Java [62] or C vs Java vs Python [63]. Any of the three/four most used programming languages is free, well supported and has a large user community, is reliable and efficient. Ease of learning can be discussed: C will have a more complicated syntax than Python. The major differences are the use of pointers (C), parameter passing by reference and value (C again), programming paradigm (procedural in C, object oriented in others), being compiled or interpreted (C and Python/Java respectively). There are few studies that attempt to outline the choice of the first programming language for novice students. Some think it is important to consider some factors such as proposed some points like Course objectives, Teacher preferences, available implementations, and relationships with other course units, as well as the "real world": students are often more motivated to study a familiar language that is known to be requested by employers [64]. Other studies, like "A project-based approach to programming language evaluation "[65] uses an evaluation method for programming languages using several items: language design and implementation (accuracy and speed), human factors (usability and ease), software engineering (portability, reliability and reuse) and application mastery specific applications). The problem of paradigms is also debated, with some considering that students should be exposed to all major paradigms through the use of a multiparadigmatic language and do not try to identify "the" correct paradigm [66, 67].

There are studies that propose multi-option choices. A document "has a design of choice with a weighted multicriteria method and where evaluation criteria are identified such as Reasonable Financial Cost, Academic/Student Version Availability, Academic Acceptance, Textbook Availability, Lifecycle Stadium, Industry Acceptance, Marketing (regional and national), Student/Academic/Full System Requirements, Operating System Dependency, Proprietary/Open Source, Development Environment, Debugging

Facilities, Fundamentals Learning Ease, Secure Code, Advanced Course Features Subsequent, More or Less Complicated Programming, Web Development Support, Teaching Support, Object Oriented Support, Support Availability, Instructor and Staff Teaching, and Expected Level of New Students [68]. Others consider that a programming language selection is intrinsically a multi-criteria decision-making problem, so we can consider that the initial programming language selection can also be a multi-criteria decision-making problem too [69]. Other studies compare the various programming languages in which various inclusion and exclusion criteria are used, such as Be suitable for teaching, Be interactive and fast, Promote correct writing, Allow you to program in "small", Provide a continuous development environment, having a good community of users, open source, good support, being free, having good teaching material, not just being used for educational purposes, being reliable and efficient [22]. We notice that there are not many studies to choose the initial programming language. The few we encounter use choices. Multicriteria and with weights assigned to each of the factors, while others use criteria that exclude some programming languages. It is a study that we would like to do to help anyone who needs to make one of these decisions to do so in a conscious and with the help of others, while making own choice.

5 Conclusion

"What is the best initial programming language?" is the million-dollar question. There is, and probably never will be, a consensus on which programming language should be chosen to introduce the student to the world of computer science. It is an ongoing problem because it depends on trends and changes (as was the case with paradigms) and even on the commercial machine.

This article develops several parts of the problem: which programming languages are most used in the "real world" and in higher education institutions, addresses the question of the importance of the initial language and lists the various attempts to outline the various points to consider when someone chooses which programming language to adopt in a given introductory course.

The choice of programming language for introductory teaching must accompany the evolution, but because it has a propaedeutic character, the choice must obey several requirements, namely pedagogical, and acceptance from the outside world. The first programming language of a future computer science professional is just the beginning of a long journey. The initial programming language is just an accompaniment (and materialization) of the beginning of the acquisition of computational thinking - more important than mere syntax or notation. It is very important to show the world (especially those who are not computer science) that the question of the initial programming language is important (because there are simpler languages than others) but what really matters is that the student "enters" computational thinking and that learning the syntax of languages or paradigm requirements, the student (future IT professional) will succeed in the programming world if he falls in love with programming and works hard to be successful.

References

1. Austing, R.H., Barnes, B.H., Bonnette, D.T., Engel, G.L., Stokes, G.: Curriculum '78: recommendations for the undergraduate program in computer science—a report of the ACM curriculum committee on computer science. Commun. ACM **22**(3), 147–166 (1979)
2. Sobral, S.R.: The first programming language and freshman year in computer science: characterization and tips for better decision making. In: WorldCist'20 - 8th World Conference on Information Systems and Technologies, Budva, Montenegro (2020)
3. Knuth, D.: The Art of Computer Programming. Addison-Wesley, Boston (1968)
4. Gries, D.: The Science of Programming, Springer, Heidelberg (1981)
5. Dijkstra, E.W.: A Discipline of Programming. Prentice Hall, Upper Saddle River (1976)
6. Aho, A., Ullman, J.D.: Foundations of Computer Science: C Edition (Principles of Computer Science Series). W. H. Freeman (1994)
7. McCracken, M., Almstrum, V., Diaz, D., Guzdial, M., Hagan, D., Kolikant, Y.B.-D., Laxer, C., Thomas, L., Utting, I., Wilusz, T.: A multi-national, multi-institutional study of assessment of programming skills of first-year CS students. In: ITiCSE on Innovation and Technology in Computer Science Education (2001)
8. Bergin, S., Reilly, R.: Programming: factors that Influence SuccessSusan. In: Proceedings of the 36th SIGCSE Technical Symposium on Computer Science Education (2005)
9. Becker, B.A., Fitzpatrick, T.: What do CS1 syllabi reveal about our expectations of introductory programming students? In: 50th ACM Technical Symposium on Computer Science Education (2019)
10. Luxton-Reilly, A.: Learning to program is easy. In: ACM Conference on Innovation and Technology in Computer Science Education (2016)
11. Dijkstra, E.W.: The Humble Programmer. Commun. ACM **15**(10), 859–866 (1972)
12. Mitchell, J.C.: Concepts in Programming Languages. Cambridge University Press, Cambridge (2003)
13. Sprankle, M.: Problem Solving and Programming Concepts. 9 edn. Pearson, London (2011)
14. Pigott, D.: Online Historical Encyclopaedia of Programming Languages (1995–2020). https://hopl.info/
15. Scott, M.L.: Programming Language Pragmatics. 3rd edn. Elsevier, Amsterdam (2009)
16. Stackoverflow.com, "Stackoverflow" (2020). https://insights.stackoverflow.com/survey/2020
17. TIOBE Software BV, "TIOBE," Set. (2019). https://www.tiobe.com/tiobe-index/
18. Smith, C., Rickman, J.: Selecting languages for pedagogical tools in the computer science curriculum. In: Proceedings of the Sixth SIGCSE Technical Symposium on Computer Science Education (1976)
19. Wexelblat, R.L.: First programming language: consequences (Panel, in Discussion) (1979)
20. Tharp, A.L.: Selecting the "right" programming language. In: SIGCSE 1982 Technical Symposium on Computer Science Education, Indianapolis, Indiana, USA (1982)
21. Duke, R., Salzman, E., Burmeister, J., Poon, J., Murray, L.: Teaching programming to beginners - choosing the language is just the first step. In: ACSE 2000 Proceedings of the Australasian Conference on Computing Education (2000)
22. Mannila, L., Raadt, M.D.: An objective comparison of languages for teaching introductory programming. In: 6th Baltic Sea Conference on Computing Education Research: Koli Calling 2006 (2006)
23. Atchison, W.F., Conte, S.D., Hamblen, J.W., Hull, T.E., Keenan, T.A., Kehl, W.B., McCluskey, E.J., Navarro, S.O., Rheinboldt, W.C., Schweppe, E.J., Viavant, W., Young Jr., D.M.: Curriculum 68: recommendations for academic programs in computer science: a report of the ACM curriculum committee on computer science. Commun. ACM **11**(3), 151–197 (1968)

24. Koffman, E.B., Miller, P.L., Wardle, C.E.: Recommended curriculum for CS1, 1984. Commun. ACM **27**(10), 998–1001 (1984)
25. Tucker, A.B., ACM/IEEE-CS Joint Curriculum Task Force: Computing Curricula 1991: Report of the ACM/IEEE-CS Joint Curriculum Task Force, p. 154. ACM Press (1990)
26. The Joint Task Force IEEE and ACM: CC2001 Computer Science, Final Report (2001)
27. Cassel, L., Clements, A., Davies, G., Guzdial, M., McCauley, R.: Computer Science Curriculum 2008: An Interim Revision of CS 2001. ACM (2008)
28. Task force ACM e IEEE: Computer Science Curricula 2013. ACM and the IEEE Computer Society (2013)
29. Giangrande Jr., E.: CS1 programming language options. J. Comput. Sci. Coll. **22**(3), 153–160 (2007)
30. Kemeny, J.G., Kurtz, T.E.: BASIC - A Manual for BASIC, the elementary algebraic language, Dartmouth College (1964)
31. Parker, K., Davey, B.: The history of computer language selection. In: IFIP Advances in Information and Communication Technology, pp. 166–179 (2012)
32. Dijkstra, E.W.: Go to statement considered harmful. Commun. ACM **11**(3), 147–148 (1968)
33. Knuth, D.: Structured programming with go to statements. Comput. Surv. **6**(4), 261–301 (1974)
34. Dahl, O., Dijkstra, E., Hoare, C.: Structured programming. Academic Press Ltd. (1972)
35. Wirth, N.: The programming language pascal. In: Pioneers and their Contributions to Software Engineering. Springer (1971)
36. Gupta, D.: What is a good first programming language?. Crossroads ACM Mag. Students **10**(4), 7 (2004)
37. Levy, S.: Computer language usage in CS1: survey results. ACM SIGCSE Bull. **7**(3), 21–26 (1995)
38. McCauley, R., Manaris, B.: Computer science degree programs: what do they look like? A report on the annual survey of accredited programs. ACM SIGCSE Bull. **30**(1), 15–19 (1998)
39. Farooq, M., Khan, S., Ahmad, F., Islam, S., Abid, A.: An evaluation framework and comparative analysis of the widely used first programming languages. PLoS ONE **9**(2), e88941 (2014)
40. Sobral, S.R.: 30 years of CS1: programming languages evolution. In: 12th Annual International Conference of Education, Research and Innovation (2019)
41. Siegfried, R., Greco, D., Miceli, N., Siegfried, J.: A longitudinal analysis of the reid list of first. Inf. Syst. Educ. J. **10**(4), 47–54 (2016)
42. Scoy, F.L.V.: The Return of the Reid List (2001). https://home.adelphi.edu/~siegfried/Rei dList/Reid%20List%2021.htm
43. Siegfried, R., Liporace, D., Herbert-Berger, K.: What can the reid list of first programming languages teach us about teaching CS1? In: 50th ACM Technical Symposium on Computer Science Education (2019)
44. Sobral, S.R.: Bachelor's and master's degrees integrated in Portugal in the area of computing: a global vision with emphasis on programming UCS and programming languages used. In: 11th Annual International Conference of Education, Research and Innovation (2018)
45. Sobral, S.R.: Introduction to programming: portrait of higher education in computer science in Portugal. In: 11th International Conference on Education and New Learning Technologies (2019)
46. Murphy, E., Crick, T., Davenport, J.H.: An analysis of introductory programming courses at UK universities. Art Sci. Eng. Program. **1**(2), 18 (2017)
47. Becker, B.A.: A survey of introductory programming courses in Ireland. In: ITiCSE 2019: 2019 ACM Conference on Innovation and Technology in Computer Science Education (2019)

48. Arcos, G., Aguirre, G.L., Hidalgo, B., Rosero, R.H., Gómez, O.S.: Current trends of teaching computer programming in undergraduate CS programs: a survey from ecuadorian universities. In: Ibero-American Symposium on Computer Programming (2018)
49. Aleksić, V., Ivanović, M.: Introductory programming subject in European higher education. Inform. Educ. **15**(2), 163–182 (2016)
50. Ezenwoye, O.: What language? - the choice of an introductory programming language. In: 48th Frontiers in Education Conference, FIE 2018 (2018)
51. Roberts, E.: The dream of a common language: the search for simplicity and stability in computer science education. In: 35th SIGCSE Technical Symposium on Computer Science Education (2004)
52. Laakso, M., Kaila, E., Rajala, T., Salakoski, T.: Define and visualize your first programming language. In: 8th IEEE International Conference on Advanced Learning (2008)
53. Stefik, A., Hanenberg, S.: The programming language wars: questions and responsibilities for the programming language community. In: 2014 ACM International Symposium on New Ideas, New Paradigms, and Reflections on Programming & Software (2014)
54. Goosen, L.: A brief history of choosing first programming languages. In: History of Computing and Education, vol. 3 (2008)
55. Guerreiro, P.: A mesma velha questão: como ensinar Programação? In: Quinto Congreso Iberoamericano de Educación Superior (1986)
56. Collberg, C.S.: Data structures, algorithms, and software engineering. In: Software Engineering Education - SEI Conference 1989 (1989)
57. Bruce, K., Freund, S.N., Harper, R., Larus, J., Leavens, G.: What a programming languages curriculum should include. In: SIGPLAN Workshop on Undergraduate Programming Language Curricula (2008)
58. Nanz, S., Furia, C.A.: A comparative study of programming languages in rosetta code. In: 2015 IEEE/ACM 37th IEEE International Conference on Software (2015)
59. Alzahrani, N., Vahid, F., Edgcomb, A., Nguyen, K., Lysecky, R.: Python versus C++: an analysis of student struggle on small coding exercises in introductory programming courses. In: 49th ACM Technical Symposium on Computer Science Education (2018)
60. Wainer, J., Xavier, E.: A controlled experiment on python vs C for an introductory programming course: students' outcomes. ACM Trans. Comput. Educ. **18**(3), 1–16 (2018)
61. McMaster, K., Sambasivam, S., Rague, R., Wolthuis, S.: Java vs. python coverage of introductory programming concepts: a textbook analysis. Inf. Syst. Educ. J. **15**(3), 4–13 (2017)
62. Farag, W., Ali, S., Deb, D.: Does language choice influence the effectiveness of online introductory programming courses? In: 14th Annual ACM SIGITE Conference on Information Technology Education (2013)
63. Sobral, S.R.: CS1: C, JAVA or python? Tips for a conscious choice. In: International Conference of Education, Research and Innovation (2019)
64. King, K.N.: The evolution of the programming languages course. ACM SIGCSE Bull. **24**(1), 213–219 (1992)
65. Howatt, J.: A project-based approach to programming language evaluation. ACM SIGPLAN Notices **30**(7), 37–40 (1995)
66. Luker, P.A.: Never mind the language, what about the paradigm? In: Twentieth SIGCSE Technical Symposium on Computer Science Education (1989)
67. Budd, T.A., Pandey, R.K.: Never mind the paradigm, what about multiparadigm languages? ACM SIGCSE Bull. **27**(2), 25–30 (1995)
68. Parker, K.R., Chao, J.T., Ottaway, T.A., Chang, J.: A formal language selection process. J. Inf. Technol. Educ. **5**(1), 133–151 (2006)

69. Mishra, A., Chandel, A., Motwani, D.: xtended MABAC method based on divergence measures for multi-criteria assessment of programming language with interval-valued intuitionistic fuzzy sets. Granular Comput. **5**, 97–117 (2020)
70. ACM - Association for Computing Machinery, Curricula Recommendations. ACM (2020) https://www.acm.org/education/curricula-recommendations

New Approaches to Studying Rodent Behavior Using Deep Machine Learning

Alexander Andreev[1,2], Eugenia Ahremenko[1], Danila Apushkin[1,2], Ilya Kuznetsov[3], Ilya Kovalenko[1], Eduard Korkotian[1,3(✉)], and Vyacheslav Kalchenko[1,3(✉)]

[1] Perm State University, Bukirev 15, 614990 Perm, Russia
[2] Perm State Pharmaceutical Academy, Polevaya 2, 614990 Perm, Russia
[3] The Weizmann Institute, Herzl 234, 76100 Rehovot, Israel
{eduard.korkotian,a.kalchenko}@weizmann.ac.il

Abstract. One of the fundamental tasks in behavioral neurophysiology is the study of movement control. During past decades, methodological arsenal in this field was limited to the expert's visual observation, especially in the open field paradigm when animals can move, groom, rare or explore objects. In the past few years, the massive adoption of machine learning and artificial intelligence has led to revolutionary changes in the field. Here we present a new and user friendly approach to the analysis of the behavior of rodents, based on a pre-trained and continuing to learn neural network, which is capable of exhibiting a high degree of flexibility and good level of performance. Open field behavior of two groups (juvenile and adult) female wistar rats has been recorded. Human expert identified several behavioral patterns for farther deep machine learning, based on open source Orange SqueezeNet, a pre-trained neural network. In addition, we used the transfer learning technique with three more layers over SqueezeNet with 10-fold cross validation. The correctness of the assessments, made by the neural network consisted $92 \pm 3\%$ based on unbiased and double-blinded expert evaluation. Moreover, younger and older rat groups revealed significant difference in such behavioral patterns as moving and hole testing, identified by the network. We conclude that our system is demonstrative, does not require high programming skills, looks intuitive, easy to use, is based on an open source resource, therefore, can be recommended as behavioral research tool and an educational platform.

Keywords: Behavior · Machine learning · Orange SqueezeNet · Pre-trained neural network · Rodents

1 Introduction

One of the most fundamental tasks of neurophysiology is the study of how the brain determines and controls the movement of skeletal muscles, which is expressed in the complex concept of "behavior". It can reflect: the functioning of local neural networks [1], interactions between individual brain structures [2], the solution of tactical and strategic tasks of the organism as a whole [3], as well as the conditional character, hormonal and psychological status of the animal: the level of stress [4], anxiety [5], curiosity or

© The Author(s), under exclusive license to Springer Nature Switzerland AG 2021
T. Antipova (Ed.): ICADS 2021, AISC 1352, pp. 365–374, 2021.
https://doi.org/10.1007/978-3-030-71782-7_32

depression [5, 6], aggression or friendliness [7], socialization (for example, selfishness or altruism) [8] and so on. In addition, behavior can serve as an objective indicator of physiological discomfort, pain [9, 10], cognitive disorders [11], the identification of which is especially important for studying models of various disorders and their treatment [11]. Throughout the development of behavioristics, its methodological arsenal was limited to the expert's visual observation of the animal's motor skills directly during the experiment or watching video recordings made on a recording device [12, 13]. This is especially true in the open field paradigm, where the animal is intended to explore free movement, spatial learning, new object exploration, curiosity, and anxiety-related patterns.

The use of the open field paradigm assumes the presence of an arena of any constant form, delimited by walls or other obstacles that are insurmountable for a rodent. Variable parameters of such an experiment can be the time, the features of the arena, and the presence of familiar or novel objects on it. During the time allotted for the experiment, the animal can move in the center or along the walls of the field, doing it at a variable speed, freeze for a while, exhibit orienting response, change the interposition of the head, body and tail, modulate the gait, interact with existing objects or walls, make supported and unsupported rear leg stance etc. [10, 13, 14]. Typical registration parameters are the distances of horizontal locomotion in the center or periphery of the field, vertical behavior (such as number and type of rearing), and the timing of grooming [15, 16]. These practically limit the possibilities of an unbiased expert assessment of the behavior and comparison of various groups of rodents, both online and offline. If any assessment of the emotional state of the animal was carried out, then it was always subjective and, in essence, depended on the personal experience and attentiveness of the expert, as well as on the quality of the examination itself.

The introduction of the first automated and commercialized behavioral analysis platforms was usually associated with the localization of the animal in space and reliable tracking of the distance traveled with an orientation to the head, nose tip, tail, or the whole body mass [17, 18]. In the past three years or so, the massive adoption of machine learning and artificial intelligence has led to fundamental and revolutionary changes in this well-established field of research. In this context, two general paths have emerged for the further development of rodent behavior. The first tendency is associated with a kind of decomposition of the bodies of a mouse or rat into a set of variable vectors describing different elements of their movable skeleton [13, 21]. This approach has undoubted advantages in the form of a thoroughly described set of trajectories, but there are also some significant disadvantages. Firstly, such a system presupposes a very cumbersome sensory and analytical apparatus, and secondly, the presence of a carefully described set of standard positions, which are compared with current data, limit the flexibility and degree of freedom of the ongoing unique analysis [19, 20]. In the current study, we present a fundamentally different and user friendly approach to the analysis of the behavior of rodents, based on a partly pre-trained and partly continuing to learn neural network, which is capable of exhibiting a high degree of flexibility, and potentially identifying and adapting new, previously unspecified behavioral patterns.

2   Materials and Methods

2.1   Animals and Behavioral Recordings

Female Wistar rats were accommodated under standard conditions, according to 5–6 animals per a Type 3 cage, supporting a standard temperature, balanced diet and a 12/12 light/dark cycle. The average age of the rats by the period of the behavioral tests was 2.5 months (10 weeks) for young rats and 10 months for old rats.

The open field as a circular experimental arena of 97 cm in diameter, divided into 19 sectors of the same area in three concentric circles and equipped with floor holes, was used to observe and classify rat behavioral patterns. Three groups of criteria were used for the classification: 1 - movement; 2 - exploratory behavior; and 3 - emotional behavior. The first pattern of behavior was expressed in the movement of the animal across an open field. The exploratory group of patterns included wall-supported climbing, unsupported rearing, and hole exploration. Emotional behavior was represented by instances of grooming (illustrations of behaviors are shown in Fig. 1). The behavioral tests were accompanied by video recording. Digital recordings were done using Canon 5D camera, 30 frames per sec, during 3 min, under 320 lux illumination.

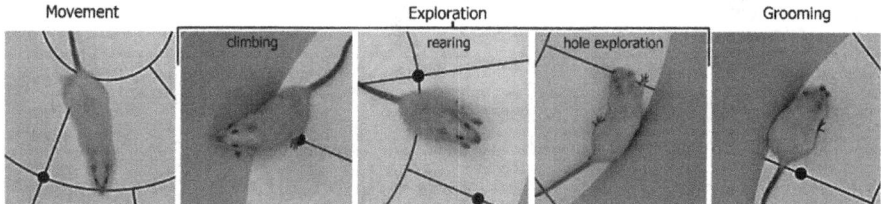

Fig. 1. Main behavioral patterns. Examples of rat behavioral patterns, identified by a human expert (from left to right): movement (1); types of exploratory behavior- climbing (2), rearing (3), hole exploration (4); and grooming (4) as an example of emotional behavior.

2.2   Deep Machine Learning

The approach of the automated analysis and classification of the behavioral patterns was based on open source Orange Data Mining (v. 3.24.0) software (University of Ljubljana) and Image Analytics Widget [22]. We have aimed to use rather simple and open source materials to investigate the applicability of convolutional deep neural networks in considering pose and/or behavioral pattern automated estimation when processing the data of the classic open field test.

An open source cv-code plugin was used for cropping the video frame, which was automatically following the moving animal in the field. The reason for cropping and tracking animals within the area of interest was to cut out all irrelevant details, which was an important adaptive preparatory step for further successful machine learning. The recorded open field video stream was a subjected of frame-by-frame human expert analysis, so that 30 to 130 single-framed examples of 5 selected behavioral patterns were extracted and saved. Original frames were of 224 × 224 pixels, saved in BMP format (Fig. 1). Each frame was a matter of augmentation procedure, made using ImageJ software. This was done to increase the variety of machine learning material. The augmentation included 18 routine procedures, consisted of: horizontal and vertical flips, Gaussian blurring, median and despeckle filtering, turns 15, 30 and 45° both to the left and right, up-left and down-right shifts, Gaussian noise additions, zooming in and zooming out by the factor of 1.25. Thus, the total number of image variations reached 19, including the initial.

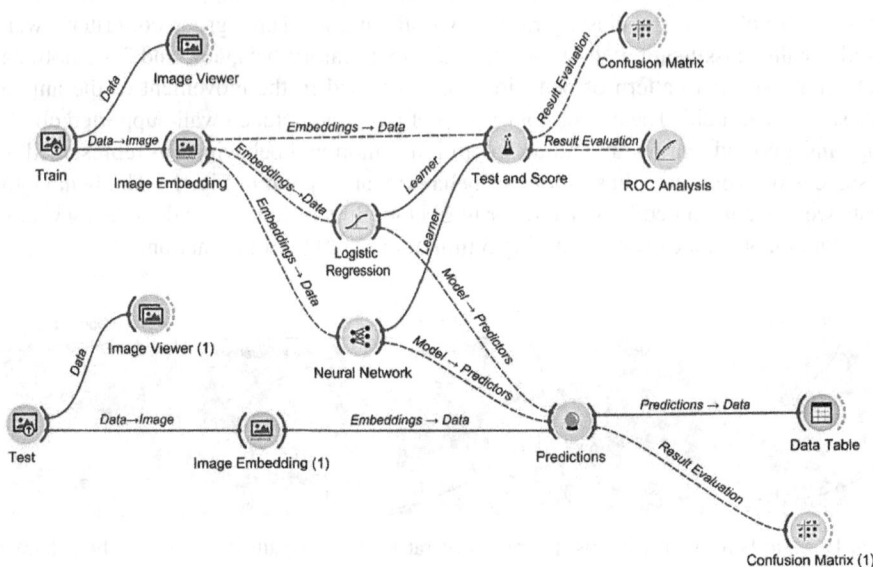

Fig. 2. The basic model created in Orange. Initial folder (Train) contains the collection of five behavioral examples, classified and preselected by an expert. Image viewer shows these static images and the Image Embedding (converting images to descriptor sets) marks images according the expert's classification and transforms them using SqueezeNet (a standard neuronal network, created for image classification) where the last layer has been modified in accordance with the particular task (see text for more details). Two classifiers of the images have been used: The Logistic Regression with standard parameters and the Neural Network based on multilayer deep learning. Test and Score terminal evaluates the efficiency of the system learning produced by each of classifiers. In each case, the result is manifested in logic regression which assigns the image to one of five classes. Some 80% of the reclassified data is used for teaching the system and another 20% - for analysis as a completely novel set of images for the farther validation of efficacy. Comparing these automatically obtained results to the expert's assessment allows the system to evaluate its robustness and create a Confusion Matrix (a matrix of recognition errors) and/or ROC Analysis presenting the probability of the correct decision made by the system. Then a novel (Test) image may be presented for Image Embedding (1) and farther evaluated in accordance with the Predictions of pre-trained network (shows the probability of compliance with a particular class for each image). At the same time, Image Viewer (1) present the image to a human expert along with a creation of Confucian Matrix (1) and output Data Table.

The workflow was based on Orange SqueezeNet, a pre-trained neural network (Fig. 2). The cropped and augmented image sets, made in accordance with the human expert classification, were embedded into Orange. Using deep machine learning, they were converted into vectors so that each image was represented as a vector consisting of 1100 values, which were invariant for particular animal pose/pattern. This embedding served an input of the operator of logistic regression. We used logistic regression supervised learning with target variables derived from the expert classification. Following these steps, the result was correlated with specific rat behavioral patterns, and the correctness of the assessments, made by the completed neural network, was concluded and reported. In addition, we used the transfer learning technique with three more layers over SqueezeNet with 10-fold cross validation.

3   Results and Discussion

3.1 Network Reliability Assessment

At the first stage, the fill-sized images have been used for farther classification and machine learning. These images contain lot of secondary or unimportant details, which increase the rate errors, made by the Network. Results of full-sized image representation to Logistic Regression (LogReg) and Neural Network (NNnet) are shown in Table 1. Each type of behavior was expressed in 15 frames, pre-identified by the human expert.

Table 1. Confusion matrices for full-sized images. Results of full-size images in the direct processing. Number of correctly defined images are in the diagonal, highlighted in violet. The number of incorrectly identified images is shown in cells marked with yellow-red, from bright to dark, depending on the number of errors. Rows, actual values, columns, predicted values. LogReg- Logistic Regression, NN - Neural Network. Behavioral patterns: C climbing, G – grooming, R – rearing, M – moving and H – hole exploration.

LogReg	C	G	R	M	O	Σ
C	12	1	0	0	2	15
G	6	4	0	2	3	15
R	7	2	2	1	3	15
M	1	1	1	8	4	15
O	3	3	0	1	8	15
Σ	29	11	3	12	20	75

NNet	C	G	R	M	O	Σ
C	11	1	0	3	0	15
G	1	11	0	3	0	15
R	0	4	9	2	0	15
M	0	2	0	12	1	15
O	0	2	1	3	9	15
Σ	12	20	10	23	10	75

Climbing and grooming have had the highest level of recognition in both LogReg and NNet, while the last method was efficient also for rearing which was the lowest category for LogReg.

Much better results have been obtained with the cropped images, as shown in Table 2.

Table 2. Confusion matrices for cropped images. Number of correctly defined images, central diagonal highlighted in violet. The number of incorrectly identified images is shown in cells marked with yellow-red, from bright to dark, depending on the number of errors. Rows, actual values, columns, predicted values. LogReg- Logistic Regression, NNet - Neural Network. Behavioral patterns: C climbing, G – grooming, R – rearing, M – moving and H – hole exploration.

LogReg	C	G	R	M	O	∑
C	15	0	0	0	0	15
G	1	13	0	0	1	15
R	2	0	13	0	0	15
M	0	0	0	15	0	15
O	0	2	0	2	11	15
∑	18	15	13	17	12	75

NNet	C	G	R	M	O	∑
C	15	0	0	0	0	15
G	1	13	0	0	1	15
R	1	0	14	0	0	15
M	0	0	0	15	0	15
O	0	0	0	2	13	15
∑	17	13	14	17	14	75

It is likely that the result of recognition of full-sized images is worse, since the neural network gives the animal's position in space more weight than its behavioral class (body shape). In addition, the recognition of full-sized images is more difficult due to the many "noisy" image elements. Besides the animal, these images contain empty space, waste products and other objects that do not carry any relevant information about the animal's behavior. Wherein, the animal itself, captured at given behavioral pattern, occupied about a quarter of the image surface of the cropped frames.

The overall correctness of the system after cropping reached $92 \pm 3\%$, based on unbiased and double-blind expert evaluation.

3.2 Network Performance Assessment

To further test the performance and reliability of our artificial intelligence system, we made an attempt to identify the age of the studied animals by the characteristics of their behavior. This rather complicated and unobvious assessment criterion was chosen by us as one of the most objective and independent of other circumstances. There is potentially no clear reason to believe that younger and older animals will behave differently in relation to basic behavioral patterns at all. However, this is possible due to natural differences in the level of curiosity, anxiety and hormonal status of animals. Our study was based solely on a pre-trained analysis of the neural network and objective indicators of the age of rats. The results are presented below.

Overall, behavioral patterns of 15 juveniles (2 months-old) and 16 adult (10 months-old) animals were recorded during 180 s. For each individual animal, a "behavioral passport" was built. The Orange-based model identified patterns of behavior for each 25th frame, i.e. one frame per second, after which the occurrence of each of the patterns was calculated. Thus, an individual five-dimensional characteristic was obtained, according to the number of recognized types of behavior. This allowed us to check the robustness and the resolution of the Orange model for distinguishing between "juvenile" and "adult" groups using multivariate analysis of variance (MANOVA).

Obviously, not each 25th frame necessarily contained one of the behavior patterns. In order to filter out such uninformative frames, we used the model's confidence characteristic in interpreting each specific frame - the "likelihood ratio". The implementation of this criterion was carried out according to the following principle: the given frame is recognized as "climbing" if the "climbing" likelihood for it exceeds a certain threshold. To determine an acceptable threshold value for likelihood, we examined the relationship between this threshold value and the MANOVA p-value, reflecting the distinguishability of groups (Fig. 3).

One can see that for frames with higher likelihood threshold, the distinguishability of the groups is also higher so the p-values decrease (red line). Of the likelihood value of about 0.6 and above, the distinguishability of groups reaches the statistically significance,

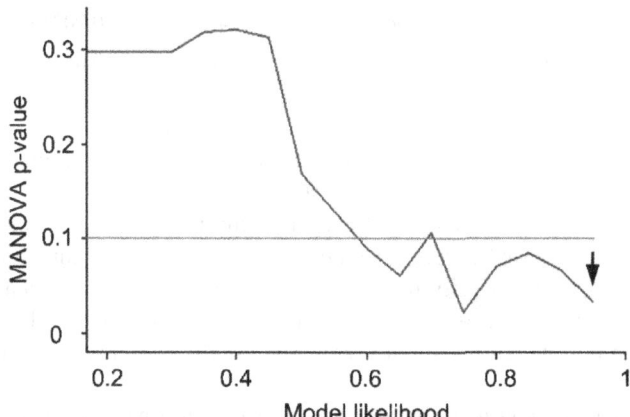

Fig. 3. Dependence of distinguishability of behavioral groups (MANOVA) on the threshold value of likelihood for the model of logistic regression created in "Orange". X-axis represents the likelihood threshold. If the likelihood value of a given frame was below the threshold, it was considered behaviorally irrelevant and not taken into account when calculating the significance of differences between groups. Y-axis represents the p-value of the analysis of variance MANOVA. The gray line indicates the threshold of statistical significance $p \leq 0.1$.

stabilizing at the level of 0.04 - 0.10, depending on the specific threshold value of likelihood.

The likelihood value 0.95 was taken as a reference threshold value (arrow on Fig. 3), which showed a high efficiency of group discrimination. At this likelihood value, MANOVA was performed and the results are shown in Table 3.

Table 3. MANOVA analysis for likelihood value of 0.95

Parameters	Df	Pillai	approxF	numDf	denDf	p-value
Age	1	0.36645	2.8921	5	25	0.03405
Residuals	29					

Since the probability of type I error in the global MANOVA test is less than 0.05, there was a reason to perform post-hoc testing (Tukey HSD) to find differences in each of the parameters. Results are presented in the Table 4 below and in Fig. 4.

Table 4. Post-hoc Tukey HSD analysis for 5 behavioral patterns.

Behavior pattern	Shift estimate	Confidence. Low	Confidence. High	p-value	Significance
Climbing	−0.529	−17	15.9	0.948	ns
Grooming	−12.8	−26.3	0.701	0.0623	ns
Rearing	−7.75	−18.3	2.83	0.145	ns
Moving	15.5	6.06	24.9	0.00219	**
Hole testing	5.59	0.121	11.1	0.0455	*

Significant differences can be noted between the two age groups for mobility (moving) and examination of holes (hole testing).

Interestingly, of all five behavioral patterns, only movement and hole exploration demonstrated a difference between juvenile and adult rats. At the same time, both great mobility and increased curiosity in the study of objects (holes) were characteristic of juvenile animals. Moreover, such independent or less dependent on age behavioral indicators as grooming and rearing, did not show statistically significant differences among age groups.

Summarizing, we can note that our pre-trained neural network based on an open access system and easy to use has demonstrated very high reliability indicators: about 92% reliability, according to an independent and unbiased expert assessment. Moreover, this system was able to identify such non-obvious and rather unexpected differences, as the difference in the behavior of juvenile and adult rats under completely identical conditions. Particularly, the difference appeared not in all, but only in certain patterns: in

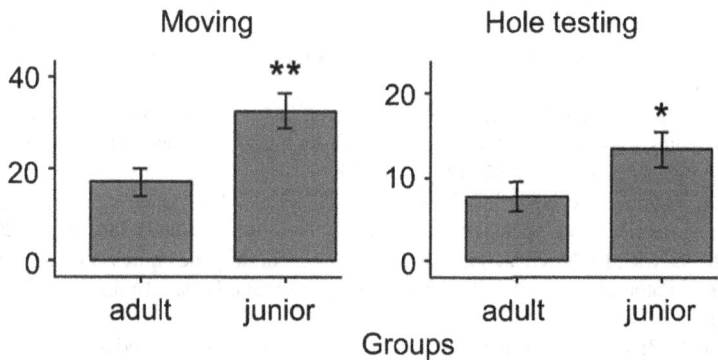

Fig. 4. Post-hoc analysis results. Statistically significant difference between age groups for moving and hole testing. X- axis – age groups, Y-axis – averaged occurrence of behavioral patterns.

moving (activity) and hole exploring (curiosity) behaviors. This is in line with general expectations, indicating that our pre-trained network is performing well.

In our opinion, this system is demonstrative, does not require high programming skills, looks intuitive, easy to use, based on an open source resource, therefore, can be recommended as behavioral research tool as well as an educational platform.

References

1. Sadaghiani, S., Kleinschmidt, A.: Brain networks and proportional to-oscillations: structural and functional foundations of cognitive control. Trends Cognit. Sci. **20**(11), 805–817 (2016)
2. Kempermann, G.: Environmental enrichment, new neurons and the neurobiology of individuality. Nat. Rev. Neurosci. **20**(4), 235–245 (2019)
3. DeChurch, L.A., Mesmer-Magnus, J.R.: the cognitive underpinnings of effective teamwork: a meta-analysis. J. Appl. Psychol. **95**(1), 32–53 (2010)
4. Antoniuk, S., Bijata, M., Ponimaskin, E., et al.: Chronic unpredictable mild stress for modeling depression in rodents: meta-analysis of model reliability. Neurosci. Biobehav. Rev. **99**, 101–116 (2019)
5. Molendijk, M.L., de Kloet, E.R.: Coping with the forced swim stressor: current state-of-the-art. Behav. Brain Res. **364**, 1–10 (2019)
6. Hughes, R.N.: Neotic preferences in laboratory rodents: Issues, assessment and substrates. Neurosci. Biobehav. Rev. **31**(3), 441–465 (2007)
7. Lischinsky, J.E., Lin, D.: Neural mechanisms of aggression across species. Nat. Neurosci. **23**(11), 1317–1328 (2020)
8. Silva, P.R.R., Silva, R.H., Lima, R.H., et al.: Are there multiple motivators for helping behavior in rats? Front. Psychol. **11**, 1795 (2020)
9. Ong, W.-Y., Stohler, C.S., Herr, D.R.: Role of the prefrontal cortex in pain processing. Mol. Neurobiol. **56**(2), 1137–1166 (2019)
10. Abdus-Saboor, I., Fried, N.T., Lay, M., et al.: Development of a mouse pain scale using sub-second behavioral mapping and statistical modeling. Cell Reports **28**(6), 1623 (2019)
11. Wyss-Coray, T.: Ageing, neurodegeneration and brain rejuvenation. Nature **539**(7628), 180–186 (2016)
12. Giancardo, L., Sona, D., Huang, H., et al.: Automatic visual tracking and social behaviour analysis with multiple mice. PLoS ONE **8**(9), e74557 (2013)

13. Geros, A., Magalhaes, A., Aguiar, P.: Improved 3D tracking and automated classification of rodents' behavioral activity using depth-sensing cameras. Behav. Res. Methods **52**(5), 2156–2167 (2020)
14. Lindsay, A.J., Caracheo, B.F., Grewal, J.S., et al.: How much does movement and location encoding impact prefrontal cortex activity? An algorithmic decoding approach in freely moving rats. Eneuro **5**(2), e0023 (2018). UNSP
15. van den Boom, B.J.G., Pavlidi, P., Wolf, C.J.H., et al.: Automated classification of self-grooming in mice using open-source software. J. Neurosci. Methods **289**, 48–56 (2017)
16. Swerdlow, N.R., Light, G.A., Trim, R.S., et al.: Forebrain gene expression predicts deficits in sensorimotor gating after isolation rearing in male rats. Behav. Brain Res. **257**, 118–128 (2013)
17. Azarfar, A., Zhang, Y., Alishbayli, A., et al.: An open-source high-speed infrared videography database to study the principles of active sensing in freely navigating rodents. GigaScience **7**(12), 1–6 (2018)
18. Richardson, A.G., Ghenbot, Y., Liu, X., et al.: Learning active sensing strategies using a sensory brain-machine interface. Proc. Natl. Acad. Sci. U.S.A. **116**(35), 17509–17514 (2019)
19. Morales, L., Tomas, D.P., Dalmau, J., et al.: High-throughput task to study memory recall during spatial navigation in rodents. Front. Behav. Neurosci. **14**, 64 (2020)
20. Alexandrov, V., Brunner, D., Hanania, T., et al.: High-throughput analysis of behaviour for drug discovery. Eur. J. Pharmacol. **750**, 82–89 (2015)
21. Mazur-Milecka, M., Ruminski, J.: Deep learning based thermal image segmentation for laboratory animals tracking. Quant. Infrared Thermography J. (2020). https://doi.org/10.1080/17686733.2020.1720344
22. Demsar, J., Curk, T., Erjavec, A., Gorup, C., Hocevar, T., Milutinovic, M., Mozina, M., Polajnar, M., Toplak, M., Staric, A., Stajdohar, M., Umek, L., Zagar, L., Zbontar, J., Zitnik, M., Zupan, B.: Orange: data mining toolbox in Python. J. Mach. Learn. Res. **14**, 2349–2353 (2013)

Advances in E-Information Systems

Modeling of Investment in It-Business

Gennady Ross$^{(\boxtimes)}$ ⓘ and Valery Konyavsky ⓘ

Plekhanov Russian University of Economics, Moscow, Russia
ross-49@mail.ru

Abstract. This paper proposes a methodology for assessing the economic efficiency of investments in the international cluster (hereinafter the consortium) of IT
development products of the Eurasian Economic Union to create a common digital
platform. The essence of this methodology is based on the financial statements
affecting the cash flow and the simulation modeling of cash flows in perspective,
the basis of which is the optimization model of evolution-simulation model (ESM).
This model includes a set of interrelated simulation models operating under uncertainties and commercial risks (overstatement risk and understatement risk), under
which the efficiency of investments in consortium activities is predicted. These
risks are the main motives of consortium behavior, and the condition of their equality is the criterion of ESM, which allows to predict the expected sales volumes,
prices, costs, quality of IT products or services, solvent demand, competition, etc.
The analytical method of pre-project marketing in predicting cash flows of the
consortium, which under strong simplifications allows to make approximate calculations, is presented. The proposed ESM, unlike the analytical model, makes it
possible to take into account the characteristics of the product, its quality, duration
of use, market saturation effect, advertising costs and advertising efficiency, competitors' prices, taxes, etc. This model, being implemented in Decision, allows to
investigate the influence of different factors on financial flows, as well as scenarios
unfolding in time, in a dialog mode.

This approach opens up the fundamental possibility of using the method of
discounted cash flows to assess the effectiveness of investments in improving the
commercial activities of the consortium. The computational experiment on realization of consortium ESM in the environment of Equilibrium module of Decision
tool system is presented.

Keyword: Simulation models · Equilibrium random processes theory ·
Investment modeling · Underestimation and overestimation risks ·
Evolutionary-simulation model · Tool system decision

1 Introduction

The socio-economic activity of all countries, in particular the countries of the Eurasian
Economic Union (EAEU), has entered the digital stage of its development today, and
the "digital" imperceptibly permeates all spheres of our life and has a growing influence
on the EAEU countries. The digitalization of the economy, as shown in [1], leads to the
disappearance of entire sectors of the economy, together with enterprises and jobs. It

T. Antipova (Ed.): ICADS 2021, AISC 1352, pp. 377–389, 2021.
https://doi.org/10.1007/978-3-030-71782-7_33

changes the social behavior of people, affects labor relations, property relations. Finally, it dilutes the tax base and creates certain threats to the existence of regulators, including public and state institutions. Therefore, for the further development of the EAEU, a single base on a common digital platform (hereinafter referred to as the Platform) is needed, said Russian Prime Minister Mikhail Mishustin at a plenary session of the Digital Future of the Global Economy forum in Alma-Ata. According to him, Russia is capable of advanced developments that will soon change the world.

The main task of the Platform is to provide tools for economic regulation of the EAEU members within a special cluster, as an independent unit that unites leading enterprises interested in development, within the framework of state regulation, implementation of cluster policy and other measures of state support. The core of the cluster can be a specially organized enterprise - a development institute (hereinafter referred to as a consortium), which requires both one-time and permanent investments. When organizing such a consortium, it is important to determine the assessment of the effectiveness of possible investments and approaches to its calculation. It is known that the efficiency of investments depends on: an increase in transactions and turnover of IT products of the consortium; expanding services and range of IT services; improvement and improvement of the advertising activities of the consortium.

The functioning of the consortium is associated with risks that are the result of real and fundamentally unavoidable uncertainty. To reduce uncertainty, EFM was used [2], within which a methodology for assessing the effectiveness of investment in a consortium was developed [3–7]. EFM includes a structural mathematical formulation of the statistical optimization problem (simulation models), optimization algorithms, as well as methodological techniques for modeling various decision-making strategies. In the proposed ESM, the condition of minimizing commercial risks (the risk of overestimation and the risk of underestimation) is used as a criterion for evaluating the optimization of the consortium's activities.

The largest commercial risks of the consortium are associated with the uncertainty of sales volumes and prices for goods and services that the consortium will be able to implement in the medium and especially long term. Volumes and prices depend not so much on the consortium itself, but on buyers, their incomes and preferences, the actions of competitors, fashion, political risks and other external random factors. These risks can be called investment risks in the sense that the money that the consortium must invest in the production and sale of goods and services may not only not bring income, but also not return. At the same time, we mean investments not only in the expansion, modernization or diversification of production [8, 9], but also in the circulating assets necessary for the delivery and storage of goods, the provision of services, and the implementation of the production process.

The proposed method of discounting cash flows in combination with the ESM of the functioning of the consortium makes it possible to predict the expected cash flows and calculate estimates of the effectiveness of investments in improving its commercial activities. A computational experiment for the implementation of EFM for the analysis of various strategies of consortium behavior in the environment of the Equilibrium module of the Decision instrumental system is presented.

2 Challenges of Building a Digital Industry Ecosystem

The principles of openness and community unification in the formation of digital industry ecosystems become the basic solution of the digital economy. A digital industry ecosystem is an environment that enables the innovative development and distribution of digital services, digital products, applications and devices in a particular sector of the digital economy. The goal of the ecosystem is to provide the public with digital services that are shaped in real time, subject to all norms and regulations, and in an environment of maximum trust.

The ecosystem creates the basis for public-private partnerships (PPP), avoiding the cost of mass development and promotion of applied services, providing an environment for the participation of small and medium-sized businesses, supporting manufacturers of services, devices and appliances. The platform should also enable disparate systems and organizations to work together, both technically and commercially [1].

The basis for the creation and development of the EACP-UDS ecosystem can be a cluster created as an independent unit based on the association of leading enterprises interested in development, within the framework of state regulation, implementation of cluster policy and other measures of state support. The core of the cluster should be a specially organized enterprise - institute of development (hereinafter the consortium), support and dissemination of the ecosystem. At the same time, the creation and improvement of the digital platform requires a one-time and ongoing investment for the effective maintenance of IT products. This raises questions such as the possible scale of the consortium, the forms of its implementation, and economic indicators. At the same time, the main problem is the problem of assessing the effectiveness of investment in the consortium in the implementation of commercial sales of IT products over time.

Thus, the functioning of the consortium implements the business process of production of modern technology business activities of various enterprises and organizations. The IT products of the consortium include system and network integration, system implementation and support, custom software development, IT consulting and IT outsourcing. There are also internal IT products provided by the IT services of the enterprises. Their set is individual and largely depends on the industry, size of the organization, etc. It can be broken down into three large groups: IT infrastructure support, business application support, and user support.

In accordance with the recommendations outlined in [1], the structure of the consortium should include the following units:

1. Project office, competence center, R&D center, which carries out: proactive design and approbation of new technologies; design transformation of existing procedures and business processes, embedding into existing business processes; testing of developments on a platform simulating the business procedures of the digital industry for service embedding; development of the ecosystem platform, including OpenAPI and infrastructure services; creation of the quality monitoring system of digital services industry at all stages of the life cycle.
2. Center for digital transformation and adaptation of products and services to the digital market, which provides: digitalization of existing processes and implementation of the best solutions within the digital industry; legitimacy of existing mobile services,

merging with other businesses, organizations or institutions to improve the quality of services; simplification of information protection procedures.

3. Application Access Center (for consumers), which supports: advanced access mechanisms that hide the complexity of procedures from clients and do not require additional hardware, and simplify access to services for different categories of consumers; publication of third-party applications.

4. A platform for quality control and distribution of developments, a competence center and a business accelerator (for development companies), which allows you to: use infrastructure services and tools for developers; create applications that can work with OpenAPI with other digital systems and services within the ecosystem; test developed applications, publish and sell them through application stores; facilitate the search for investors and the creation of the initial reputation of startups; participate in the expert community for the development of the semantic core, standardization of requirements for applications and services.

However, for the implementation of this proposal, the EAEU needs a solid financial and economic justification for assessing investment in an international consortium of IT products. We need models that would take into account the possible commercial risks of the consortium and make it possible to model various strategies for investing in its business.

When predicting the expected sales volume and prices, as well as the economic parameters of the consortium's activities, the criterion of equality of commercial risks is one of the defining elements of the EFM [10–13]. The commercial risks are twofold, i.e. there is both a risk of overstatement and a risk of understatement, and both of these risks are present at the time of the decision. The risk of overestimation lies in the fact that the decision may turn out to be too optimistic, not realistic enough and, therefore, impracticable in full. For example, it is a risk to develop a number of IT products and not implement all of them. The risk of underestimation is that the decision is too pessimistic, does not make enough use of available opportunities. An example of this risk is the development of an insufficient range of IT products (versus demand).

This approach forms the basis of the technology for changing the value of the consortium as a result of evaluating the effectiveness of investment investments. The paper presents an optimization forecasting model, which is based on simulation models and dialogue procedures with the Equilibrium module of the Decision system [2, 3].

3   Evolutionary-Simulative Model and Its Main Characteristics

For a consortium, the amount of investment for the successful development and implementation of IT products is important. The consortium management plans the values of important economic indicators, which generally characterize the results of its work or plans for the future [5].

To formalize the task of planning the consortium activity let us introduce the following notations. Let *PL* denote the planned indicator of the consortium, which is a deterministic value, which value is set by the consortium management at the beginning of the planning period, and Fa - actually obtained result, which is a random value, which depends on a large number of internal and external random factors, its specific value is found out at the end of the planning period.

Random factors, on which Fa depends, are related to sales volume, prices, revenues, prices on foreign markets, ruble exchange rate, labor productivity, etc. Let f_i, $i = 1,...,$ I be the factors, which we will consider as scalar random variables, and $\bar{f} = (f_1, ..., f_I)$ be a random vector. Besides random factors, the parameter Fa also depends on some number of conditionally constant quantities, such as, for example, tax and payment rates, expenditure norms, etc., which we will refer to as initial indicators. Let p_j, $j = 1,..., J$ be the initial indicators, scalar quantities, and $\bar{p} = (p_1, ..., p_J)$ be the vector of initial indicators.

Analyzing the consortium activity we can define sets of factors and initial indicators and develop a simulation model $Fa = \rho\left(\bar{f}, \bar{p}\right)$, which will enable us to carry out statistical tests and obtain Fa realizations. We will mark the realization Fa, i.e. the one obtained in a statistical test (a one-time play on computer of the plan fulfillment process), by the upper index ^{e}: Fa^e. If ^{e} has a specific numerical value, it indicates the number of the statistical test.

Since PL is deterministic and Fa is a random variable with a common area of values, one of two situations is possible: either it turned out that $PL > Fae$, or that $PL < Fae$. Let us consider each of them.

If $PL > Fa^e$ then it means that the plan was overestimated, so it could not be fulfilled completely and there were costs. Depending on the specific conditions the content of costs can be very different. Let us denote the value of these costs. If it had been possible to foresee the circumstances and set the plan $PL = Fa^e$ in advance, the costs could have been avoided.

The situation $PL < Fa^e$ means that the plan was underestimated. In this case there is also a cost, but a completely different kind - it is a missed opportunity. Let us denote these costs by $\Psi_2(PL, Fa^e)$. They could have been avoided if the plan would have been accepted in the size $PL = Fa^e$.

When calculating the costs of overestimation and the costs of underestimation, some differences in the composition of factors and changes in calculation procedures must be taken into account. For example, if the PL sales plan was overestimated, then part of the volume $(PL - Fa^e)$ can be sold through other channels, on other, less favorable terms. Taking this into account, the total sales volume will be Fa_1^e. In calculating the costs of underestimation there is no such a need and the total sales volume will be different: Fa_2^e. Thus, in the most general case 4 simulation models are needed:

1. $Fa_1 = \rho_1\left(\bar{f}, \bar{p}\right)$ - model for calculating the implementation of the planned indicator Fa_1^e used in calculating the costs of overestimation;
2. $Fa_2 = \rho_2\left(\bar{f}, \bar{p}\right)$ - model for calculating the realizations of the planned indicator Fa_2^e used in calculating the costs of understatement;
3. $\Psi_1(PL, Fa_1) = \rho_3\left(PL, Fa_1, \bar{f}, \bar{p}\right)$, if $PL > Fa_1$ - a model for calculating overestimation costs;
4. $\Psi_2(PL, Fa_2) = \rho_4\left(PL, Fa_2, \bar{f}, \bar{p}\right)$, if $PL < Fa_2$ - a model for calculating the costs of understatement.

The risk of overestimation is the mathematical expectation of overestimation costs $M\{\Psi_1(PL, Fa_1)\}$, and the risk of underestimation is the mathematical expectation of underestimation costs $M\{\Psi_2(PL, Fa_2)\}$, where M is a sign of the mathematical expectation.

Summarizing all the above, we obtain the formulation of ESM [2]. Notes: f_i, $i = 1,..., I$ - factors (random scalar quantities); $\bar{f} = (f_1, ..., f_I)$ - vector of factors; $p_j, j = 1,..., J$ - initial indicators (conditionally-constant scalar quantities); $\bar{p} = (p_1, ..., p_J)$ - vector of initial indicators.

Relationships of the model:

$$Fa_1 = \rho_1\left(\bar{f}, \bar{p}\right) \tag{1}$$

$$Fa_2 = \rho_2\left(\bar{f}, \bar{p}\right) \tag{2}$$

$$\Psi_1(PL, Fa_1) = \rho_3\left(PL, Fa_1, \bar{f}, \bar{p}\right), \text{ if } PL > Fa_1 \tag{3}$$

$$\Psi_2(PL, Fa_2) = \rho_4\left(PL, Fa_2, \bar{f}, \bar{p}\right), \text{ if } PL < Fa_2 \tag{4}$$

Let's consider the features of simulation models (1)–(4) as a part of ESM.

The simulation model $Fa_1 = \rho_1\left(\bar{f}, \bar{p}\right)$ (Eq. (1)) reflects solvent demand for the firm's products. In contrast to (11), the model $\rho_1\left(\bar{f}, \bar{p}\right)$ includes adjustment parameters and allows to determine the composition and interrelation of factors and initial indicators accordingly. This simulation model is implemented in EquilibriumFullRus. Indicators as Function Usl_zav_MM2.

The simulation model $Fa_2 = \rho_2\left(\bar{f}, \bar{p}\right)$ (Eq. (2)) reflects solvent demand in the same way as the model $\rho_1\left(\bar{f}, \bar{p}\right)$, but without taking into account sales of some goods outside the market, i.e. $Fa_2 = Fa_2 - f_{13}$. The simulation model $\rho_2\left(\bar{f}, \bar{p}\right)$ is implemented in EquilibriumFullRus. Indicators as Function Usl_zan_MM2. The difference between Fa1 and Fa2 is explained by the fact that Fa1 is used in the calculation of overestimation costs. In this case it is necessary to take into account the possibility of selling the product outside the market, and Fa2 - when calculating the costs of understatement - when it is not necessary.

The simulation model $\Psi_1(PL, Fa_1) = \rho_3\left(PL, Fa_1, \bar{f}, \bar{p}\right)$ (Eq. (3)) expresses the overestimation costs of the firm, i.e. the costs that arise in the situation when the sales plan PL turns out to be greater than the demand Fa1 (taking into account out-of-market sales). This simulation model takes into account the peculiarities of the setting and automatically calculates the cost of production taking into account the sales volume and fixed costs. The model $\left(PL, Fa_1, \bar{f}, \bar{p}\right)$ is implemented in EquilibriumFullRus. Indicators as Function Izd_zav_MM2!(PL, Fa2).

The simulation model $\Psi_2(PL, Fa_2) = \rho_4\left(PL, Fa_2, \bar{f}, \bar{p}\right)$ (Eq. (4)) expresses the understatement cost, that is, the cost of the consortium that occurs in the situation when the sales plan PL turns out to be less than the demand Fa2 (not including out-of-market sales). This simulation model, like the previous one, takes into account the specifics of the setting and automatically calculates the cost. The model is implemented in EquilibriumFullRus. Indicators as Function Izd_zan_MM2!(PL, Fa2).

4   Pre-project Marketing

The ultimate success of an investment project depends to a greater extent on the quality of pre-project marketing research (PMR), the goals and objectives of which were formulated in the work of Nikolai Fetyukhin [16]. The ultimate goal of PPM is to get a forecast of expected financial flows from the sale of certain goods and services in the market. Having forecasts, you can choose the best scenario and try to implement it. The financial result of production and realization of any good or service is the difference between the sum of all receipts and the sum of expenditures connected with the production and realization of the goods for a certain period of time, in other words, it is a Cash flow (CF) indicator [6, 7]. The CF indicator is the basis of all integral indicators used in justification of investment projects (NPV, IRR, PB, etc., Fig. 1), as well as most other indicators reflected in the balance sheets of economic agents (EA).

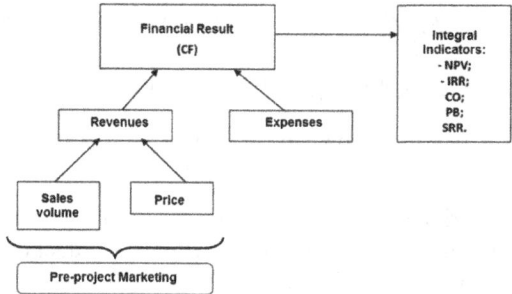

Fig. 1. Tasks of pre-project marketing in cash flow forecasting.

The CF indicator comprehensively takes into account the totality of factors determining the conditions of consortium functioning: sales volumes, prices, cost price, quality of goods or services, effective demand, competition, etc. The main way to estimate sales volumes and prices is the so-called analytical marketing, the essence of which consists in drawing up a simple model that expresses the mathematical dependence of expected sales volume on the factors that determine effective demand and prices of the offered product or service. Analytical marketing is most convenient for demonstration of ESM capabilities to reduce error [4]. When substantiating investment projects it is reasonable to consider the following main factors: the number of customers, customer income, consumption patterns, and competition.

Let's illustrate the effect of error reduction with the data given in Table 1.

Table 1. Demand factors and price.

Price, (rub)	Unit cost (rub)	Factors									
		Number of clients (quantity)		Average annual income (rub)		Share of expenditure on information technology		Share of service costs		Competitor's market share	
30	10	min	max	min	max	min	max	min	max	min	max
		10000	30000	5000	20000	0,1	0,35	0,35	0,75	0,50	1,00

In this case we have:

$$Fa^{max} = f_1^{max} * f_2^{max} * f_3^{max} f_4^{max} * (1 - f_5^{min})/C$$
$$= 30000 * 20000 * 0,35 * 0,75 * (1 - 0,5)/30 = 2625000$$

$$Fa^{min} = f_1^{min} * f_2^{min} * f_3^{min} f_4^{min} * (1 - f_5^{max})/C$$
$$= 10000 * 5000 * 0,1 * 0,3 * (1 - 1)/30 = 0$$

Uncertainty Interval:

$$\Delta Fa = Fa^{max} - Fa^{min} = 2625000 - 0 = 2625000$$

$$PL = Fa^{min} + (Fa^{max} - Fa^{min}) * \frac{1}{1+\sqrt{\frac{S}{C-S}}}$$
$$= 0 + (2625000 - 0) * \frac{1}{1+\sqrt{\frac{10}{30-10}}} = 1855875$$

Computational experiments on Decision are needed to obtain an estimate of ΔPL. These calculations show that $\Delta PL \sim 131390$. Thus, the error reduction is almost 20 times: $\frac{\Delta Fa}{\Delta PL} \approx \frac{2625000}{131390} = 19,9$. This error is 7% of the nominal, $\frac{\Delta PL}{PL} \approx \frac{131390}{1855855} * 100 = 7\%$, which for the situation under consideration is quite acceptable.

Having calculated the sales plan PL, we can determine the planned costs (r1),

$$\left.\begin{array}{l} r_1 = S * PL \\ r_2 = C * PL \\ r_3 = (C - S) * PL \end{array}\right\}.$$

revenues (r2) and profits (r3):

The error of the values r1, r2 and r3 calculated in this way, as well as the reliability P0 is the same 7%.

5   Evolutionary-Simulative Model of Consortium Activity

This model, when implemented in Decision, allows us to investigate the influence of different factors on financial flows, as well as scenarios unfolding over time, in a dialog mode. In addition, the model, which we consider below, allows for further improvement. In particular, a variety of additional sources of damage to the firm can in principle be taken into account when developing algorithms for calculating overestimation and underestimation costs. Analysts point to:

- Ineffective management of working capital and capital structure;
- insufficient control over the ratio of fixed and circulating assets, borrowed funds and own capital;
- lack of optimization of investments by factors of importance;
- lack of calendar planning of investments in order to ensure the rhythm of payments;
- the shortcomings in the organization of control over the structure of investments, correlation of parts of the financial portfolio in terms of riskiness and profitability;
- the methods of competition;
- price competition without regard to the consumer properties of goods;
- industrial espionage and the practice of lobbying the interests of the firm in governmental bodies;
- the influence of subjective evaluations on the quotation of securities.
- How appropriate to take into account these features should be decided by the developer of the model [14].

Factors: f1 - buyer group (category); f2 - buyer income of a particular category; f3 - cost (the proportion of income that buyers of a given category are willing to spend on a given product); f4 - cost (for a product subset); f5 - cost (for a particular product); f6 - propensity to buy (buyer's desire to purchase goods or save money); f7 - outside market (ability to market goods outside that market sector, such as under pre-arranged contracts); f8 - propensity to buy; f9 - inflation; f10 - strengthening of the ruble; f11 - quality; f12 - non-payments; f13 - out of market; f14 - competitors (market share occupied by competitors); f15 - average price of goods from a competitor; f16 - price of components (in unit of goods); f17 - supplier's price (wholesale unit price for merchants); f18 - material price (in unit of goods); f19 - advertising costs; f20 - number of customers covered by one ad; f21 - probability that a potential customer becomes a real customer.

Indicators: p1 - price of components (costs of the consortium to purchase the components included in the unit of product being produced); p2 - own investment (investment in producing the unit of product); p3 - transportation and storage costs (per unit of product); p4 - planned price (per unit of the product); p5 - average rate of deposit (the rate at which the consortium receives credit from the bank); p6 - loss of quality during storage; p7 - taxes in the cost of production; p8 - profit tax; p9 - fixed costs (management, etc.); p8 - tax on profit. Costs per unit of a good or service); p10 - investment (investments planned for the planned period in the expansion of production or sale of a given good or service); p11 - the share of return on investment; p12 - the share of profits on advertising; p13 - time; p14 - the share of investment in advertising; p15 - the share of investment in increasing the volume.

It is also possible to take into account the parameters of model adjustment, which allow in each case to take into account all or some factors and initial indicators.

Depending on the scope of the consortium, the specifics of the product, methods of sales, advertising, factors and baselines are present in different combinations and have different relationships. Therefore, the consortium model has a convenient interface that allows you to configure the model in a dialog mode.

6 Computational Experiments

Suppose that the EquilibriumFulRus. Indicators module is installed. To load the consortium model "Firm" from the menu, follow the dialog procedure:

Library of equilibrium models → Firm.

Then you will be asked to change the composition of factors and initial indicators. If you answer "Yes", a dialog box appears, which shows a menu for selecting the type of consortium activity. In order to be able to study the course of the processes in time it is necessary to select any of these options, taking into account the advertising.

Further the system asks some more qualifying questions (ways of distribution of advertising, whether there is a brand, etc.) and gives the possibility to enter the initial data of the conditional example if you wish. After that appears "Form 1" (Fig. 2), which shows a fragment with a list of factors, estimates of their marginal values, initial indicators and their values.

D6		× ✓ fx	500			
A	B	C	D	E		G
1	Joint mold 1. Direct settlement					FIRM
2		Initial data			Research	The results of the optimization
3		factors and	Dimension	Concern		plan
4	№	indicators			plan	Optimum
5	1 Group 1, min		pieces	100,00		overstatement/understatement
6	2 Group 1, max		pieces	500,00		reliability for increased
7	3 revenue 1, min		rub/m	1 240,00		reliability for understatement
8	4 revenue 1, max		rub/m	2 750,00		НОРМАТИВ
9	5 costs, min		%	4,00		Optimum
10	6 costs, max		%	10,00		overstatement/understatement
11	7 propensity to buy, min		%	85,00		reliability for increased
12	8 propensity to buy, max		%	87,00		reliability for understatement
13	9 market size, min		pieces	5,00		
14	10 market size, max		pieces	9,00		Calculated indicators
15	11 other competitors, min		%	45,00		name
16	12 other competitors, max		%	55,00		profi (plan)
17	13 average competitor price, min		rub	910,00		costs (plan)
18	14 average competitor price, max		rub	970,00		financial reserves
19	15 advertising costs, min		rub	5 300,00		advertising costs
20	16 advertising costs, max		rub	5 700,00		tax deductions (plan)
21	17 ad price, min		rub	330,00		average unit yield
22	18 ad price, max		rub	450,00		◇
23	19 number of customers, min		pieces	1 260,00		◇
24	20 number of customers, max		pieces	1 750,00		◇
25	21 the probability potential client, min		per.unit	0,01	◇	◇
26	22 the probability potential client, max		per.unit	0,03	◇	◇
27	23 the price of components		rub	120,00	◇	◇
28	24 own investments		rub	90,00	◇	◇

Fig. 2. Fragment of the Form 1 "Firm" with the initial data.

In the process of computational experiments, the dependences of calculated indicators and the main characteristics of ESM (optimum, expressing sales volume, Over/under, reliability) from the minimum or maximum value of any factor or any initial indicator, in particular from time, can be built. When constructing the dependence on time, you can also take into account the expected changes in the limiting values of the factors and the initial indicators.

In this case the initial indicator "Investments" reflects capital investments in production and sales of products, and the indicator "Equity investments" shows current assets, which are included in the cost of production. Investments are covered by profits over time. In addition, profits may be partly spent to cover running costs, particularly advertising.

One of the most important indicators of the safety of a consortium [14–16] is the change over time of its financial reserves and the change in the average risk-adjusted specific return. In order to construct these dependencies, a dialog procedure is performed, as a result of which the dependency graph shown in Fig. 3. The graph reflects the expected change in financial reserves due to the implementation of a particular IT product or service. And the graph is based on the interaction of all factors, including revenue and preferences of expected customers, price competition, advertising and its effectiveness, etc. The graph clearly shows that over the next 3 periods (months) there will be fluctuations in financial stocks, and then for another 6 periods there will be a steady and steep decline in stocks. Based on these data alone, it is obvious that supplying the market with this IT product is unreasonable .

For a more reasonable conclusion, you can perform a number of other calculations: consider a longer period, to see what will happen in the most likely expected changes, such as inflation, changes in income in the target group (the factor "Income"). It is possible that the situation could be reversed by a change in advertising costs. To see how these or those changes will affect the situation, it is necessary to make the desired changes in Form 1 "Firm" (Fig. 2) and get Fig. 3.

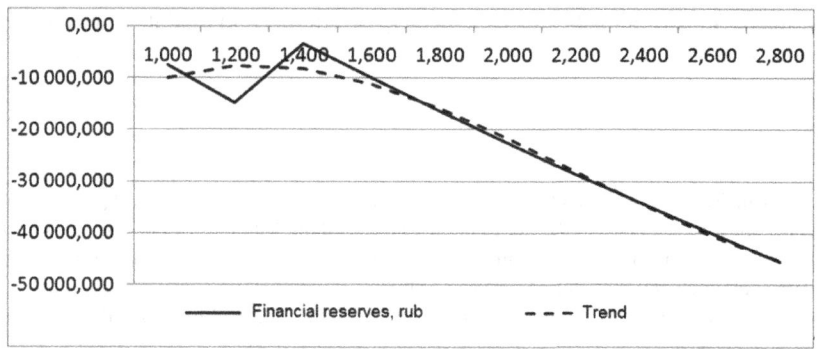

Fig. 3. Change in financial reserves.

With the help of the Firm model, the estimates of these indicators can be refined, and the indicators themselves can be built into the model, if necessary. In this case it is necessary to decide on the binding of these or those indicators to the firm as a whole and to the individual commodity. Let us consider the current liquidity ratio (CLR) as an example. It is calculated by the formula:

CLR = Current assets/Current liability.

Current assets and current liabilities in this formula refer to the consortium as a whole, and in the "Firm" model all indicators refer to a specific commodity. From a practical point of view, it is important for decision-making by consortium management to calculate the refined values of working capital, short-term credit and CTL precisely in relation to a particular product in order to make marketing sales plans, to determine the advisability of removing a product from production or, conversely, to build up. In order to characterize the consortium as a whole, it is possible to make calculations according to the "Firm" model for the main goods and on this basis to clarify the values of the indicators and to predict their values.

7 Conclusions

As a result of the study, it was found that within the framework of the created consortium, it is possible to effectively implement the business process of production of modern IT business activities of various enterprises and organizations of the Eurasian Economic Union. To justify investment in the activities of the consortium, an ESM was developed, which includes a set of interrelated simulation models to evaluate the business of the consortium to simulate the cash flows in the future.

The proposed ESM uses a minimax functional of the commercial risk objective (the risk of overestimation and the risk of underestimation of investments) as an optimization criterion to optimize the sales volume and prices of IT products and economic parameters of the consortium. It is possible to consider different scenarios of consortium functioning.

The rationale for using the method of assessing the cash flows of the consortium using ESM, which allows you to predict the expected cash flows after making investments in the activities of the consortium. The computational experiment on realization of some scenarios of consortium behavior with the help of Equilibrium module of Decision tool system is performed.

References

1. Akatkin, Yu.M., Karpov, O.E., Konyavsky, V.A., Yasinovskaya, E.D.: Digital economy: conceptual architecture of the digital industry ecosystem. Bus. Inform. 4(42), 17–26 (2017)
2. Lichtenstein, V.E., Ross, G.V.: Information Technologies in Business. Practicum. Textbook, Part 1. Finance and Statistics, Moscow (2008)
3. Lichtenstein, V.E., Pavlov, V.I.: Economic and Mathematical Modeling. Prior Textbook, Moscow (2001)
4. Avdiisky, V.I., Bezdenezhny, V.M., Lichtenstein, V.E., Ross, G.V.: Economic Justice and Security of Economic Agents. Finance and Statistics, Moscow (2016)
5. Melnik, M., Antipova, T.: Organizational aspects of digital economics management. In: Antipova, T. (eds.) Integrated Science in Digital Age. ICIS 2019. Lecture Notes in Networks and Systems, vol. 78, pp. 148–162. Springer, Cham (2020). https://doi.org/10.1007/978-3-030-22493-6_14
6. Yendovitsky, D.A.: Comprehensive Analysis and Control of Investment Activity. Finance and Statistics, Moscow (2001)
7. Lichtenstein, W.E., Ross, G.V.: Equilibrium Random Processes: Theory, Practice, Info Business. Finance and Statistics, Moscow (2015)
8. Ross, G.V.: Modeling of Production and Socio-Economic Systems using the Apparatus of Combinatorial Mathematics. MIR, Moscow (2001)
9. Bulygina, O.V., Emelyanov, A.A., Yashin, E.S., Ross, G.V.: Investments, innovations, import substitution: simulation modeling and fuzzy logic in risk management. Appl. Inform. 15(01), 85 (2020)

10. Rozhkova, N., Blinova, U., Rozhkova, D.: The concept of management accounting based on the information technologies application. In: Antipova, T., Rocha, Á. (eds.) Information Technology Science. MOSITS 2017. Advances in Intelligent Systems and Computing, vol. 724. Springer, Cham (2018).https://doi.org/10.1007/978-3-319-74980-8_8

11. Mingaleva, Z., Mirskikh, I.: Small innovative enterprise: the problems of protection of commercial confidential information and know-how. Middle East J. Sci. Res. **13**(SPLISSUE), 97–101 (2013)

12. Voskanyan, Y., Shikina, I., Kidalov, F., Davidov, D.: Medical care safety - problems and perspectives. In: Antipova, T. (eds.) Integrated Science in Digital Age. ICIS 2019. Lecture Notes in Networks and Systems, vol. 78. Springer, Cham (2020). https://doi.org/10.1007/978-3-030-22493-6_26.

13. Vukovic, N., Pobedinsky, V., Mityagin, S., Drozhzhin, A., Mingaleva, Z.: A study on green economy indicators and modeling: Russian context. Sustain. (Switz.) **11**(17), 4629 (2019)

14. Lichtenstein, V.Y., Ross, G.V., Los, V.P., Tyshuk, E.D.: Methodology for analysis of financial and economic security: management of capital flows (Article), pp. 58–59 (2018). Volume 7, Issue 4.36 Special Issue 36

15. Ross, G. Konyavsky, V. Multi-agent Systems as the Basis for Systemic Economic Modernization, Lecture Notes in Networks and Systems, vol. 136, pp. 177–187 (2021)

16. Fetyukhin, N.: Pre-project marketing research. https://www.cmsmagazine.ru/library/items/internet-marketing/nf_2011-06-08/

Investigating Social and Economic Components of Strategic Business Development Conditions in an Information Economy: A Case of Ukraine

Svitlana Tkalenko[1]([✉]) [iD], Zhanna Derii[2] [iD], Sergii Zakharin[3] [iD],
Maryna Zakharina[4] [iD], and Petro Viblyi[5] [iD]

[1] Department of European Economics and Business, Kyiv National Economic
University named after Vadym Hetman, pr. Peremogy 54/1, Kyiv 03057, Ukraine
sv.tkalenko@gmail.com
[2] Department of Theoretical and Applied Economics, National University «Chernihivska
Politekhnika»,
Shevchenko Street 95, Chernihiv 14000, Ukraine
jannet_d@ukr.net
[3] Department of Tourism,
Kyiv National University of Culture and Arts, Konovaltsa, 36, Kyiv 01133, Ukraine
z0679330105@gmail.com
[4] Department of Social Work, National University «Chernihivska Politekhnika», Shevchenko
Street 95, Chernihiv 14000, Ukraine
zakharina_m@ukr.net
[5] Department of Finance, National University «Lviv Polytechnic», Mytropolyta Andreia Street,
Building 4, Lviv 79016, Ukraine
lider577@gmail.com

Abstract. The present time is characterized by constant economic, social and cultural, political changes that cover the whole world. In the conditions of globalization there are cardinal shifts in the sphere of international relations and international business, characterized by transformational, modernizing changes, and have a direct impact on the development of economies around the world, including Ukraine. Globalization has also contributed to the formation of a new era - the era of information society. By the transition to the information society, the level of social and economic development is related to the access to information resources. A new economic sector is being created, where intellectual, social and investment capital place an increased role. The aim of the study is to assess the relationship between FDI attraction in Ukraine and its main macroeconomic indicators in the context of socio-economic development of business and globalization. The novelty is the substantiation of the directions of attraction of investments in the economy of Ukraine. The study is based on the analyzed macroeconomic indicators for the period 1998–2018, which made it possible to select those that have the greatest impact on FDI. The method of modeling economic processes is used to identify the effect of global influence on attracting foreign investment as a socio-economic component of strategic business development in the economy of Ukraine. The results of the simulation showed: achieving a high level of dynamics of FDI attraction in Ukraine in recent years; the existence of a strong link between FDI and foreign trade.

T. Antipova (Ed.): ICADS 2021, AISC 1352, pp. 390–402, 2021.
https://doi.org/10.1007/978-3-030-71782-7_34

Keyword: Globalization · Information society · International business · Information technologies · Strategy · Social capital · Intellectual capital · Foreign direct investment · Foreign trade

1 Introduction

In the context of globalization, the basis of strategic business development is no longer an extensive industrial method of mass production of goods and services and spontaneous self-regulation, but accumulated information resources, advanced information and communication technologies, innovative and synergistic way of producing new scientific knowledge and information products and services. The modern era is characterized by constant transformation and transition of the society from the outdated entropy-market system to a new highly organized management system [1]. This is a system in which knowledge along with other types of economic resources is the source of growth. A dominant factor is the process of the formation and use of intellectual capital, ensuring the creation of high-tech and scientific products, provision of highly qualified services, which leads to the increased competitiveness of business and the economy as a whole.

In such conditions of today, the formation of information society takes place, which reflects the main challenges of a rapidly changing world. Social, economic theory has revised its own preconditions, categorical structures, changes in the status and objectives of theoretical knowledge [2–4]. Thus, Wallerstein notes that "the society of the first half of the XXI century in its complexity, instability and at the same time openness far exceeds everything we saw in the XX century. The modern world system as a historical system has entered a stage of final crisis, and is unlikely to exist in fifty years" [4].

It should be noted that the modern period has a global character and affects basic principles, foundations of business organization, economy of most countries. The epoch of the consciously planned, programmed, purposeful sustainable development is coming. The current crisis is a crisis of the industrial-market paradigm, ideology, concept, model of the world economy development, which should be replaced by a new innovative-synergistic information paradigm, a model of the social and economic development, formation of knowledge economy [5]. In such a model of the development, the main role is played by a highly qualified specialist, a carrier of intellectual capital and creative and innovative abilities, and who acts as the main driving force in ensuring the business competitiveness.

In the era of information economy, as noted by leading experts, it is worth talking about the emergence of a new type of competition - hypercompetition or global innovation competition [6]. That is, the controlled hypercompetitive development of global, regional markets in the conditions of advanced dominant innovations leads to the integration into global structures and includes new advanced methods of the programmed, controlled influence on goals, motives, interests, needs and economic behavior of people in order to get benefits and effects.

In such conditions, nation-states compete more and more diversely and fiercely with each other for new scientific knowledge, the right to control and regulate resources, information and financial flows, place and share in world markets, ownership of intellectual and information capital, the right to control and manage economic processes, that determines their leadership and high competitiveness in world markets. However, at the same time, new global (supranational) institutions and centers of management, coordination and control of the national, regional and world economy as a whole are being formed. Methods, mechanisms and forms of economic regulation and competition at the global, regional, national, sectorial and local levels are changing qualitatively, becoming more flexible and active.

Undoubtedly, the leading place in the modern information economy is occupied by huge companies - multinational corporations. It is these corporations that offer innovative products, services, maintenance and management that are characterized by global innovation.

2 Literature Review

Among the publications related to the issues of this article, the following research vectors should be noted:

1. Research of theoretical and methodological aspects of the information economy formation.

New approaches to the characterization of the globalized world, in which modern business is developing, are laid by Wallerstein, who defined globalization as part of objective reality. According to his conception, capitalism immediately began developing as a holistic system of world relations, the individual elements of which were national economies. Globalization, according to Wallerstein, also looks like a "misleading concept". Therefore, speaking about the discourse of globalization, the scientist argues: "discourse is in fact a gigantic misunderstanding of current realities - a deception imposed on us by influential groups and, worse, often imposed on us by ourselves in despair. This discourse forces us not to notice real problems before us and not to understand the historical crisis in which we find ourselves".

That is, Wallerstein focuses on risks and problems of the modern world, and challenges that will face all participants in world economic relations [5].

Theoretical substantiation of the formation and development of information economy was originally accumulated in the theory of post-industrialism, in which the analysis of knowledge industry and the role of information in modern economic relations occupies a prominent place.

One of the founders of the concept of post-industrial society is Bell, who, based on the analysis of economic and social processes in the United States in the postwar period, formulated the basic elements of the post-industrial theory. The author places considerable emphasis on the formation of service economy, origin of scientific knowledge, positioning of post-industrial society in the general model of social progress [6].

D. Bell identifies priority services in information society, such as: transition from post-industrial to service society; codified theoretical knowledge for the implementation of technological innovations; transformation of new "intelligent technology" into a key tool of the system analysis and the decision-making theory.

In addition, he argues that now the formation of a new social system based on telecommunications gets a significant importance for economic and social life, for the methods of knowledge production and for the nature of human labor. Moreover, in fact there is a parallel formation of post-industrial society with a revolution in the organization and processing of information and knowledge [7]. This analysis is gradually separated into a new scientific discipline - the theory of information economy.

According to Toffler, post-industrial civilization has become a product of new technologies, and the third wave of civilization marks the emergence of "super-industrial society" [8]. In this order, all previous rules and principles of economic models are leveled and the opposite begins. Production also has a different nature, it is possible to work not in offices, but at home, a self-service economy emerges when production and consumption are concentrated in one place.

Among the Ukrainian scientists who study the problems of information economy, it is worth noting Shkarlet, who in particular provided the author's interpretation of information economy, the essence of which is to consider it as a type of economic system, in which information plays a key role in the development of major areas and industries of national production, is an integral part of the implementation of the processes of production, distribution, exchange and consumption [9].

2. Identification of the essential content and levels of the information economy manifestation in terms of the impact on business.

The leading researcher of the modern information society is Castells, who makes an attempt to comprehend a fairly rapid and uneven, in pace and space, the development of communication capabilities of society, assessing the widespread use of the Internet. The new social form - network society - acquires a global form, differing in specific manifestations, and has significant differences in its impact on people's lives depending on the history, culture and social institutions [10–12].

Castells critically analyzes the Internet as a major factor in globalization. Reducing the number of traditional workers increases the flexibility of economy, which methods of the activity organization spatially cover the entire labor market, forming a new social structure, which Castells calls a network society. The modern economy is characterized by a significant increase in productivity, which decisive growth factors are investment in information technology and high productivity in the computer industry. The main asset of entrepreneurs is the mind and experience in terms of the minimum necessary "infrastructure" of production: a computer, telephone, workplace, which can be outside the office.

One of the representatives of the information economy platform is Joseph Stiglitz, who formulated theoretical foundations of its development and, in addition, practically tested his theory.

Joseph Stiglitz with George Akerlof and Michael Spence were awarded the Nobel Prize for developing the market theory with asymmetric information. The American scientist was able to prove that the understanding of asymmetric information allows a better understanding of many economic phenomena. The information technique of asymmetry is called screening. One of Stiglitz's best-known studies is screening, a method used by one economic agent to extract personal information from another [13–17].

The asymmetry of information not only leads to transaction costs for data retrieval, but is a structural problem that can change markets for the worse and even lead to their decline. In his work, the scientist analyzed the impact of asymmetric information on markets and different levels of economic development.

In modern society, information technology is becoming the most important factor that has a growing impact on social and legal relations. However, if in the Western scientific literature, the last decades have been marked by increased attention to legal problems of information society, then in the domestic science there are not enough holistic concepts that focus on information issues. Works of Vitlinsky and Katunina are very prominent, within which, a scheme of signs and markers of digital economy is provided, among them they named the following:

- conditionality of economic processes by rapidly implemented systemic changes in the essence, format and a number of aspects of the behavior of the economic environment;
- development of forecasting tools based on intellectual and mathematical apparatus and the latest model and information technologies aimed at deepening the degree of the events predictability;
- significant transformations taking place in the real sector of the economy and practice of the managing complex of social and economic systems [18].

Bairachna analyzes modern specifics of the information society formation and reveals the problem field of information society. According to the scientist in the information society, information is the main tangible and intangible value and the most important factor in the development of the individual and society. The second most important value is knowledge as the ability to obtain reliable, objective and comprehensive information and the ability to interpret it critically. According to her, the legal field of information society, which is currently being formed, is the most proportional to the problems of today; it contributes to the humanization of information society and its daily functioning [19].

For a modern understanding of science, it is important to consider that the leading trend is the cooperation of research between different countries, which contributes to the development of Internet technologies; different states, individual scientists, businesspersons, business entities join the network.

Despite extensive research on theoretical and applied developments in international business development, general problems of combining the economic component of strategic business development and socially oriented business in the information economy remain open, which determines the relevance and direction of further research.

3   Materials and Methods

Taking into account the peculiarities of TNC business development, we study the movement of foreign investment in Ukraine. The method of modeling economic processes to identify the effect of global influence on attracting foreign investment using the software product E-Views was tested. This study is based on the analyzed economic indicators for the period 1998–2018, which allowed the authors to choose those of them that have the greatest impact on the inflow of foreign investment into the economy of Ukraine, which in particular include: GDP (shows the scale of economic activity), ES (shows the competitiveness of domestic services in the world market), IS (shows the volume of foreign services), LM – number of employed population (reflects human resources, potential of Ukraine. Statistical data for analysis are taken for the period 1998–2018 on the basis of data from the State Statistics Service of Ukraine [20].

The authors built a multifactor regression model, namely: the hypothesis on the influence of selected economic factors on foreign direct investment in the current conditions of the information economy development.

A general view of the FDI model illustrating the relationship of selected variables is described by the following equation:

$$FDI = f(GDP, ES, IS, LM)$$

Where, FDI – direct foreign investments, US $ million;
GDP – GDP of Ukraine, US $ million;
ES – export of services, US $ million;
IS – import of services, US $ million,
LM – employed population, thousands of people;

In the process of analyzing the model, the authors constructed a correlation matrix that explains the relationship between variables. Correlations between the variables of this model have been identified. The model is adequate if it reflects current trends in economic processes. The constructed model was tested for autocorrelation and tests for heterosedacticity, i.e. classical assumptions and testing of the predictive quality of the regression model. The results of the author's model proved its adequacy and high quality.

4 Results

4.1 Strategic Business Development: Place of Intellectual and Social Capital

Global innovation is, first of all, the offer of advanced innovative highly competitive goods and services with qualitatively new, and in most cases, universal functions and consumer properties which have steady demand in world markets and will receive the status of global novelties, brands. Changes in the global market also affect changes within organizations, which must adapt as quickly as possible to new conditions in order to thrive in the market. One way they can follow global trends is to make changes to their business strategy [21].

It is worth noting that large corporations do not focus on the single global market, but in regional blocs. Due to government actions and cultural differences, the world is now divided into three blocks and is developing in the EU-US-Japan triad. In order to achieve success and a certain level of competitiveness, most top managers of large enterprises, including industries such as chemicals or the automotive business, today focus on national and regional strategies. Stagnation of a global strategy can only be justified in a few areas, including home appliances; in others, such as the automotive industry, regional services are required.

Proponents of globalist theories argue that goods are produced and distributed evenly around the world. However, we'd like to note that goods produced in the United States, the EU or Japan, respectively, are distributed in the North American Free Trade Area (NAFTA), Europe and a small number of countries in Asia and Oceania (the main economic centers are the United States, the EU and Japan, respectively). As proof, here are the facts: more than 90% of cars produced in Europe are sold in the same area; about 90% of paint, steel, electronic devices, energy, transport services are produced and used according to the regional principle of the triangle [22].

In general, a significant proportion of TNCs sell an average of 80% of their share of the triad in the domestic market. Therefore, the world of international business is not global, but regional; and only a small proportion of corporations operate in all regions of the triad. Thus, world trade is developing within the triad, as evidenced by indicators of exports and imports. Only a small proportion of goods and services end up on another continent. Centers of this triad do not rely on their partners from other parts of the world, but rely mainly on neighboring countries in their region.

In addition, today not only the economic component is important for business competitiveness, but also social responsibility. Socially responsible business is preferred by all stakeholders. For example, one of the world's leading companies with a multibillion-dollar capitalization is Google's search and service system, which can be described as global, innovative and competitive. It is the leading, global TNCs, such as IBM, Coca-Cola, General Motors and others that provide innovation, high dynamism, competitiveness, leadership in global markets and at the same time are socially responsible.

For host economies such as Ukraine, TNCs have a number of advantages, in particular: growth of foreign investment inflows, growth of export-import operations, in particular in the field of services, growth of investments in intellectual assets, in the social sphere, etc. The dynamics of FDI attraction is the basis of long-term development of the economy, the basis of the reproduction process, which will contribute to the expansion of high-tech forms of reproduction of fixed capital and the accumulation of highly intellectual capital [23].

Through the use of intellectual and social capital, creative abilities of specialists and company management, ICT, the position of business in world markets increases. Intellectual and social capital is becoming one of the important properties of competition.

We consider intellectual capital as intellectual abilities of people used by corporations in the process of their intellectual work. The concept of intellectual capital is introduced into scientific circulation by the Canadian-American economist of the twentieth century Galbraith, who uses this term in the sense of intellectual activity [24]. Stewart was the first to study the nature of intellectual capital: "intellectual capital is intellectual material that includes knowledge, experience, information, intellectual property and participates in the creation of values" [25].

Social capital, which affects the competitiveness of corporations, determines the access to resources and information through appropriate channels, does not have a single clear definition due to its specific properties. Thus, today within the transition process to a new economy, scientific and technological revolution and formation of information society, intellectual and social capital is the most important competitive advantage of business entities, and above all large corporations, as well as the most important resource of the country, able to ensure its future.

4.2 Econometric Modeling of the Factors Influencing the FDI Growth

The hypothesis is that in the conditions of information economy and socio-economic orientation of business at the present stage, TNC investments in Ukraine will contribute to GDP growth, increase in exports and imports of services (as a leading position in information economy), employment growth. We will analyze the data under the test hypothesis. Let's consider the function FDI = f (GDP, ES, IS, LM) and check it with the help of the econometric model.

The correlation matrix that explains the relationship between our chosen variables (GDP, ES, IS, LM) and the results of multifactor regression are given in Tables 1 and 2, respectively.

Table 1. Correlation

	FDI	GDP	ES	IS	LM
FDI	1.000000	0.193852	0.965015	0.952300	−0.301381
GDP	0.193852	1.000000	0.232229	0.187561	−0.212306
ES	0.965015	0.232229	1.000000	0.977364	−0.323867
IS	0.952300	0.187561	0.977364	1.000000	−0.331188
LM	−0.301381	−0.212306	−0.323867	−0.331188	1.000000

Source: authors' development
Table 1 presents a correlation matrix that explains the relationship between selected variables and shows their influence on foreign direct investments and incurred costs for innovations. Matrix constructed by us confirms the success of the model.

There is a strong positive correlation between FDI and ES, IS (correlation coefficient 96.5 and 95.2%, respectively), a slight positive correlation between FDI and GDP, and a small negative correlation between FDI and LM. With a negative correlation, a decrease in the value of a variable leads to an increase in another variable. With a positive correlation, an increase in the value of a variable leads to an increase in another variable. We can also observe a small correlation between all variables; an acceptable result of correlation between variables actually confirms that there is no multicollinearity (i.e. close interdependence between variables); therefore, our model is successful.

Regarding the quality requirements of the factors, we will apply the accepted average of 5–10% of the significance level. According to F-statistics, all the coefficients of the regression equation are not equal to 0 at the same time. In our equation, only exports of services are less than 5%; therefore, the most significant in this case is ES = 2.26%.

As a rule, there are no strict requirements for the constant, but in our case, it is also significant (0.04%). The variables GDP, IS, LM are insignificant, since their prob. Are more than 5–10%.

R^2 shows the extent to which our chosen criteria and their number explain changes in FDI, i.e. our selected criteria explain 93.4% of changes in FDI growth. The correlation coefficient is 0.92, which indicates that there is a strong enough connection. The probability of accepting the null hypothesis of F-statistic $= 0.00$, i.e. close to zero, which confirms the need to take an alternative hypothesis, which speaks of the significance of the equation as a whole.

Table 2. The Results of Multi-factor Regression of FDI

Dependent Variable: FDI Method: Least Squares
Sample: 1998 2018 Included observations: 21

Variable	Coefficient	Std. Error	t-Statistic	Prob.
GDP	−0.004079	0.012456	−0.327482	0.7475
ES	3.852375	1.526452	2.523744	0.0226
IS	1.584201	2.621942	0.604209	0.5542
LM	0.005447	0.031089	0.175222	0.8631
C	−14422.12	4086.610	−3.529116	0.0028
R-squared	0.933784	Mean dependent var		25508.43
Adjusted R-squared	0.917231	S.D. dependent var		18724.95
S.E. of regression	5387.109	Akaike info criterion		20.22566
Sum squared resid	4.64E+08	Schwarz criterion		20.47436
Log likelihood	−207.3695	Hannan-Quinn criter		20.27964
F-statistic	56.40873	Durbin-Watson stat		1.143300
Prob(F-statistic)	0.000000			

Source: authors' development
Table 2 shows the results of multi-factor regression using the least squares method and different coefficients, which as a whole prove the significance of the equation.

Information criteria - Akaike, Schwarz, Hannan-Quinn, Durbin-Watson are given in Table 2. We check the equation for the presence of autocorrelation using the Durbin-Watson test. This Durbin-Watson criterion d $= 1.14$. Using Durbin-Watson statistics in the table, we'll determine significant points d_L and d_U. For the number of observations 21 and 4 of variables at the significance level $\alpha = 5\%$ $d_L = 0,72$ and $d_U = 1,55$; at the significance level 1% $d_L = 0,93$ and $d_U = 1,81$. The interval between it is a zone of uncertainty: it means that we cannot unambiguously say whether there is an autocorrelation or not. Thus, the presence of higher-order autocorrelation should be checked using the Breusch-Godfrey test (Table 3).

Breusch-Godfrey test results: when lag $= 2$, the probability of accepting the null hypothesis is more than 5%, which means that this lag is not significant; when lag

Table 3. Autocorrelation test results

Correlation LM Test	Prob		Prob	
	RESID(−2)	RESID(−1)	F(2,14)	F(1,15)
lag 2	0,2195	0,2519	0,1099	-
lag 1	-	0,0854	-	0,0854

Source: authors' development
Table 3 shows the presence or absence of autocorrelation.

= 1, the probability of accepting the null hypothesis is 0.08, which is higher at 5% significance level, and means that we can accept the null hypothesis regarding the absence of autocorrelation, i.e. no autocorrelation. Probability of accepting the null hypothesis is 17.3%, i.e. we claim that in our model there is no autocorrelation of random deviations.

The following tests – are to check heteroskedacticity. Among these, we use the test White, Glejser, Breusch-Pagan-Godfrey, Harvey, ARCH to check the heteroskedasticity (i.e., non-constant dispersion) of random errors of the linear regression model. The test results are shown in Table 4.

Table 4. Heteroskedasticity test results

Test results	F-statistic	Obs*R-squared	Scaled explained SS	Prob. F(14,6)
White	2.214700	17.59513	13.93959	0.1678
Glejser	1.604601	6.012313	5.098407	0.2215
Breusch-Pagan-Godfrey	1.079095	4.461620	3.534681	0.3995
Harvey	2.146780	7.334308	4.875642	0.1220
ARCH	1.085220	1.137236	-	0.3113

Source: authors' development
Table 4 shows the presence or absence of heteroskedasticity.

Let's check for heteroskedasticity using the White test, which is universal. By the value of accepting the null hypothesis, all variables are not statistically significant, because the probability of accepting the null hypothesis. In addition, prob. F = 16.8%, which indicates the absence of heteroskedasticity.

Next check for heteroskedasticity is the case using the Glejser test for our variables. All variable models were also tested separately by the test. Not all variables are statistically significant, as the probability of accepting the null hypothesis is more than 5%, i.e. it can be accepted. Presence of regression as a whole is also high, i.e. in our model there is no heteroskedasticity, that means, the model residues do not depend on the selected variables.

Another test for the presence of heteroskedasticity is the Breusch-Pagan-Godfrey test. The probability of accepting the null hypothesis is 0.3995, which is more than 5%

and confirms the absence of heteroskedasticity. In all tests, the null hypothesis is the absence of heteroskedasticity, the alternative presence.

The next step of checking the model for quality is RESET Test, its coefficient is more than 10%, which testified to the sufficient quality of the proposed model. We check the model for explanatory ability, because the constructed model should accurately reflect the FDI growth using the available independent variables (Fig. 1).

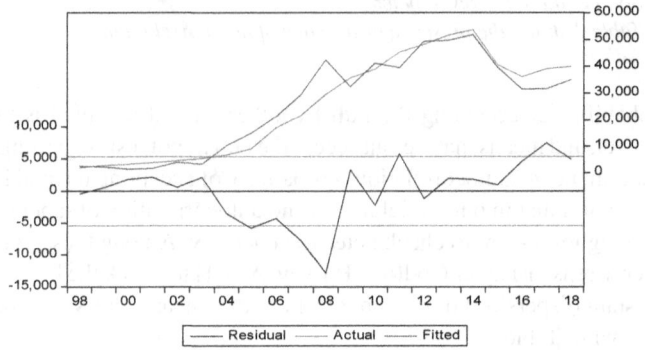

Fig.1. Explanatory ability of the model. *Source: author's development.*

As can be seen from the graph, the simulated Fitted values fairly accurately reflect Actual values, so by this criterion, the model is completely acceptable. Thus, our Eq. (1) is statistically significant, with a high coefficient of determination. The general view of the multi-regression model of the dependence of FDI growth on independent variables is as follows:

Substituted Coefficients:

$$FDI = -0.00407912787279 \cdot GDP + 3.85237526917 \cdot ES + 1.5842014639 \cdot IS + 0.00544746124051 \cdot LM - 14422.1226435 \tag{1}$$

5   Conclusion

The current stage of development of the world economy is characterized by constant changes, which occur not only under the influence of globalization, but also Covid-19. The basis of strategic business development is information resources, ICT, innovation resources and investment capital. Covid-19 has made adjustments to the business process, which today suffers losses. At the same time, intellectual and investment capital will contribute to the further development of the business and its entry to a new level of organization and profitability. For the next two or three years the forecasts for the development of world business remain disappointing, with the exception of the pharmaceutical, logistics and IT businesses.

Global challenges facing the world, TNCs will lead to a new round of economic and business development.

Social and economic component of the strategic business development in information economy contributes to economic growth. In particular, for Ukraine as a result of regression it was found that the growth of FDI by 1% in the formation and development of information economy is most positively affected by exports and imports of services (385 and 158%, respectively); with an increase in GDP by 1%, the growth of FDI is almost unchanged; the level of the employed workforce also has almost no effect on the FDI growth.

In addition, the social and economic component of the business development contributes to the environment protection, infrastructure development, and not only of developed countries but also other regions and countries where the world's leading TNCs develop their business. In turn, informatization destroys national and state frameworks and leads to the increased internationalization. In such circumstances, corporations can use more tools in organizing their business and get additional opportunities, as well as gain more free access to any information.

References

1. Dyatlov, S.: Innovative and synergistic dominant of anti-crisis policy in the 21st century. The state and the market: mechanisms and methods of regulation in the conditions of transition to innovative development. SPb: Asterion, T.2. 374 p. (2010). [in Russian]
2. Castells, M.: The Rise of the Network Society. 2nd edn. 625 p. (2010)
3. Beck, U.: What is Globalization? 182 p. Polity Press, Cambridge (2009)
4. Wallerstein, I.: The end of the familiar world. Sociology of the XXI century, 368 p. Moscow (2003). [in Russian]
5. Wallerstein, I.: Globalization or the Age of Transition? A Long-Term View of the Trajectory of the World System (2000). https://www.iwallerstein.com/wp-content/uploads/docs/TRAJWS 1.PDF
6. Bell, D.: The Coming of Post-Industrial Society: A Venture in Social Forecasting, 507 p. Basic Books, New York (1999)
7. Bell, D.: Social framework of the information society. New technocratic wave in the West, pp. 330–342. Progress, Moscow (1986). [in Russian]
8. Toffler, A.: The Third Wave, 781 p. (2004). [in Russian]
9. Sergiy, S., Maxim, D.: Identification of the essence of information economy. Problems and prospects for economy and management, vol. 1, no. 9, pp. 99–105 (2017). [in Ukrainian]
10. Castells, M.: The Internet Galaxy: Reflections on the Internet, Business, and Society, 292 p. November 2001
11. Castells, M.: The Information Age: Economy, Society and Culture, vol. 3, January 2000
12. Castells, M., Himanen, P.: The Information Society and the Welfare State: The Finnish Model, 197 p. October 2011
13. Stiglitz, J.E., Hoff, K.: Equilibrium fictions: a cognitive approach to societal rigidity. Am. Econ. Rev. **100**(2), 141–146 (2010). https://www-wds.worldbank.org/servlet/WDSContentSe rver/WDSP/IB/2010/02/26/000158349_20100226083837/Rendered/PDF/WPS5219.pdf
14. Stiglitz, J.E.: Information and the Change in the Paradigm in Economics. Noble Prize Lecture (2001). https://www.nobelprize.org/nobel_prizes/economic-sciences/laureates/2001/stiglitz-lecture.pdf
15. Arnott, R., Greenwald, B., Stiglitz, J.E.: Information and economic efficiency. Inf. Econ. Pol. **6**(1), 77–82 (1994). https://www.nber.org/papers/w4533.pdf

16. Akerlof, G.A., Kranton, R.E.: Identity and the economics of organizations. J. Econ. Perspect. **19**(1), 9–32 (2005)
17. Akerlof, G.A.: The market for «Lemons»: quality uncertainty and the market mechanism. Q. J. Econ. **84**(3), pp. 488–500 (1970). Published by: The MIT Press Stable. https://www.jstor. org/stable/1879431
18. Vitlinsky, V.V., Katunina, O.S.: Methodological aspects of modeling the development and viability of digital economy systems and counterparties. Probl. Econ. **1**(35), 333–341 (2018). [in Ukrainian]
19. Bairachna, L.: The problem field of the information society. Inf. Law **2**(17), 77–84 (2016). [in Ukrainian]
20. Website State Statistics Committee of Ukraine. www.ukrstat.gov.ua/
21. Stjepić, A.M., Ivančić, L., Vugec, D.S.: Mastering digital transformation through business process management: investigating alignments, goals, orchestration, and roles. J. Entrep. Manage. Innov. (JEMI) **16**(1), 41–74 (2019). https://doi.org/10.7341/20201612
22. Rugman, A.M., Alain, V.: A perspective on regional and global strategies of multinational enterprise. J. Int. Bus. Stud. **35**(1), 3–18 (2004)
23. Tkalenko, S., Sukurova, N., Honcharova, A.: Determinants of the Foreign Direct Investments in Terms of Digital Transformation of the Ukrainian Economy. DSIC 2019. Advances in Intelligent Systems and Computing, vol. 1114, pp. 148–164. Springer, Cham (2019). https:// link.springer.com/chapter, https://doi.org/10.1007/978-3-030-37737-3_14
24. Galbraith, J.K.: The New Industrial State, 518 p. Princeton University Press (2007)
25. Stewart, T.A.: Intellectual Capital. The New Wealth of Organizations. Doubleday, New York (1997)

Right to Be Forgotten in the Age of Machine Learning

Quang-Vinh Dang(⊠)

Industrial University of Ho Chi Minh City, Ho Chi Minh City, Vietnam
dangquangvinh@iuh.edu.vn

Abstract. The right to be forgotten (RtbF) is considered as one of the fundamental human rights in many legal systems. However, given the popularity of the computer systems in our daily life, and particularly the rapid development of the machine learning techniques, the RtbF need to be considered again. In this study we review the definitions of RtbF in several major legal documents and the application of the right in practice. We then discuss the differential privacy as a framework to support the RtbF.

Keywords: Ethics of AI · Law of machine learning · Privacy preserving machine learning

1 Introduction

The right to be forgotten (RtbF) is defined by the European Commission as "the right of individuals to have their data no longer processed and deleted when they are no longer needed for legitimate purposes" [11]. RtbF has its deep root in the European legal system. For instance, the French civil law has defined "le droit à l'oubli" (*right of oblivion* in English) [18] in 2010.

Since the beginning when the RtbF to be introduced, there exists a debate about the conflict of the RtbF and the freedom of expression. Many European countries support the RtbF from the Constitution [33], except for some countries like United Kingdom or Ireland. It is understandable as these countries use the common-law system rather than the civil laws as the continental Europe. However, we still can find the support to the right in the expression and interpretation of the supreme courts of these countries [21]. To be note, the right to be forgotten has been declared since the French revolution in 1789 and The United States Declaration of Independence [16].

In practice, after the decision of Spanish law in 2010 in the case Google vs Spain [18], there are a lot of requests on URL removal submitted to Google. According to a report [1], Google denies 75% of the requests. The company mostly approved requests concern to privacy.

While the RtbF has been defined, even not at 100% of clear, in the legal documents [15], the recent development of machine learning techniques and the usage of these techniques in our daily life have raised a lot of new questions regarding the RtbF. The problem has been addressed firstly by the paper of [7] (Figs. 1, 2 and 3).

© The Author(s), under exclusive license to Springer Nature Switzerland AG 2021
T. Antipova (Ed.): ICADS 2021, AISC 1352, pp. 403–411, 2021.
https://doi.org/10.1007/978-3-030-71782-7_35

No	Country	% of total URLs sent	# of URLs sent
#1	Germany	28.1%	34 474
#2	UK	23.6%	28 918
#3	France	14.5%	17 741
#4	Netherlands	7.3%	8 918
#5	Belgium	3.7%	4 533
#6	Spain	2.2%	2 653
#7	Romania	2.0%	2 497
#8	Sweden	1.8%	2 165
#9	Austria	1.6%	1 999
#10	Switzerland	1.6%	1 919

Fig. 1. Number of link removal requests approved by Google [1]

Type of refusal	% of URLs
Concerns your professional activity	29.7%
You are at the origin of this content	19.9%
The information is about another soon	9.4%
No name on page	8.4%
Your profile on a social network	7.6%
Topical and in the public's interest	7.4%
Relevant, topical and in the public's interest	6.3%
Does not refer to a physical person	5.8%
You are a public personality	3.5%
Government data	1.6%
Others	0.5%
Total	100%

Fig. 2. The reasons for Google to remove links due to RtbF [1]

On the other hand, the work in differential privacy [10] shed a new light to the problem. Dwork does not introduce a single problem to deal with privacy protection, but she defines a completely new framework to understand the privacy in a generic data mining system.

As Dwork does not describe a particular technique to guarantee the RtbF, the recent studies in *machine unlearning* [6–8, 30] tell us an efficient way to implement the RtbF in practice.

In this paper, we study the RtbF and its implementation in the era of machine learning.

Type of requests	% of URLs
Invasion of privacy	62.1%
Damage to reputation	9.6%
Damange to image	4.2%
Legal proceedings	3.0%
Identity theft	2.9%
Violation of the presumption of innocence	1.3%
Deceased persons	1.1%
Homonyms	0.8%
Other	14.9%
Total	100%

Fig. 3. The reasons for Google to remove links due to RtbF [1]

2 The Definition of the RtbF

2.1 In the Law

Memory includes two sides of a coin: remembering and forgetting [25]. While the right to be remembered are somehow clear: for instance, an employee usually requires their employer to "remember" the payday, the right to be forgotten indeed raises a lot of controversy. The main concern is, while "remembering" is easily to verify, how can we know if someone actually forgot something. In fact, trying to forget means trying to remember.

The RtbF has the origin from the idea that, after a certain amount of time, any criminal records should be vested [34]. It is to protect not only the person who perform the criminal acts, but also the victims and the society. There is no good to repeat everyday about the evil actions have been made, say fifty years ago. Furthermore, the victims should be released from the act to come back to the normal life. For instance, a victim of a sexual attack should require everyone to "forget" the attack so they can move forward.

Probably by anticipating the controversy of the definition, European countries have limited the RtbF to the right of erasing personal data [35]. It is understandable because the requirement is easier to define and verify.

Similar requirements of the RtbF can be found in Argentina, USA or Asian countries [35]. For instance, Section 1798.105 of the CCPA (The California Consumer Privacy Act, 2018) states, "A consumer shall have the right to request that a business delete any personal information about the consumer which the business has collected from the consumer", and that "A business that receives a verifiable request from a consumer… Shall delete the consumer's personal information from its records." [13]. However, it is worthy to note that in general, the US law does not favor the privacy but the freedom of expression, particularly for the media [18, 31].

2.2   In the Era of Machine Learning

In the digital age, the RtbF addresses a very core problem of the technology today: it is almost impossible to fully erase any piece of data [28]. There are several backup systems such as archive.org, as displayed in Fig. 4, stored almost everything has been published on the Internet. Hence, even the website owner removes some information from their own website, the information can be retrieved somewhere else.

Fig. 4. Websites such as archive.org allow the Internet users to surf through the history of the Internet.

European commission defined the core fundamental of the RtbF as "If an individual no longer wants his personal data to be processed or stored by a data controller, and if there is no legitimate reason for keeping it, the data should be removed from their system" [27]. However, several questions needed to be answered:

- How to define the removal action? Does the data keeper need to remove completely the data or un-linking the data is enough?
  - In many situations, the record of the data and the actions on the data needed to be kept for auditing purpose as required by law. How can the data keeper satisfy both requirements?
  - We would like to note that the term "erasure" is used throughout the General Data Protection Regulation (GDPR) but has not been explained [31].
  - Furthermore, most of the legal systems currently treat human and computer alike, hence require the same kind of forgotten applied to both of them [31]. However, it is not clear how "minds" work for both human and computers.
- How to define the personal data?
  - It is common that a piece of data involving more than one person. While a person want this information to be removed, others might not agree.
  - What is the boundary between personal and public data? For instance, a transaction on Bitcoin network should be visible to public, but a transaction between A and B is a private business [12].
- How to actually remove data in machine learning?
  - In machine learning [29], the information of a single data instance is not only kept in the raw format, but rather to be learnt and embedded in the machine learning models. Here, the information collected from multiple data points are mixed into a single piece of compressed representation. How can we actually remove a single point of data from the established knowledge. It is similar to ask a person to forget a single lesson from the entire education background.

- While there exists some pioneering work to understand the impact of a single data point to the model [19], it is still an open question to fully interpret how to undo the learning process [6]. Nevertheless, the work of understanding the contribution of each single data point is a very important task. The task is visualized in Fig. 5.

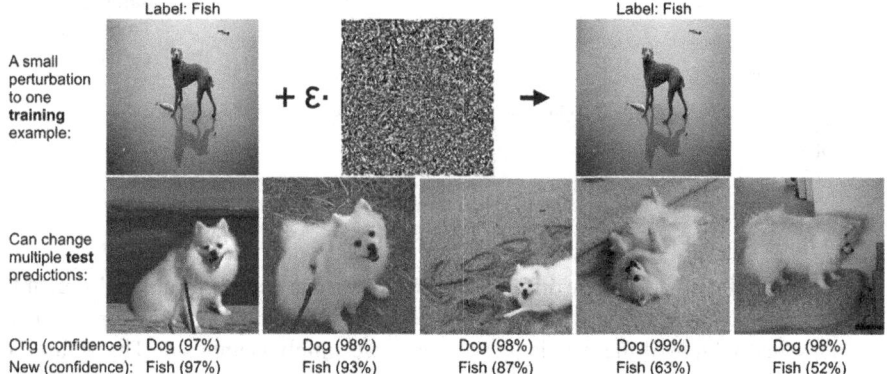

Fig. 5. Influence of single data point to the training process [19]

In the next section, we will discuss the above question in the new light of the differential privacy [10] and machine unlearning [20]. We will show that, differential privacy provides us a much stronger framework to define the forgotten than the definition of the European commission.

3 Differential Privacy as the Definition of Forgotten

Differential Privacy or DP for short is defined in the work of [9, 10]. Intuitively the DP requires that, any attacker cannot distinguish if whether or not a single data point is included in the training dataset or not.

Formally, the differential privacy is defined as:

$$Pr(A(D_1) \in S) \leq Pr(A(D_2) \in S) * exp(\epsilon) \tag{1}$$

wherein:

- A is a randomized algorithm.
- D_1 and D_2 are two most identical dataset, different in only one single data point.
- S is a range of the output of A.
- is a pre-defined real value, stated how much privacy the system will support.

We visualized the differential privacy in Fig. 6.

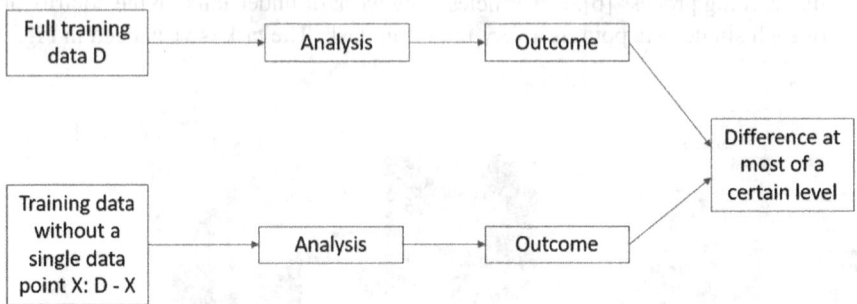

Fig. 6. Differential privacy

From the point of view of the differential privacy, we can define "forgotten" as unable to distinguish the presence of the data. It is a very general definition of the forgotten, as we do not really know if the data is there or not - we simply cannot say. The definition is much stronger than the traditional definition of the forgotten, as we have to explicitly check if the data exists, hence the process of checking itself will reveal the data.

4 Machine Unlearning

Machine unlearning is the term coined by [7] and gained the momentum recently [3, 8, 14, 24]. The core idea of machine unlearning is to let a machine learning model to forget a single particular data point on-demand [7].

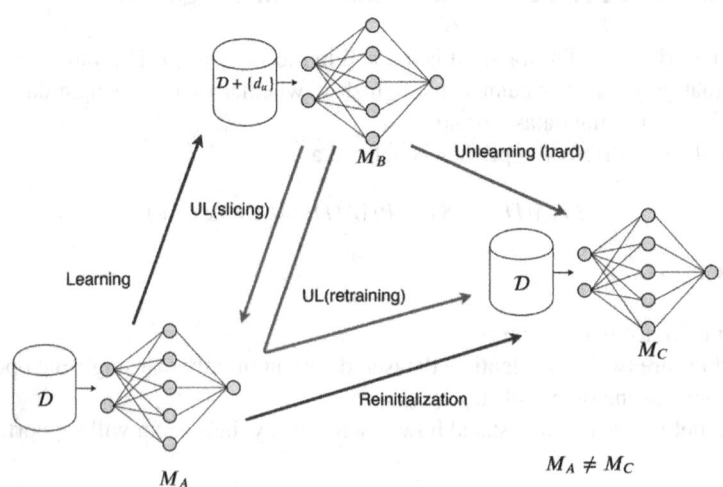

Fig. 7. Machine unLearning [6].

Machine unlearning is visualized in Fig. 7. Indeed, the naive approach of machine unlearning is just to remove the desired data point from the training dataset and training everything else from scratch. However, it is very inefficient and time-consuming.

In the work of [6], the authors proposed the framework called SISA to unlearn more efficiently. In fact, the core idea of SISA is to slice the data into multiple parts so we only to retrain a single part that contains the data point to be removed.

The authors of [14] defined four principles of data deletion:

- Linearity
  - It requires that simple unlearning should be done in linear time. It is not clear at the time of writing if it is achievable. However, for some simple algorithms such as clustering, the linear time complexity is feasible [14, 17, 24, 26].
- Laziness
  - Laziness means that the deletion could be postponed until it actually need to be done [2, 5, 32].
- Modularity
  - The idea of modularity is to isolating the different parts of the data, hence reducing the dependence level between data points. By doing so, we can safely remove a data point without affect the model that based on the others [4, 22].
- Quantization
  - Quantization seems to be surprised at first. However, if the model space is a continuous space, any small change on the dataset might lead to the dramatic change of the model [23].

In the work of [14], the authors defined the data deletion in a stronger requirement than the requirement of the differential privacy. The authors require that, the result of deleting a data point x from the training set is a model that is the exactly the same if we train the model from scratch without x. While the authors of [14] showed that it is possible for some simple algorithms such as k-means, there is a doubt that we can achieve the results for other complicated models such as deep neural networks.

5   Conclusions

The right to be forgotten is indeed a very important human right in the settings of the modern era. Despite the fact that the computer systems never forget and there is no way to actually verify if someone forget something, there are still some recent techniques that allow us to define the forgotten. In this paper, we argue that the differential privacy can be considered as the framework to define the forgotten, and machine unlearning is a usable technique to practice the RtbF.

References

1. Report: 2 years in, 75 percent of right to be forgotten asks denied by google (2016). https://sea rchengineland.com/report-2-years-75-percent-rightforgotten-asks-denied-google-249424

2. Atkeson, C.G., Moore, A.W., Schaal, S.: Locally weighted learning for control. In: Lazy Learning, pp. 75–113. Springer (1997)
3. Baumhauer, T., Schöttle, P., Zeppelzauer, M.: Machine unlearning: linear filtration for logit-based classifiers. arXiv preprint arXiv:2002.02730 (2020)
4. Berman, O., Ashrafi, N.: Optimization models for reliability of modular software systems. IEEE Trans. Softw. Eng. **19**(11), 1119–1123 (1993). https://doi.org/10.1109/32.256858
5. Birattari, M., Bontempi, G., Bersini, H.: Lazy learning meets the recursive least squares algorithm. In: Kearns, M.J., Solla, S.A., Cohn, D.A. (eds.) Advances in Neural Information Processing Systems 11, NIPS Conference, Denver, Colorado, USA, 30 November–5 December 1998, pp. 375–381. The MIT Press (1998). https://papers.nips.cc/paper/1507-lazy-lea rning-meets-the-recursive-leastsquares-algorithm
6. Bourtoule, L., Chandrasekaran, V., Choquette-Choo, C., Jia, H., Travers, A., Zhang, B., Lie, D., Papernot, N.: Machine unlearning. arXiv preprint arXiv:1912.03817 (2019)
7. Cao, Y., Yang, J.: Towards making systems forget with machine unlearning. In: 2015 IEEE Symposium on Security and Privacy, pp. 463–480. IEEE (2015)
8. Chen, M., Zhang, Z., Wang, T., Backes, M., Humbert, M., Zhang, Y.: When machine unlearning jeopardizes privacy. arXiv preprint arXiv:2005.02205 (2020)
9. Dwork, C., McSherry, F., Nissim, K., Smith, A.: Calibrating noise to sensitivity in private data analysis. In: Theory of Cryptography Conference, pp. 265–284. Springer (2006)
10. Dwork, C., Roth, A., et al.: The algorithmic foundations of differential privacy. Found. Trends Theor. Comput. Sci. **9**(3–4), 211–407 (2014)
11. European Commission: A Comprehensive Approach on Personal Data Protection in the European Union. Communication From The Commission To The European Parliament, The Council, The Economic And Social Committee and The Committee of The Regions (2010)
12. Franco, P.: Understanding Bitcoin: Cryptography. Engineering and Economics. Wiley, Hoboken (2014)
13. Garg, S., Goldwasser, S., Vasudevan, P.N.: Formalizing data deletion in the context of the right to be forgotten. In: Annual International Conference on the Theory and Applications of Cryptographic Techniques, pp. 373–402. Springer (2020)
14. Ginart, A., Guan, M., Valiant, G., Zou, J.Y.: Making AI forget you: data deletion in machine learning. In: Advances in Neural Information Processing Systems, pp. 3518–3531 (2019)
15. Gollins, T.: The ethics of memory in a digital age: interrogating the right to be forgotten (2016)
16. Heyman, S.J.: Free Speech and Human Dignity. Yale University Press, London (2008)
17. Hinneburg, A., Gabriel, H.: DENCLUE 2.0: fast clustering based on kernel density estimation. In: Berthold, M.R., Shawe-Taylor, J., Lavrac, N. (eds.) Advances in Intelligent Data Analysis VII, 7th International Symposium on Intelligent Data Analysis, IDA 2007, Ljubljana, Slovenia, 6–8 September 2007, Proceedings. Lecture Notes in Computer Science, vol. 4723, pp. 70–80. Springer (2007). https://doi.org/10.1007/978-3-54074825-0_7
18. Jones, M.L.: Ctrl+ Z: The Right to Be Forgotten. NYU Press, New York (2018)
19. Koh, P.W., Liang, P.: Understanding black-box predictions via influence functions. In: ICML. Proceedings of Machine Learning Research, vol. 70, pp. 1885–1894. PMLR (2017)
20. Kwak, C., Lee, J., Park, K., Lee, H.: Let machines unlearn–machine unlearning and the right to be forgotten (2017)
21. Lambert, P.: Understanding the new European data protection rules. CRC Press, Boca Raton (2017)
22. Mondal, A.K., Roy, B., Roy, C.K., Schneider, K.A.: Micro-level modularity of computaion-intensive programs in big data platforms: a case study with image data. CoRR abs/1910.11125 (2019). https://arxiv.org/abs/1910.11125

23. Mukherjee, S., Niyogi, P., Poggio, T.A., Rifkin, R.M.: Learning theory: stability is sufficient for generalization and necessary and sufficient for consistency of empirical risk minimization. Adv. Comput. Math. **25**(1–3), 161–193 (2006). https://doi.org/10.1007/s10444-004-7634-z

24. Neel, S., Roth, A., Sharifi-Malvajerdi, S.: Descent-to-delete: gradient-based methods for machine unlearning. arXiv preprint arXiv:2007.02923 (2020)

25. Pereira, A.G., Vesnić-Alujević, L., Ghezzi, A.: The ethics of forgetting and remembering in the digital world through the eye of the media. In: The Ethics of Memory in a Digital Age, pp. 9–27. Springer (2014)

26. Rahimi, A., Recht, B.: Random features for large-scale kernel machines. In: Platt, J.C., Koller, D., Singer, Y., Roweis, S.T. (eds.) Advances in Neural Information Processing Systems 20, Proceedings of the Twenty-First Annual Conference on Neural Information Processing Systems, Vancouver, British Columbia, Canada, 3–6 December 2007, pp. 1177–1184. Curran Associates, Inc. (2007). https://papers.nips.cc/paper/3182-random-features-for-large-scale-kernelmachines

27. Reding, V.: The EU data protection reform 2012: making Europe the standard setter for modern data protection rules in the digital age. In: Innovation Conference Digital, Life, Design Munich, vol. 22 (2012)

28. Rosen, J.: The right to be forgotten. Stanford Law Rev. **64**, 88 (2011)

29. Shalev-Shwartz, S., Ben-David, S.: Understanding Machine Learning: From Theory to Algorithms. Cambridge University Press, Cambridge (2014)

30. Sommer, D.M., Song, L., Wagh, S., Mittal, P.: Towards probabilistic verification of machine unlearning. arXiv preprint arXiv:2003.04247 (2020)

31. Villaronga, E.F., Kieseberg, P., Li, T.: Humans forget, machines remember: artificial intelligence and the right to be forgotten. Comput. Law Secur. Rev. **34**(2), 304–313 (2018)

32. Vinh, N.X., Epps, J., Bailey, J.: Information theoretic measures for clusterings comparison: Variants, properties, normalization and correction for chance. J. Mach. Learn. Res. **11**, 2837–2854 (2010)

33. Waldock, C.H.M.: The European convention for the protection of human rights and fundamental freedoms. Brit. Yb Int'l L. **34**, 356 (1958)

34. Walker, R.K.: The right to be forgotten. Hastings LJ **64**, 257 (2012)

35. Werro, F.: The Right To Be Forgotten: A Comparative Study of the Emergent Right's Evolution and Application in Europe, The Americas, and Asia, vol. 40. Springer Nature (2020)

Reducing the Influence of the Cognitive Miser Phenomenon and Cognitive Biases on the Selection Process in Exploratory Research Using a Reflective Approach

Olga Popova[(✉)] [iD], Boris Popov, Vladimir Karandey, and Viktor Afanasyev

Kuban State Technological University, Krasnodar, Russian Federation
popova_ob@mail.ru

Abstract. The aim of the study was to show how the reflective approach helps to reduce the influence of the cognitive phenomenon and cognitive biases on the selection process in exploratory research. For this, it was supposed to use the visualization of the subject area, for example, optimization methods, on any information medium. Visualization can relieve unnecessary stress on the brain, which is expensive to store all the details of a problem situation. To represent the subject area, the structure of the binary tree of the system of questions and answers was chosen. The structure and methodology for obtaining it allowed us to present knowledge in a form that completely removes the known cognitive biases - false attribution, zero risk effect, Stockholm shopping syndrome, Barnum effect, Dunning-Kruger effect, "false consensus effect" and the "cognitive curmudgeon" phenomenon. The effects were considered in the aspect of choosing one of the most suitable alternatives from the whole variety of alternatives. As a result of the study, it was found that structured knowledge devoid of cognitive distortions makes it possible to find the most powerful solution to a problem, draw up an invention formula and obtain a new solution method. An example of the emergence of the geometric programming method as a new optimization method was considered. The reflexive approach proposed by the authors allows obtaining structured information about the studied subject area without distortion and finding new solution methods for solving new problems. Now the subject area can be obtained without heuristic errors with significant savings in mental energy. This will significantly reduce research time, problem solving time and increase the accuracy of decisions.

Keywords: Cognitive biases · Phenomenon · Cognitive miser · Exploratory research · Subject area

1 Introduction

Today, research in psychology makes it possible to make the process of choosing an alternative from a set of alternatives as close as possible to the process of choosing in exploratory research. With the right approach to the study of these phenomena, this process can be made as efficient as possible. Build a mathematical model that takes these phenomena into account. Automate this process by making it as close as possible to human intelligence. Earlier author the main stages of exploratory research were taken into account and the selection process was automated [1–3]. Now it is necessary to combine the latest research in psychology and those obtained by author. Results from the methodology for evaluating the time of exploratory research [4]. We will be able to get a new understanding of the usefulness of information, taking into account the phenomena identified by psychologists and make the necessary additions to the theoretical foundations of computer science. For this, we will continue to work with information quality indicators, so that to reduce the influence of the cognitive phenomenon and cognitive biases on the selection process in exploratory research. The aim of the study was to apply a reflexive approach to visualizing a domain, such as optimization methods. Visualization can relieve unnecessary stress on the brain, which is expensive to store all the details of a problem situation. To represent the subject area, the structure of the binary tree of the system of questions and answers was chosen. The structure and methodology for obtaining it can allowed us to present knowledge in a form that completely removes the known cognitive biases - false attribution, zero risk effect, Stockholm shopping syndrome, Barnum effect, Dunning-Kruger effect, "false consensus effect" and the "cognitive curmudgeon" phenomenon. The effects will be considered in the aspect of choosing one of the most suitable alternatives from the whole variety of alternatives. We might gets structured knowledge devoid of cognitive distortions makes it possible to find the most powerful solution to a problem, draw up an invention formula and obtain a new solution method. We will considered an example of the emergence of the geometric programming method as a new optimization method. We will verify the proposed reflexive approach. Can it obtaining structured information about the studied subject area without distortion and finding new solution methods for solving new problems? Can be obtained the subject area without heuristic errors with significant savings in mental energy?

2   Cognitive Miser Phenomenon and Cognitive Biases

The concepts of mental or cognitive distortion were first introduced by Daniel Kahneman and Amos Tversky [5]. They proved that the choice of an alternative is most often dictated by subjective factors, and not by exact calculation, and by various stereotypes in the perception of incoming information [6]. Therefore, cognitive distortions are considered systematic deviations in behavior, perception and thinking, due to both prevailing stereotypes and structural features of the human brain. Mankind faced the problem of cognitive distortions at the turn of the Middle Ages and the New Time. Consider an illustrative example of distortion found by Bertrand Russell. In 1610, Galileo made the astronomical discovery of four moons around the planet Jupiter. For this, a new scientific instrument was used - a telescope, in which a new lens system was used. A completely new optical glass processing technology was applied to them. This discovery contradicted university science in Europe, so for a long time the professors did not even want to hear about new astronomical discoveries obtained with the help of a telescope. For several hundred years, it was believed that the number of planets should exist only seven, because seven is the correct number. The orbits along which the planets revolve should represent a perfect circle, not an ellipse. Scientists needed to abandon the old experience of solving problems, to obtain and accept new actual experience from solving new complex problems, using new technical means. After all, only new knowledge helps to solve new more complex problems. Earlier knowledge helps to solve past simpler problems.

It was during this period in the history of the development of science that philosophers proved that the flow of new knowledge is endless and ideas about the world will constantly change. Despite this proven fact, today in solving problems, most are guided by their personal experience and do not seek to find the best and leave their comfort zone due to the saving of brain energy. It is energetically more beneficial for a person to consider the world understandable. Therefore, the most common example of cognitive bias is disappointed expectations. At the moment, there are a lot of cognitive biases.

Consider cognitive biases from a theoretical perspective. First, they are associated with false attribution - the attribution of non-existent qualities and properties to something. Second, with incorrect estimates of probabilities and with memory errors. The simplest example of the manifestation of an erroneous assessment of probabilities from our daily life should be considered a player's mistake, which is expressed in an erroneous opinion about the influence of a previous event on the probability of subsequent ones. Under the influence of cognitive distortion, the player is inclined to believe that the probability of getting "heads" increases each time if, when flipping a coin, "heads" fall out several times. In fact, the probability of getting "heads" or "tails" is 1:2, so from several tosses the probability of getting a fixed combination will be the same. Therefore, multiple bad luck is not evidence of future gain. Despite the mathematical evidence, many act irrationally, under the influence of cognitive distortion. Consider a few more common cognitive biases today. The zero risk effect is a special case of player error, where instead of reducing the high probability of a serious risk, they prefer to reduce the low probability of a fatal risk. Such distortions are the reasons for the appearance of various phobias, for example, aerophobia. Airplane crashes are rare fatal events that cause fear. Although fatal accidents happen daily. Another group of cognitive biases arises from the influence of the psyche: the Hawthorne effect, the placebo phenomenon and the sympathetic partner. Consumer cognitive biases are common today, such as Stockholm buying syndrome. A striking example is the situation when the buyer of an expensive car of a certain brand and color begins to see the same cars everywhere. In this way, his choice after the purchase is justified. There are also useful cognitive biases that can help you make faster decisions when speed of decision is more important than accuracy. To save resources, people tend to trust the first information. Let's give an example. If the first review of a movie novelty on the site's list is negative, but most of the reviews are positive, then anyway a negative opinion about the film may become dominant for us. This also includes the Barnum effect. In horoscopes written "especially for us", we see descriptions of our personality in general phrases that are applicable to most people. The last group of cognitive biases is influenced by the quality and way of thinking of the individual. These include the Dunning-Kruger effect and the "false consensus effect". The first of them shows that people with low qualifications tend to make erroneous decisions, are unable to realize the erroneousness of their decisions and they have an overestimated idea of their abilities, and highly qualified specialists suffer from insufficient confidence in their own strengths and knowledge. The second of the latter group, distortion forces you to project your own way of thinking onto others. People think that their usual way of reasoning is natural for their society, although there is no reason to think so. These cognitive biases come in unpleasant surprises.

People often misjudge the possibilities of their thinking. For example, they do not pay attention to information suggesting the right decision and choose the simplest one, because they automatically strive to save mental resources. This phenomenon was introduced into cognitive and behavioral psychology by the psychologists Susan Fiske and Shelley Taylor in 1984 under the name "cognitive miser". They described the strategy of cognition and behavior, which consists in reducing new knowledge to already existing knowledge in this phenomenon. The errors, or errors of heuristics, obtained as a result of the action of such a phenomenon, were proposed by them to be considered as failures in the processing of the available information. Daniel Kahneman continued to investigate heuristics. An example of a common heuristic in decision making is the recognition heuristic. If it is necessary to make a choice from several alternatives, then a person will make his choice based on the information that he has about the properties of one or more alternatives, when he does not know anything about the properties of other alternatives. Let's take an example from social communication. To the question: "Which of the two cities is larger?", The respondent will choose a city that is known to him as big. If, among the two indicated cities, he saw the names of a city that he knows as small, then he will choose another, unknown to him, alternative. The strategy does not take into account the additional information revealed by the current situation with its own set of alternatives, which are not known to us. The consequences of such a mistake can be serious if the choice of the alternative must be precise. The "cognitive curmudgeon" tends to use less time and energy for thinking, but does not consider the situation from different points of view, so the result of his activities in the end is not optimal. This thinking strategy is common to all people. Let us explain why it takes place. Psychologists distinguish two types of thinking. The fast type of thinking works automatically, does not require energy and turns on instantly. And the slow type of thinking takes up a lot of energy and is included in solving problems that require concentration. For solving the problems of choosing an alternative from a variety of alternatives, quick and emotional reactions, which belong to the first type of thinking, are not suitable. Slow thinking can bring the wrong decision with a small amount of information and the occurrence of cognitive distortions. That is why the slow type of thinking is divided into algorithmic and reflective types. Algorithmic thinking is based on the substitution of general formulas, it helps to divide each problem into several elements and solve them sequentially. Reflective thinking spends energy on understanding the problem, its solution and calls into question the conditions of the task, which in everyday life a person does very rarely. It is easier to quickly find a solution to a problem than to reformulate it and find the optimal solution. Here is an example from the life of professor of medicine Peter Ubel, based on his research. In Ubel's experiment, subjects were required to distribute 100 organs for transplantation among 200 children. The children were divided into two groups. The first group includes 100 children with an 80% chance of recovery, and the second, this probability is only 20%. Most of the participants in the experiment distributed 50 organs between the two groups, as they wanted to give fair hope to everyone for recovery. Further, Peter

Ubel removed groups from the problem statement. The participants in the experiment allocated organs in accordance with the probability of patient survival, and her decision changed fundamentally. The conclusion was drawn. The division of children into groups forced the participants in the experiment to act not rationally and incorrectly to use the concept of justice arising from the condition of the problem - the division of children into two equal groups. Due to cognitive stinginess, a heuristic error occurred. The legitimacy of the distribution of organs equally between groups cannot be proved by reasonable arguments. The subjects acted under the conditions of the task, without changing them, and allowed the conditions of the task to control their thoughts and the selection process. Professor of the Department of Human Development and Behavioral Psychology at the University of Toronto Keith Stanovich [7] proved in her works that you need to be able to question the given conditions of the problem. Otherwise, the conditions of the problem will control us, and we will not control the choice of the method for solving this problem. Keith Stanovich suggested introducing the concept of "cognitive generosity," which requires energy and increased alertness that can be learned.

3 Theory and Research Methods

Today, visualization is an affordable tool for eliminating various types of cognitive biases described in the first chapter. Let us prove this statement. Drawing a problematic situation helps to remove the harmful effect that occurs in the process of choosing an alternative from among many alternatives, caused by energy savings during brain activity. The performed visualization on any information carrier relieves the unnecessary burden on the brain, which is expensive to store all the details of a problem situation. In the process of reflective thinking, the necessary arrangement and connection between the elements of the problem condition are found, which will differ from the initial conditions. Therefore, it is easier to perform any operations with elements of the problem condition using the necessary tools and techniques for representing knowledge about the problem conditions. Earlier, the author of the article was Popova O.B. the structure of a binary tree of the system of questions and answers was proposed, as well as a method of its visualization The proposed structure and methodology can be used to represent knowledge in a form that completely removes the cognitive distortions listed above - false attribution, zero risk effect, Stockholm shopping syndrome, Barnum effect, Dunning-Kruger effect, "false consensus effect" and the "cognitive curmudgeon" phenomenon. All the listed distortions should be considered in the aspect of working with alternatives, when one of them is chosen the most suitable alternative.

When constructing a sketch of a problem situation, the maximum effect is achieved from a reflective approach. Let us explain why this is true. Finding a solution to the problem by enumerating already available solutions from the base of previously solved problems is, of course, easier than reformulating it and finding the optimal solution, but there is no guarantee that the conditions of the current problem are formulated correctly. Under the influence of the above cognitive biases, heuristic errors arose when choosing an alternative, which negatively influenced the accuracy of the choice result. Therefore, when constructing the subject area, the initially specified conditions of the problem were questioned and the subject area was built in relation to completely different conditions. As a result, the conditions of the current problem ceased to control the process of choosing a method for solving this problem. Although the result of choosing an alternative may depend on the set of conditions for the current task. When constructing a subject area, a problem situation is not considered the problem itself with its conditions, but methods of solution or a set of alternatives, among which a solution method or a chosen alternative is sought. Reflectivity is achieved through finding the correspondence of all input conditions from the problem with the properties of the desired method for solving the problem. Sometimes additional information can be found from the available problem conditions. Author a sketch of the subject area "optimization methods" was obtained and analyzed. On it you can find the most powerful solution for the current problem and the area of the most powerful inventive solutions. This became possible due to the use of a binary tree to represent the subject area. The process of moving through the tree helps to visualize the choice of the method for solving the problem. Here the path along the tree is the path to solving the problem of choosing a method. In this tree, sheets are methods for solving optimization problems, and intermediate nodes are properties of problems that made it possible to structure knowledge about the subject area "optimization methods". The knowledge about all optimization methods known today of varying degrees of complexity and solving different types of problems was structured. Information about their properties was carefully selected and structured, distinguishing them from methods. The pre-structured knowledge was the result of the effect of "cognitive bounty" and required energy and increased attention to get a sketch of the subject area of "optimization techniques" (Fig. 1).

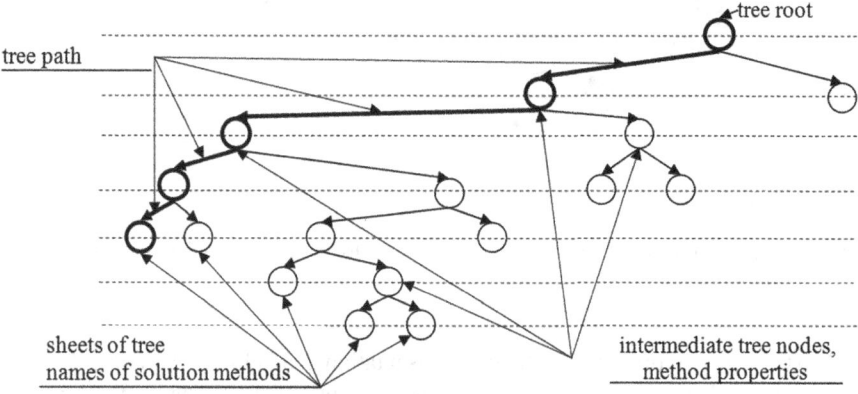

Fig. 1. Basic definitions.

For further research, the following facts were taken into account. The closer a property is to the root of a tree, the more general it is. The upper properties first divide the solution methods into large groups, then subgroups, then separate the methods of one subgroup from each other (see Fig. 2). This is how knowledge is structured.

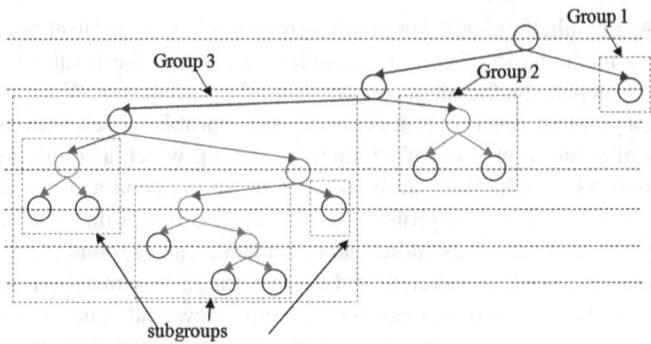

Fig. 2. Properties structured knowledge.

The full path along the tree from the root to the leaf of the tree points to the strongest solution for the current problem (see Fig. 3), since it collects all the necessary properties for solving the current problem.

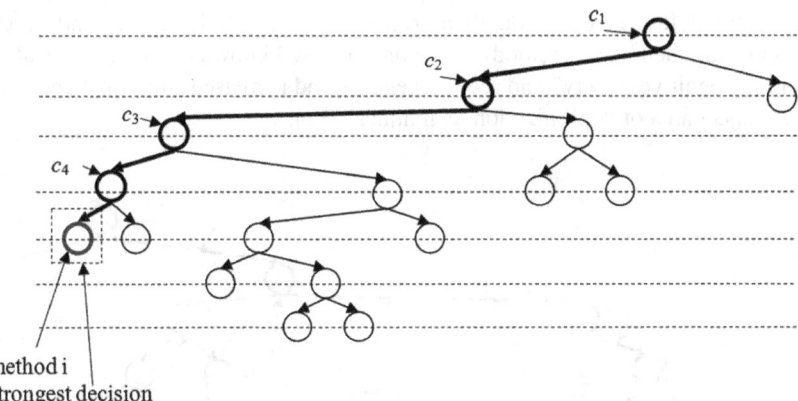

Fig. 3. The path along the tree to the strongest solution to the current problem with the properties c_1, c_2, c_3 and c_4.

Next, we give a form of knowledge representation that helps to overcome the cognitive distortions listed in the introduction, to find potentially new knowledge. Here, these should be considered the found new properties of potentially new solution methods used to solve new more complex problems. Earlier knowledge helps to solve past and simpler problems. An incomplete (complete) path along the tree helps to draw up a "claim" for a new solution method, which will be included in a new group of methods or a subgroup of methods. The formula is shown below.

NAME of the METHOD that has the properties $\{x_i\}$, **characterized in that it has the property** z_i, which does not match the properties $\{y_i'\}$ and $\{y_i''\}$,

where $\{x_i\}$ - full tree path or not full tree path to the prototype of the resulting method;

$\{y_i'\}$ - many properties that are located to the left of the specified path along the tree;

$\{y_i''\}$ - many properties that are located to the right of the specified path along the tree.

An incomplete path through the tree indicates the receipt of a solution method from a new group or subgroup of solution methods, which is still one in this group or subgroup (see Fig. 4). That is, a new direction is being formed in the methods of solving, for example, optimization problems. To find it, it is necessary to identify a property that the current group or subgroup of solution methods does not have in the specified path along the tree, which was not previously known and does not correspond to the properties $\{y_i'\}$ and $\{y_i''\}$.

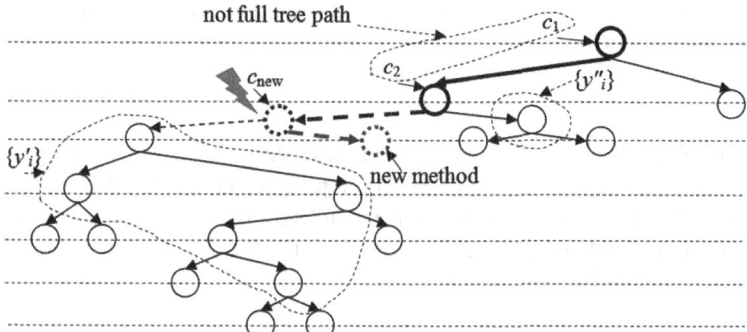

Fig. 4. How to draw up a "formula of invention" for a new solution method, which will be included in a new group of methods or a subgroup of methods, according to the sketch of the obtained subject area "optimization methods".

The full path along the tree allows you to find a solution method that belongs to an already known group or subgroup of methods (see Fig. 5). To find it, it is necessary to identify a property that the final solution method does not have in the specified path along the tree, which was not previously known and does not correspond to the properties $\{y_i'\}$ and $\{y_i''\}$.

The process of finding the strongest solution for the current problem was automated and presented in the form of software "Optimel" (The Support System of Choice the Optimization Method), which was obtained under an open license - GNU General Public License 3, in English for distribution among scientific community conducting exploratory research and software development.

Consider the search for the strongest solution for the current problem and obtaining the strongest inventive solution. Let's show the operation of the Optimel software using one example. It is necessary to find an optimization method that will minimize the volume of the end motor. The required electrical parameters at the output are set - power, voltage, current and shaft rotation speed. It is necessary to determine the optimal geometrical dimensions of the electric motor stator so that its volume is minimal. In the process of answering software questions, information about the problem being solved was collected. There is one objective function in the task, which reflects the engine displacement. And several restrictions on power, current, voltage and some geometric

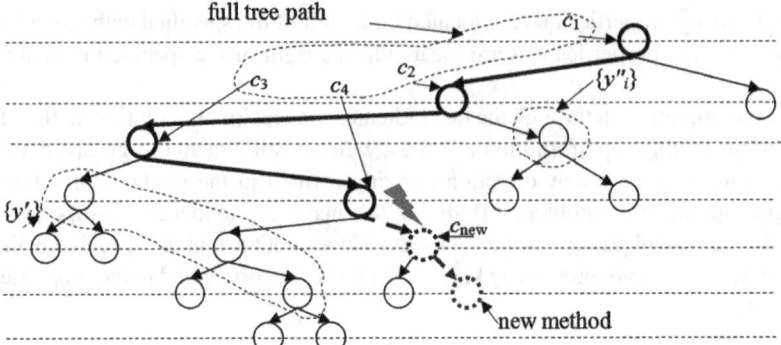

Fig. 5. How to draw up a "invention formula" for a new solution method, which belongs to an already known group or subgroup of methods, according to the sketch of the obtained subject area "optimization methods".

parameters. Since this construction has been little studied and there is no information about the optimal relationships between certain parameters or about their optimal sizes, the number of variables to be optimized is much larger than the number of equations in the optimization problem. Optimel software gave the following sequence of questions and a suggested solution (see Fig. 6).

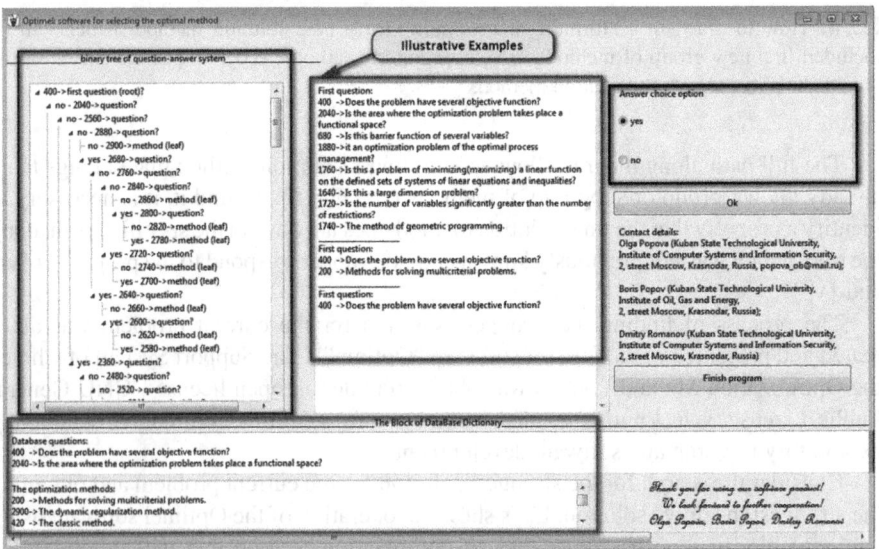

Fig. 6. An example of the Optimel program.

Now let's consider an algorithm for obtaining a new optimization method, for example, the appearance of a geometric programming method. To obtain a new solution method, it is necessary to determine an incomplete or complete path through the tree using a sketch of the tree and a list of properties. List properties that are part of an incomplete or full tree path − $\{x_i\}$. Now it is necessary to conduct a study of the method (if the full path along the tree is taken) or a group of methods (if not the full path along the tree is taken), in order to identify the property z_i, which a method or method group does not have. Found property z_i should not match properties $\{y_i'\}$ and $\{y_i''\}$,

where $\{y_i'\}$ - many properties that are located to the left of the specified path along the tree;

$\{y_i''\}$ - many properties that are located to the right of the specified path along the tree.

Let us write down the formula of the invention for the geometric programming method, as for a newly discovered method. Although it is already known and used for a long time.

METHOD OF GEOMETRIC PROGRAMMING, **which has the following properties**

{Does the task have one objective function? **Yes**

Are the areas in which the optimization problem occurs - are they functional spaces? **no**

Function to minimize multiple variables? **Yes**

Is your optimization task a task of optimal process control? **no**

The problem to be solved is the problem of minimizing (maximizing) a linear function on the sets specified by systems of linear equalities and inequalities? **no**

High-dimensional optimization problem? **yes**}, **characterized in that it has the property** {Is the number of variables much greater than the number of constraints? **yes**}.

4 Conclusions

The aim of the study was to show how the reflective approach helps to reduce the influence of the cognitive phenomenon and cognitive biases on the selection process in exploratory research.

For this, the method of visualizing the subject area was used, using the structure of a binary tree of the system of questions and answers. The methodology for obtaining structured knowledge and the resulting structure of the subject area were tested for the presence of cognitive biases - false attribution, zero risk effect, Stockholm shopping syndrome, Barnum effect, Dunning-Kruger effect, "false consensus effect" and the "cognitive curmudgeon" phenomenon. For the study, data were used - subject area optimization methods.

The effects were considered in the aspect of choosing one of the most suitable alternatives from the whole variety of alternatives. As a result of the study, it was found that structured knowledge devoid of cognitive distortions makes it possible to find the most powerful solution to a problem, draw up an invention formula and obtain a new solution method. An example of the emergence of the geometric programming method as a new optimization method was considered. The reflexive approach proposed by the authors allows obtaining structured information about the studied subject area without distortion and finding new solution methods for solving new problems. Now the subject area can be obtained without heuristic errors with significant savings in mental energy. This will significantly reduce research time, problem solving time and increase the accuracy of decisions.

The structural approach not only significantly reduces the identified effects, but also fully complies with the principle of the reflective approach, where mental energy is completely and one-time spent in the process of visualizing the subject area. Further, quick thinking is used when moving through the constructed tree, taking into account the existing conditions of the current task. Thus, there is a significant saving in mental energy. The subject area is obtained without heuristic errors.

Acknowledgments. The reported study was funded by RFBR [Project title: Development of the theory for qualitative assessment of the information taking into account its structural component, No. 19-47-230004 from 19.04.2019]. All the work on compiling the paper and obtaining calculated and experimental data was evenly distributed among its authors.

References

1. Popova, O., Popov, B., Romanov, D., Evseeva, M.: Optimel: software for selecting the optimal method. SoftwareX **6**, 231–236 (2017)
2. Popova, O., Popov, B., Karandey, V., Evseeva, M.: Intelligence amplification via language of choice description as a mathematical object (binary tree of question-answer system. Proc. – Soc. Behav. Sci. **214**, 897–905 (2015)
3. Popova, O., Popov, B., Karandey, V., Gerashchenko, A.: Entropy and algorithm of obtaining decision trees in a way approximated to the natural intelligence. Int. J. Cogn. Inform. Nat. Intell. **13**(3), 50–66 (2019)
4. Popova, O., Popov, B., Karandey, V., Afanasyev, V.: To the problem of developing a methodology for estimating the time of exploration research. Mod. High Technol. **11**, 88–96 (2019)
5. Curved mirror. Popular Mechanics, 6 (2016)
6. Mental distortion: why do we think irrationally? https://emosurf.com/post/8747
7. Stanovich, K.: Rational Thinking. What Aptitude Tests Don't Measure. Career Press, Wayne (2016)

Modern and Historical Aspect of Automation of Accounting

Olga A. Zubrenkova⬤, Sergey N. Kozlov⬤, Zinaida A. Mishina⬤,
Yulia Y. Sysoeva⬤, and Ekaterina N. Zubenko(✉)⬤

State Budgetary Educational Institution of Higher Education "Nizhny Novgorod State
Engineering and Economic University", Oktyabrskaya Str., 22 a, 606340 Nizhny Novgorod
Region, Knyaginino, Russia
zubenkoen@yandex.ru

Abstract. In modern conditions, the success of the economic entities activities
largely depends on their awareness and ability to use information about their internal resources and external market environment effectively. The preparation and
adoption of economic decisions at various levels of a management system requires
processing of a large amount of accounting information, its comparison, analysis,
interpretation. Accounting system is faced with problems of integrating various
types of accounting: management, tax, social, environmental, etc. Along with the
data of accomplished facts of economic life, functions of an accountant-analyst are
expanded through formation and preparation of forecast information about future
financial results and directions of business entities development. Modern discoveries in the field of information technologies (open technological platforms, cloud
technologies, unified international format for submitting financial statements in
electronic form, etc.) significantly expand the possibilities of building an accounting system that integrates data on internal business processes of an economic entity
and external market environment. Electronic reference and information systems
are becoming more and more widely used in accounting, the development perspective of which is accumulation and systematization of professional knowledge
in the field of accounting, analysis, and audit in specialized knowledge bases.

Keywords: Automation · Accounting process · Information technology

1 Introduction

Over the past 15–20 years, as a result of the process of informatization of society,
computers and Internet technologies have become an integral part of a modern person's
life as a global information resource and means of communication.

In modern conditions, business partners, suppliers and consumers learn a significant part of information about accounting and financial activities of an economic entity
through the Internet, including through a corporate website. For example, in the field of
educational services via the Internet, users receive more than 60% of information about
a company, a little more than 20% of information comes from magazines, and 10% from
reference books [1].

© The Author(s), under exclusive license to Springer Nature Switzerland AG 2021
T. Antipova (Ed.): ICADS 2021, AISC 1352, pp. 423–432, 2021.
https://doi.org/10.1007/978-3-030-71782-7_37

The level of information security and ability to manage it automatically are becoming the most important conditions for achieving innovative competitive sustainable economic development in the context of global competition. For this in Russia, an ecosystem of the digital economy is being formed, operating with digital data, which is becoming a key means of financial management in all spheres of socio-economic activity, a necessary tool for effective interaction between businesses in Russia and abroad, scientific and educational communities, state and citizens based on the so-called smart (intelligent) technologies using cloud services [2].

In the economic literature, it has been repeatedly noted that viability and efficiency of the economic entity functioning in many ways depends on the nature and quality of the process of automation of accounting and financial activities. At present, in Russia, as in the whole world, there is a tendency towards an increasing demand for information systems that allow automating accounting and financial activities of an economic entity [3].

The purpose of the work is to determine the role of accounting and financial activities in the system of automation of the organization's activities. Abstract-logical and computational-constructive methods, the method of comparison are used in the article, historical and modern aspects of automation of accounting and financial activities are analyzed, the conclusion about the development of information technologies in the Russian Federation is made.

2 Materials and Methods

2.1 Researchers

Today, many scientists in their works are actively discussing advantages and disadvantages of automation of accounting and financial activities: Bagdanov, V. S. [7], Rasputin, A. P [9], Khotina, N. V. [13], Sidorova, M. I. [14], Fedorova, E. I. [15], Korneva, O. V. [16], Gulin, D., Hladika, M., Valenta, I. [17], Wilson, R., Sangster, A [19], H. Rkein, Z. A. Issa, F. J. Awada, H. J. Hejase [18].

2.2 Information Base Research

The basis for research is works of both domestic and foreign authors. In their works, they reveal the results of theoretical and practical research in the field of automation of accounting and financial activities in Russia. These questions are raised in scientific conferences materials, scientific journals articles, dissertations, monographs.

2.3 Research Method

The study uses various approaches and methods, such as the abstract-logical, monographic, comparative method.

2.4 Purpose of the Study

In order to determine advantages and disadvantages of using automation in accounting and financial activities, it is necessary to resolve the following issues:

1) to determine capabilities of organizations to use automation in accounting and financial activities;
2) to study possible risks when using automation in accounting and financial activities;
3) to assess the impact of automation on the effective development of the organization.

3 Result and Discussion

Introduction of automation into accounting and financial activities is directly related to additional costs. Not every organization is ready for additional costs. It is the factor that is holding back active dissemination of information technology in all production

Table 1. Dynamics of costs distribution of organizations for information and communication technologies by type, billion rubles

Indicators	2015	2016	2017	2018	2019	Variation, (+, −)
Information and communication technology costs - total	516	1175	1153	1249	1488	1273
Including: purchasing of computers and office equipment	113	260	234	249	297	218,5
Purchasing of telecommunication equipment	...	154	156	145	162	162
Purchasing of software	81,2	162	203	280	282	260,1
Payment for communication services	168	279	256	241	262	205,3
Including payment for Internet access	39,2	73,0	68,1	69,3	69,8	69,8
For employee training related to development and use of information and communication technologies	3,7	12,2	6,8	6,5	6,6	4,4
Payment for services of external organizations and specialists in information and communication technologies (except for communication and training services)	98,9	200	232	254	376	341,1
Other costs	51,1	107	65,1	74,2	102	80,9

processes. According to the Federal State Statistics Service for 2019, the cost of services for creation of solutions for automation of accounting and financial activities amounted to 1,488 billion rubles. (Table 1) [4].

The data presented in Table 1 indicate that the largest share of costs falls on payment for the services of external organizations and specialists in information and communication technologies (except for communication and training services) – 376 billion rubles, as well as for the purchase of computing hardware and office equipment and software – 297 billion rubles and 282 billion rubles respectively.

It should be noted that consideration of sectoral features of the process of automation of accounting and financial activities of economic entities should be preceded by a fundamental clarification of generally accepted terminological (conceptual) apparatus of the process of automation of accounting and financial activities. This applies primarily to the fundamental term "automation". This clarification is necessary to develop a unity of scientific interpretation of economic essence and feasibility of automating activities.

The term "automation" is common knowledge. At the same time, it should be noted that existing interpretations of this term are rather vague. In domestic practice, there is no unified approach to interpretation of the term automation. Here are a small number of definitions of this term most often found in scientific literature.

So, for example, in a modern economic dictionary, automation is understood as the use of machines, machine technology and technology in order to facilitate human labor, displace its manual forms, increase its productivity, and is also aimed at using computers and other technical means of processing and transmitting information in production management, economic processes [5].

At the legislative level, namely in GOST R 54862-2011, a set of technical means, software, maintenance for automatic control, monitoring, optimization of personnel work during operation in order to ensure energy efficient, economical and safe operations for maintenance of engineering equipment of a building [6].

Automation, as V. S. Bagdanov notes, is one of directions of scientific and technological progress, the use of self-regulating technical means, economic and mathematical methods and control systems that free a person from participation in processes of obtaining, transforming, transferring and using energy, materials or information that significantly reduce the degree of this participation or complexity of the operations performed [7].

Despite the opinions pluralism, most scientists are unanimous that the use of technical means and special control systems partially or completely frees a person from direct participation in process of production, receipt, transformation, etc. energy, materials and information.

Over the centuries, accounting and financial activities methods have been formed, accuracy of calculations has increased, and tools have been improved. The history of computing technology introduction in this area is much less known. In recent years, in connection with constantly increasing business requirements for information technology, IT resources have been increasing, the number of highly qualified specialists has been increasing, but at the same time new types of information threats, financial risks, and psychological problems have arisen. Adequately assess the influence of technical means on information support of control system, reliably determine the prospects for development of an automated form of accounting, it is possible by summarizing the experience gained, starting from the origins of machine information processing.

The carried out studies indicate that in domestic practice, information technology has gone through several stages of historical development associated with the development of technical tools.

Based on the review of scientific works devoted to automation of accounting activities of economic entities, periodization of the evolution of automation of accounting activities is systematized (Table 2).

Table 2. Historical stages of development of information technologies in domestic practice

№	Stage name	Stage characteristics	Note
1	1st stage. Manual (until the second half of the 19th century)	"Manual" technologies: pen, inkwell, book, elementary manual means of counting	The main goal of technology is the presentation and transmission of information in a desired form
2	2nd stage Mechanical (late 19th century–40s of the 20th century)	"Mechanically" technologies: typewriter, arithmometer, telegraph, telephone, voice recorder	The main goal of technologies is to present information in the required form in more convenient means, to reduce cost of correcting losses and distortions
3	3rd stage Electric (40s–60s of XX century)	"Electric" technologies: the first vacuum tube computers and the corresponding software, electric typewriters, teletypes (telexes), copiers, portable voice recorders	The main goal of technologies is to generate information that meets the needs of the main users
4	4th stage Electronic (70s–mid 80s)	"Electronic" technologies: computers, control systems (ACS), information retrieval systems	The main goal of the technologies is formation of information that meets the needs of main users, organizations of further analytical work
5	5th stage Computer (from the mid 80s–1990s)	"Computer" technologies: personal computer	The main goal of technologies is to use network technologies and telecommunications to work with information
6	6th stage Internet (from mid-90s–2000s)	"Internet/Intranet" accounting technologies: global, regional and local computer networks	The main goal of technology is introduction and widespread use of computerized accounting
7	7th stage Modern (since 2000–Present)	"Cloud" accounting technologies: software complex located on remote servers	The main goal of technology is to use work over the Internet

At present, development of process of automation of accounting and financial activities is included in priority directions of the Concept of long-term socio-economic development of the Russian Federation. The strategy of development of the information society in Russia provides for implementation of existing cultural, educational, scientific and technological potential of the country in the medium-term perspective and provide the Russian Federation with a worthy place among the leaders of the global information society [8].

It should be noted that the basis of the process of automation of accounting and financial activities is the use of information. Information is a random memorized choice of a variant from many possible and equal, as well as new information that allows you to improve the processes associated with the transformation of matter, energy and information itself.

In turn, economic information reflects facts of accounting and financial activities of an economic entity using a system of natural and cost indicators.

The system of economic information in the country covers both control information (planned and regulatory) and informative: operational production accounting, book-keeping, banking information, statistics (Fig. 1).

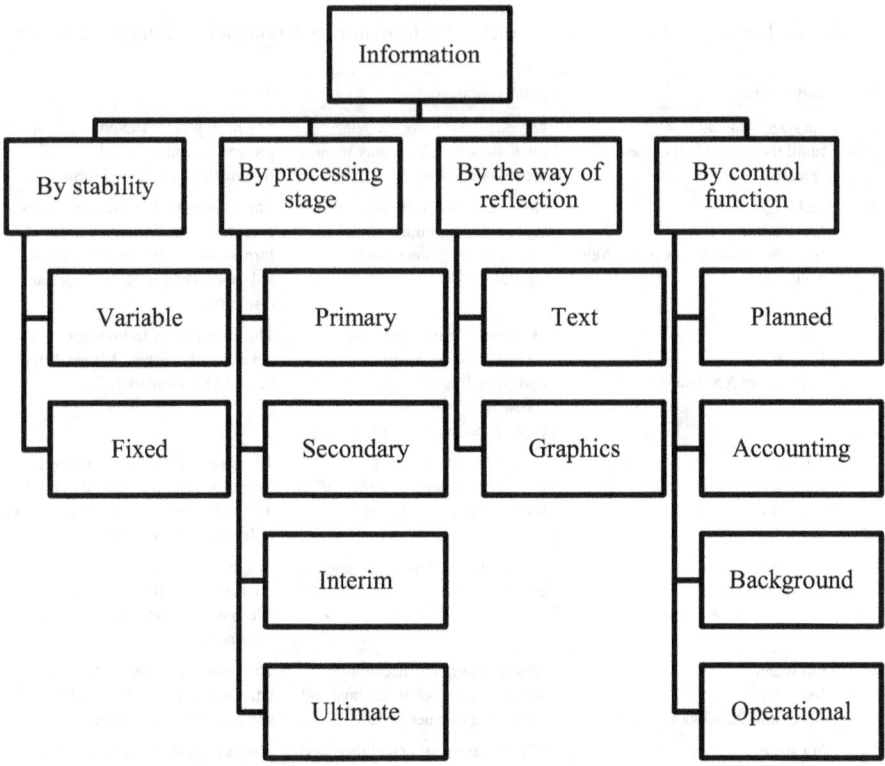

Fig. 1. Types of economic information

Economic information is characterized by a large volume, multiple use, renewal and transformation, a variety of its sources and consumers, a large number of logical operations and specific mathematical calculations to obtain many types of ultimate information [9].

It should be noted that the development of theory and practice of accounting and financial activities is directly influenced by internal and external factors. The results of the survey made it possible to group six main groups of such factors (Fig. 2).

From the data in Fig. 2, it follows that 35.3% of respondents indicate the need to introduce information technologies, 23.6% note the emergence of new technologies for conducting accounting and financial activities.

It should be noted that fundamental changes have taken place in economic processes over the past decades, which have had a significant impact on the ratio of accounting and reporting. This became possible due to automation of accounting and financial activities, in which accounting data are detailed to the information of accounts of the fifth and higher order; and, finally, the revolutionary development of information technology, which made it possible to develop and implement numerous computer programs that automate accounting, management, inventory control and other areas of accounting [10].

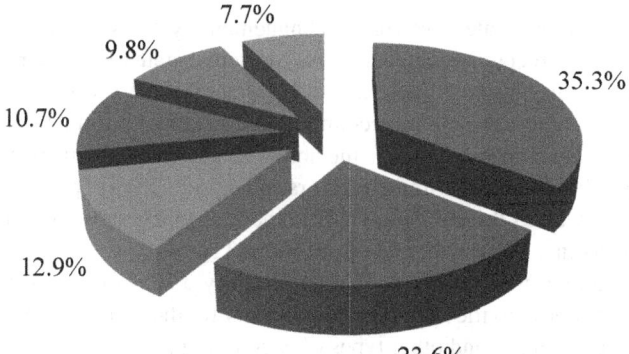

- understanding the need to implement information technology
- emergence of new technologies for implementation of accounting
- understanding the basics of building information technology
- readiness for implementation by an organization
- willingness to provide qualified resources
- readiness for a clear organization of the survey and implementation

Fig. 2. Factors influencing implementation of information technology in the organization's management system*

It is also important that development of information technology provides new opportunities for presenting information, including on a daily basis. Financial accounting continues to rely on periodic review and analysis of financial information, while stakeholders, auditors and others expect comprehensive real-time reporting. The transition from a periodic basis, focused strictly on financial information, to a more complex presentation of financial statements requires a transition period with a phased implementation. All this will entail a violation and displacement of current and familiar roles. Not many organizations are ready for this [17].

It should also be noted that with the use of automation in accounting and financial activities, employees will lose their acquired useful skills. The loss of this kind of skills certainly negatively affects thinking activity of employees, however, introduction of automation of accounting and financial activities will make it possible to speed up production processes without spending much time, for example, on calculations, which previously required a lot of effort and concentration of attention, allowing you to use the freed up time for development of an organization or self-development [18].

Back in his work, Alan Sangster said that not all accountants are willing to use automation in their work. As a rule, they were more willing to use automation in accounting and financial activities of accountants working in an industrial sphere (since the tasks performed by them are based on an algorithmic nature). Accountants of smaller organizations were reluctant to implement automation, which caused industry imbalances and often misunderstandings [19].

4 Results

The effective use of automated enterprise management systems allows you to collect information quickly, process it, analyze, assess the actual state of an enterprise, predict and, on the basis of this, promptly make any management decisions. The introduction of systems for automated processing of accounting and financial information has become the norm for modern enterprises. This is the ability to automatically display documents and transactions, elimination of arithmetic errors and reproduction of primary and reporting documents, as well as the ability to analyze the financial condition of an enterprise and provide results in a convenient and visual form [11].

In turn, accounting information is characterized by a set of data on the results of financial and economic activities of an economic entity for the reporting period, obtained from the data of accounting and other types of accounting [12].

Historically, the need to accumulate and organize a large amount of data manually on paper was the basis for improving forms of accounting. At the same time, development of automation also had a positive effect on the change in forms of accounting.

Thus, for example, the use of automation of accounting and financial activities at initial stages was reduced to reduction of errors in arithmetic calculations, simplifying work of accounting services, on the one hand, by automating grouping and generalization of data, complicating work due to the need for additional verification, with the other side. At the same time, the role of an accountant did not change, only the function of entering data into automated control systems was added, if an accounting employee did this, very often additional operators were involved.

In modern conditions of diversity of socio-economic ties in the sectors of national economy and their management, the role of information is increasing, especially for planning, budgeting and activities of all sectors of national economy. This is ensured by information support of a single information base. This allows you to obtain uniform data for planning, budgeting, analysis of the results obtained and, accordingly, making appropriate decisions to improve production process.

Organization of collection, storage, processing and dissemination of information on activities of an economic entity is carried out in an information base, which provides for an indication of a set of admissible information structures, operations on data and a set of restrictions for stored data values.

The main role of information technology in accounting remains collection, analysis, systematization, interpretation, control and storage of data. This information significantly influences management decisions, therefore, the automated accounting system significantly affects the degree of efficiency of management automation [13].

At the same time, at present, information technologies are becoming not only a means of implementing various models of accounting, but the strongest factor in the development of theory and methodology of accounting and financial activity. Removing restrictions on the complexity and speed of performing mathematical calculations, the possibility of using statistical methods of information processing expand capabilities of accounting and analytical services and stimulate development of new methods and principles of accounting that are impossible in the conditions of "paper" technology [14].

The above, on the one hand, creates conditions for continuous development of the theory of accounting and financial activity, on the other hand, it is the basis for improving automated systems. It should also be noted that even in the conditions of high automation of accounting and financial activities, it is impossible to completely remove a person from this process, since in many cases professional judgment is required to correctly identify individual facts and, accordingly, generate objective information for users.

In the context of reforming accounting and financial activities, one of the primary tasks for training qualified personnel is to consider and improve the features of accounting in organizations of various fields of activity. At the same time, due attention should be paid to issues of complex automation of an accounting process [15].

At the present stage of development of information technologies, the following classes of tools are used in automation of accounting and financial activities and economic analysis:

- non-specialized software packages with analytical capabilities;
- specialized software for creating an information storage of accounting data;
- integrated enterprise management systems [16].

5   Conclusion

Summing up the above, we can conclude that the information system of an organization consists of various subsystems, among which the economic one, which mainly uses the information of an accounting system, occupies the largest share. The accounting system of an organization consists of operational, statistical, tax and book-keeping, which differ in the composition of the reflected phenomena, purpose and methods of implementation.

Modern approaches to automation of accounting and analytical activities enable organizations to free up time while performing current and routine operations. But at the same time, costs of implementing automation are still a limiting factor.

The use of modern automated information processing systems and economic and mathematical methods will expand possibilities of accounting based on integration: once registered and entered into a computer memory, data can be repeatedly used in a single accounting system of an organization. At the same time, each of the types of accounting in the process of integration retains its methodological features and its purpose. In addition, the development of information technologies in the Russian Federation has a direct impact on economic aspects of the life of society, including entire system of accounting and financial activities.

References

1. Rozanova, N.M.: Information technology: a means of survival or a competitive advantage? World Econ. Int. Relat. **7**, 74–81 (2010)

2. Chernov, V.A.: Implementation of digital technologies in financial management of economic activities. Econ. Region **16**(1), 283–297 (2020)
3. Kurashova, M.V., Zadorozhny, V.N.: The effectiveness of information support for management decisions. Econ. Region **3**(7), 83–97 (2006)
4. Federal State Statistics Service. https://rosstat.gov.ru/
5. Raizberg, B.A., Lozovsky, L.Sh., Starodubtseva, E.B.: Modern Economic Dictionary, 6th edn. INFRA-M, Moscow (2011)
6. GOST R 54862-2011. National standard of the Russian Federation. Energy efficiency of buildings. Methods for determining the impact of automation, management and operation of a building (approved and put into effect by the Order of Rosstandart dated 15.12.2011 N 1567-st)
7. Bagdanov, V.S.: Development of automation systems. Design. Customization. Implementation. BHV-Petersburg, SPb. (2005)
8. Order of the Government of the Russian Federation of November 17, 2008 No. 1662-r. Strategy for the development of the information society in Russia, approved by the President of the Russian Federation on 07.02.2008 No. Pr-212
9. Rasputin, A.P.: Concept of economic information, its properties, value and features of processing on a computer. Sociol. Econ. Sci. **5**, 60–63
10. Ageeva, O.A., Rebizova, A.L.: Trends in the development of accounting methods in the conditions of information technologies. Eurasian Union Sci. **4-1**(13), 11–13 (2015)
11. Sadykova, L.G., Geimasheva, G.R.: Trends in the development of information technologies used in accounting. Bull. Mod. Res. **11.8**(26), 411–413 (2018)
12. Guseva, T.A.: Analysis and Diagnostics of the Financial and Economic Activities of the Enterprise. Textbook. Publishing House TRTU, Taganrog (2015). 225 p.
13. Khotina, N.V.: Information technologies in the field of accounting and the prospects for their development. In the Collection: Problems of Managing Sustainable Development of business Structures in Different Areas of Activity Collection of Scientific Papers of the International Economic Forum. Under the General Editorship of N. A. Lytneva, pp. 149–152 (2017)
14. Sidorova, M.I.: Information technologies as an integral element of the modern accounting model. Account. Anal. Audit **3**, 82–92 (2015)
15. Fedorova, E.I.: Accounting automation in budgetary institutions. Success. Mod. Sci. (7), 77–79
16. Korneva, O.V.: Features of the organization of accounting in a computer environment. Labor Soc. Relat. (7), 129–134
17. Gulin, D., Hladika, M., Valenta, I.: Digitalization and the challenges for the accounting profession. In: Proceedings of the ENTRENOVA - ENTerprise REsearch InNOVAtion Conference, Rovinj, Croatia, 12–14 September 2019, IRENET - Society for Advancing Innovation and Research in Economy, Zagreb, vol. 5, pp. 502–511. https://hdl.handle.net/10419/207712
18. Rkein, H., Issa, Z.A., Awada, F.J., Hejase, H.J.: Impact of Automation on Accounting Profession and Employability: A Qualitative Assessment from Lebanon. Faculty of Business Administration, Al Maaref University, Beirut, Lebanon. https://doi.org/10.21276/sjbms.2019. 4.4.10
19. Wilson, R., Sangster, A.: The automation of accounting practice. J. Inf. Technol. **7**, 65–75 (1992). https://doi.org/10.1057/jit.1992.11

Development of Network System for Connection PLC to Cloud Platforms Using IIoT

Chuquimarca Jiménez Luis[(✉)] [ID], Asencio Gonzabay Alba[ID],
Torres Guin Washington[ID], Bustos Gaibor Samuel[ID], and Sánchez Aquino José[ID]

Facultad de Sistemas y Telecomunicaciones, Universidad Estatal Península de Santa Elena,
Santa Elena, Ecuador
{lchuquimarca,wtorres,sbustos,jsanchez}@upse.edu.ec,
albaasenciogonzabay@hotmail.com

Abstract. The IIoT is a reality being part of Industry 4.0, this technology integrates intelligent systems to the industrial area adapting the connection between many devices and production machines, allows process monitoring using a communication network between PLC S7-1200 and Siemens IOT2040 Intelligent Gateway, this is an Industrial Gateway that collects, processes and transfers production data to the Freeboard platform. However, a fundamental part is to analyze and compare latency of production data between an automated process in real time and its connectivity with the outside world. The access method used is the OPC-UA communication technology for collects data from the connected devices inside the production line, also to IIoT the tool Node-Red as the gateway, to connected OPC-UA client and Freeboard platform for a monitoring and analysis in real-time, the delay time in communication from the production area to the web platform and vice versa is between 2 ms and 10 ms.

Keywords: IIoT · OPC-UA · MQTT · PLC · IOT2040 · Freeboard

1 Introduction

The Industry Internet of Thing (IIoT) improves manufacturing by the easy accessibility of data, producing a foreseeable maintenance, improved safety and more operational efficiency. The networks of IIoT can improve interconnection, scalability, interoperability, time savings, and cost savings for industrial organizations [1]. One of the problems with IIoT is the use different protocols for sending and receiving process data. The communication protocols which are currently in use is Open Platform Communications Unified Architecture (OPC-UA). However, the MQ Telemetry Transport (MQTT) transmission protocol becomes the standard for IIoT [2, 3].

This work proposes an experimental methodology to investigate the impact of quality of connection parameters on the latency from the production line to the Cloud and vice versa, using gateways with OPC-UA. An OPC Server/Client configuration uses to exchange data between Programmable Logic Controller (PLC) and IIoT device via Profinet using TCP/IP protocols [4].

© The Author(s), under exclusive license to Springer Nature Switzerland AG 2021
T. Antipova (Ed.): ICADS 2021, AISC 1352, pp. 433–443, 2021.
https://doi.org/10.1007/978-3-030-71782-7_38

In the industry, the most accepted protocol is the machine to machine (M2M) inter-action with OPC-UA. Therefore, M2M can become "smart" enabling flexible interface where machines could be automatized [5, 12].

For increased performance in the industry, the software Supervisory Control and Data Acquisition (SCADA) are completed with several levels of data gathering and manipu-lation structures. Vertical data integration is augmented with horizontal interoperation, is important to implement IIoT [3]. The driver used is Profinet considering protocol OPC-UA, although is important refer to software advanced Totally Integrated Automa-tion (TIA) Portal optimizes the processing, machine operation and planning procedures [6, 13].

The PLC first the centralized OPC was introduced on PCs (e.g. KepserverEx), added with OPC-UA server, and then OPC-UA servers on the Human-Machine Interface (HMI) and PLC levels [3]. Also, the laboratory tests based on S7-1200 with Profinet, using as hardware platforms Raspberry Pi 3 (RPi3) and IOT2040, with a software WinSCP which transfer program files [6, 7].

Horizontal data can be accumulated in PLC or SCADA, but they cannot be stored in long-term due to limit in storage space [8]. The possible solution is using Cloud platform combined with PLC and IoT2040 device [6]. Cloud Platforms are used for data analysis and visualization (e.g. Freeboard). The data production can be taken using embedded sensors with new connectivity options [5]. Then, data are sent in the Cloud from every place of production and then be analyzed in Real-time [9, 10]. The communications between devices can be performed through wired and wireless networks, using different technologies such as RS485 and WiFi [2].

2 Proposed System

This project allows the device to be remotely controlled due to transferring of process data between the PLC and IoT Gateway, using the tool is called Node-RED. Therefore, the main objective is the process communication of systems and sensors for networking technologies IIoT. Node-Red is an editor that makes it easy to link communication via a wide range of nodes for MQTT protocol in a multiplatform solution (see Fig. 1) [1].

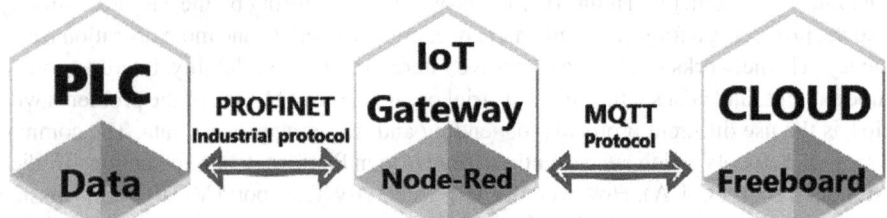

Fig. 1. Transfer of process data from PLC to Freeboard Cloud platform.

Transferred data to the Cloud platform, backups of data can be achieved. Also, Freeboard allows to control variables (Bool, Word) directly from the Cloud platform based on visualization elements in Dashboard [1].

2.1 Hardware Selection

The IoT Gateway communicate data from a PLC S7-1200, is after to stores the process data into the Freeboard which controls device remotely. IIoT Gateway from company Siemens (Simatic IoT 2040) was selected for acquisition, processing, and transmission of process data. The Simatic IoT2040 can be remote control device of peripherals via MQTT (see Fig. 2) [1].

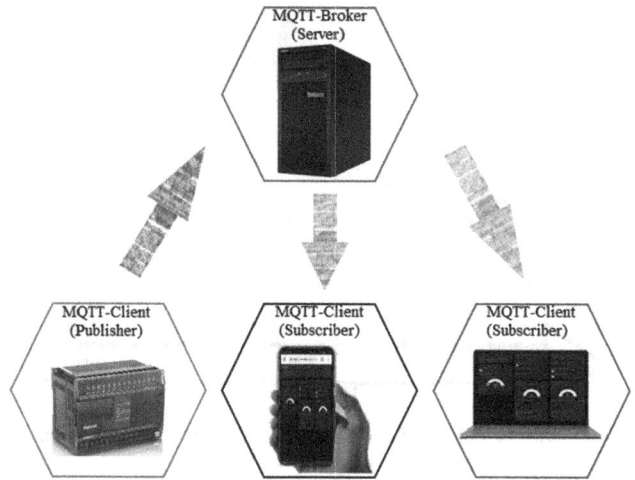

Fig. 2. Data exchange between clients and servers using the MQTT standard.

2.2 Software Selection

The proposed cloud architecture for the remote control and monitoring of devices consists of an S7-1200 PLC connected to the Gateway together with the graphical node programming tool and Freeboard for the interface design. In addition to SCADA data acquisition software and Virtual Private Network (VPN) [9, 14].

2.3 Communication System

There are several ways to communicate between different industrial devices, it is very to use a PROFINET connection, because leads to increased reliability of the network system (see Fig. 3) [2].

The data is exchanged between industrial engineering systems use OPC UA, it is interface to export information of the automation systems, which at field level are implemented with other fieldbus networks. Also, the OPC UA functions without being constrained to a specific communication protocol, it is possible the communication between heterogeneous systems, thus achieving flexibly harmonizes different industrial networks [10, 15].

Using point-to-point network communication architecture to decrease the problems associated with Internet latency and industrial network system stability [9].

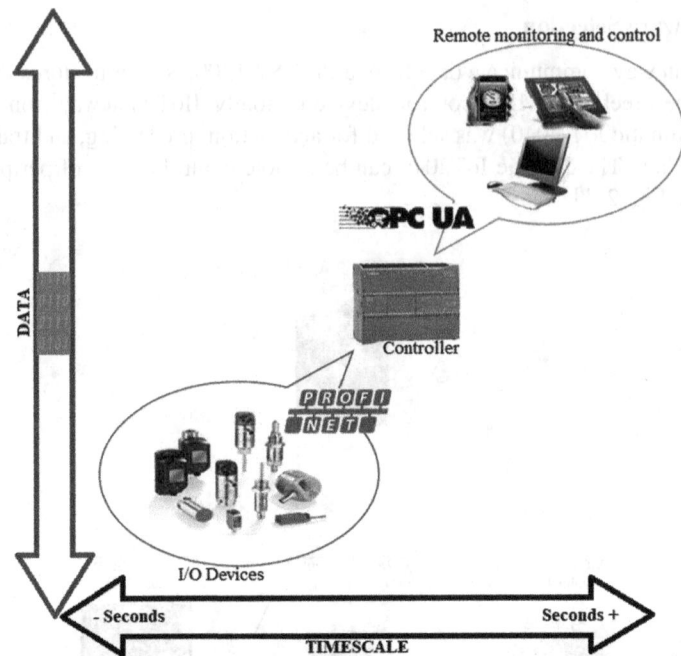

Fig. 3. System with Profinet and OPC UA protocol reduces communication latency.

3   Content Development

Devices are available in development for real-time control, including tasks controlled by the S7-1200 PLC. Each PLC must contain the fundamental program under "Data". Data from the automated process is collected and transferred to the cloud through the IoT with its programming defined in Node-Red and the configuration in the Freeboard platform (see Fig. 4), the main objective of the program changes the states of the structures of predefined variables through the IIoT Gateway or on the Cloud platform [1].

There are different elements that must be controlled and monitored, devices such as sensors, actuators, motors, PLC, HMI, SCADA systems are connected. The communication links of the IoT gateway function as an interface between different networks and support different communication protocols. The SIMATIC IOT2040 has a reliable open platform for data collection, processing and transfer and is used as a gateway between the data acquisition controller and the cloud [2].

The connection between the PLC and the IoT Gateway can be via the official of the PLC S7-1200 libraries of the Node-RED. These libraries categorize of required PLC items and decide what data blocks will be used. After, it groups a request together into one or more packets at maximum speed. The libraries through "rack slot 1" and within programmable interface TIA Portal access at blocks used. Also, the PUT/GET access in the hardware configuration must be enabled within the PLC S7-1200 [1].

For the exchange of information across the hierarchical systems is used OPC UA, also applied at the enterprise resource planning (ERP) or manufacturing execution systems

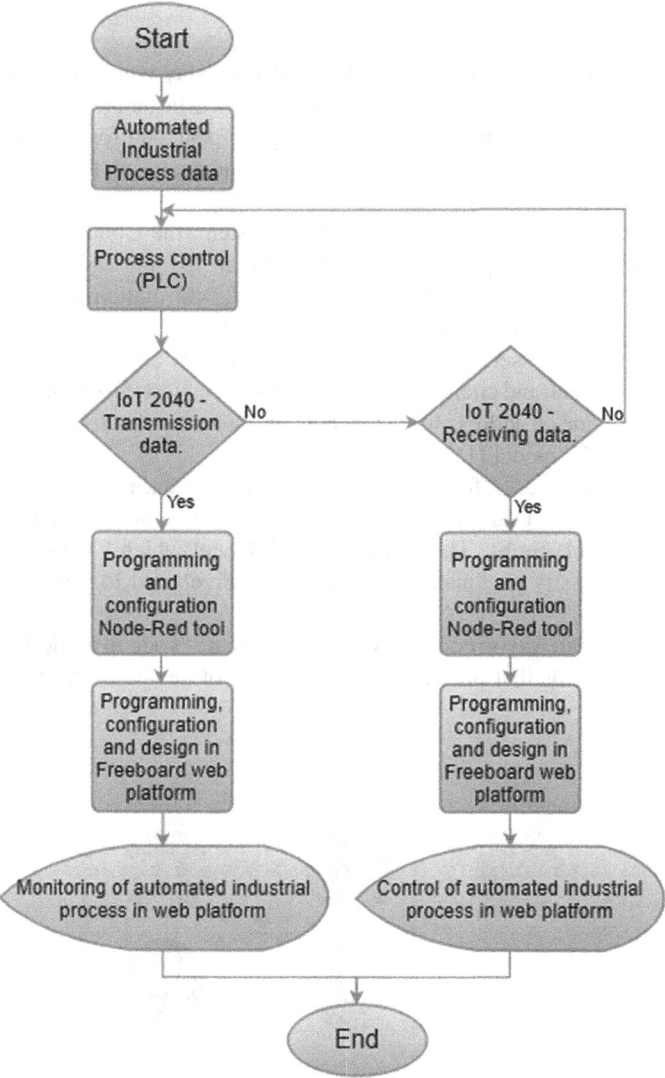

Fig. 4. Logical diagram of tasks that are controlled by the PLC and send to the web platform.

(MES) and to the field devices. The OPC UA client API (application program interface) is used for send/receiving the OPC UA service requests/responses to/from the OPC UA server. Also, can be used to find available OPC UA servers and research their address space. The OPC UA is used for exchanged a varies of data structures "nodes", including the variables that can be remotely called [10].

4   Methodology

The information of a communication system IoT has delay which depends on the network. In this work, the experimental method approach in define a test methodology adaptive to situations. The proposed methodology is designed for IIoT services when device sends data to the Cloud where are processed and analyzed; the result is sent back to be used by automatized system device [10, 16, 17].

There are three parts of the communication system, the Control Device, the IIoT Gateway, and the Cloud Application, which using an IoT messaging protocol called MQTT. After a complete roundtrip, is acquired the information the time which are useful for the estimation of the delays (see Fig. 5) [10].

OPC-UA uses client-server architecture and Servers are applications which indicate the information model, and clients are applications that request information from servers. In each server, define an address space that containing nodes of software or real physical objects the OPC-UA model [11].

Node-RED tool is used for wiring together different hardware devices, APIs and online services, where a sample set of nodes that permit communicate between different protocols and platforms. The Node OPC UA Client which has the address of the PLC server, and finally is connect the Node Client Freeboard IoT which has the data information [11].

The IOT2040 gateway is connect with the Freeboard Node in Node-RED, where is create different boards that present data information of devices, sensors, actuators, or other [11].

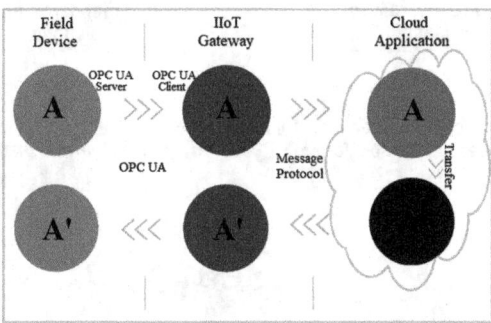

Fig. 5. Proposed experimental setup for the measurement of the communication delay between machine with OPC UA interface and Cloud platform.

5   Results

The machine controller with OPC UA communication Siemens S7-1200 using the gateway toward the Cloud platform is implemented using the embedded device IOT2040 which is placed in the same communication network of the PLC. Both the IIoT gateway and the Cloud application adopted in this use case have been deployed using tool Node-RED.

For the PLC the synchronization uncertainty with IOT2040 is derived from Siemens internal reports that state a maximum delay of 6 ms [10]. The results on the standard uncertainty of communication and synchronization from PLC to IoT and vice versa are shown in Table 1 and Table 2, and communication from IoT to cloud platform and vice versa is shown in Table 3 and Table 4.

Table 1. Data transfer from PLC S7-1200 to IoT2040 and the of response time the data packet.

PLC S7-1200 to Gateway IoT2040			
DATA	Sent	Received	Lost
	60	60	0
TIME	Minimum	Maximum	Medium
	3 ms	14 ms	6 ms

Table 2. Data transfer from IoT2040 to PLC S7-1200 and the response time of the data packet.

IoT 2040 to PLC S7-1200			
DATA	Sent	Received	Lost
	35	35	0
TIME	Minimum	Maximum	Medium
	1 ms	8 ms	1 ms

Table 3. Data transfer from IoT2040 to Cloud and the of response time the data packet.

IoT 2040 to Cloud platform			
DATA	Sent	Received	Lost
	30	30	30
TIME	Minimum	Maximum	Medium
	2 ms	9 ms	2 ms

Table 4. Data transfer from Cloud to IoT2040 and the of response time the data packet.

Cloud platform to IoT 2040			
DATA	Sent	Received	Lost
	30	30	30
TIME	Minimum	Maximum	Medium
	1 ms	5 ms	1 ms

With the data obtained, graphs are created as shown in Fig. 6 and Fig. 7, the packages are sent from the S7-1200 PLC to the IoT2040 and vice versa, also from the IoT2040 to the Cloud platform and vice versa, when carrying out the respective tests the changes over time. Certainly, the data packet upload time is much longer than the download time.

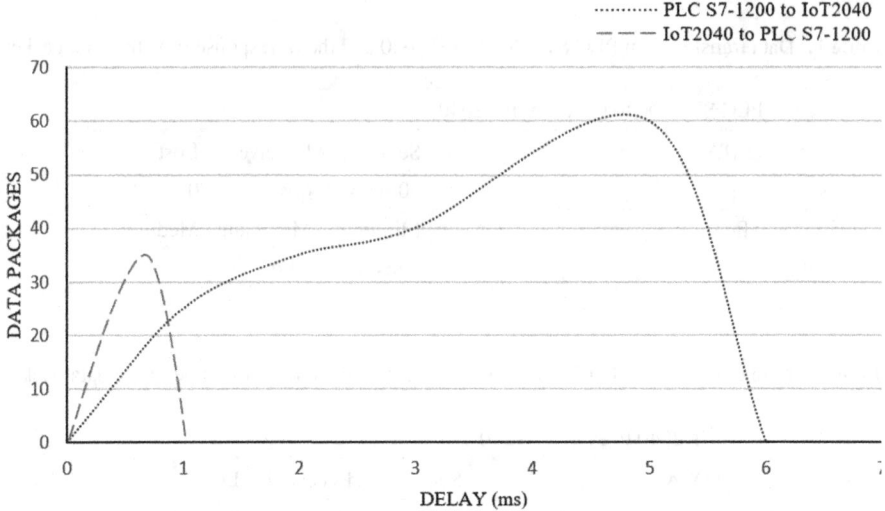

Fig. 6. Comparison of data latency in PLC and IoT communication.

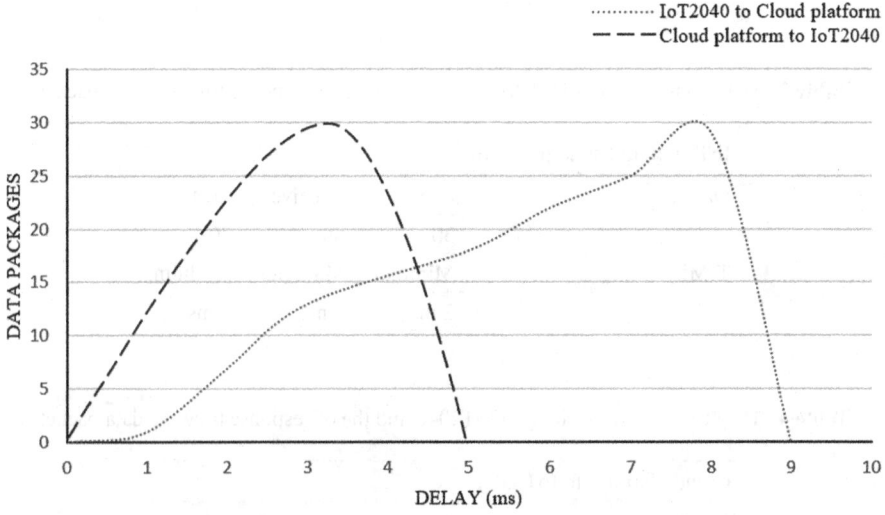

Fig. 7. Comparison of data latency in IoT and Cloud Platform communication.

The Cloud data systems can be realized on the Node-RED platform. S7 nodes are programmed to receive Boolean data when activating or deactivating a variable in the

PLC, also a node as counter for receives information about the quantity of a variable from PLC, finally the Dweetio nodes send the data to the Freeboard web platform. The most important parameters are the IP address, variable/structure selection and unique identification name. The saving data on cloud platform Freeboards takes the form of protocol MQTT. The control of PLC is possible either directly from IIoT Gateway (see Fig. 8) [1].

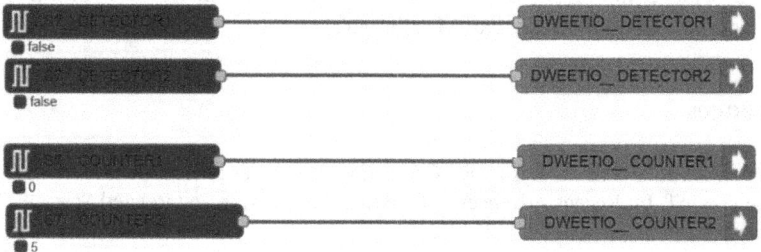

Fig. 8. The nodes programmed in the Freeboard web tool.

The platform Freeboard can be used control panel and visualization elements in the graphical interface. Everything is continuously synchronized in Node-RED (see Fig. 9).

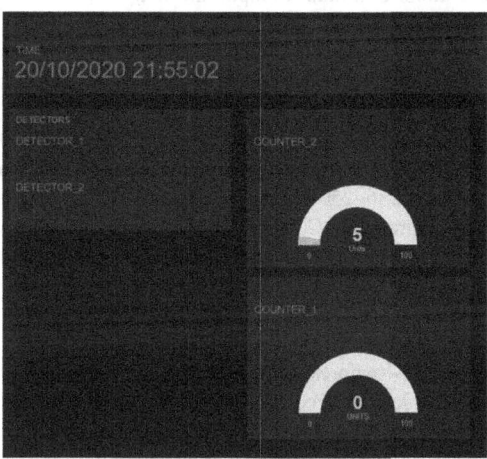

Fig. 9. Data monitoring on Freeboard web platform.

6   Conclusion

In the future of the Industry 4.0 services will require short response time. Therefore, this project was used protocol MQTT, which worked excellent. Freeboard Cloud is a platform which offers a high response speed, obtaining a time less than 1 s with changing process data and the security works as a fuse in the event of failure on the communications or device. The results demonstrate the feasibility of OPC UA and IIoT gate-ways, with a hardware and software (Siemens S7 1200 PLC controller, IOT 2040 Gateway, and Freeboard platform). Therefore, the proposed monitoring architecture based on cloud complies with the challenges of latency in data processing, data storage and requirement of analytic information.

For a better comparison, it would be advisable in the future to extend this article adding another Cloud platform call MindSphere.

References

1. Gavlas, A., Koziorek, J., Rakay, R.: Comparing of transfer process data in PLC and MCU based on IoT. In: Recent Advances in Electrical Engineering and Related Sciences: Theory and Application, pp. 390–399 (2020)
2. Mateoiu, A.-M., Korodi, A.: OPC-UA based small-scale monitoring and control solution for android devices case study for water treatment plans. In: 4th International Conference on Control, Automation and Robotics, pp. 190–195 (2018)
3. Ferrari, P., Flammini, A., Rinaldi, S., Sisinni, E., Maffei, D., Malara, M.: Evaluation of communication delay in IoT applications based on OPC UA. In: Workshop on Metrology for Industry 4.0 and IoT, pp. 224–229 (2018)
4. Salhaoui, M., Arioua, M., Guerrero-González, A., García-Cascales, S.: An IoT control system for wind power generators. Commun. Comput. Inf. Sci. **855**, 469–479 (2018)
5. Bellagente, P., Ferrari Paolo, P., Flammini, A., Rinaldi, S., Sisinni, E.: Enabling PROFINET devices to work in IoT: characterization and requirements. In: IEEE International Instrumentation and Measurement Technology Conference Proceedings, pp. 1–6 (2016)
6. Combs, L.: Control Engineering, Downers Grove, 15 December 2011. https://www.contro leng.com/articles/cloud-computing-for-scada/. Accessed 10 Aug 2020
7. Lee, B., Kim, D.-K., Yang, H., Oh, S.: Model transformation between OPC UA and UML. Comput. Stand. Interf. **50**, 236–250 (2017)
8. Mijovic, S., Shehu, E., Buratti, C.: Comparing application layer protocols for the Internet of Things via experimentation. In: IEEE 2nd International Forum on Research and Technologies for Society and Industry Leveraging a Better Tomorrow (RTSI), pp. 1–5 (2016)
9. Forsström, S., Jennehag, U.: A performance and cost evaluation of combining OPC-UA. In: Global Internet of Things Summit (GIoTS), pp. 1–6 (2017)
10. Gavlas, A., Zwierzyna, J., Koziorek, J.: Possibilities of transfer process data from PLC to cloud platforms based on IoT. Int. Fed. Autom. Control - PapersOnLine **51**, 156–161 (2018)
11. Ioana, A., Korodi, A.: VSOMEIP - OPC UA gateway solution for the automotive industry. In: IEEE International Conference on Engineering, Technology and Innovation (ICE/ITMC), pp. 1–6 (2019)
12. Coito, T., Viegas, J.L., Martins, M.S.E., Cunha, M.M., Figueiredo, J., Vieira, S.M., Sousa, J.M.: A novel framework for intelligent automation. Int. Fed. Autom. Control - PapersOnLine **52**, 1825–1830 (2019)
13. Ferrari, P., Flammini, A., Rinaldi, S., Sisinni, E., Maffei, D., Malara, M.: Impact of quality of service on cloud based industrial IoT applications with OPC UA. MDPI - Electron. **7**(7), 109 (2018)
14. Huan, H., Tian, H., Yan, Z., Yao, Z.: Group control elevator dispatching system based on S7-1200 PLC. In: Chinese Automation Congress (CAC), pp. 3853–3857 (2019)

15. Zidek, K., Pitel, J., Pavlcnko, I., Lazorick, P., Hosovsky, A.: Digital twin of experimental workplace for quality control with cloud platform support. In: 4th EAI International Conference on Management of Manufacturing Systems, EAI/Springer Innovations in Communication and Computing, pp. 135–145 (2019)
16. Dhandapani, G., Veilumuthu, R.: Cloud based Real-time condition monitoring model for effective maintenance of machines. In: Proceedings of the 31st International Congress and Exhibition on Condition Monitoring and Diagnostic Engineering Management, pp. 118–124 (2018)
17. Cavalieri, S., Salafia, M.G., Scroppo, M.S.: Integrating OPC UA with web technologies to enhance interoperability. ScienceDirect - Comput. Stand. Interf. **61**, 45–64 (2019)

Some Current Aspects of Big Data Evolution

Artem A. Balyakin⬡, Marina V. Nurbina⁽⊠⁾ ⬡, and Sergey B. Taranenko

NRC Kurchatov Institute, 1, ac. Kurchatov sq., Moscow 123182, Russia
Nurbina_MV@nrcki.ru

Abstract. The paper discusses the current trends in digitalization process. A number of technologies that are critical for the digital ecosystem formation (such as big data and artificial intelligence technologies) are discussed. The current approach to digital technology regulation is described. A number of characteristics of the big data life cycle are performed. The complementary connection between science and digital technologies based on it and the social life is shown. The state policy that promotes innovative development is proved to be based on setting general norms and prospects. The methods and approaches from the scientific field are supposed to be the basic tool to form digital society. A number of examples of the introduction of digital technologies in the scientific field are discussed.

Keywords: Big data · Digitalization · Megascience facilities · Ecosystem · Life cycle · Science

1 Introduction

Digitalization in modern society is considered as a classical example of breakthrough technologies. In the current century, digital technologies would play the same role as electricity did for the XXth century, enabling the authorities to solve emerging challenges. For instance, in the Russian Federation (according to the Strategy for the Development of the Information Society), the introduction of digital technologies is considered as one of the tools for solving the major challenges facing the country [1].

In addition to its obvious implementation into the technological sphere, digital technologies are also being applied in the social and economic life. The corresponding arising changes cannot be attributed to exclusively positive ones [2]: new risks and challenges contribute to the formation of digital inequality, unavoidable in the near future [3], and contradictions between existing and emerging institutions, including the degradation of social practices [4].

In this paper, the term "digital technologies" is attributed mostly to big data and data handling methods (aiming at obtaining "smart results"). We note that this approach often refers to artificial intelligence and decision-making systems that accompany big data technologies. The outcomes of this technologies development would result in a global digital infrastructure and its penetration into all spheres of human life.

Despite the variety of big data sources [5], similar approaches and algorithms are used for their processing in various fields [6]. This trait (replication property) can be treated as one of the basic characteristics of big data technologies.

T. Antipova (Ed.): ICADS 2021, AISC 1352, pp. 444–450, 2021.
https://doi.org/10.1007/978-3-030-71782-7_39

2 Big Data Concept

From a legal point of view, "big data" is not a legal term at the moment, but rather a collection of approaches with many different implications for various disciplines such as economics, natural science, legal and social sciences, etc. [7].

The Russian Federation emphasizes the position that the main source of big data is the Internet and therefore "big data" are defined as Big User Data [8]. According to the conclusion of the US Federal Trade Commission, "big data" are arrays of structured or unstructured data characterized by large volume, variety, high rates of change, and real-time processing [9]. In European Union main attention is paid to socio-economic aspects of big data turnover, and no specific legal definition has been accepted. In fact, big data can be defined as a set of data and information that defies ordering and sorting at the present stage of human development.

This approach can lead to legal uncertainty in this area. In our opinion, the most applicable definition should emphasize a new qualitative property of large amounts of information (rather than quantative ones), considered as "big data", in comparison with the usual datasets. Thus, big data are perceived as a complex that includes both the data itself and the approaches and methods of their analysis. In this case, big data are rather a process, not an object.

3 Perception of Digital Technologies

The abovementioned shows the high role of digital technologies in general (and, more narrowly, big data technologies) for the social sphere [10]. The growth of technical capabilities greatly impacts the social space. Contrarily, at the same time, the changes caused by innovative technologies are institutionally dependent [4, 7]: their possible negative and positive consequences are tied to the proper and timely development of social institutions.

In numerous works related to the field of big data, technical issues of organizing the collection, storage and processing of information prevail. The socio-economic and humanitarian aspects of high technologies are presented much worse. For example, the author [11] gives big data technology a "humanitarian tint", proposing to use the concept of cultural and social capital, developed by P. Bourdieu, in relation to big data. In his opinion, big data, like cultural and social capital, simplifies interaction with the information environment, which creates conditions for the conversion of capital into economic goods, the last to generate net income.

General philosophical questions of the algorithms creators responsibility are discussed, for example, in [12, 13]. The authors proceed from the fact that algorithms silently structure our life, removing responsibility from officials, "depersonalizing" decisions, believing it to be a common good [10]. Instead of reliability and objectivity to other uncertain procedures, the sensitivity of algorithmic models to human bias increases [13]. The solution is seen as a deeper empirical study of algorithmic models used in practice [13], and the growth of "internal responsibility", when it is proposed to increase the involvement of employees who develop algorithms, making them ethically responsible for the solutions developed [12]. This approach looks very controversial, since it does not mention the responsibility of the "customer" who sets technical tasks. As a rule, it is the state [14], which, depending on its goals and objectives, implements different approaches (and, as a result, demonstrates the demand for different algorithms). For example, building a digital economy based on blockchain technology in Estonia is more successful and meets less resistance in society than in the UK, where cultural traditions, coupled with a larger population, create problems for building the so-called "algorithmic government" [14].

As a rule, many studies touch on "borderline" issues related to improving the technical characteristics of systems, leading to its qualitative growth. In particular, the authors of [15, 16] showed that an increase in the number of data used does not mean an improvement in the output results. The problem is associated with both nonlinear saturation [16] and an increase in the influence of noise with increasing data set [15]. On the other hand, the algorithm used in big data technologies itself generates "new" big data [11]: the collection of more and more data makes the information environment more and more controlled and manageable, which in turn creates the opportunity to collect even more data. This trend does not lead to qualitative changes yet, and the redundant data is actually not of interest. They are sometimes necessary however for the formation of more fine-tuned decision-making mechanisms (for example, in the field of sales and service delivery).

The solution to emerging difficulties is seen both in new algorithms and approaches [15, 16], and in the development of new ethical rules and norms related not only to the developers of algorithms (as suggested in [13]), but also to the system as a whole. Thus, in [17], using the example of the work of international scientific collaborations (including megascience-class facilities), it is proposed to use the rules for protecting the emerging intellectual property of project employees as a starting point, which are previously developed and approved by the project participants.

The authors stress the complementary relationship between science and digital technologies (rooted from the scientific approaches and tasks) and the social life. There is a bilateral influence of one on the other [10].

On the one hand, the algorithms used are a priori biased, and on the other hand, the results obtained seriously change the socio-cultural landscape (therefore, provoking changes in algorithms in return). For example, in the legal field, there is a deformation of law due to the delegation of decision-making from a person to algorithms [4]. In a practical aspect, this is realized by training algorithms on existing practices (precedents). As a result, the emerging artificial intelligence (decision-making device) can accumulate both the norm and negative practices. This can be avoided only by setting rules of a higher level than the algorithms themselves [5, 17].

4 Evolution of Big Data Technologies

4.1 From Big Data to Smart Content

Understanding digitalization phenomenon requires studying the evolution of technologies involved in this process. So, in relation to big data and artificial intelligence, there is a process of "data intellectualization", which has received the name of the phenomenon of "smart content" (vs "habitual" big data). It has the following features:

1. Knowledge vs new data. The task of such a process is to get away from a simple search for correlations, which is typical of most methods of working with big data. The results obtained should be categorized as "knowledge" (as a result of data interpretation) and not "new data".
2. Knowledge vs code. Data is thought of as an "active participant" in the processing and decision-making process. The idea was expressed to combine "data" and "code". In particular, in accordance with the FAIR principle [18], algorithms are included in the "circle" of stakeholders and participate in decision-making (as an independent object, whose "vote" - the decision obtained - is counted with a certain weighting factor).
3. An priori assignment of good and evil. The social component of big data technologies is taken into account: solutions based on digital technologies are not socially neutral, and must be "adjusted" in advance in the direction necessary for society. Accordingly, the contradiction between personal data and open data could be removed. On the other hand, it opens wide opportunities for fake results and eliminates the unwanted results (thus inclining into socially committed results). To manage that process the human control proves to remain inevitable.
4. Quality vs quantity. On the "technical" side, smart data will be characterized by fewer variables than currently, along with the ability to fold and recover data: the amount of data will decrease without loss of quality. This thrusts the need of new algorithms development, and smart data storage.

In our opinion, smart content is the data of the subject area based on big data, on which, despite their immensity, an adequate interpretation is built due to immanent intelligent procedures. From the user's point of view, smart data is in some way an intermediary between digital technology and civil society (an understandable representation of data set).

4.2 Big Data Life Cycle

The life cycle as an analytical concept is characterized by life time, and understood as a characteristic time of significant changes. Hereinafter we list a number of the big data life cycle features in the aspect of legal regulation.

- Taking into account the dynamics of the context (not that all is big data that seems to be). The very concept of "big data" is dynamic. The current development of deep learning technology, and then the expected development of universal AI, as well as the possible development of cognitive (nature-like) AI, will significantly change what we can design (create, process, use) as big data. At the same time, some of what we previously understood as big data will lose this status, becoming a traditional database.
- Accelerating social change. Big data is conceived according to current trends (fashion-like style), and as social life is tremendously unstable and ever altering, so does the big data differs. In practical aspect, we mention the requirements regarding the "right to be forgotten", etc., that may change dramatically within a short period of time.

- Integration of big data. Big data has both an opportunity and a tendency to be part of larger sets (a kind of fractal structure). The latter is clearly manifested, first of all, in retail and banking.
- Multiplicity of characteristic life cycle times (presence of several characteristic times for one phenomenon).
- Dynamism of definitions and legal mechanisms during the life cycle. According to the authors, the resilience of big data to these changes, including the need to support them (functioning, updating, legal adaptation, etc.) should be included in the analysis of the big data life cycle.
- Taking into account consideration of the upstream and downstream stages of the big data life cycle. In the early stages, big data may not meet the big data requirements. Moreover, at this stage they are collected and processed. An independent legal regime is needed. In the final stage (refusal of support, modification, etc.), not only atomized rights and needs (necessities) may be affected, but the functioning of systems and infrastructure may also be disrupted. With the possible absence of an alternative (both physically and economically possible), this is a matter of national security.

To our viewpoint, the big data life cycle is a set of stages, various forms and pace differenced processes in the pregenesis (initial stage, as a creation of the idea of big data and corresponding issues), genesis and post-genesis (functioning of big data) as a systemically connected and organized institution. By now the big data technology has not reached the stage of mature technology, and came to a stand at the position of the tool (for other scientific and industrial areas). Its further evolution is still under question, and the course has not been settled yet.

5 Conclusion

Big data technologies have already outgrown the role of a tool for solving a narrow circle of strictly applied problems to an almost independent scientific and technological field, which in the minds of most people is rigidly linked with artificial intelligence and decision-making systems. This complex phenomenon is better treated as a process, rather than an object. Thus, being a process, it gives rise to a number of risks and challenges. To overcome them a number of actions is required.

Adequate legal regulation in the digital sphere should be one of the first steps [19]. Understanding the essence of the processes caused by digitalization leads to the idea that it is society (represented by the state) that sets the rules and regulations for business, and not vice versa. To preserve social well-being and development, it is necessary to ensure the creation and development of a well-thought-out and effective regulation system through the law of relationships between man, science, state, society and nature [20].

It is necessary to understand the dynamics of development of digital technologies, taking into account the peculiarities of the life cycle of big data and other technologies in strategic planning. This leads to the need to separate legal mechanisms at different stages of building the digital economy.

For example, in the scientific field, digitalization processes are clearly represented in the work of megascience class facilities [21]. Particularly, the coronavirus pandemic has given an additional impetus to the widespread introduction of the remote access regime and the creation of digital twins of real objects [20]. This phenomenon is still under development.

According to the authors, there is no need for immediate legal regulation of all emerging technologies at the time of their creation. The main task of states interested in the development of advanced digital technologies is to formulate meanings and set a perspective. An example is national strategies for the development of artificial intelligence, digital technologies, etc. Perhaps part of such a regulatory system belongs to the newly formed complex branch of law - scientific research law (the law of science) [22].

In this regard, the authors actively support the idea of developing a set of supranational/international principles of human interaction in connection with the creation of AI systems, including the ethics of the application of advanced digital technologies and the separation of the law of science into a separate industry. We believe that such a set of rules can be based on the Report on the Ethics of AI (2019) of the World Commission on the Ethics of Scientific Knowledge and Technology of UNESCO [23].

Acknowledgments. This work was supported by the RFBR grant 18-29-16130.

References

1. Decree of the President of the Russian Federation of 09.05.2017 No. 203. On the Strategy for the Development of the Information Society in the Russian Federation for 2017–2030. https://pravo.gov.ru/proxy/ips/?docbody=&firstDoc=1&lastDoc=1&nd=102431687. Accessed 20 Oct 2020
2. Insight Report. The Global Information. Technology Report 2016. Innovating in the Digital Economy. https://www3.weforum.org/docs/GITR2016/WEF_GITR_Full_Report.pdf. Accessed 01 Nov 2020
3. Zobova, L., Shcherbakova, L., Evdokimova, E.: Digital spatial competition in the global information space. Fundam. Res. **5**, 64–68 (2018). [in Russian]
4. Zhulego, V.G., Balyakin, A.A., Nurbina, M.V., Taranenko, S.B.: Digitalization of society: new challenges in the social sphere. Bull. Altai Acad. Econ. Law **9–2**, 36–43 (2019). [in Russian]
5. Balyakin, A.A., Malyshev, A.S., Nurbina, M.V., Titov, M.A.: Big Data: Nil Novo Sub Luna. In: Antipova T. (ed.) Integrated Science in Digital Age. ICIS 2019. Lecture Notes in Networks and Systems, vol. 78, pp. 364–373. Springer, Cham (2020)
6. Grigorieva, M., Golosova, M., Ryabinkin, E., Klimentov, A.: Exabyte repository of scientific data. In: Open Data Systems. DBMS, Moscow (2015). https://www.osp.ru/os/2015/04/13047963
7. Perspectives in Law, Business and Innovation, p. 341. Springer Nature Singapore Pte. Ltd. (2017)
8. National Program "Digital Economy of the Russian Federation". https://government.ru/info/35568/. Accessed 02 Sept 2020
9. Big Data: A Tool for Inclusion or Exclusion? Understanding the Issues (FTC Report). https://www.ftc.gov/reports/big-data-tool-inclusion-or-exclusion-understanding-issues-ftc-report. Accessed 04 Oct 2020
10. Balyakin, A.A., Taranenko, S.B., Nurbina, M.V., Titov, M.A.: Social aspects of big data technology implementation. ICS. J. Digit. Sci. **1**(1), 15–24 (2019)

11. Sadowski, J.: When data is capital: datafication, accumulation, and extraction. Big Data Soc. **6**(1), 1–12 (2019)
12. Kirsten, M.: Ethical implications and accountability of algorithms. J. Bus. Ethics **160**(4), 835–850 (2019)
13. Kemper, J., Kolkman, K.: Transparent to whom? No algorithmic accountability without a critical audience. Inf. Commun. Soc. **22**(14), 2081–2096 (2019)
14. Engin, Z., Treleaven, P.: Algorithmic government: automating public services and supporting civil servants in using data science technologies. Comput. J. **62**(3), 448–460 (2019)
15. Triguero, I., García-Gil, D., Maillo, J., Luengo, J., García, S., Herrera, F.: Transforming big data into smart data: An insight on the use of the k-nearest neighbors algorithm to obtain quality data. WIREs Data Min. Knowl. Discov. **9**(2), e1289 (2018)
16. Succi, S., Coveney, P.V.: 2019 Big data: the end of the scientific method? Phil. Trans. R. Soc. **377**(2142), 20180145 (2019)
17. Kahn, M.: Co-authorship as a proxy for collaboration: a cautionary tale. Sci. Public Policy **45**(1), 117–123 (2017)
18. The FAIR Data Principles. https://www.force11.org/group/fairgroup/fairprinciples. Accessed 03 Aug 2020
19. Balyakin, A.A., Nurbina, M.V., Taranenko, S.B.: Comparative legal features of the formation of a digital ecosystem. Int. Legal Courier **1–2**(37–38), 42–45 (2020)
20. Nurakhov, N.: The basic processes of creating a "Megascience" project. In: International Conference on Integrated Science, ICIS 2019: Integrated Science in Digital Age, Batumi, Georgia, pp. 329–339 (2019)
21. Nurbina, M.V., Nurakhov, N.N., Balyakin, A.A., Tsvetus, N.Yu.: Mega science projects for business. In: Ahram, T., et al. (eds.) Human Interaction, Emerging Technologies and Future Applications III Proceedings of the 3rd International Conference on Human Interaction and Emerging Technologies: Future Applications (IHIET 2020), Paris, France, pp. 488–492 (2020)
22. Kashkin, S.Yu.: The formation of the law of science as a new integrated branch of law, Bull. Univ. Named After O.E. Kutafina, Moscow (5), 16–27 (2018)
23. Report of COMEST on robotics ethics. UNESDOC. https://unesdoc.unesco.org/ark:/48223/pf0000253952. Accessed 09 Nov 2020

The Impact of Big Data on Firm Performance

Myriam Ertz[(⊠)], Shouheng Sun, and Imen Latrous

LaboNFC, University of Quebec in Chicoutimi, Saguenay, Canada
Myriam_Ertz@uqac.ca

Abstract. Big data applies statistical processing, analytical techniques, and sophisticated algorithms to large quantities of structured and unstructured data. Since the processed data is increasingly of higher volume, higher velocity, of higher variety, and potentially of higher veracity, organizations may gather key information and knowledge about extra-organizational aspects such as consumers and markets, but also from intra-organizational aspects including refined understandings of processes, systems, procedures, bottlenecks and other issues. Overall, this increased intelligence could lead firms to function more efficiently and be more effective in attaining their strategic and commercial objectives. Organizations may therefore develop competitive strengths which could then translate into competitive advantages and improved firm performance. This study is based on the assumption that big data analytics could improve organizational processes to the point that such improvements may significantly improve firm financial performance. However, in contrast to past research which explored the impact of big data on firm performance as a whole, this study provides an exploratory research on the impact of specific big data analytics (i.e., descriptive, predictive, and prescriptive) on the financial performance of 560 organizations listed in the S&P500 and in the S&P/TSX60 stock indices. The preliminary results of this study suggest three key contributions. First, big data analytics has a significant and extensive impact on corporate performance. Second, descriptive analytics contribute positively to the profit-related performance (i.e. share price) whereas prescriptive analysis contributes positively to both revenue and profit-related performance. Furthermore, the contribution of BDA to the revenue performance of manufacturing industry is greater than in other industries.

Keywords: Big data analytics · Firm performance · Linear regression

1 Introduction

Big Data is defined as high-volume, high-velocity, and high variety information assets that demand cost-effective, innovative forms of capturing, storing, distributing, managing and analyzing that information [1, 2]. Big Data Analytics (BDA) refers to the "application of statistical processing, and analytics techniques to big data for advancing business" [3].

Increasing need, availability and affordability of data, as well as culture change towards evidence-based management, support the growing interest to BDA [4]. Currently, although for lots of companies, big data is no longer a hyped trend, but rather a

© The Author(s), under exclusive license to Springer Nature Switzerland AG 2021
T. Antipova (Ed.): ICADS 2021, AISC 1352, pp. 451–462, 2021.
https://doi.org/10.1007/978-3-030-71782-7_40

necessity, more than 87% of organizations have low business intelligence (BI) and analytics maturity, according to a survey by Gartner, Inc. [5]. While major (e-)retailers such as Walmart and Amazon and others are often over-emphasized as best-in-class users of BDA, the reality is that there are many other companies that use these technologies. For example, The North Face uses both artificial intelligence and machine learning in its recommendation system Watson by IBM, to deliver custom recommendations and improve overall shopping experience [6]. Likewise, Starbucks uses artificial intelligence to personalize marketing campaigns and promotions [6].

In the academic field, despite much interest to the concept of Big Data and analytics generated from researchers in several domains (e.g. finance, marketing, information sciences, transportation, agriculture, healthcare, social networking, telecommunications, and so on), there is a lack of empirical evidence of the benefits associated with BDA. More surprisingly, little attention has been paid to the empirical investigation of the impact of BDA on financial performance, despite the critical importance of such a relationship to reach strategic objectives. This lack could be due to the fact that the investigation of such a relationship requires the integration of several research areas, disciplinary frameworks and expertise in a cohesive inter-disciplinary endeavor.

This project aims at filling that void in the literature by adopting an inter-disciplinary perspective to assess the impacts of BDA for marketing purposes on firm financial performance. More specifically, this project consists in a large-scale study of organizations that are part of the S&P 500 (Standard & Poors 500) in the USA, and of the S&P/TSX 60 (Standard & Poors/Toronto Stock Exchange 60), in Canada, to identify to what extent the implementation of BDA, in the marketing function, forms a competitive advantage that materializes through financial performance. More precisely, the project aims at achieving the following specific objectives: (1) determine the BDA methods and techniques. This objective aims at determining to what extent three groupings of BDA are used by the studied companies; (2) identify the main applications of BDA. That is, the identification of the areas of benefits, or goals of BDA usage in marketing research and analytics projects of the organizations; (3) estimate the impact of BDA methods and techniques on the financial performance of organizations. This objective refers to the estimation of a predictive model to assess the impact of the BDA on main financial indicators of a firm; and (4) estimate the impact of BDA methods and techniques on the sustainability. This implies discovering how BDA contributes to creating a sustainable competitive advantage of a firm.

2 Literature Review

2.1 Big Data Analytics in Marketing

Many organizations are already applying conventional or advanced BDA, in order to achieve strategic marketing objectives such as improving understanding of customers, retaining customers, improving customer experiences, sales forecasting, price optimization, website or search optimization, and supply chain optimization, among many others [7]. The benefits of BDA to develop and sustain a competitive advantage [8], to predict customers' propensity to buy certain products [3], to guide decision-making and improve marketing processes and outcomes [9], have been well-documented in the literature. BDA appears therefore of critical importance to sustain firm competitiveness and growth in an increasingly globalized and competitive economy. But does it?

In light of the rapidly changing evolutions in the realm of information sciences, the application of BDA in marketing appears essential to remain consumer-centered, differentiated and well-positioned. The trend is probably going to be even stronger with a more generalized transition to connected objects, and autonomous objects that will produce even larger quantities of data to analyze. However, although the concept of BDA raised a lot of interest from scholars in different research domains including marketing [10], only a handful academic studies provided an initial empirical assessment of the impact of BDA in marketing on organizations' financial performance [11, 12]. Yet, no study provides empirically-supported guidelines as to which method or technique or combination of method and technique improves most significantly financial performances. This study aims at filling the gap in the literature in this regard by adopting an inter-disciplinary approach combining both marketing and finance to contribute to both disciplines.

2.2 Impact of Big Data Analytics on Financial Performance

There are several ways in which BD creates value for the organization [13]:

- Creating transparency, achieved through easier accessibility of data to stakeholders and integrating data from different organizational units;
- Enabling experimentation to discover needs, expose variability, and improve performance, which means collecting sufficient and accurate performance data, its further analysis and setting controlled instruments to find ways to upgrade performance;
- Segmenting populations to customize actions, enabled by specific segmentation instruments and fulfilled with targeting;
- Replacing/supporting human decision making with automated algorithms, applying sophisticated analytics to large datasets and therefore optimizing decision-making processes;
- Innovating new business models, products and services, which covers both enhancing and creating new products, services and business models, including analysis of the data obtained from the current products' usage.

This value is easy to monetize, and therefore helps companies to achieve better financial performance. It can be measured in different indicators, such as sales revenues, profits, market share, stock prices, etc. [14]. Most studies show positive dependence between investments in BDA and improvements of financial indicators of the firm. As McKinsey discovered, BDA usage has potential to increase operating margin up to 60% and reduce expenditure up to 8%. Müller, Fay and Vom Brocke discovered that live BDA assets in average contribute to 3–7% improvement in firm productivity in terms of financial performance [15]. They also found out that firms in information technology-intensive and/or highly competitive industries tend to gain higher value from investment in BDA assets. Announcements about BDA investment also boost market capitalization of the firm through rising of stock prices [16]. However, in this case, since investors tend to expect more returns from non-knowledge-intensive firms investment announcements than from knowledge-intensive, the value for the first type of companies will be bigger. From the practical point of view, it can be observed that not only traditional technology giants such as Google, Amazon, Facebook, Microsoft, IBM etc. keep investing in BDA capabilities, but also retailers, financial and ICT corporations, car manufacturers, energy and oil companies, etc., implying that BDA has potential to bring value across a broad scope of industries, what we are going to test.

Overall, no wonder that BDA leads to improvement of financial performance, as it generates new insights, supports decision making and enhances productivity of business processes. Specifically, it leads to "more efficient planning (e.g. schedule and cost variance, capacity utilization), manufacturing process (e.g. process downtime, machine efficiency, waste reduction), and quality assurance (e.g. defective units, rejected units)" [17]. Also, integrated with traditional marketing analytics, BDA can lead through knowledge fusion to the new product success [18].

2.3 Impact of Big Data on Sustainability

Regarding the ways in which BD enhances sustainability, researchers highlight a variety of processes. For example, digital technologies such as Internet of Things, BD and Analytics enable functioning of Product-Service Systems (PSS) Business Models, which help to overcome Circular Economy (CE) challenges [19]. If IoT contributes mainly to product upgradability and material tracking, also providing huge amount of real-time data to companies, BDA, in its turn, enables better and more sustainable decision making and helps in the provision of advanced services such as preventive and predictive maintenance. All in all, IoT, Big Data and Analytics help companies to overcome CE challenges such as operational risks (through monitoring of users' activities), loss of ownership and users' willingness to pay (through advanced services such as preventive and predictive maintenance or the optimization of the usage phase), technology improvement (by digital upgrade of devices) and return flow uncertainties (by estimation of products and components residual life) [19]. BDA also has environmental and social impacts on innovation and performance of the sustainable supply chains [20]. In this case, BDA plays a facilitating role for achieving shared sustainability goals through mutual support and information sharing within the supply chain network [21]. Thus, big data analytics plays an important role in implementing sustainable way of doing business, which then leads to a long-term competitive advantage.

3 Methodology

This study draws on two theories: resource-based theory (RBT) and Knowledge Based View (KBV) [22, 23] to postulate that the use of BDA may create superior value for business. Therefore, we conduct an empirical analysis by selecting relevant variables and building corresponding models to estimate the impact of BDA methods and techniques on the financial performance of organizations.

3.1 Data Collection

For the purposes of the study, three databases were built: the database of financial indicators, the database of big data analytics techniques and the sustainability database. First, the data about 162 types of financial indicators was collected for the companies of S&P500 and S&P/TSX 60 list (as of May 2019) for the 2009–2018 time period. The indicators referred to the income statement, balance sheet and financial ratios of each company reported at the end of the financial year, as well as the companies' share price and market capitalization reported at the end of the calendar year. Also, three control variables such as the sector the company operates in the industry and the type of organizational structure (holding or not) were used. According to both SIC and NAICS classification, this study divides the sample companies into four categories: FMCG (fast-moving consumer goods), (Semi)-durable goods, personal services, and professional services. All the data was obtained from Mergent Online and Yahoo! Finance databases.

Second, the database of big data analytics methods and techniques used by mentioned organizations was built. To obtain the information about these specific analytical instruments, we analyzed 11 569 academic publications written in English, published in the period of 2016–2018 and extracted from 4 international databases: ABI/Inform, Business Source Complete, Scopus and Web of Science. Aggregating information from several databases enabled covering of both business and highly technical publications, as well as articles from different industries and multidisciplinary studies. Starting from extraction of articles containing the terms "big data" and the name of the company either in the title, or abstract, or keywords, we then conducted their filtering, leaving 2 735 relevant publications, and finally the content analysis of the remaining articles. The results were put in the SPSS database for further analysis.

3.2 Variable Design and Model Construction

The dependent variables in this paper is enterprise performance. There are many indicators to measure corporate performance. Each indicator reflects the performance and development ability of the enterprise from different perspectives. In order to avoid the one-sidedness of using single indicator, we selected eleven indicators including the ROI (return on investment), ROE (return on equity), ROA (return on assets), EBITDA (earnings before interest, taxes, depreciation and amortization), Net income, Operating income, Gross profit, Total revenue, Sales revenue, Share price, and Earning before tax, which basically cover all characteristics of corporate performance. The value of each variable is set as the actual average value of the corresponding performance indicator of the sample enterprise in 2016, 2017 and 2018.

As to the independent variables, given the breadth of existing BDA methods and techniques, we use the classification of INFORMS (Institute for Operations Research and Management Science) of three hierarchical groupings for BDA [24]:

– Descriptive analytics refer to "creating report to summarize business activities, to answer the questions of 'what happened?' or 'what is happening?'" [24, p. 8]. It includes Association Analysis, Sequence Analysis, Cluster Analysis, Similarity Matching, Link Analysis.

- Predictive analytics "is the process of making intelligent/scientific estimates about the future values of some variables like customer, demand, interest rates, stock market movements" [24, p. 9], to answer the question "what will happen?". It includes Decision Trees, Neural Networks, Partial Least Squares Regression, least-angle regression (LARS).
- Prescriptive analytics: is the highest echelon in analytics [...] where the best alternatives among many – that are usually created identified by predictive and/or descriptive analytics – courses of action is determined using sophisticated mathematical models" [24, p. 9]. This type of analytics tries to answer the question "what should I do?" The corresponding techniques include optimization, simulation, and heuristics-based decision modelling techniques. It includes Stochastic Optimization, Optimization, Multi Criteria Decision Making Techniques, Decision Modeling, Network Science, Simulation Techniques, Deep Learning, Artificial Intelligence.

For a more comprehensive and in-depth investigation of the impact of BDA methods and techniques on the financial performance of the corporation. We conduct two sets of research at different levels in this paper. One group uses the broad category of BDA (i.e. Descriptive analytics, Predictive analytics and Prescriptive analytics) as the first level independent variables for modeling and analysis. The purpose is to observe the impact of each broad category of big data analytics on corporate performance as a whole. Another group uses the specific analysis methods and techniques as the secondary level variables for modeling and analysis. The purpose is to observe and evaluate the impact of each specific big data analysis technology on enterprise performance. Such research design can make a more comprehensive and detailed analysis of the impact of big data analysis technology on enterprise performance. In addition, by comparing the analysis results of the two hierarchical variables, it can better verify the validity and the reliability of the research results. The two sets of hypotheses proposed are as follows.

- Set1_ H_{ij}: The adoption of big data analysis BDA_j (j = broad category, i.e. descriptive analytics, predictive analytics and prescriptive analytics) has a positive impact on the corporate performance indicator DV_i.
- Set2_ H_{ij}: The adoption of big data analysis BDA_j (j = specific method and technique) has a positive impact on the corporate performance indicator DV_i.

Based on the above two hypotheses, this paper uses the regression analysis method and general linear model to carry out empirical research. The corresponding model is shown in Eq. (1)

$$DV_i = \alpha_{i,0} + \sum \alpha_{i,j} \times BDA_j + \sum \beta_{i,k} \times CV_k + \varepsilon_i \tag{1}$$

Among the terms, $\alpha_{i,0}$ is the constant term, DV_i refers to the dependent variables, that is, the indicator of corporate performance. BDA_j refers to the big data analytics (broad category and specific method and technique), CV_k refers to the control variables. ε_i and is the residual term, which is the noise generated by the omitted variables that influence dependent variable in the model. $\alpha_{i,j}$ and $\beta_{i,k}$ are the coefficients of the independent variables and control variables.

4   Results and Discussion

4.1   Descriptive Statistics

As to the data collection, we received the feedback from 101 companies. We counted the number and distribution of all S&P 500 and the TSX60 companies in different sectors, and also calculate the proportion of selected sample companies in all companies of S & P 500 and the TSX60 in different sectors. The distribution of samples is shown in Table 1.

Table 1. Sample distribution

Sector	Number and proportion of companies in the S&P 500 and the TSX60 in each sector		Number of samples	Sample proportion in each sector
FMCG	82	15%	12	15%
(Semi)-durable goods	130	24%	23	18%
Personal services	137	25%	36	26%
Professional services	200	36%	30	15%
Sum	549	100%	101	18%

As can be seen from Table 1, the sample distribution is basically reasonable and can be used to analyze the true situation of the population. Next, we observe the difference of application of big data analysis technology in different sectors. The adoption of big data technology in different sectors are shown in Fig. 1 and Fig. 2.

It can be seen from Fig. 2 that the application of big data analysis technology has a significant difference among different sectors. Currently, descriptive analytics and prescriptive analytics are widely used in enterprises, especially prescriptive analysis. It shows that with the development of big data technology, more and more advanced data analysis technology is widely used by enterprises. The value of big data has been recognized by enterprises, and the emphasis on BDA has raised unprecedented levels. As the exponential growth in the volume of corporate data, more and more companies are constantly seeking deeper data mining by using the more sophisticated and advanced data analysis techniques such as Artificial Intelligence, Simulation Techniques, Optimization, and Network Science.

From the perspective of specific industry application, big data analysis technology is less adopted by the FMCG industry and the technology category is relatively single. The most used is the optimization algorithm, which mainly applied to the process optimization of production and supply to reduce related costs and improve the operating efficiency of the whole product cycle, so as to improve the enterprise revenue and profit. Compared to FMCG, the application of big data technology in durable goods

Prescriptive Analytics
Descriptive Analytics
Predictive Analytics
FMCG
Professional services
(Semi)-durable goods
Personal services

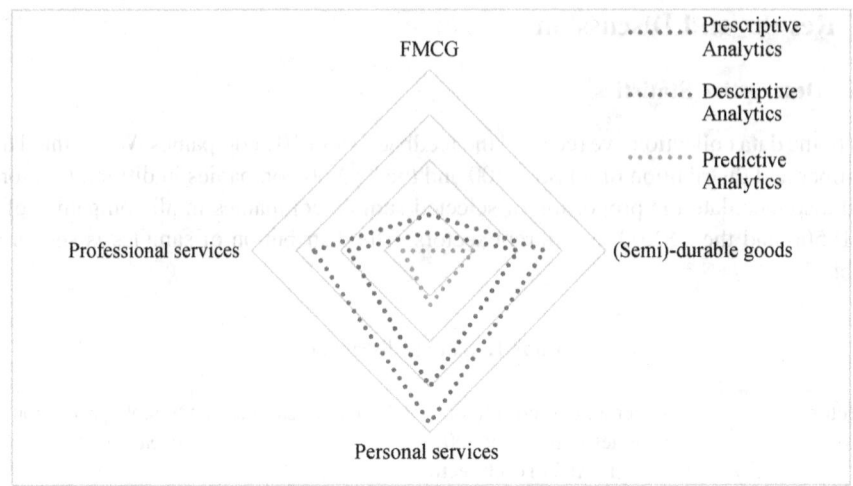

Fig. 1. Usage of the Broad category of BDA in different sectors

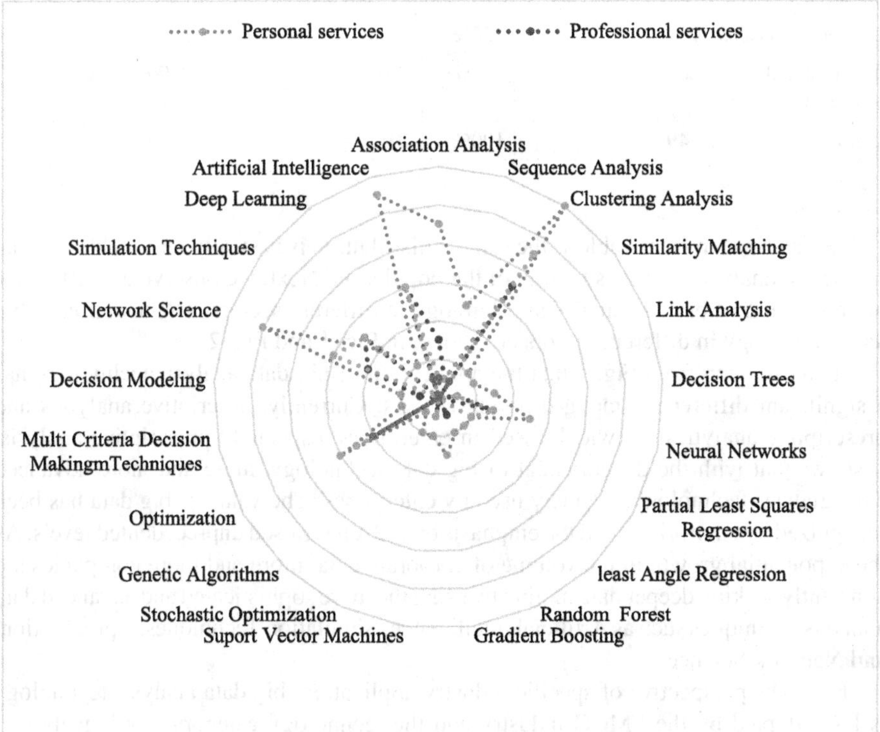

Personal services
Professional services
Association Analysis
Artificial Intelligence
Sequence Analysis
Deep Learning
Clustering Analysis
Simulation Techniques
Similarity Matching
Network Science
Link Analysis
Decision Modeling
Decision Trees
Multi Criteria Decision MakingmTechniques
Neural Networks
Optimization
Partial Least Squares Regression
Genetic Algorithms
least Angle Regression
Stochastic Optimization
Random Forest
Suport Vector Machines
Gradient Boosting

Fig. 2. Specific method and technology usage in different sectors

manufacturing enterprises is much more extensive. This is mainly because of the different characteristics of the products between them. The problems faced and that need to be solved by durable goods manufacturers in the operation and management are more systematic, complicated and long-term. Big data analysis technologies can be applied to industrial design, R&D, manufacturing, sales, service and other aspects of enterprise operation management. Because the product life cycle is long and costs of production are generally relatively high, it is very necessary for enterprises to strengthen the development and application of data value from the aspects of analyzing customer needs and preferences, improving production technology, improving the internal management level of enterprises, and improving after-sales service to improve customer loyalty. The professional services sector is very similar to the durable goods industry in the application of big data analysis technology. In addition, among these 4 sectors, big data analysis technology is most widely used in the personal service industry, and almost all forms of big data analysis technology are adopted. This also fully reflects the social attributes and behavioral complexity of human beings. The main big data analysis technologies used are cluster analysis (e.g., customer or market segments, identification and classification of unknown behavior patterns), network analysis (e.g., the exploration and excavation of social relations and network effects), and artificial intelligence.

4.2 Regression Analysis

Based on the sample data, we conduct the regression analysis and finally obtain some model equations with relative strong explanatory power (The F value of the model is significant at the 0.05 level). The statistically significant results are summarized in Table 2 and Table 3.

Table 2. Standardized Beta and significance testing results for first level independent variables

	Net income	Operating income	Gross profit	Total revenue	Sales revenue	Share price	Earning before tax
Size	0.187	0.252^*	0.555^{**}	0.806^{***}	0.861^{***}	0.053	0.218^*
Sector (Semi)-durable goods	0.114	0.120	0.251	0.196^{**}	0.213^{**}	−0.152	0.102
Descriptive Analytics	−0.020	−0.010	−0.079	−0.042	−0.095	0.659^{***}	0.002
Prescriptive Analytics	0.444^{***}	0.427^{**}	0.497^{**}	0.221^{**}	0.194^*	0.070	0.415^{**}

Note: $^{***}p < 0.001$, $^{**}p < 0.01$, $^*p < 0.05$. The numbers in the table are the standardized Beta. Each column corresponds to a dependent variable equation.

As shown in Table 2 and Table 3, descriptive analytics and prescriptive analytics contribute positively to the corporate performance to some extent. To be specific, descriptive

Table 3. Standardized Beta and significance testing results for secondary independent variables

	Net income	Operating income	Gross profit	Total revenue	Sales revenue	Share price	Earning before tax
Size	0.178	0.233*	0.540*	0.780***	0.866***	−0.004	0.206*
Sector (Semi)-durable goods	0.046	0.044	0.364	0.145*	0.234*	−0.157	0.037
Association Analysis	0.084	0.101	0.080	0.049	−0.059	0.403***	0.091
Sequence Analysis	−0.078	−0.067	0.000	−0.023	0.039	0.324***	−0.072
Network Science	0.382**	0.352**	0.563**	0.193**	0.284**	0.014	0.307**
Artificial Intelligence	0.408***	0.382**	0.101	0.158*	0.037	0.165	0.404***

Note: $^{***}p < 0.001$, $^{**}p < 0.01$, $^*p < 0.05$. The numbers in the table are the standardized Beta. Each column corresponds to a dependent variable equation.

analytics contribute positively to share price. Within the descriptive analytics, association analysis and sequence analysis show significant positive effects on the share price of the company. Compared to the descriptive, the influence of prescriptive analytics on corporate performance is broader. Its contribution involves various revenue and profit-related performance indicators such as Net income, Operating income, Gross profit, Total revenue, Sales revenue, Earning before tax. However, this is mainly because Network Science and Artificial Intelligence provide significant positive contributions, other method and technology belonging to prescriptive analytics do not appear to contribute as much to the results.

5   Conclusions

This study is unique in that it provides key insights into the impact of specific families and types of Big Data Analytics techniques on firm performance. As such, the study contributes meaningfully to the literature in several ways. First, the study shows that BDA has a significant overall impact on firm performance. Second, descriptive analytics contribute positively to the profit-related performance (i.e. share price) whereas prescriptive analysis contributes positively to both revenue and profit-related performance. Finally, the contribution of BDA to the revenue performance of manufacturing industry is greater than in other industries.

In sum, descriptive analytics contribute positively to the profit-related performance (i.e. share price) whereas prescriptive analysis contributes positively to both revenue and profit-related performance. Interestingly, we found in the results that descriptive analytics and prescriptive analytics does not form an overlap in the affected performance indicators of the enterprises, that is, different big data analysis technologies may have different influence mechanisms on enterprise performance.

Therefore, it can be seen that some hypotheses in Sect. 3 are tenable, big data analysis has a significant and extensive impact on corporate performance. Different types of methods can contribute positively to different corporate performance indicators. In particular, it is worth noting that the impact of the application of big data analysis on enterprise performance is also significantly different in different sectors. Sector (Semi)-durable goods is markedly different from other industries, which mainly reflect on the indicators of Total revenue, Sales revenue. It shows that the contribution of BDA to the revenue performance of manufacturing industry is greater than other industries.

In addition, through the cross-validation of the analysis results of the two sets of models, it can be concluded that the results obtained in this study are effective and reliable, and can explain the impact of the application of BDA on the financial performance of organizations to a certain extent.

References

1. Gartner IT Glossary. https://www.gartner.com/it-glossary/big-data/. Accessed 21 Sept 2019
2. Tech America Foundation's Federal Big Data Commission, Demystifying bigdata: A practical guide to transforming the business of Government. https://www.techamerica.org/Docs/fileMa nager.cfm?f=techamerica-bigdatareport-final.pdf. Accessed 21 Sept 2019
3. Grover, V., Chiang, R.H., Liang, T.P., Zhang, D.: Creating strategic business value from big data analytics: a research framework. J. Manag. Inf. Syst. **35**(2), 388–423 (2018)
4. Erevelles, S., Fukawa, N., Swayne, L.: Big Data consumer analytics and the transformation of marketing. J. Bus. Res. **69**(2), 897–904 (2016)
5. Gartner: Gartner shows 87 percent of organizations have low BI and analytics maturity. https://www.gartner.com/en/newsroom/press-releases/2018-12-06-gartner-data-shows-87-percent-of-organizations-have-low-bi-and-analytics-maturity. Accessed 21 Sept 2019
6. Sykes, N.: 5 companies using big data and AI to improve performance. https://www.kol abtree.com/blog/5-companies-using-big-data-and-ai-to-improve-performance/. Accessed 11 Dec 2020
7. Rexer, K., Gearan, P., Allen, H.: Data science survey 2015. https://www.rexeranalytics.com/data-science-survey.html. Accessed 21 Sept 2019
8. Nicola, S., Ferreira, E.P., Ferreira, J.J.P.: A quantitative model for decomposing & assessing the value for the customer. J. Innov. Manag. **2**(1), 104–138 (2014)
9. Gopalkrishnan, V., Steier, D., Lewis, H., Guszcza, J.: Big data, big business: bridging the gap. In: Proceedings of the 1st International Workshop on Big Data, Streams and Heterogeneous Source Mining: Algorithms, Systems, Programming Models and Applications, pp. 7–11. ACM (2012)
10. Deighton, J.: Big data. Consum. Mark. Cult. **22**, 1–6 (2018)
11. Martin, K.D., Borah, A., Palmatier, R.W.: The dark side of big data's effect on firm performance. In: Marketing Science Institute Working Paper Series, pp. 16–104 (2016)
12. Colicev, A., Malshe, A., Pauwels, K., O'Connor, P.: Improving consumer mindset metrics and shareholder value through social media: the different roles of owned and earned media. J. Mark. **82**(1), 37–56 (2016)
13. McKinsey Global Institute, Big Data: The next frontier for innovation, competition, and productivity. https://www.mckinsey.com/~/media/McKinsey/Business%20Functions/McK insey%20Digital/Our%20Insights/Big%20data%20The%20next%20frontier%20for%20i nnovation/MGI_big_data_exec_summary.pdf. Accessed 21 Sept 2019

14. Martin, D.M., Schouten, J.: Sustainable marketing. Pearson Prentice Hall, Upper Saddle River (2011)
15. Müller, O., Fay, M., vom Brocke, J.: The effect of big data and analytics on firm performance: an econometric analysis considering industry characteristics. J. Manag. Inf. Syst. **35**(2), 488–509 (2018)
16. Zhang, T., Wang, W.Y.C., Pauleen, D.J.: Big data investments in knowledge and non-knowledge intensive firms: what the market tells us. J. Knowl. Manag. **21**(3), 623–639 (2017)
17. Popovič, A., Hackney, R., Tassabehji, R., Castelli, M.: The impact of big data analytics on firms' high value business performance. Inf. Syst. Front. **20**(2), 209–222 (2018)
18. Xu, Z., Frankwick, G.L., Ramirez, E.: Effects of big data analytics and traditional marketing analytics on new product success: a knowledge fusion perspective. J. Bus. Res. **69**(5), 1562–1566 (2016)
19. Bressanelli, G., Adrodegari, F., Perona, M., Saccani, N.: Exploring how usage-focused business models enable circular economy through digital technologies. Sustainability **10**(3), 639 (2018)
20. El-Kassar, A.N., Singh, S.K.: Green innovation and organizational performance: the influence of big data and the moderating role of management commitment and HR practices. Technol. Forecast. Soc. Change **144**, 483–498 (2019)
21. Gupta, S., Kumar, S., Singh, S.K., Foropon, C., Chandra, C.: Role of cloud ERP on the performance of an organization. Int. J. Logist. Manag. **29**(2), 659–675 (2018)
22. Barney, J.B.: Firm resources and sustained competitive advantage. J. Manag. **17**, 99–120 (1991)
23. Lee, R.P., Grewal, R.: Strategic responses to new technologies and their impact on firm performance. J. Mark. **68**(4), 157–171 (2004)
24. Delen, D., Ram, S.: Research challenges and opportunities in business analytics. J. Bus. Anal. **1**(1), 2–12 (2018)

Advances in Public Administration

Development of the e-Government in the Context of the 2020 Pandemics

Darya Rozhkova[1]([✉]) [iD], Nadezhda Rozhkova[2] [iD], and Uliana Blinova[1] [iD]

[1] Financial University under the Government of the Russian Federation,
125993 Moscow, Russia
rodasha@mail.ru
[2] State University of Management, 109542 Moscow, Russia

Abstract. The epidemic caused by the coronavirus COVID-19 influenced the world and had a major economic impact. A sudden and dramatic increase in online activity and necessity to stay at home increased a demand for relevant and reliable information accounting and its fast diffusing. Some of these changes will continue beyond this pandemic. In this article, the author examines the creation, development and implementation in the work of state and municipal institutions of the Government Platform in Russian health care system, including creation of the information platform (COVID-19). Using pandemics as a research context, we have evaluated the role of information and communication technologies (ICTs) in emergency situations which includes technologies to inform, prevention technologies, technologies to engage. This paper focuses on a case study. We made a conclusion about results of implementation the Government Platform in health care system during crises caused by pandemic 2020 through the lenses of technologies implementation. Our results show that largely the information platform (COVID-19) used technologies to inform and prevention technologies, however engaging mechanisms are quite poor.

Keywords: e-Government · Government as a Platform · Orchestration

1 Introduction

In recent times, major technological changes, including the appearance of Internet technologies, brought unique characteristics to the global economic system. The term "digital economy" describes economic and social activities in the context of information and communication technologies, using the Internet and devices (Negroponte 1995). The subject of the research in this work is the ongoing digitalization of public services, as well as the peculiarities of digitalization during the crisis.

According to many scientists, the global world is rapidly entering the era of the digital platform economy, in which the tools and mechanisms used are based on the Internet and online platforms form the foundation of the economic and social life (Kenney and Zysman 2015, 2016; Miller 2020). The main principles of digital platform economy include: the network effect principle, the principle of digital potential, the principle of

T. Antipova (Ed.): ICADS 2021, AISC 1352, pp. 465–476, 2021.
https://doi.org/10.1007/978-3-030-71782-7_41

infinity and super speed, the value-building principle, the principle of information asymmetry reduction (Johnson et al. 2008; Rozhkova 2017). One of the main characteristics of a digital economy is the high role of the information and knowledge in the production of products and services and the active use of digital methods of storing, processing and transmitting data.

The European Commission also defines online platforms through their functional design as "search engines, social media, e-commerce platforms, online stores, price comparison sites" (The European Commission 2016). Moreover, some companies initially built their enterprise as online platforms (for example, Aliexpress), while others "grew" into platforms thanks to the progressive involvement number of participants, increasing the proposed business opportunities (for example, Facebook).

The idea of a Government as a Platform (GaaP) was put forward by the writer Tim O'Reilly as a formation of an open platform by the state in order to involve various partners in cooperation with authorities (O'Reilly 2011). The concept of Government as a Platform includes an idea that the information produced by and on behalf of citizens is a national asset.

A Government as a Platform concept has been implemented by governments around the world, including USA (Orszag 2011), UK (Brown et al. 2017), Italy (Cordella and Paletti 2019) and Russia (Petrov et al. 2018; Styrin et al. 2019).

Platforms connect providers and consumers of information and services, organizing network interactions, thereby acting as a tool in the public administration system (Janssen and Estevez 2013; Mukhopadhyay et al. 2019), increase overall efficiency of the service production and delivery (O'Reilly 2011; Walravens and Ballon 2013), extend different values associated with public services (Cordella and Bonina 2012; Bannister and Connolly 2014).

Note, however, that the context of all studies is devoted to the "normal" functioning of platforms. However, it remains unclear how these mechanisms can be implemented in the context of crisis component. There is a research gap in understanding of the GaaP, main technologies that is used and enhanced during crises.

This research sheds light on the impact that crises have on the creation of GaaP, features of its operation. The global economy faces a variety of challenges: different kinds of crises (economic, social or terrorism) and natural disasters (eruption, flooding or epidemics). The state should respond to any change faster than any other participants of economic activity.

We have chosen the novel coronavirus infection (COVID-19) which became a research context in many investigations (Niyazbekova 2020). From one hand, the main consequence of COVID-19 is a medical impact (a dramatic effect on a health of people), from the other hand – economic one (reductions in income, a rise in unemployment, a bankruptcy of many organizations) (Antipova 2021). COVID-19 have impacted almost all industries. It boosts a further digitalization (Rozhkova et al. 2019). Thus, it is important to use the role of information and communication technologies (ICTs) in emergency situations in order to minimize all negative repercussions.

The paper proceeds as follows. An introduction includes a relevance of the study and a research context. Section 2 reviews a Government as a Platform concept and a main types of crises communications. Section 3 provides a general background of research and the experience of creating the Government platform in health care system of the Russian Federation. In Sect. 4 we made an attempt to evaluate results, summarize main problems and prospects. Summary and conclusions describe main findings and prospects of future research.

2 Theoretical Framework

Nowadays, technologies' development plays an important role today in finance (Ross and Liechtenstein 2018), government auditing (Antipova 2018). Digital platforms are phenomena which includes enlarging of traditional principal-agent relationships (Ghazawneh and Henfridsson 2013) and network effects (Belleflamme and Peitz 2018).

A Government as a Platform concept demonstrates all the complexity of large-scale digital transformation (Evans and Basole 2016). The introduction of digital platforms requires government agencies to conduct serious legal, organizational, moral and ethical transformation. A lot of resources are required to transform previously created information systems into state digital platforms.

To digitalize public administration based on a platform approach it is necessary to carry out organizational and technological changes including fixing and framing to tackle the variety of barriers (Meijer 2015). When introducing digital technologies into the public administration system, significant budgetary expenditures are required, which are difficult to calculate in advance. Moreover, the introduction of technological innovation does not automatically lead to changes in the organizational culture (Oreg and Goldenberg 2015).

A Government as a Platform concept is a broader instrument than platforms in commercial organizations in terms of functionality and impact on public, industry and market processes (Jamieson et al. 2020). Government as a Platform forms and digitally regulates industry processes, makes transparent relations between existing industry players through secure participation technologies, assess mechanism of generating added value for each participant and the industry as a whole.

A Government as a Platform operates not only within one industry, but also to organize a cross-industry interaction due to open architectural standards. It requires more coordination of various authorities, as well as the choice of the operator. The operator will coordinate relationships between different groups of platform participants, enhance collaboration and co-creation (Janssen and Estevez 2013).

In the crises context governments can use the role of information and communication technologies (ICTs) in emergency situations which includes (Asmolov 2016):

1. Technologies to inform. The main aim of such technologies is to provide a general information picture of the crisis. The important role of IT in overcoming an information deficit is obvious (Qayyum 2014).
2. Prevention technologies. They target to transmit information that may be important to specific users to improve their personal safety, and recommend specific steps to reduce the level of risk. It includes as well collaboration of different participants (Brown 2004) and creation of different information to reduce uncertainty, describe unpredictable environments etc. (McKinney and Yoos 2016).

3. Technologies to engage. They offer users active forms of participation in crisis response through digital mechanisms for mobilizing their resources. The use of this kind of technology is popular at a corporate level through co-creation of value (Bengtsson and Kock 2014; Gnyawali 2016) and quite limited in the context of government (Fuglsang 2008). Nevertheless, government interaction with their citizens will bring more efficient and effective public services (Kokkinakos et al. 2012).

3   Research Setting and Methodology

3.1   General Background of Research

The method that we use is an explanatory case study (Yin 2009). The essence of the case method used in different fields of science is that it gives an idea of a solution or a set of solutions, describes why these decisions were made, how they were introduced, and to what resulted in (Schramm 1971).

The process of conducting a case study includes the stages that are traditional for any scientific research: defining the goal and objectives, collecting, analysing and interpreting the data obtained, preparing a result on the study. The main methods of collecting data during the case study are document analysis, observation, interviewing. Features of the above methods allow us to consider the studied case in broader historical, social, cultural aspects. We sourced data through interviews and documents.

As an object of the research, we have chosen Russian Government Platform in health care system and the information system (platform) for accounting information in order to prevent the spread of a new coronavirus infection (COVID-19). These platforms targeted to form the possibility of quick access to primary health care, increasing high-speed information exchange in medical organizations. Some authors highlighted a need for reconfiguration and optimization of the health care infrastructure in Russia (Antipova and Shikina 2017).

This is one of the most demanded recent projects. Never before state had a program which covers such a huge number of participants, includes mobile application, convenient interfaces for recording on the Internet, SMS notification of an upcoming visit to the doctor, an electronic medical record, an Internet portal for doctors and many other convenient functions of the system. Government Platform in health care system is an important step to create and implement the concept of a Government as a Platform (GaaP) (O'Reilly 2011).

3.2   The Government Platform in Health Care System

In Russian practice, the term "platform" is often utilized as a synonym of an information system. An information system is a system designed to store, search and issue information at the request of users (information service system).

The Unified Government Information System (the Government Platform in health care system) in the health care sector is a national information system which has been created to provide effective information support to government agencies and organizations of the healthcare system, as well as citizens within the framework of the management of medical care. Its development took the first step in 2011 and was completed by 2017.

The principles of building the Unified Government Information System are determined by the specifics of financing medical institutions. Hospitals, clinics and other medical organizations working in the mandatory health insurance system receive the main funding from the budget, which, in turn, consist of insurance premiums deducted by employers. In order to be able to quickly monitor the activities and reporting of health facilities from the centre in order to efficiently spend the funds, it was decided to centralize the IT system. The general architecture of the Unified Government Information System consists of a segment of centralized system-wide components and a segment of applied components.

The first, according to the concept, include subsystems for integrating application systems, maintaining a directory of system users, maintaining a register of reference information, dictionaries of medical terminologies and a register of electronic documents, a subsystem for managing a certification centre, operating management, maintaining e-mail, etc. All components are operated by the Ministry of Health.

The segment of applied components includes transactional (formation of primary information about the activities of medical institutions, automation of information exchange), managerial (integrated electronic medical record), personalized accounting of services provided) and reference (information support of the population, medical staff, students) subsystems.

Applied information systems are subdivided into federal (created by the Ministry of Health) and regional. According to the concept, Russian regions create and operate regional application systems, integrate them with federal application systems and centralized services. Treatment-and-prophylactic institutions report on their own activities to the health authorities with the help of implemented medical information systems. All these innovations in a health care industry improve a medical care safety (Voskanyan et al. 2020, 2021), decrease costs (Somkin et al. 2021) and increase a satisfaction of a population (Buzin 2020).

The global experience of 2020 in the fight against the Covid-19 virus have shown the problem that the states in any moment can face the conditions caused by natural and technogenic factors that may threaten the life of the population for a long time. An important step concerning pandemic was an introduction of the Decree of the Government of the Russian Federation, issued March 31, 2020 No. 373 "On Approval of the Interim Rules for Recording Information to Prevent the Spread of a New Coronavirus Infection (COVID-19)".

According to this decree, the Ministry of Health of the Russian Federation is the operator of an information system (platform) for accounting information in order to prevent the spread of a new coronavirus infection (COVID-19). Information platform keeps records of information about persons with a confirmed diagnosis of a new coronavirus infection (COVID-19) and hospitalized persons with signs of pneumonia, as well as about persons in contact with patients. Information platform (COVID-19) includes federal and regional segments.

The functions and participants of the information platform (COVID-19) are presented in Fig. 1 (Decree of the Government of the Russian Federation 2020).

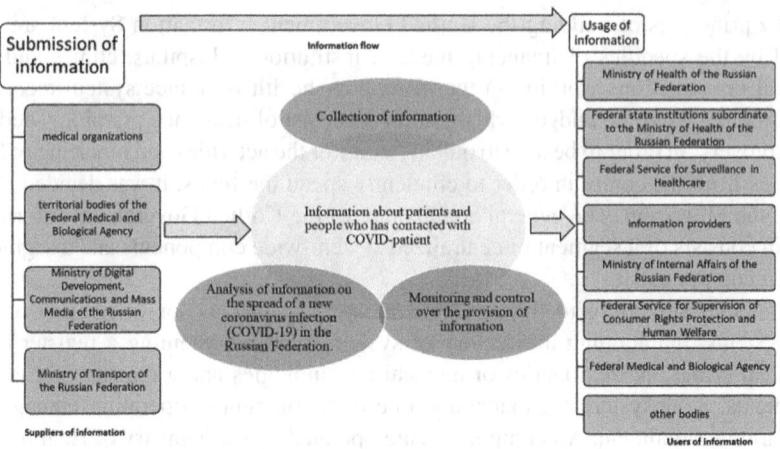

Fig. 1. The function and participants of the information platform (COVID-19)

The objectives of the information platform (COVID-19) are:

a) collection and recording of up-to-date information about patients and people who has contacted with COVID-patient;
b) organizing the exchange between users the information about patients and contact persons in compliance with the requirements of the legislation of the Russian Federation in the field of personal data.

The Ministry of Health of the Russian Federation (the information platform operator):

a) maintain an information platform (COVID-19) in accordance with the legislation of the Russian Federation on information, information technology and information protection;
b) develop and coordinate federal executive authorities (information providers), regulate the transfer of information in a protected form;
c) develop and communicate to all interested participants instructions and forms according to which information is submitted to an information platform (COVID-19);
d) coordinate the submission of information to the information platform (COVID-19);
e) protect of information contained in the information platform (COVID-19) in accordance with the requirements established by the Federal Security Service of the Russian Federation and the Federal Service for Technical and Export Control;
f) establish requirements and measures to protect the information contained in the information platform (COVID-19);
g) provide an access to the information platform (COVID-19) with the differentiation of the corresponding access rights;
h) provide a methodological support about technical use and content of the information platform (COVID-19).

4 Analysis of the Case

Developing of the information platform (COVID-19) can be divided into 2 different steps. Step 1 includes an implementation of a personalized accounting system for all cases of coronavirus infection (March 2020).

The Russian Ministry of Health has introduced a system of personified registration of all cases of coronavirus infection and pneumonia. In accordance with the Decree of the Government of the Russian Federation No. 373 of March 31, 2020, all medical organizations providing medical care to such patients, the federal executive authorities enter information about them into the information platform. Each of the participants, within the framework of their competence, has the opportunity to use all the information accumulated in the system and necessary to carry out work to prevent the spread of a new coronavirus infection both at the regional and federal levels. Based on the data received from the information system, a report is formed, which is available to federal remote consulting centres for anaesthesiology and resuscitation on the diagnosis and treatment of the new coronavirus infection COVID-19 and pneumonia.

Currently, about 6 thousand employees from almost 1 thousand medical organizations in 84 federal subjects of the Russian Federation enter information into a unified information system for recording patients with COVID-19 and pneumonia, which makes it possible to monitor the situation with the incidence of coronavirus infection COVID-19 and pneumonia in real time, to analyse the spread of COVID-19 throughout the entire territory of the Russian Federation. After the completion of updating the data in the information system and full involvement in the process of entering data on patients of all constituent entities of Russia, the information platform (COVID-19) will become a single source of reliable data on patients with coronavirus infection and pneumonia.

Step 2 includes a creation of information about health workers and students who are entitled to incentive payments for work with coronavirus infection. As well an information about sick doctors has been implemented (June 2020). The information platform (COVID-19) significantly expanded due to information about sick medical workers, as well as the establishment of incentive payments to doctors and students working with patients with coronavirus infection. Now the information platform (COVID-19) contains information about profile patients or those who have been in contact with patients, it is necessary to separately enter information about sick doctors, as well as about students and health workers who are entitled to incentive payments for working with infected people, and about medical organizations in which they are working.

The data of infected health workers must be uploaded to the system within one day of the confirmation of COVID-19. Responsible for this are the regional ministry of health or the institutions of the Federal biomedical agency. The same amount of time is given to a medical organization to inform medical and other workers, including students, about the establishment of "COVID" bonuses from the moment of the beginning of the activity for which they are encouraged.

Compared to the March 2020 (government decree No. 373), the list of departments that have gained access to this information has grown. These are the Compulsory Medical Insurance Fund (including its territorial branches), the Social Insurance Fund, the Rosgvardia, the government apparatus and the presidential administration, the Ministry of Internal Affairs, "other bodies or organizations" that are determined by the federal operational headquarters. According to experts, in the context of multitasking, a large number of participants, many data transmission channels, the information platform of the Ministry of Health of Russia for monitoring COVID-19 is a tool necessary for making timely and well-grounded management decisions, modelling the situation, and making forecasts for the development of the epidemic process. It is very important that the use of the system can significantly reduce the volume of reporting documents, summaries, notifications that medical organizations, already experiencing a shortage of personnel, are still forced to collect and send to various instances. The data accumulated in the system on the methods of diagnosis and treatment helps to quickly develop and implement in practice the most effective algorithms for the provision of medical care.

We have studied that the information platform (COVID-19) had been built according to the principle of maximum possible unification and integration of functional areas of management (centralized approach). The advantages of information platform (COVID-19) include: the use of uniform requirements, development of interdepartmental interaction, optimization of expenses for information, telecommunication and transport infrastructure, energy and other resources. At the same time, the system is less adaptive to external changes and user requests, require significant costs to ensure the reliability of operation and safety of information resources, as well as compliance with formal procedures for their application. Thus, there is a room for improvements in order to increase the key performance of the public system in general and information platform in particular (Antipova 2017).

Largely the information platform (COVID-19) used technologies to inform. One of the main challenges of the global crisis, the picture of which is changing, is monitoring information and presenting a dynamic, holistic picture of what is happening. This task has become especially difficult because of the number of unverified facts, rumours and misinformation was constantly growing. The information platform (COVID-19) allowed to familiarize with the picture created on the basis of relatively big data in an accessible form.

If we evaluate prevention technologies embedded to the information platform (COVID-19), we can say that the level is quite high. Unlike tools that offer a general picture of events based on an analysis of big data, prevention technologies seek to convey information about risks in the most targeted way. Here the task is to determine what information is directly related to a specific target audience, which, on the basis of this information, should take steps to increase the degree of their personal safety, as well as the safety of those around them. The information platform (COVID-19) has reliable epidemic information and is a base of prevention measures (sending SMS, Telegram-channel etc.). In response to coronavirus, government agencies of Russian Federation quickly began to use messengers as an effective communication channel. In Singapore, there is already a system for receiving information via a smartphone about possible contacts with patients with coronavirus infection. The TraceTogether application allows you to identify everyone with whom the owner of the smartphone was in close contact. Unfortunately, such initiatives are not used in Russia. One of recommendation can be a step to enrich and to enlarge a usage of prevention technologies. For example, state can add additional information about COVID-19 in widely used governmental portal ("gosuslugi") or based on smartphone applications send all recent news to citizen.

Take notice that the information platform (COVID-19) does not use technologies to engage. Engagement technologies suggest that users have the opportunity to actively participate in the response to a crisis. But due to centralized approach and the vision of government authorities the information platform (COVID-19) does not involve people. The role of technologies to engage is certainly underestimated in Russia. A creation of a collaborative nature through public and private organizations collaboration can boost the process of solutions co-creation (Criado et al. 2020). Further research is clearly needed (interviews, analysis of foreign experience) that will provide an understanding of why the Russian authorities underestimate the potential of technologies to engage, possibilities of co-creation and motivation to implement the idea of Government as a Platform through citizens involvement.

5 Summary and Conclusions

The next challenge for the world community today is the outbreak of coronavirus infection (COVID-19). This virus had an impact not only on the healthcare sector, but also became a trigger for large-scale digitalization, starting the process of adaptation of all spheres of society to the conditions of a new "remote" reality. All this has led to the transformation of the methods and types of interaction between public authorities and citizens, implemented through the integration of an increasing number of state institutions into the digital space. The work of state information platforms is carried out by creating new automated processes for collecting and processing information necessary for the implementation of the powers of state bodies and ensuring the exchange of information between these bodies, as well as for other purposes established by federal laws, as well as increasing the efficiency of information exchange and forecast of its development.

The introduction of restrictive measures in connection with the spread of COVID-19 made it difficult to provide state and municipal services through the physical appeal of citizens to the office of the relevant service. In such circumstances, government platforms have become the key to maintaining the services provided to citizens at the appropriate level. Thus, state information platforms are a necessary component to ensure the stable functioning of state infrastructure in a pandemic.

References

Antipova, T.: Human-computer interaction in the public sector performance evaluation analysis. In: Rocha, Á., Correia, A., Adeli, H., Reis, L., Costanzo, S. (eds.) Recent Advances in Information Systems and Technologies. WorldCIST 2017. Advances in Intelligent Systems and Computing, vol. 570. Springer, Cham (2017). https://doi.org/10.1007/978-3-319-56538-5_56

Antipova, T.: Using blockchain technology for government auditing. In: 13th Iberian Conference on Information Systems and Technologies (CISTI), Caceres, pp. 1–6 (2018). https://doi.org/10.23919/CISTI.2018.8399439

Antipova, T.: Coronavirus pandemic as black swan event. In: Antipova, T. (ed.) Integrated Science in Digital Age 2020. ICIS 2020. Lecture Notes in Networks and Systems, vol. 136. Springer, Cham (2021). https://doi.org/10.1007/978-3-030-49264-9_32

Antipova, T., Shikina, I.: Informatic indicators of efficacy cancer treatment. In: 12th Iberian Conference on Information Systems and Technologies (CISTI) Lisbon, Portugal, 21–24 June 2017, pp. 1–5 (2017).https://doi.org/10.23919/CISTI.2017.7976049

Asmolov, G.: Subject, crowd and the governance of activity: The role of digital tools in emergency response Doctoral dissertation, The London School of Economics and Political Science (LSE) (2016)

Bannister, F., Connolly, R.: ICT, public values and transformative government: a framework and programme for research. Gov. Inf. Q. **31**(1), 119–128 (2014). https://doi.org/10.1016/j.giq.2013.06.002

Belleflamme, P., Peitz, M.: Platforms and network effects. In: Handbook of Game Theory and Industrial Organization, vol. II. Edward Elgar Publishing (2018)

Bengtsson, M., Kock, S.: Coopetition - Quo vadis? Past accomplishments and future challenges. Ind. Mark. Manag. **43**(2), 180–188 (2014)

Brown, H.G., Poole, M.S., Rodgers, T.L.: Interpersonal traits, complementarity, and trust in virtual collaboration. J. Manag. Inf. Syst. **20**(4), 115–138 (2004)

Brown, A., Fishenden, J., Thompson, M., Venters, W.: Appraising the impact and role of platform models and government as a platform (Gaap) in UK government public service reform: towards a platform assessment framework (PAF). Gov. Inf. Q. **34**(2), 167–182 (2017)

Buzin, V.N., Mikhailova, Y.V., Chukhriyenko, I.Y., Buzina, T.S., Shikina, I.B., Mikhailov, A.Y.: Russian healthcare through the eyes of the population: Dynamics of satisfaction over the past 14 years (2006-2019): review of sociological studies. Profilakticheskaya Meditsina **23**(3), 42–47 (2020). https://doi.org/10.17116/profmed20202303142

Communication from the commission to the European parliament, the council, the European economic and social committee and the committee of the regions. Online Platforms and the Digital Single Market Opportunities and Challenges for Europe, European commission, 18 p. (2016)

Cordella, A., Bonina, C.M.: A public value perspective for ICT enabled public sector reforms: a theoretical reflection. Gov. Inf. Q. **29**(4), 512–520 (2012). https://doi.org/10.1016/j.giq.2012.03.004

Cordella, A., Paletti, A.: Government as a platform, orchestration, and public value creation: the Italian case. Gov. Inf. Q. **36**(4), 101409 (2019)

Criado, J.I., Guevara-Gómez, A., Villodre, J.: Using collaborative technologies and social media to engage citizens and governments during the COVID-19 Crisis. The case of Spain. Digit. Gov.: Res. Pract. **1**(4), 1–7 (2020)

Decree of the Government of the Russian Federation, issued March 31, 2020 No. 373 "On Approval of the Interim Rules for Recording Information to Prevent the Spread of a New Coronavirus Infection (COVID-19)" (2020)

Evans, P.C., Basole, R.C.: Revealing the API ecosystem and enterprise strategy using visual analytics. Commun. ACM **59**(2), 23–25 (2016)

Fuglsang, L.: Capturing the benefits of open innovation in public innovation: a case study. Int. J. Serv. Technol. Manag. **9**(3–4), 234–248 (2008)

Ghazawneh, A., Henfridsson, O.: Balancing platform control and external contribution in third-party development: the boundary resources model. Inf. Syst. J. **23**(2), 173–192 (2013)

Gnyawali, D.R., Madhavan, R., He, J., Bengtsson, M.: The competition–cooperation paradox in inter-firm relationships: a conceptual framework. Ind. Mark. Manag. **53**, 7–18 (2016)

Jamieson, D., Wilson, R., Martin, M.: Is the GaaP wider than we think? Applying a sociotechnical lens to Government-as-a-Platform. In: Proceedings of the 13th International Conference on Theory and Practice of Electronic Governance, pp. 514–517 (2020)

Janssen, M., Estevez, E.: Lean government and platform-based governance - doing more with less. Gov. Inf. Q. **30**, 1–8 (2013)

Johnson, M.W., Christensen, C.M., Kagermann, H.: Reinventing your business model. Harv. Bus. Rev. **86**(12), 57–68 (2008)

Kenney, M., Zysman, J.: Choosing a future in the platform economy: the implications and consequences of digital platforms. In: Kauffman Foundation New Entrepreneurial Growth Conference, vol. 156160 (2015)

Kenney, M., Zysman, J.: The rise of the platform economy. Issues Sci. Technol. **32**(3), 61 (2016)

Kokkinakos, P., Koussouris, S., Panopoulos, D., Askounis, D., Ramfos, A., Georgousopoulos, C., Wittern, E.: Citizens collaboration and co-creation in public service delivery: the COCKPIT project. Int. J. Electr. Gov. Res. (IJEGR) **8**(3), 33–62 (2012)

McKinney, E.H., Yoos, C.J.: Information about information: a taxonomy of views. MIS Q. **34**(2), 329–344 (2016)

Meijer, A.: E-governance innovation: barriers and strategies. Gov. Inf. Q. **32**(2), 198–206 (2015)

Miller, V.: Understanding Digital Culture. SAGE Publications Limited, Thousand Oaks (2020)

Mukhopadhyay, S., Bouwman, H., Jaiswal, M.P.: An open platform centric approach for scalable government service delivery to the poor: the aadhaar case. Gov. Inf. Q. **36**(3), 437–448 (2019)

Negroponte, N.: Being Digital, pp. 65–67. Hoder and Stoughton, London (1995)

Niyazbekova, S.U.: Foreign banks management in the context of a pandemic: problems and solutions. E-Management, **3**(3), 4–12 (2020). (in Russ.). https://doi.org/10.26425/2658-3445-2020-3-3-4-12. Ниязбекова Ш.У. Менеджмент зарубежных банков в условиях пандемии: проблемы и пути их решения. E-Management, 3(3) с. 4–12 (2020)

Oreg, S., Goldenberg, J.: Resistance to Innovation: Its Sources and Manifestations. University of Chicago Press, Chicago (2015)

O'Reilly, T.: Government as a platform. Innov. Technol. Gov. Global. **6**(1), 13–40 (2011)

Orszag, P.: Open government directive. https://www.whitehouse.gov/open/documents/open-government-directive. Accessed 10 Oct 2020

Petrov, M., Burov, V., Shklyaruk, M., Sharov, A.: State as a platform. (Cyber) state for the digital economy. Digital transformation. M.: Foundation CSR (2018). (in Russian) Петров М., Буров В., Шклярук М., Шаров А. Государство как платформа. (Кибер) государство для цифровой экономики. Цифровая трансформация. М.: Фонд ЦСР (2018)

Ross, G., Liechtenstein, V.: Management of financial bubbles as control technology of digital economy. In: Antipova, T., Rocha, Á. (eds.) Information Technology Science. MOSITS 2017. Advances in Intelligent Systems and Computing, vol. 724. Springer, Cham (2018). https://doi.org/10.1007/978-3-319-74980-8_9

Qayyum, A., Thompson, K.M., Kennan, M.A., Lloyd, A.: The provision and sharing of information between service providers and settling refugees. Inf. Res. **19**(2) (2014)

Rozhkova, D., Rozhkova, N., Blinova, U.: Digital universities in Russia: prospects and problems. In: Antipova, T., Rocha, Á. (eds.) Digital Science 2019. DSIC 2019. Advances in Intelligent Systems and Computing, vol. 1114. Springer, Cham (2020). https://doi.org/10.1007/978-3-030-37737-3_23

Rozhkova, D.: Digital platform economy: definition and operating principles. Manag. Econ. Syst. **10**(104), 32 (2017). https://uecs.ru/component/content/article/4582. Accessed 19 Oct 2020. (in Russian) (Рожкова Д.Ю. Цифровая платформенная экономика: определение и принципы функционирования. Управление экономическими системами, **10**(104), 32 (2017). Режим доступа. https://uecs.ru/component/content/article/4582 (дата обращения 19.10.2020)

Schram, W.: Notes on case studies of instructional media process. Institute for Communication Research Stanford University (1971)

Somkin, I., Rozhkova, D., Orlov, E.: Peculiarities of standard cost-based accounting. In: Popkova, E.G., Sergi, B.S. (eds.) "Smart Technologies" for Society, State and Economy. ISC 2020. Lecture Notes in Networks and Systems, vol. 155. Springer, Cham (2021). https://doi.org/10.1007/978-3-030-59126-7_46

Styrin, E.M., Dmitrieva, N.E. Sinyatullina, L.H.: Gosudarstvennye tsifrovye platformy: Ot kontsepta k realizatsii (Government Digital Platform: From Concept to Implementation). Public Adm. Issues, (4), 31–60 (2019). (in Russian). Стырин Е.М., Дмитриева Н.Е., Синятуллина Л. Х. Государственные цифровые платформы: от концепта к реализации. Вопросы государственного и муниципального управления, (4) (2019)

Voskanyan, Y., Shikina, I., Kidalov, F., Davidov, D.: Medical care safety - problems and perspectives. In: Antipova, T. (ed.) Integrated Science in Digital Age. ICIS 2019. Lecture Notes in Networks and Systems, vol. 78. Springer, Cham (2020). https://doi.org/10.1007/978-3-030-22493-6_26

Voskanyan, Y., Shikina, I., Kidalov, F., Andreeva, O., Makhovskaya, T.: Impact of macro factors on effectiveness of implementation of medical care safety management system. In: Antipova, T. (ed.) Integrated Science in Digital Age 2020. ICIS 2020. Lecture Notes in Networks and Systems, vol. 136. Springer, Cham (2021). https://doi.org/10.1007/978-3-030-49264-9_31

Walravens, N., Ballon, P.: Platform business models for smart cities: from control and value to governance and public value. IEEE Commun. Mag. **51**(6), 72–79 (2013)

Yin, R.K.: Case Study Research: Design and Methods (Fourth ed.). SAGE Publications, Inc., London (2009)

Penta-Helix Model in Sustaining Indonesia's Tourism Industry

Eko Priyo Purnomo[1], Aqil Teguh Fathani[2(✉)], Deni Setiawan[1],
Mochammad Iqbal Fadhlurrohman[1], and Dwi Heru Nugroho[3]

[1] Jusuf Kalla School of Government, Universitas Muhammadiyah
Yogyakarta, Yogyakarta, Indonesia
eko@umy.ac.id, deni.setiawan.psc19@mail.umy.ac.id,
Fadlurrohmani84@gmail.com
[2] Department of Government Affairs and Administration, Jusuf Kalla School of Government,
Universitas Muhammadiyah Yogyakarta, Yogyakarta, Indonesia
aqil.teguh.psc19@mail.umy.ac.id
[3] Magister of Law, Universitas Muhammadiyah Yogyakarta, Yogyakarta, Indonesia
dherunugroho@gmail.com

Abstract. This study analyzes and sees Penta-helix actors' role and contribution in maintaining the tourism sector during the COVID-19 pandemic. The involvement of these actors before the pandemic succeeded in advancing the Indonesian tourism sector. Since the COVID-19 pandemic, a large contraction in the tourism sector has been inevitable. This research was conducted using data analysis from NVivo 12 Plus software. The research data consisted of files and documents over three months (August, September, October) and supported by online media data related to the pandemic's tourism sector. The study results showed that the government and business/private had the highest scores to maintain the tourism sector, with an average of 27.88% and 28.11%. Compared to other actors (academic, community, media), these high values were caused by a close relationship between government and business/private in the economy, employment, mobility, and tourism policy. Second, academic and community actors were in third and fourth place with 21.31% and 13.44%. Third, the actor with the lowest score was the media, with a score of 9.27%. During the COVID-19 pandemic, the media prioritized information about the spread and development compared to information on the tourism sector.

Keywords: Penta-Helix model · Sustainability · Tourism industry · Tourism activities · COVID-19

1 Introduction

The research aims to describe how Penta-helix collaboration can enhance Indonesia's tourism area during the Corona Virus Disease 19 (COVID-19) pandemic. The emergence of COVID-19 at the end of 2019 has a significant impact on Indonesia's global activities, namely the decrease in human mobility to carry out all activities, including tourism

© The Author(s), under exclusive license to Springer Nature Switzerland AG 2021
T. Antipova (Ed.): ICADS 2021, AISC 1352, pp. 477–486, 2021.
https://doi.org/10.1007/978-3-030-71782-7_42

activities [1]. The tourism industry is a leading sector for the Indonesian government in improving the national economy because it provides a considerable contribution [2]. The high contribution of tourism to the national economy is also driven by other actors' involvement, namely academics, business/private, society, and the media (Penta-helix), supporting tourism progress. However, since the emergence of the COVID-19 pandemic in Indonesia (March 2020) has put pressure on tourism, contractions and large losses have been inevitable due to the pandemic [3]. Therefore, this research also assesses Penta-helix actors' role and influence to maintain Indonesian tourism's sustainability during the COVID-19 pandemic.

The tourism industry's high contribution to the national economy before the COVID-19 pandemic is inseparable from the Penta-helix collaboration model's various interest actors [4]. In the last five years (2014–2019), the implementation of the Penta-helix collaboration was considered successful in advancing the development of the Indonesian tourism industry [5]. The Penta-helix collaboration model has five (5) actors and is interrelated in development efforts in the tourism sector, namely academics (A), business (B), community (C), government (G), and media (M) [6]. The Penta-helix collaboration model is a refinement of the collaboration model between the government, private sector, and society by including other important actors, namely academics and media [7]. The concept of tourism development using the Penta-helix collaboration model is an innovation towards change in the form of cross-stakeholder [8]. Actor mapping shows how significant the role and influence of actors are in developing tourism [9]. In general, the Penta-helix concept tends to see developments in the export-import industry activity sectors. As time goes by, the concept of Penta-helix collaboration is increasingly shifting to other industries, such as the tourism industry [10].

Coronavirus Disease-19 (COVID-19) presence in early 2020 in Indonesia hurt national industries [5]. The rapid spread of the COVID-19 pandemic is a big surprise for the global economy, including affecting the Indonesian economy. One of the most influential in COVID-19 pandemic is the tourism industry [11]. The tourism industry is one of the industries that has been badly affected by the spread of COVID-19, seen from the massive decline in foreign tourist arrivals, resulting in a large contraction of income from the tourism sector [12]. Before the COVID-19 pandemic, the tourism industry continued to increase every year (2015–2019), and it is targeted that tourist attendance in 2020 will reach 20 million tourists (Fig. 1) and (Table 1).

Based on Fig. 1 and Table 1, the tourism crisis during the COVID-19 pandemic is a shared responsibility. In this case, it is necessary to conduct coordination and cooperation by all stakeholders. Coordination and cooperation within the Penta-helix are needed to sustain and restore tourism that has been destroyed by the pandemic. Coordination in these efforts must be carried out in a structured and systematic manner so that the implementation of the recovery runs well. Each actor's role in the Penta-helix contributes to maintaining and restoring the tourism sector.

This study analyzes tourism sustainability efforts using the Penta-helix model approach and can assess the most influential actors in maintaining tourism sustainability during the COVID-19 pandemic. This research is qualitative research using secondary data. Secondary data were documents and files discussing the tourism sector. The data came from the Ministry of Tourism and other related institutions. Data from online media also supported this article with 30 news stories from August-September-October or the

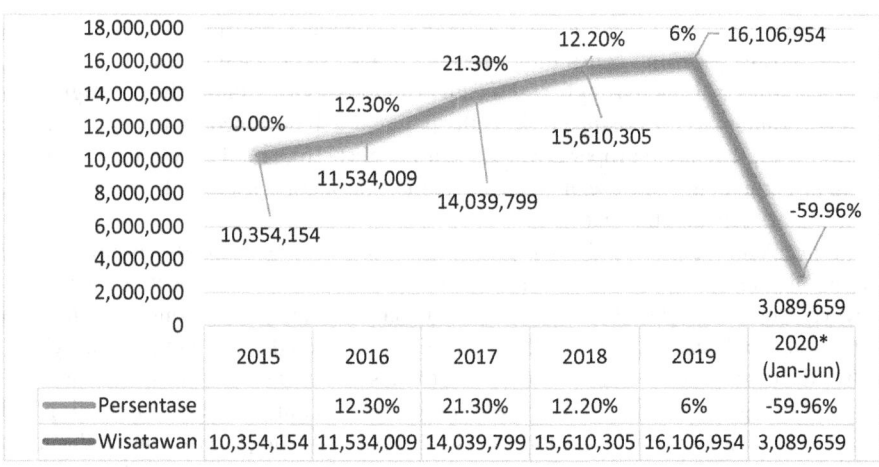

Fig. 1. Foreign tourist visit

Table 1. Contribution of tourism industry to the national economy

	2015	2016	2017	2018	2019
Contribution to the National Economy	4.3%	4.2%	4.11%	4.5%	5.5%
Foreign Exchange (USD)	12.2 Billion	13.6 Billion	15.24 Billion	19.8 Billion	20 Billion
Labour (Million)	11.4	11.8	12.4	12.7	13

Source: Central Bureau of Statistics, 2020

new life order policy (New Normal Era) that began in Indonesia. The data coding was carried out using NVivo 12 Plus software and a Crosstab Query. The NVivo 12 Plus in this study can analyze deeply and distinguish each actor's role and involvement [13]. The Nvivo12 Plus software obtained maximum results following the study's title, seeing the most influential actors in maintaining tourism during the COVID-19 pandemic.

2   Finding and Discussion

Collaboration between government administration stakeholders to solve is a limitation of dealing with them [14]. The collaborative governance approach is a relevant step to be developed because it explains the collaboration system that will introduce various concepts that unite the government, private sector and society. In this collaboration, a dialogue will occur in solving a problem [15]. In formal policies, the government is dominated by actors who influence the interests of public affairs. Although governments no longer exercise centralized control over public policies, they still can influence them. The power used by the state today is to negotiate with actors in policy networks. Members of this network are increasingly accepted as equal partners in the policy process, public and private actors collaborating to obtain resources they cannot access independently.

For example, using private companies for policy implementation allows the government to avoid costly and time-consuming procedural and accountability issues. Currently, governance theory is increasingly significant in public service affairs. The paradigm shift from government to governance reflects the political will designed to drive governance reform by implementing proper governance principles [16]. Good governance currently has high complexity because it involves supporting actors, such as business/private, academia, society and the media.

So far, Penta-helix actors' involvement in the tourism sector positively impacts the national economy because they can maximally carry out their respective roles [17]. However, since COVID-19 emerged, all actors have been unable to control the tourism sector's impact because the COVID-19 pandemic has also damaged other vital sectors such as the mining industry, export-import activities, trade, and others [18]. Researchers' data in looking at Penta-helix actors who play a role and influence in maintaining and trying to restore the tourism sector are as follows. First, the government as the main actor seeks to control the spread of the COVID-19 pandemic by making policies, in this case also including policies in implementing health protocols in every tourism activity, and always paying attention to the losses of the tourism economic sector caused by COVID-19. Second, Business/Private has always played a role in encouraging the development of the tourism industry, making an immense contribution to investment and development to improve tourism's quality and facilities further. However, when COVID-19 business/private suffers the most losses. The tourism sector's economy has fallen by 90% since COVID-19 entered Indonesia [19].

Third, a community in the Penta-helix model has been the leading partner in developing the tourism sector. Groups and communities always welcome the presence of ideas in tourism development during the pandemic. All groups, associations, institutions, and the community always follow appeals and government policies, especially closing access to tourism destinations or destinations only for local communities [4, 20]. Fourth, in the Penta-helix collaboration model, academics' role is considered a drafter, providing new ideas and concepts in developing national or regional tourism. Academics are a source of knowledge with concepts, new theories relevant to tourism development with scientific studies, and research to improve tourist destinations' quality and quantity [21]. Besides, they play a role in providing standardization, product certification, and human resources skills (HR). Fifth, in the Penta-helix collaboration model, media are expenders to support the publication, promote and create a brand image through websites, online and offline media. In other words, the media are tools for marketing and attracting tourists' attention [22].

In the current situation, these actors' roles and contributions are needed to restore the economy, society, health, mobility, and the environment. In the preliminary discussion, the tourism industry is a leading sector that supports the national economy. The government's priorities for the tourism sector, which began in 2014, gave significant results to make an enormous contribution to the national economy (Fig. 1 and Table 1). However, since the COVID-19 pandemic in Indonesia, the tourism sector has experienced a large contraction, affecting the national economy [23, 24].

Figure 2 explains that COVID-19 has a significant influence on tourism activities. The outline of Fig. 2 is also connected to the food and beverages business, hotels, jobs, shopping, consumers and producers, exhibitions, travel, transportation, and others. This

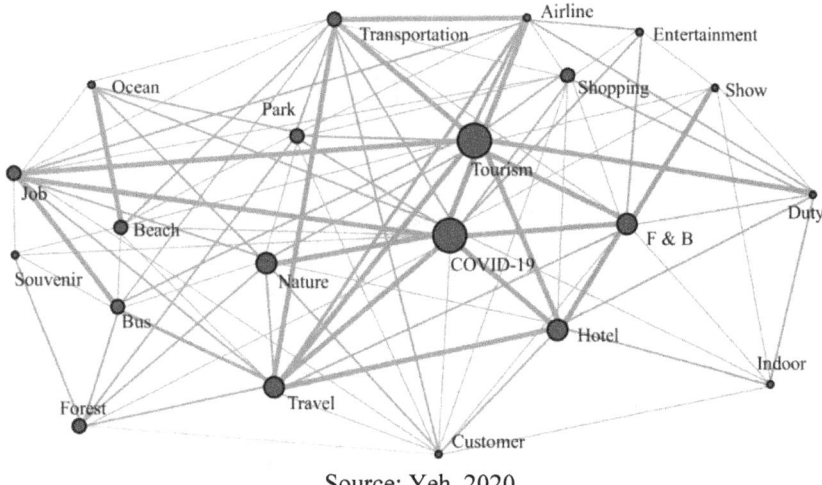

Source: Yeh, 2020

Fig. 2. The effect of COVID-19 on tourism

condition proves that COVID-19 has a significant effect on all holiday activities and tourism activities. This pandemic's enormous effect requires a joint role, especially the helical actors, to maintain and slowly rebuild tourism.

Based on data findings and analysis results using the Nvivo 12 plus software, the five actors in the Penta-helix model have different roles and influence on the tourism sector. This study used several indicators of great concern in rebuilding tourism, the economy [25], employment [26], the health of tourists [27], health policy in tourism [28], and human mobility in tourism activities [29]. The indicator used is a benchmark for coding and data analysis to get results.

Based on Fig. 3 and Table 2, each actor has their value to maintain and rebuild Indonesia's tourism sector. Of the overall indicators, the highest score was achieved by a business or private, with an average of 28.11%. The second place is the government with 27.88%, and the third is academies with 21.31%. The fourth is the community with 13.44%, and the fifth is media with 9.27%. The discussion of each indicator is as follows.

First, on economic indicators, the government is the most influential actor in the tourism economic sector than other Penta-helix actors. This situation cannot be avoided because the tourism sector is the leading sector in Indonesia. After all, it has an enormous contribution and has been helpful to the national economy. Since the COVID-19 pandemic in Indonesia, the country's foreign exchange earnings from the tourism sector have fallen by 90% [19]. The magnitude of the declining value has become the government's concern in the tourism economic sector. The government has taken several strategic steps to slowly revive the tourism sector, including providing a tourism grant of IDR 3.8 trillion [30]. The grants are given in subsidies for hotels, restaurants, and businesses directly related to tourism throughout Indonesia to provide industry opportunities to continue tourism economic activities during the COVID-19 pandemic.

Second, in job indicators, business or private is the actor with the most role and influence on continuing work in the tourism industry. The business/private sector must bear large losses due to the cessation of tourism activities caused by COVID-19. This

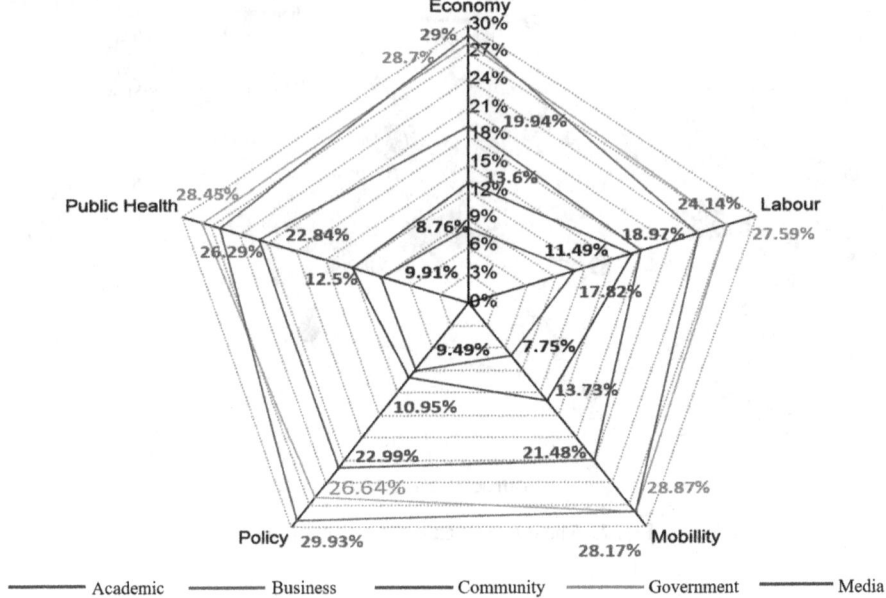

Fig. 3. The role and influence of Penta-helix actors

Table 2. The value of Penta-helix actors during the pandemic

	Economy	Labour	Mobility	Policy	Public health	Total
Academic	19.94%	18.97%	21.48%	22.99%	22.84%	21.31%
Business or Private	28.7%	27.59%	28.87%	26.64%	28.45%	28.11%
Community	13.6%	17.82%	13.73%	10.95%	12.5%	13.44%
Government	29%	24.14%	28.17%	29.93%	26.29%	27.88%
Media	8.76%	11.49%	7.75%	9.49%	9.91%	9.27%
Total	100%	100%	100%	100%	100%	100%

condition continues to mass layoffs of employees/termination of employment in the scope of tourism. Business/private must also provide severance pay for these former employees' survival during the pandemic. The number of layoffs in Indonesia's tourism sector reached 82% [31]. The business/private tourism sector must carry out this figure to maintain the tourism sector's sustainability, which is currently in crisis.

Third, business/private and government are the leading actors on the mobility indicator due to the close relationship. Clashes and differences in views on human mobility are a big problem in the current pandemic situation. First, during the COVID-19 pandemic, people's movement from one place to another becomes a calculation and analysis to see and assess the number of COVID-19 spread caused by this mobility [32]. Second, to maintain and improve the tourism sector, human mobility is needed so that the tourism sector can survive during a pandemic. Business/private endeavors to bring out all innovation and creativity to attract tourists' attention during the COVID-19 pandemic, such as promoting destinations at low prices, applying health instruments during tourism activities, and providing health insurance for tourists during tourism activities. The magnitude of the two actors' roles and concerns is an essential point in preventing the spread of COVID-19 and seeking tourism sustainability during the pandemic, even though on a low scale.

Fourth, in the tourism policy indicator, the government and business/private are the leading actors seen by the reciprocal relationship and mutual support of each policy program. The government issued a policy to maintain the sustainability of tourism during the pandemic. This policy is supported by businesses/private by implementing tourism activities. The government's new policy is to prioritize and implement health protocols to prevent the spread of COVID-19 in tourism activities. Health instruments in the implementation of tourism activities refer to the tourism ministry's programs, namely Cleanliness, Health, Safety and Environmental Sustainability (CHSE) [33] as well as the Circular of the Minister of Tourism and Creative Economy/head of the Tourism and Creative Economy Agency Number 2 of 2020. Fifth, on indicators health of tourists, business/private has a high value compared to other actors. Even though the government issues every policy, business/private actors significantly contribute and pay attention to tourists' health in its implementation.

The three other actors are academics, community, and media, with low tourism sector scores. Still, the roles and contributions of these actors are vital in maintaining the sustainability of tourism. (1) During the COVID-19 pandemic, academics have always provided concepts, ideas, strategic steps to the government and business/private sector to suppress the spread of COVID-19 and strive to ensure that tourism activities do not stop completely. One of them is to provide input to the government and business/private sector to continue to carry out local scale tourism activities to keep the economy running even on a small scale. (2) Even though it does not have a high score, the community has a significant role in maintaining tourism by implementing every policy and regulation. People can still take vacations only to the local area while still fulfilling the health protocols.

Third, business/private and government are the leading actors on the mobility indicator due to the close relationship. Clashes and differences in views on human mobility are a big problem in the current situation. First, during the COVID-19 pandemic, people's movement from one place to another becomes a calculation and analysis to see and assess the number of COVID-19 spread caused by this mobility [32]. Second, to maintain and improve the tourism sector, human mobility is needed so that the tourism sector can survive during a pandemic. Business/private endeavors to bring out all innovation and creativity to attract tourists' attention during the COVID-19 pandemic, such as promoting destinations at low prices, applying health instruments during tourism activities, and providing health insurance for tourists during tourism activities. The magnitude of the two actors' roles and concerns is an essential point in preventing the spread of COVID-19 and seeking tourism sustainability, even though on a low scale.

Of the five indicators, the Penta-helix actors who are more visible in their role, contribution, and influence in maintaining tourism sustainability during the COVID-19 pandemic are the government and business/private because the two actors have a close and interrelated relationship with the tourism industry. The government needs business/private to maintain and encourage tourism during a pandemic. Business/private also needs government assistance to reduce the number of COVID-19 spread caused by tourism mobility. Even though other actors (Academic, Community, Media) only had low scores, they also have contributed significantly to implementing tourism activities during the pandemic. The five Penta-helix actors are running correctly to maintain the tourism sector and rebuild the tourism industry.

3 Conclusion

Indonesia has lost one of the primary income sources, which has a multiplier effect, namely the tourism industry. The Penta-helix collaboration has succeeded in advancing the tourism industry before the COVID-19 pandemic. Still, during the COVID-19 pandemic, the tourism sector has suffered significant damage but can still survive even on a low scale. The involvement of interest actors, in this case, the Penta-helix, must be carried out maximally to maintain the tourism sector. Further, these actors must immediately encourage the tourism sector's awakening with a new style, namely implementing health protocol instruments in every tourism activity. The goal is that the tourism sector can survive in the future, vulnerable to external disturbances. Therefore, it is necessary to maintain and increase collaboration, in this case, the Penta-helix, to create innovations so that the tourism sector can continue to survive even in a crisis.

References

1. Yazid, S., Lie, L.D.J.: Dampak Pandemi Terhadap Mobilitas Manusia di Asia Tenggara. Jurnal Ilmiah Hubungan Internasional, 75–83 (2020). https://doi.org/10.26593/JIHI.V0I0.3862.75-83
2. Budimanta, A.: Sektor Unggulan Pemerintah, Komine Ekonomi Industri Nasional (KEIN), Jakarta (2019)
3. Nasution, D.A., Erlina, E., Muda, I.: Dampak Pandemi Covid-19 Terhadap Perekonomian Indonesia. J. Benefita 5(2), 212–224 (2020). https://doi.org/10.22216/jbe.v5i2.5313
4. Hardianto, W.T., Sumartono, S., Muluk, K., Wijaya, A.F.: Tourism investment services in Batu city with penta helix perspective. Int. J. Manag. Adm. Sci. 5(05), 17–22 (2017)
5. Sugihamretha, I.D.G.: Respon Kebijakan: Mitigasi Dampak Wabah Covid-19 Pada Sektor Pariwisata. J. Perenc. Pembang. Indones. J. Dev. Plan. 4(2), 191–206 (2020). https://doi.org/10.36574/jpp.v4i2.113
6. Halibas, S.A., Sibayan, R.O., Maata, R.: The pentahelix model of innovation in Oman: an HEI perspective. Interdiscip. J. Inf. Knowl. Manag. 12, 159–174 (2017)
7. Amrial, A., Muhammad, A., Muhamad, E.: Penta helix model: a sustainable development solution through the industrial sector. Soc. Hum. Sci. (November), 152–156 (2017). HISAS 14th Proceedings of Conference
8. Calzada, I.: Local entrepreneurship through a multi-stakeholders' tourism living lab in the post-violence/peripheral era in the Basque Country. Reg. Sci. Policy Pract. 11(3), 451–466 (2019). https://doi.org/10.1111/rsp3.12130

9. Williams, A.M., Baláž, V.: Tourism risk and uncertainty: theoretical reflections. J. Travel Res. **54**(3), 271–287 (2015). https://doi.org/10.1177/0047287514523334

10. Sulistyo, A.: Strategi Pengembangan Objek Wisata Minat Khusus Dalam Upaya Menciptakan Pariwisata Berkelanjutan Di Kabupaten Bantul (Studi Kasus: Karst Tubing), 11th Univ. Res. Colloqium 2020, pp. 1–8 (2020). https://repository.urecol.org/index.php/proceeding/article/view/876/851

11. Wishnutama, W.: Kerugian Pariwisata Indonesia Akibat Corona, Jakarta (2020)

12. Tauhid, T., Argubi, A., Ramadhon, R., Kamaluddin, K.: Revitalisasi Kebijakan Pengembangan Pariwisata Dalam Menghadapi Pandemi Covid-19 di Kota BIma. Sadar Wisata **3**(1), 13–24 (2020)

13. Zahra, A.A., Purnomo, E.P., Kasiwi, A.N.: New democracy in digital era through social media and news online. Humaniora **11**(1), 13 (2020). https://doi.org/10.21512/humaniora.v11i1.6182

14. Renn, O.: Stakeholder and public involvement in risk governance. Int. J. Disaster Risk Sci. **6**(1), 8–20 (2015). https://doi.org/10.1007/s13753-015-0037-6

15. Ansell, C., Gash, A.: Collaborative governance in theory and practice. J. Public Adm. Res. Theory **18**(4), 543–571 (2008). https://doi.org/10.1093/jopart/mum032

16. Mannaa, M.T.: Halal food in the tourist destination and its importance for Muslim travelers. Curr. Issues Tour., 1–12 (2019). https://doi.org/10.1080/13683500.2019.1616678

17. Muhyi, H.A., Chan, A.: The penta helix collaboration model in developing centers of flagship industry in Bandung City. Rev. Integr. Bus. Econ. Res. **6**(1), 412 (2017). https://buscompress.com/journal-home.html

18. Uğur, N.G., Akbiyik, A.: Impacts of COVID-19 on global tourism industry: a cross-regional comparison. Tour. Manag. Perspect. **36**(September), 100744 (2020). https://doi.org/10.1016/j.tmp.2020.100744

19. Karunia, A.M., Jatmiko, B.: Akibat Pandemi, Pendapatan Devisa Sektor Pariwisata Turun hingga 90 Persen, Kompas.com (2020). https://money.kompas.com/read/2020/09/25/135500926/akibat-pandemi-pendapatan-devisa-sektor-pariwisata-turun-hingga-90-persen. Accessed 10 Dec 2020

20. Zulkhibri, M., Sinay, J.B.: Assessing ASEAN economic policy responses in a pandemic. ASEAN Policy Br., vol. 02, no. May 2020

21. Yuningsih, T., Darmi, T., Sulandari, S.: Model Pentahelik Dalam Pengembangan Pariwisata. JPSI (J. Public Sect. Innov.) **3**(2), 84 (2019). https://doi.org/10.26740/jpsi.v3n2.p84-93

22. Nugraha, Y.M.: Analisis Potensi Promosi Pariwisata Halal Melalui E-Marketing di Kepulauan Riau. J. Penelit. dan Karya Ilm. Lemb. Penelit. Univ. Trisakti **3**(2), 63–68 (2018). https://www.trijurnal.lemlit.trisakti.ac.id/lemlit/article/view/2990

23. BPS, Kunjungan Wisatawan Mancanegara, bps.go.id (2020). https://www.bps.go.id/subject/16/pariwisata.html#subjekViewTab3. Accessed 25 July 2020

24. Trading Economics, Indonesia GDP Growth Rate, Trading Economics (2020). https://tradingeconomics.com/indonesia/gdp-growth

25. Ozili, P.K., Arun, T.: Spillover of COVID-19: impact on the global economy. SSRN Electron. J. (2020). https://doi.org/10.2139/ssrn.3562570

26. Olivia, S., Gibson, J., Nasrudin, R.: Indonesia in the time of Covid-19. Bull. Indones. Econ. Stud. **56**(2), 143–174 (2020). https://doi.org/10.1080/00074918.2020.1798581

27. Eslami, S., Khalifah, Z., Mardani, A., Streimikiene, D., Han, H.: Community attachment, tourism impacts, quality of life and residents' support for sustainable tourism development. J. Travel Tour. Mark. **36**(9), 1061–1079 (2019). https://doi.org/10.1080/10548408.2019.1689224

28. Lee, S., Hwang, C., Moon, M.J.: Policy learning and crisis policy-making: quadruple-loop learning and COVID-19 responses in South Korea. Policy Soc., 1–19 (2020). https://doi.org/10.1080/14494035.2020.1785195

29. Bonaccorsi, G., et al.: Economic and social consequences of human mobility restrictions under COVID-19. Proc. Natl. Acad. Sci. U. S. A. **117**(27), 15530–15535 (2020). https://doi.org/10.1073/pnas.2007658117

30. Jyestha, V.: Pemerintah Harus Selektif Salurkan Dana Hibah Pariwisata Rp 3,3 Triliun, Kompas.com (2020). https://www.tribunnews.com/nasional/2020/10/05/pemerintah-harus-selektif-berikan-dana-hibah-pariwisata-rp-33-triliun. Accessed 10 Dec 2020

31. Putri, C.A.: Survei: Karena Covid-19, 35% Pekerja di Indonesia Kena PHK, CNBC Indonesia (2020). https://www.cnbcindonesia.com/news/20201007145144-4-192535/survei-karena-covid-19-35-pekerja-di-indonesia-kena-phk. Accessed 10 Dec 2020

32. Prawoto, N., Purnomo, E.P., Zahra, A.A.: The impacts of Covid-19 pandemic on socio-economic mobility in Indonesia. Int. J. Econ. Bus. Adm. **8**(3), 57–71 (2020)

33. Wishnutama, W.: Panduan Pelaksanaan Kebersihan, Kesehatan, Keselamatan dan Kelestarian Lingkungan Pada Penyelenggaraan Kegiatan Pariwisata. Kementrian Pariwisata dan Ekonomi Kreatif, Jakarta (2020)

Public Communication of Local Government Leaders: A Case Study of Three Major Governors in Indonesia

Syamsul Bahri Abd. Rasyid[1]([✉]) [iD], Achmad Nurmandi[2] [iD], Suswanta[2] [iD],
Dyah Mutiarin[2] [iD], and Salahudin[3] [iD]

[1] Department of Government Affairs and Administration, Universitas Muhammadiyah
Yogyakarta, Yogyakarta, Indonesia
syamsulbahri100296@gmail.com
[2] Department of Government Affairs and Administration, Jusuf Kalla School of Government,
Universitas Muhammadiyah Yogyakarta, Yogyakarta, Indonesia
[3] Department of Government Science, Social and Political Science Faculty, Universitas
Muhammadiyah Malang, Malang, Indonesia

Abstract. For public purposes and political contact, political insiders, regional leaders, and state officials are increasingly utilizing Twitter. To the full extent that it is a lobbying method for issues that have not become a public concern, Twitter can relay messages. This paper intends to review the use of Twitter from the official Twitter accounts of the three governors with the most followers, namely; Anies Baswedan (DKI Jakarta), Ganjar Pranowo (Central Java), and Ridwan Kamil (West Java), in the six months, with indicators; empowering citizens, building public trust in public institutions, civic engagement, providing significant insights, and strengthening transparency, which researchers draw from previous research. This study used descriptive qualitative research methods to describe the comparison and connectivity between Twitter usage indicators by @aniesbaswedan, @ganjarpranowo, and @ridwankamil. Nvivo 12 plus software with crosstab query features, word frequency, cluster analysis, and project map was used to analyze tweet intensity, conversation topics, relations between indicators, and indicator relations with each Twitter account. This study's findings show that Twitter usage by @aniesbaswedan, @ganjarpranowo, and @ridwankamil is outstanding by always tweeting oriented towards the indicators in question and having a relationship between indicators on each Twitter account. This paper's contribution is to depict the use of Twitter as a means of communication and public interaction between the three governors' accounts and their communities. It is essential to conduct studies on the usage of Twitter with more subjects of the governor account using a mixed-method approach.

Keywords: Social media · Twitter · Governors · Local government · Public communication · Indonesia

1 Introduction

Twitter is one of the most followed social networks in the world. Twitter has been one of the fastest-growing media networking sites in the last few years since its release in 2006 [1]. A survey result from a survey institute, Statista, said that up to the third quarter of 2018, there were 326 million Twitter users. Twitter's active users in Indonesia are the third-largest after America and India, which reached 24.34 million. Apart from being used as a social media to establish relationships with other users, history also notes that Twitter is often used as a media for social movements in developed countries, for example, the Egyptian Revolution to overthrow President Husni Mubarak. At the same time, it is known as events of social movements protesting Iran's election results. From 2009 to 2010. While in Indonesia itself, Twitter is also widely used as a medium for political communication, both by political figures [2].

For the number of followers on Twitter, Ridwan Kamil has the most followers, followed by Anies and Ganjar. Since February 2010, Anies Baswedan has 4 million followers. Meanwhile, Ganjar Pranowo's Twitter account has 1.7 million followers since January 2010. Meanwhile, Ridwan Kamil has 4.3 followers since October 2009. With the use of Twitter and the massive followers of the three governors, as well as its benefits that can speed up the flow of information from local government to the community, then the use of social media to address complaints faced by the community can be resolved if social media is utilized correctly. This situation is in line with what Ganjar Pranowo and Ridwan Kamil said, "public complaints will be known and resolved if channeled through social media". Utilization of social media Twitter in the Governor of Central Java government bureaucracy, Ganjar Pranowo and Gurbernur West Java, Ridwan Kamil to improve excellent service, has implemented service standards, namely ability, attitude, appearance, attention, action, and responsibility [3].

The indicators in this paper are a collection of indicators that the authors took from previous studies: engaging citizens and strengthening transparency [4], empowering citizens [5], building public trust in public institutions [6], and providing significant insights [7]. As far as the quest of the previous studies is concerned. The authors have not found @aniesbaswedan, @ganjarpranowo, and @ridwankamil's social media use on Twitter with indicators: inspiring people, building public trust in public institutions, engaging citizens, providing valuable insights, and reinforcing accountability in a single research paper. Therefore, the authors refer more to studies [4, 5, 6]; and [7] to compile a paper on the official Twitter accounts of the three governors with the indicators above.

2 Literature Review

2.1 Social Media and Government

Social media use by government agencies is extensive across various disciplines, focusing on various topics [8]. Social media can be defined as a collection of technological instruments that enable public bodies to promote citizen engagement [4]. The use of social media by government agencies and citizens has now become mainstream. Platforms such as Facebook, Twitter, and WeChat represent global tools that can transform the interaction between government agencies, businesses, and citizens [9].

In Al-Masaeed 's study, the Jordanian government identified e-participation as the focus of its strategic initiatives. Also, it vowed to create an environment that empowered people to be more involved in government activities [5]. In 2015, the China Cyberspace Administration Office issued statements on the central government's social media strategy and the commitment of the government's social media function, namely censoring government information, engaging with people, reacting to popular perception, and supporting public services [7]. The key findings from [10] also found: (1) the significant influence of disaster information came from the probability of participants requesting additional disaster data from the newspapers, local government websites, and federal government websites; (2) respondents defined the most crucial intention to link immediately to the disaster via offline interpersonal forms rather than online organizational and personal forms; and (3) if requested by the government, participants reported a firm intention to relocate.

2.2 Challenges of Social Media

Some of the substantial challenges of social media are (1) lack of implementation plan, making the use of social media more ad hoc and challenging in the long term to sustain; (2) low quality of interaction of an institution which may not encourage citizens to take part, thus damaging the image of the institution; (3) the potential to disseminate gossip or question authority; (4) difficulty understanding the details on different social media sites that are processed every day; and (5) Exclusion, due to impairment or financial loss, of non-Internet users [11].

Moreover, disinformation on chat apps such as WhatsApp, Signal, or Telegram is an essential medium by which individuals share news and information, coordinate political activity, and discuss politics [12]. Social media users can assist policy-makers and decision-makers (media and press agencies) and providers of journalism services in Egypt to consider the essence of those that use social media as a source of awareness and attract researchers and media practitioners interested in information sources. Government journalists are often described as people working under government finance and administration control and are considered mediators and information carriers between the state and individuals [13].

2.3 Social Media Strategies

Understanding the value of government perceived by a government service through social media and social media synchronicity under government services' characteristics can provide significant insight into the government's social media [6]. Experts claim that designing a social media strategy will help governments reinforce accountability policies or promote a transparent government picture to the public [7]. In this research, the authors will discuss social media use by the government through the Twitter accounts of @aniesbaswedan, @ganjarpranowo and @ridwankamil with indicators: empowering citizens, building public trust in public institutions, citizen engagement, providing significant insights, and strengthening transparency. The research questions are: (1) What is the intensity of the tweets carried out by the @aniesbaswedan, @ganjarpranowo and @ridwankamil accounts in October 2020? (2) What are the most dominant themes in the tweets of @aniesbaswedan, @ganjarpranowo and @ridwankamil? (3) How do @aniesbaswedan, @ganjarpranowo and @ridwankamil accounts carry out the relationship between the indicators' tweets in the regional head's leadership in utilizing social media?

3 Research Method

The approach used in this research was descriptive qualitative, which described Twitter in empowering citizens, building public trust in public institutions, involving citizens, providing significant insights, and strengthening transparency. Descriptive analysis in this study used the NVivo 12 plus software application. This study's data came from the Twitter accounts of @aniesbaswedan, @ganjarpranowo, @ridwankamil, and supported by previous studies related to the government's social media use. Data were obtained using the Ncapture feature in Nvivo 12 plus by capturing tweets @aniesbaswedan, @ganjarpranowo, and @ridwankamil written for six months (July 2020–15 December 2020).

Data were analyzed using crosstab query features, word frequency, cluster analysis, and project map on the Vivo 12 Plus. The crosstab query was used to determine the intensity of tweets made by the Twitter accounts @aniesbaswedan, @ganjarpranowo, and @ridwankamil. Word frequency found out the topics of conversation or words most often used in tweets. Cluster analysis was used to determine the relationship between indicators in the use of social media Twitter. Meanwhile, the project map determined the relationship between Twitter accounts (@aniesbaswedan, @ganjarpranowo, and @ridwankamil) and their tweets on social media. The data from the three Twitter accounts of the regional heads can be seen in the table below.

Table 1. Position, Twitter Account, Followers, and Tweets during October 2020

Name	Position	Twitter Account	Followers	Tweets July-15 December 2020
Anies Baswedan	Governor of DKI Jakarta	@aniesbaswedan	4 M	356
Ganjar Pranowo	Governor of Central Java	@ganjarpranowo	1,7 M	481
Ridwan Kamil	Governor of West Java	@ridwankamil	4,3 M	145

This research aims to look at the usefulness of social media in the hands of the government, with indicators: empowering citizens, building public trust in public institutions, civic engagement, providing significant insights, and strengthening transparency, based on the tweets of the three most active governors on Twitter, namely the Governor of DKI Jakarta (@aniesbaswedan), the Governor of Central Java (@ganjarpranowo), and the Governor of West Java (@ridwankamil). The object of this research study is to answer the following three research questions.

(1) What is the intensity of the tweets carried out by the @aniesbaswedan, @ganjarpranowo and @ridwankamil accounts in October 2020?
(2) What are the most dominant themes in the tweets of @aniesbaswedan, @ganjarpranowo and @ridwankamil?
(3) How do @aniesbaswedan, @ganjarpranowo and @ridwankamil accounts carry out the relationship between the indicators' tweets in the regional head's leadership in utilizing social media?

Data analysis was carried out in three stages; namely, the first stage is NCapture text Twitter account; the second stage was data analysis with the Nvivo 12 Plus software reading text and content with similarities between one account and another, finding related items, and looking for interrelated meanings, words, and contexts between the three accounts; The third stage was discourse analysis, which was the analysis and meaning of the three Twitter accounts and relevant articles to the government's social media.

4   Result and Discussion

4.1 Tweets Intensity

In Table 1, regarding the comparison of Twitter accounts based on the number of followers, Ridwan Kamil is in the highest position with 4.3 million followers since joining Twitter in October 2009, followed by Anies Baswedan with 4 million followers February 2010, and Ganjar Pranowo with 1.7. Million followers since January 2010. However, this condition is inversely proportional to the tweet's intensity. Ganjar Pranowo tweets more frequently and far surpass Anies Baswedan and Ridwan Kamil.

On the @ganjarpranowo account, the indicator citizen involvement with a value of 479, followed by builds public trust in public institutions with a value of 472, strengthening transparency with a value of 472, empowering citizens with a value of 433, and providing significant insights with a value of 378. Meanwhile, on the @aniesbaswedan account, citizen involvement indicators, building public trust in public institutions, and empowering citizens, have a value of 355, 352, and 349, followed by indicators of strengthening transparency with a value of 330 and indicators of providing significant insight with a value of 268. Finally, on the @ridwankamil account, citizen involvement indicators and building public trust in public institutions have the same value of 150, followed by indicators of empowering citizens, strengthening transparency, and providing significant insights with a value of 143, 127, and 112.

Data sourced from the Twitter account of the Governor of DKI Jakarta (@aniesbawedan), the Governor of Central Java (@ganjarpranowo), and the Governor of West Java (@ridwankamil) in July-December in Fig. 1, shows that @ganjarpranowo's tweets more frequently than @aniesbaswedan and @ridwankamil. In a row, @ganjarpranowo and @anisbaswedan tend to be an indicator of citizen involvement. Meanwhile, @ridwankamil share a tendency towards two indicators, namely citizen involvement and building public trust in public institutions. The data above is how the governors of DKI Jakarta, Central Java, and West Java use social media in carrying out their duties as regional heads, with indicators of citizen involvement, building public trust in public

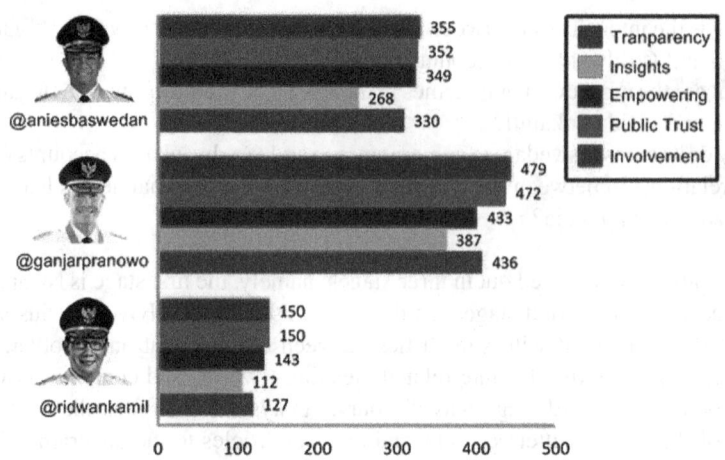

Fig. 1. Tweet intensity of @aniesbaswedan, @ganjarpranowo and @ridwankamil

institutions, and empowering citizens, providing significant insights, and strengthening transparency.

In the tweet intensity data above, it is following research conducted by [4–6]; and [7], that the use of social media government can engage citizens, build public trust in public institutions, empower citizens, provide significant insights, and strengthen transparency. In the study [4], online openness, mood, the activity level on social media, and interactivity provided by local government websites were considered essential to citizen engagement. This study's findings contribute significantly to understanding how the form of social media adopted affects citizen interaction. The research conducted in Spain is in line with the indicators of citizen involvement and strengthening transparency, which the author did on the Twitter accounts @aniesbaswedan, @ganjarpranowo, and @ridwankamil, where the use of social media by local governments involves citizens in their tweets as well as a form of strengthening of transparency online.

In [5] in Jordan, the government's latest published strategy identified electronic engagement as the focus of its strategic initiatives and promised to create an environment that empowers individuals to be more involved in government activities. In line with this research, this research also shows the government's use of social media via tweets on their Twitter accounts, demonstrating its desire to empower the citizens. In line with research at the Office of Investment and One Stop-Integrated-Services (Dinas Penanaman Modal dan Pelayanan Terpadu Satu Pintu Provinsi DKI Jakarta) [6], the results of my research are also that the use of social media can enhance government trust as long as the public feels transparency and interactivity of social media communication. The government's transparency through social media has an effect on public trust in them.

Meanwhile, social media's practical use, particularly knowledge acquisition and participation, is an essential predictor of perceptions of government accountability, responsiveness, and citizen satisfaction in research [7] in Beijing, China. In contrast, the position of public services is urgent. The increasing perception of the government's response is also in line with the current research on the government's social media use, namely @aniesbaswedan, @ganjarpranowo, @ridwankamil, which also provides significant insights to the public in every tweet.

4.2 The Intensity of the Tweet Topic

Based on word frequency analysis, the accounts @ridwankamil and @ganjarpranowo always tweet about covid, unlike @ridwankamil, which prioritizes the province's name DKI Jakarta. The most frequently discussed topics from the @aniesbaswedan, @ganjarpranowo, and @ridwankamil accounts can be seen in the pictures below, which the author has analyzed using the word frequency feature in the Nvivo Plus 12 software (Figs. 2, 3 and 4).

Fig. 2. Word Frequency of @aniesbaswedan   **Fig. 3.** Word Frequency of @ganjarpranowo

Fig. 4. Word Frequency of @ridwankamil

The word frequency from the @aniesbaswedan social media accounts results from *Jakarta* is the word most often used. This situation is in line with his position as governor of DKI Jakarta. Other words whose visualization is also remarkable are *#jagajakarta, dki, #covid19, data,* and *virus,* which means that Anies Baswedan uses his Twitter account as governor of DKI Jakarta. Meanwhile, *covid* words are most often used by the @ganjarpranowo account, which means it is in line with his position as governor of Central Java, where he uses his Twitter account to inform about health protocols against COVID-19. Other words with great visualization are *kesehatan, masker, desa, jawa,* and *pandemi.* This result illustrates that Ganjar Pranowo concerns his citizens' health by frequently making tweets related to COVID-19. Similar to @ganjarpranowo, *covid* is also the most dominant word on the @ridwankamil account. This result means that the @ridwankamil also has grave concerns for the health of their citizens. Other words with good visualization are *jawa, kota, ekonomi, bandung* dan *warga.* Thus, the tweets from the @aniesbaswedan, @ganjarpranowo, and @ridwankamil accounts are in line with the indicators of research from [4–6]; and [7].

4.3 Connectivity and Similarity of Tweets Between Twitter Accounts

The relationship between indicators in the accounts @aniesbaswedan, @ganjarpranowo, and @ridwankamil is defined based on cluster analysis results, connectivity between indicators, namely, empowering people and building public confidence in public institutions, citizen engagement, providing essential insights, and strengthening transparency in each Twitter account.

On the @aniesbaswedan account, indicators of empowering citizens, providing significant insights, and strengthening transparency have the most substantial relationship with details of the respective values, with 12.05%, 11.94%, and 11.84%, followed by indicators building public trust in public institutions of 11.20% and citizen involvement 9,09%. On the @ganjarpranowo account, the indicator of empowering citizens has the most substantial relationship with a value of 15.55%, followed by indicators providing significant insights of 15.32%, indicators of building public trust in public institutions 14.15%, indicators of strengthening transparency 14.05%, and citizen involvement 12.56%. While on the @ridwankamil account, the indicator empowers citizens and provides significant insights, has the most substantial relationship with the details of the same value, namely 5.30%, followed by indicators of strengthens transparency of 5,05%, indicators of building public trust in public institutions 4.48%, and indicators of citizen involvement 3.95% (Figs. 5, 6 and 7).

The cluster analysis results above show that each indicator: empowering citizens, building public trust in public institutions, citizen engagement, providing significant insights, and strengthening transparency, correlated with each other in the orientation of the use of social media by the Twitter accounts of @aniesbaswedan @ganjarpranowo, and @ridwankamil, in line with the research of [4] in Spain, [5] in Jordan, [6] at the Office of Investment and One Stop-Integrated-Services (Dinas Penanaman Modal dan Pelayanan Terpadu Satu Pintu Provinsi DKI Jakarta), and [7] in Beijing, China.

4.4 Relationship Between Twitter Accounts

Project Map data is the last stage of the data analysis process using Nvivo 12 plus. This process is a sub-process that is on the Map. A map is a visualization tool used to explore ideas and display the connections between one data and another. Meanwhile, the project map is a graphic representation of various items that have been made in the study. In this stage, the authors created an analysis map from coding, case, and related source data to

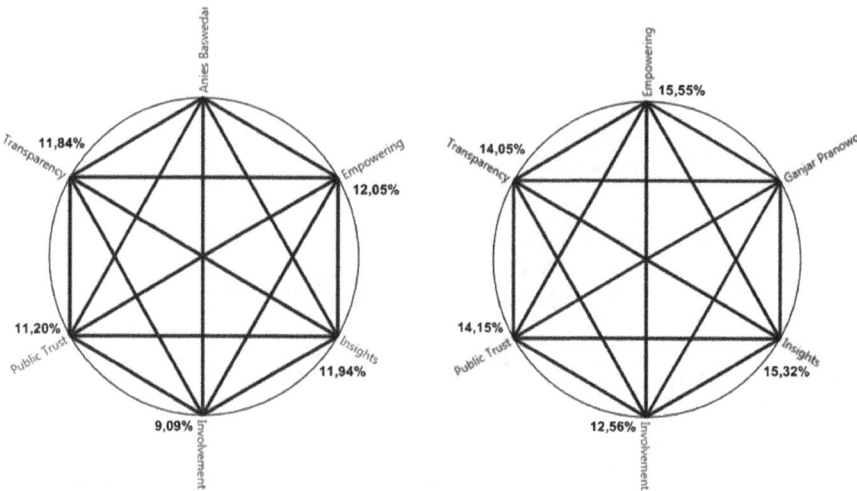

Fig. 5. Cluster analysis of @aniesbaswedan   **Fig. 6.** Cluster analysis of @ganjarpranowo

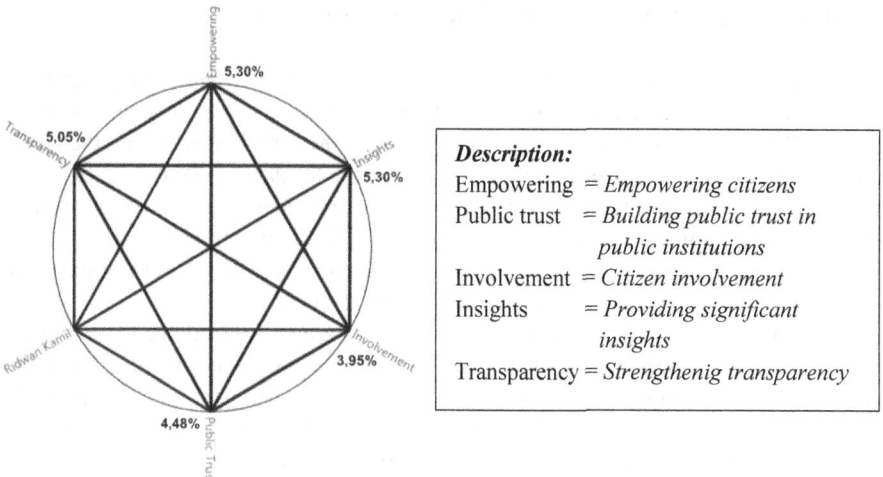

Description:

Empowering   = *Empowering citizens*
Public trust   = *Building public trust in
                    public institutions*
Involvement = *Citizen involvement*
Insights   = *Providing significant
                    insights*
Transparency = *Strengthenig transparency*

Fig. 7. Cluster analysis of @ridwankamil

display the data processing flow and the relationship of each data that the researcher has done from the beginning to the end (Fig. 8).

The results of the project map above show the Twitter accounts @aniesbaswedan, @ganjarpranowo, and @ridwankamil have a mutual relationship because the accounts @aniesbaswedan, @ganjarpranowo and @ridwankamil, can utilize social media (Twitter), to write tweets related to empowering citizens, building public trust in public institutions, citizen engagement, providing significant insights, and strengthening transparency.

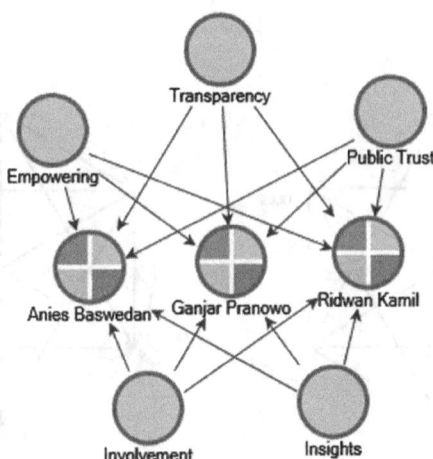

Fig. 8. Project map of @aniesbaswedan, @ganjarpranowo and @ridwankamil

5   Conclusion, Contribution, Research Limitations and Recommendations for Further Research

The use of social media by the Twitter accounts @aniesbaswedan, @ganjarpranowo, and @ridwankamil as DKI Jakarta, Central Java, and West Java governors is considered perfect. Concerning the tweets' frequency from the three Twitter accounts, the tweets from @ganjarpranowo are more severe than @aniesbaswedan and @ridwankamil. The debate topics from these three accounts also appear to be identical, namely prioritizing each tweet the names of provinces and COVID-19. This finding shows that @aniesbaswedan, @ganjarpranowo, and @ridwankamil aim to make the best use of the social media of Twitter to connect and communicate with the citizen.

This paper helps explain similarities between governors to stimulate Twitter social media use by governors across Indonesia so that government and public information exchange can run faster and community concerns can be resolved quickly due to the rapid flow of information through Twitter social media. While this research contributes to the definition of @aniesbaswedan, @ganjarpranowo, and @ridwankamil's use of Twitter social media in disseminating knowledge and engaging with the group, a significant triangulation study between qualitative data and quantitative data has not yet been demonstrated by the data analysis approach used. For this reason, in subsequent studies with a mix-method approach, problems about the government's use of Twitter social media need to be analyzed more thoroughly and concentrate on more comparisons of official government accounts.

References

1. Habibi, M.: Analisis Konten Jejaring Sosial Twitter Dalam Kasus Pemilihan Gubernur Dki 2017, Teknomatika, no. March 2018. https://tekno.kompas.com/read/2015/03/26/16465417/ Pengguna.Twitter.di.Indonesia.Capa

2. Juditha, C.: Dukungan Sosial Warganet Di Twitter Terhadap Gaya Komunikasi Pasangan Calon Presiden Pada Debat Pemilu 2019. J. Stud. Komun. dan Media **23**(1), 87 (2019). https://doi.org/10.31445/jskm.2019.1982

3. Suminto, A., Al Farizi, A.: Analisis Pemanfaatan Media Sosial Twitter oleh Ganjar Pranowo dan Ridwan Kamil. Sahafa J. Islam. Commun. **2**(2), 191 (2020). https://doi.org/10.21111/sjic.v2i2.4394

4. Haro-de-Rosario, A., Sáez-Martín, A., del Carmen Caba-Pérez, M.: Using social media to enhance citizen engagement with local government: Twitter or Facebook? New Media Soc. **20**(1), 29–49 (2018). https://doi.org/10.1177/1461444816645652

5. Al-Masaeed, S.: Social media use by the government: adoption and efficiency. Int. J. Electron. Gov. **11**(2), 205–216 (2019). https://doi.org/10.1504/IJEG.2019.101497

6. Marpianta, D.A.: Influence of use of social media of government agencies on trust to the government: study on social media Owned by Dinas Penanaman Modal dan Pelayanan Terpadu Satu Pintu Provinsi DKI Jakarta. J. Komun. Indones **VIII**(2) (2019)

7. Jia, Z., Liu, M., Shao, G.: Linking government social media usage to public perceptions of government performance: an empirical study from China. Chin. J. Commun. **12**(1), 84–101 (2019). https://doi.org/10.1080/17544750.2018.1523802

8. Nadzir, M.M.: Proposed e-government 2.0 engagement model based on social media use in government agencies. In: 2019 IEEE Conference on e-Learning, e-Management e-Services, IC3e 2019, pp. 16–19 (2019). https://doi.org/10.1109/IC3e47558.2019.8971778

9. Medaglia, R., Scholl, M.C., Loukis, E.: Introduction to the minitrack on social media and government. In: Proceedings of the 51st Hawaii International Conference on System Sciences, no. January, pp. 3–4 (2018). https://doi.org/10.24251/hicss.2018.318

10. Liu, B.F., Fraustino, J.D., Jin, Y.: Social media use during disasters: how information form and source influence intended behavioral responses. Commun. Res. **43**(5), 626–646 (2016). https://doi.org/10.1177/0093650214565917

11. El-taliawi, O.G.: Global encyclopedia of public administration, public policy, and governance. In: Global Encyclopedia of Public Administration, Public Policy, and Governance, no. January 2018 (2020). https://doi.org/10.1007/978-3-319-31816-5

12. Bradshaw, S., Howard, P.N.: Challenging Truth and Trust: A Global Inventory of Organized Social Media Manipulation. Comprop.Oii.Ox.Ac.Uk, p. 26 (2018). https://comprop.oii.ox.ac.uk/wp-content/uploads/sites/93/2018/07/ct2018.pdf

13. Mansour, E.: The adoption and use of social media as a source of information by Egyptian government journalists. J. Librariansh. Inf. Sci. **50**(1), 48–67 (2018). https://doi.org/10.1177/0961000616669977

The Influence of Social Media on the Omnibus Law-Making Process in Indonesia

Herpita Wahyuni[1](✉) ⓘ, Achmad Nurmandi[2] ⓘ, Dyah Mutiarin[2] ⓘ, Suswanta[2] ⓘ, and Salahudin[3] ⓘ

[1] Department of Government Affairs and Administration,
Universitas Muhammadiyah Yogyakarta, Yogyakarta, Indonesia
herpitawahyuni@yahoo.com
[2] Department of Government Affairs and Administration, Jusuf Kalla School of Government,
Universitas Muhammadiyah Yogyakarta, Yogyakarta, Indonesia
dyahmutiarin@umy.ac.id
[3] Department of Government Science, Social and Political Science Faculty,
Universitas Muhammadiyah Malang, Malang, Indonesia

Abstract. Social media offers governments new approaches to increasing transparency and accountability, involving increased citizen participation and collaborating in decision-making to improve information management and access as a public service. This study aims to determine the influence of social media, namely Twitter, in making Omnibus Law and using descriptive qualitative research methods, namely, research that describes the results of the study more broadly with NVivo 12 Plus software's help, which is useful and effective in helping qualitative research efficiently, helping logic consumption and research design, and providing facilities to analyze content. The results show that Twitter social media is an active media included in the Omnibus Law discussion. From the five indicators, it can be concluded that the transparency indicator has provided information that the debate on the Omnibus law will be postponed and states that the Omnibus Law consists of 79 rules, composed of 15 articles with 174 articles to be discussed in the House of Representatives of Indonesia (DPR RI). Conversation indicators that discuss always to maintain good communication. Involvement is an activity that involves the community in finding alternative solutions to problems; participation and communication have been carried out well, but community involvement in making Omnibus Law is still lacking. It needs to be involved openly.

Keywords: Social media · Policy · Omnibus Law · Government · Indonesia

1 Introduction

Many people prefer to express their views on social media [1]. Social networking links, quizzes, surveys, ratings, and other online engagement types are more than just entertaining: building confidence and faith in online social relations and giving local communities

T. Antipova (Ed.): ICADS 2021, AISC 1352, pp. 498–510, 2021.
https://doi.org/10.1007/978-3-030-71782-7_44

a rare glimpse into their citizens' direct everyday perspectives. As many city governments strive to balance accountability, accessibility, breadth of services, and competition from the private sector, citizen participation is essential to preserving a democracy representing its citizens' will [2].

Governments embrace social media to provide complementary information dissemination, connectivity, and engagement platforms through which people can reach government and government officials and make informed decisions [3]. Social media enables the government to access citizens' information, understanding, and views, making government processes more efficient and successful. By analyzing the content of comments made on social media, governments may classify and forecast people's opinions. In this way, the government will discover how people feel about the problem and behave accordingly [4].

Omnibus Law carried out by the government is an excellent plan as "The Main Means of Regulatory Structuring [5]. Indonesia's Omnibus Law creates investment legal instruments that can increase investment interest in Indonesia [6]. Omnibus Law is advantageous in terms of cost and time, but there are various requests. The rejection occurred after President Joko Widodo made a speech at the Plenary Session of the People's Consultative Assembly of the Republic of Indonesia. In the context of the Inauguration of the President and Vice President Elected 2019–2024, there were various legal arguments in addressing the contents of the speech delivered by President Joko Widodo. The implementation of the omnibus law still has to go through the House of Representatives' supervision to create checks and balances and public participation [7]. The Omnibus Law referred to in the president's speech is the Employment Creation Law and the Law on Empowerment of Micro, Small and Medium Enterprises, which aims to overcome all forms of regulations that are currently being obeyed by Indonesia so that rules must be simplified, cut and called orders [8]. The objectives of the Omnibus Law are, among others: Resolving regulatory conflicts quickly, effectively, and efficiently; Uniform government policies both at the central and regional levels to support the investment climate; Licensing management is more integrated, efficient and effective; Able to break the long bureaucratic chains; The increased coordination relationship between related agencies is regulated in an integrated regulatory Omnibus policy. There is a guarantee of legal certainty and legal security for policymakers [9].

Weaknesses of the Omnibus Law: The statutory position of the Omnibus Law concept has not been regulated. Omnibus Law has the potential to ignore the formal provisions of the formation of laws. Omnibus Law narrows openness and public participation in the shape of laws. The Omnibus Law can add to the regulatory burden if it fails to be implemented [10]. There are five steps that lawmakers must fulfill in drafting the Omnibus Law, namely: The House of Representatives (DPR) together with the government must involve the public in every stage of its drafting, the DPR and the government must be transparent in providing any information on the progress of the Omnibus Law formulation process. Must map the relevant regulations in detail, compilers must harmonize vertically with higher regulations or horizontally with equivalent regulations, and compilers must preview before being passed [9]. Social media is a means of channeling information sharing that can be connected in the government policymaking process, so this research focuses on social media's influence in making the Omnibus Law, which looks at the making process that involves the public and transparency.

The difference between this study and previous research is that this research focuses more on the direct influence of social media and discusses current issues in the Omnibus Law with NVivo 12 Plus's help so that the resulting data opens up updates of knowledge. Previous studies mostly analyzed e-government in public services, so that this research has differences in collecting and analyzing data.

2 Literature Review

2.1 Social Media as a Means of Distribution of Information

Social media is a series of new technologies that support the sharing of richer data in a highly decentralized, dynamic, and loosely structured virtual environment. Social media is supposed to facilitate engagement, learning, knowledge creation, aligning, and even challenging borders and traditional authority [10]. Social media provides policymakers with a new approach to increasing openness and accountability, increasing user involvement, and collaborating in decision-making to enhance the control of information and access to public resources [11]. Social media's key features are engagement, openness, discussion, contact, and collaboration [4].

Social media are on the rise everywhere in developed nations like Indonesia. Young people with wireless access are increasingly prevalent; Indonesians are one of the world's biggest social media consumers. To engage the public further in the policymaking process, LAPOR is an example of how the government creates custom-made social media networks. It is used for the announcement of public service records by residents. Data approval from this study will be canceled for details on data-based decision making at the national level [12]. Social media is a space for expression and communication, social media accounts that are already actively posting information so that communication integration can integrate public policy [13]. Citizen interaction is an essential component of transparency that helps produce ideas and capital and strengthen decision-making. Provide timely and reliable data through suitable networks by promoting more significant contact between individuals in the policymaking process [4]. Twitter has been generally embraced as it offers the latest coverage with instant and broad access while also reporting on the incident [14]. Twitter is a website owned and run by Twitter that provides microblogging on a social network that helps its users send and read messages called tweets [15]. By minimizing misdirected messages, speeding participation, and integrating various channels, the government's public resources supply can upgrade [12].

2.2 Optimizing the Use of Social Media as Information Connectivity for the Government

Different social media, such as blogs, microblogs, service sharing, text messages, discussion forums, and social networks that can directly engage their citizens, are currently used by government departments in different countries [16]. Regulating and enforcing government policy for the good of society by rising productivity and public service as a whole is the core of coordinating e-gov [17]. Smart Twitter is a technology used in organizations to promote working relationships with the community [18]. As observers of political policy and opponents of government behavior, the media may also play a part [19]. Social media is defined as a group of internet-based applications built based on ideology and technology that allows data exchange [20]. Various internet activities through social media can be carried out anywhere; people can engage in public discourse discussed and encourage cyber democracy so that democracy can run well [19]. The phenomenon of social media in the development of information and communication technology. Social media's presence impacts communication in all fields, such as marketing communication, political communication, and communication in learning systems [22].

3   Research Methods

This research used descriptive qualitative research. Data analysis used *NVivo 12 plus software*, data retrieval through the DPR RI Twitter account via *NCapture* from NVivo 12 Plus with Web Chrome. The data were processed with the *Crosstab feature* to automatically calculate the main statistical tests required with notable comparisons and indirect variables analyses. The *Crosstab Query feature* is to enter code (manual, generated, etc.), text data, and numeric data in variables and data patterns. At this stage, automatic calculation between all data was found regarding the media's influence on the Omnibus Law-making process. Use the *Word Tree* feature to see patterns of relationships in the discussion.

Furthermore, using the *Word Cloud feature* can find words that often appear from the results of data findings or see terms that are often discussed. This study reveals the Influence of Social Media on the Omnibus Law-Making Process. Measuring indicators of participation, transparency, conversation, interaction, and communication [4].

4   Results and Discussion

Social media can express themselves, form public opinion, and various points of view. Multiple benefits can be felt directly on different access to information and openness in various public policies. The existence of social media is considered capable of influencing a process in government policymaking. The Omnibus Law is debated, socialized, and responded to by the public before the Law is passed; transparency and participation are essential factors. A law was born based on the Law on the Formation of Legislative Regulations, Law no. 12 of 2011 in conjunction with Law No. 15 of 2019, which is the implementation of the constitutional mandate of Article 22A of the 1945 UUD Omnibus Law The Work Creation Act should be debited from the planning, drafting, discussion, promulgation and socialization stages to the public [23]. The rapid dissemination of information has had a significant impact on government and community life [24]. The Omnibus Law Bill was first announced by Jokowi in October 2019 when he only served as president for the second term; Jokowi immediately made the regulation. He then asked the Chairman of the DPR, Puan Maharani, that the job creation Bill's deliberation process be completed within three months. Shortly after that, the #GagalkanOmnibusLaw social media on Twitter became trending. The demonstration against the Omnibus Law Bill was enlivened by various workers and students who are members of the Mobile People's Alliance [25]. Social Media Twitter is a medium often used in sharing information, such as the DPR RI Twitter account, in making the Omnibus Law, which is involved and influences in decision-making. The data results show social media's impact from various indicators: transparency, conversation, participation, communication, and involvement. The results of this analysis can be seen in Fig. 1 below.

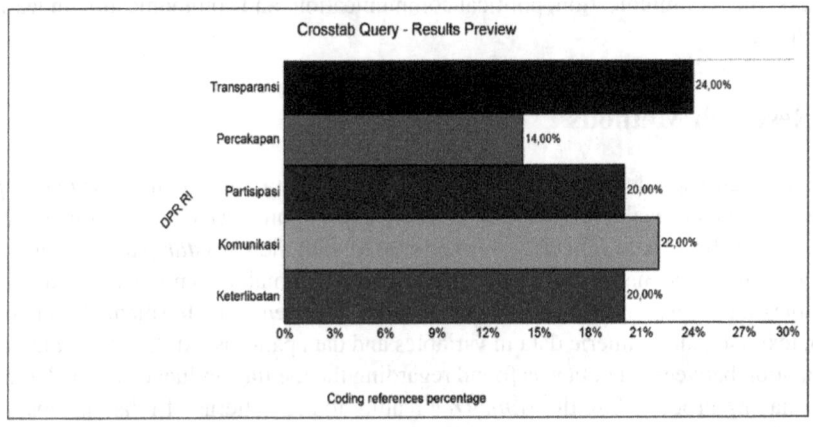

Fig. 1. Processed Crosstab Query by researchers using NVivo 12 Plus (2020)

4.1 Transparency

Transparency is the government's openness in providing information related to public resource management activities to parties who need information [18]. Transparency is a significant factor in the making and stipulation of a policy to maximize the common interest. The DPR RI Twitter account provided clarity in sharing information about the Omnibus law after the data was processed with NVivo 12 Plus obtained a score of 24.00%. The DPR RI Twitter account shared various activities regarding the Omnibus Law, namely that on April 24, 2020, the Chairman of the DPR RI Dr. (H.C) Puan Maharani said that he and the DPR RI leadership would ask the DPR RI Legislation Body to postpone the discussion of the Employment Cluster Work Creation Omnibus. The postponement period can re-absorb the aspirations of the community related to the Employment Cluster Work Creation Omnibus. We will ask to postpone the debate for clusters that need further discussion and waiting for the city's aspirations. Tweet February 13, 2020, explains that the DPR's existing mechanism will implement the Omnibus Law on Job Creation through the legislature or the Special Committee because it involves seven commissions related to the discussion of 11 clusters of 15 chapters and 174 articles. The various Tweets shared by the DPR RI Twitter account regarding transparency with citizens in making the Omnibus Law, which scores 24.00%, are considered transparent to activities and need to increase openness to the public community involvement.

4.2 Conversation

A conversation is a language event between two or more participants; the dialogue is a forum that allows the principles of cooperation and courtesy to occur in language events [26]. An open discussion in discussing the Omnibus Law will provide a broad understanding so that conversation is an essential element that must be present in the policymaking process. Data concerning conversations in the process of social media engagement, in this case, obtained a score of 14.00%, which was acquired by the data obtained by NVivo 12 Plus by discussing that the process of making Omnibus Law is in the process of creating, so that accurate and expected coordination is needed.

This element explains that in the conversation indicators from the results of the DPR RI Twitter account from the delivery of Puan Maharani on February 12, 2020, that the Coordinating Minister for the Economy along with other Ministers present said that the Omnibus Law copyright work would later consist of 79 laws, which comprised of 15 chapters with 174 Articles to be discussed in the DPR. The Omnibus's making was also explained on March 10, 2020, by the deputy chairman of the DPR RI, Sufmi Dasco Ahmad, that, in principle, the Indonesian Parliament always involves elements from the community that is guaranteed by Law and are open. The results of the conversation that was given to the DPR RI involving social media in making the Omnibus Law were concluded to be still insufficient and considered not contributing to the community concerned; the results of data analysis from the conversation indicators obtained a score of 14.00% which captured needs to be improved so that between the two can be connected more broadly. The conversation that is discussed about the Omnibus Law can be seen in Fig. 2 below.

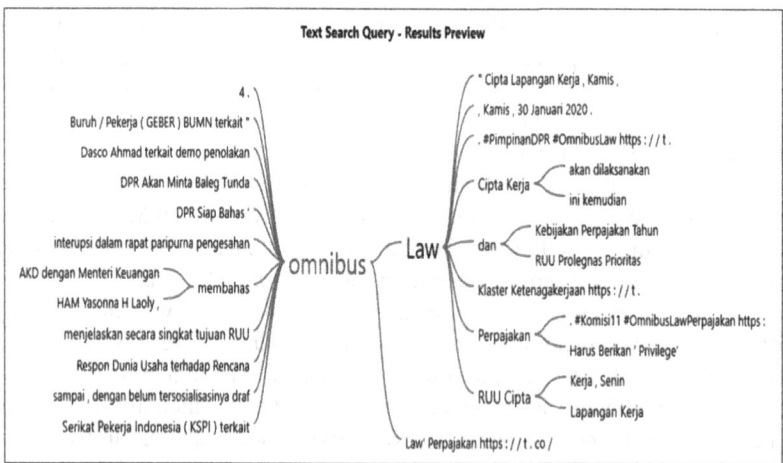

Fig. 2. The conversation about the Omnibus Law, Processed by researchers using Word Tree NVivo 12 Plus (2020)

The conversation discussed by the DPR RI account on the Twitter account was discussing the Omnibus Law involved in the picture above that the discussion there were several discussions about the Omnibus Law which had been concerned that the Indonesian Parliament delivered Puan Maharani on February 12, 2020, the Coordinating Minister for the Economy with the Another present stated that the Omnibus Law copyright work would later consist of 79 laws, which comprised of 15 chapters with 174 Articles to be discussed in the DPR. The Omnibus's making was also explained on March 10, 2020, by the deputy chairman of the DPR RI, Sufmi Dasco Ahmad, that, in principle, the Indonesian Parliament always involves elements from the community that is guaranteed by Law and are open. The results of the conversation that was given to the DPR RI involving social media in making the Omnibus Law were concluded to be still insufficient and considered not contributing to the community concerned; the results of data analysis from the conversation indicators obtained a score of 14.00% which captured needs to be improved so that between the two can be connected more broadly. The conversation that is discussed about the Omnibus Law can be seen in Fig. 2 below.

Seventy-nine laws composed of 15 articles with 174 articles were discussed in the DPR. The preparation of an open Omnibus was also presented on March 10, 2020, by the Deputy Chairperson of the Indonesian Parliament, Sufmi Dasco Ahmad, that, in principle, the Indonesian Parliament always involves the guaranteed community by Law and is of nature. So that the process of making Omnibus Law always involves various elements involved.

4.3 Participation

Participation is participation in running programs in activities to achieve a goal [22]. Participation in the question is in the following discussion. On January 16, 2020, the participation was carried out by Commission IX of the DPR RI RDPU together with the Chairperson of the Federation of Chemical, Energy and Mining Workers Union, General Chairperson of the National Welfare Movement, and General Chair of the Joint Labor/Workers Movement BUMN related to Omnibus Law. Activities carried out by interest elements in this process do not involve or openness to social media so that, in this process, social media has less influence. Participation is indispensable in the implementation of the Omnibus Law so that every element is involved in it.

The implementation of making the Omnibus Law also involved the Chairman of the Indonesian Parliament Puan Maharani accompanied by Deputy Chairman of the Indonesian Parliament Rachmat Gobel Consultation Meeting with Minister of Finance Sri Mulyani and Minister of Law and Human Rights Yasonna H Laoly, at the Nusantara III Building of the Indonesian Parliament on January 30, 2020, in a meeting to discuss Omnibus Law and equating perceptions about the Omnibus Law which will later be submitted to the government to the Indonesian Parliament. The results of data analysis obtained from the DPR RI Twitter account regarding participation in social media influence making the Omnibus Law received a score of 20.00%, which means that it has an impact even though it is not too big. The DPR RI Twitter account's various participations, which are discussed, can be seen in Fig. 3 below.

Based on the data above, the DPR RI Twitter account is active. Tweets conducted various discussions in conveying multiple information needed by community members.

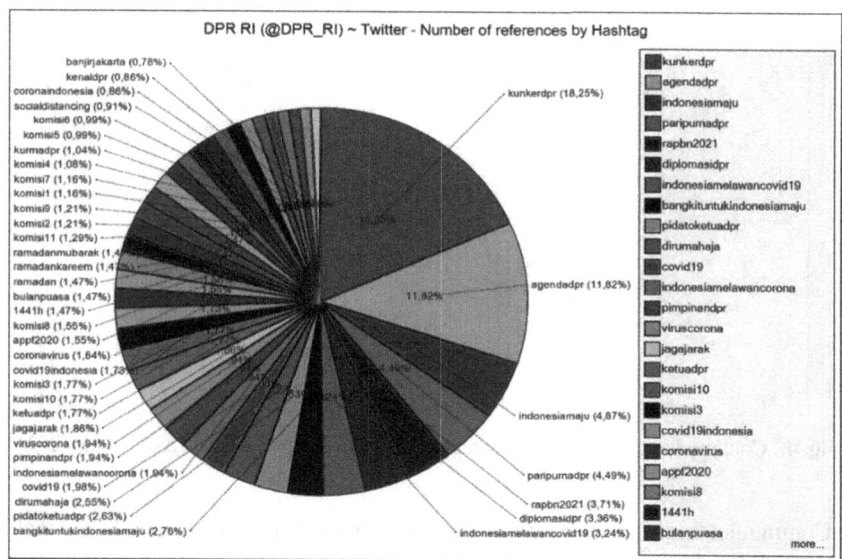

Fig. 3. Participation in the DPR RI Twitter Account, by Hashtag NVivo 12 Plus (2020)

The DPR RI Twitter account activity from the hashtag free hashtag is the highest from the hashtag kunkerdpr, agendadpr, Indonesiamaju, paripurnapdr, rapbn2021 to the Bulan-fast hashtag; various existing hashtags show the delivery of information and results that influence decision making. Hashtag activity from the DPR RI Twitter account shows the influence in different processes of delivering information and making policies.

4.4 Communication

Communication is everything regarding the arrangements and practices of communication within the government's scope; communication manages public opinion and the entire process of communication that takes place in government in achieving a goal [28]. The Twitter account social media's communication assessment from data analysis results with NVivo 12 Plus software from the DPR RI Tweet obtained 22.00%. The score was obtained from the activities of the DPR RI at the Secretary-General of the DPR RI, Indra Iskandar, explaining the microphone turned off incident when members of the Democratic Faction delivered an interruption in the plenary meeting of the ratification of the Omnibus Law on the Job Creation Bill on October 5, 2020. Communication is expected to run well because, according to the DPR RI Chairman, Puan Maharani, he said he should not create suspicions that could raise suspicion against the DPR RI. The DPR RI Twitter account also informs that the Omnibus Law will be implemented with the DPR's existing mechanism through the legislature or select committee because it involves seven commissions related to reforming 11 clusters, 15 chapters 174 articles. On this indicator, obtaining a score of 22.00% is considered to influence making the Omnibus Law because of the openness through various connected communications. The communication carried out by the DPR RI with multiple discussions can be seen in Fig. 4 below.

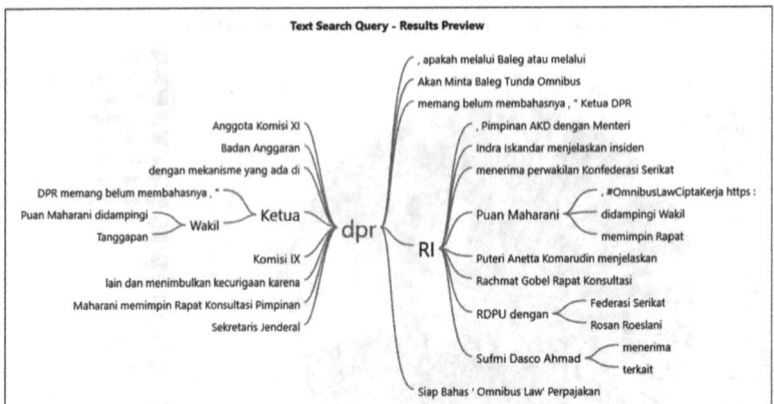

Fig. 4. Communication, Processed by researchers using Word Tree NVivo 12 Plus (2020)

Communication is an activity that involves various related elements; In the making of the Omnibus law, the implementation of communication carried out by the Indonesian Parliament from the Word Tree analysis results resulted in words that were often discussed from the products shown. Based on the data, the implementation of communication towards making Omnibus Law has an associated communication network. The communication seen from the analysis results indicates a process involved in making the Omnibus Law, which includes delivering information from the Chairman of the DPR and his deputies regarding implementing the Omnibus Law.

4.5 Involvement

Involvement is an activity involving the community in finding alternative solutions to an existing problem and providing an evaluation or assessment of the authorities' performance so that later they will be able to change the pattern of government towards better governance [24]. Making the Omnibus Law Act involves various essential elements in the government and society, hoping that each group's aspirations can be fulfilled. Analysis of data from the DPR RI Twitter account after processing it got a score of 20.00%. The implementation of the process of making the Omnibus Law involves various essential elements as explained by the Chairman of the Indonesian Parliament Puan Maharani that the DPR RI Leadership Consultation, AKD Leaders, and the Minister of Finance discussed the Omnibus Law and the 2020 Priority National Legislation Program Bill related to financial conditions, macro-fiscal and economic developments. Country on December 16, 2019. The process of making the Omnibus Law also involved Commission IX of the DPR RI RDPU with the Federation of Chemical, Energy and Mining Workers Unions, the National Welfare Movement, and the Joint Movement for Workers' Workers BUMN related to Omnibus Law Employment that will be held on January 16, 2020. Tweets' results from the DPR RI Twitter account show that making the Omnibus Law has been carried out.

Further analysis using the *Word Cloud feature* is to find words that often appear and are discussed in research topics. Using the Word Cloud can see what types of stories appear on the DPR RI Twitter account's social media account about social media's influence in making the Omnibus Law. The results of the 100 words that seem the most on this topic sourced from the DPR RI Twitter account can be seen in Fig. 5 below.

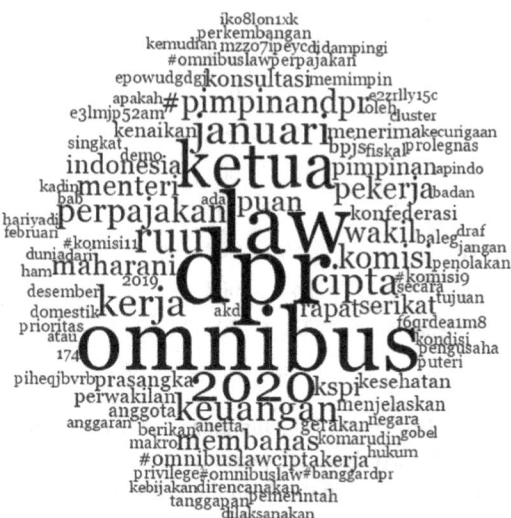

Fig. 5. Processed by researchers using World Cloud NVivo 12 Plus (2020)

The ten most important words that appear in the *story cloud* results are DPR, Omnibus, Law, Chairperson, 2020, Draft law, January; the ten most prominent names that appear in the word cloud results are DPR, Omnibus, Law, Chairperson, 2020, Draft Law, January, Create, work, and Finance. These words are closely related to the making of the Omnibus law. The word DPR is the frequency of words that appear the most, in this case, the DPR, which has an essential role in various processes and policies. The emergence of existing stories from the *Word Cloud* results offers a synchronization in the works of research on making the Omnibus Law and Finance. These words are closely related to the making of the Omnibus law.

The word DPR is the frequency of words that appear the most, in this case, the DPR, which has an essential role in various processes and policies. The second big word that appears is Omnibus, in the word Omnibus is a topic discussed in the DPR RI Twitter account. The emergence of existing stories from the *Word Cloud* results offers a synchronization in the works of research on making the Omnibus Law. The comprehensive data that discusses the challenge of the Omnibus Law is seen from five indicators, namely: Transparency, Conversation, Participation, Communication, and Engagement, which are discussed on the DPR RI Twitter account after being processed with the *Summary* NVivo 12 Plus, bringing up words that often appear in discussions and can be seen in Table 1 below.

The words that appear in the *Word Frequency* from the summary results indicate the terms used most frequently in the discussion and expect synchronization. The words most often appear the word dpr with a weight percentage of 005%, then omnibus 004%,

Table 1. Word Frequency of NVivo 12 Plus Analysis

Word	Length	Count	Weighted percentage (%)
dpr	3	74	005
omnibus	7	64	004
law	3	61	004
ketua	5	45	003
2020	4	36	002
ruu	3	29	002
januari	7	27	002
cipta	5	25	002
kerja	5	25	002
keuangan	8	23	001

Law 004%, Chairman 003%, ruu 002%, January 002%, creativity 002%, work 002%, and 001% finance. From the words that appeared, it can be seen that the Twitter account stated that the DPR played an essential role in the process of making the Omnibus Law. The emergence of words from the analysis results shows that in the discussion, the Omnibus becomes the main topic based on the Law by the regulations on the creativity of policymaking, which is expected to have proper synchronization.

5   Conclusion

Social media is sophisticated media that plays an essential role in making various policies; social media accessing multiple information. Social media Twitter is currently making the Omnibus Law; different information can be accessed quickly. Making the Law impacts various angles, namely transparency, conversation, participation, communication, and involvement; the results show that Twitter social media is an active media included in the Omnibus Law discussion. From the five indicators, it can be concluded that the transparency indicator has provided information that the debate on the Omnibus law will be postponed and states that the Omnibus Law consists of 79 rules, composed of 15 articles with 174 articles to be discussed in the DPR. Conversation indicators that discuss always to maintain good communication. Involvement is an activity that involves the community in finding alternative solutions to problems; he implementation of the process of making the Omnibus Law involves various essential elements as explained by the Chairman of the Indonesian Parliament Puan Maharani that the DPR RI Leadership Consultation, AKD Leaders, and the Minister of Finance discussed the Omnibus Law and the 2020 Priority National Legislation Program Bill related to financial conditions, macro-fiscal and economic developments; participation and communication have been carried out well, but community involvement in making Omnibus Law is still lacking. It needs to be involved openly.

This research has a research limitation on social media in giving influence in the process of making the Omnibus Law so that the activities of Twitter social media can be assessed for their openness, and further research, it is expected to be able to examine social media strategies in attracting interest in the information which was raised considering that many residents did not care about communication.

References

1. Aziz, M.N., Firmanto, A., Miftah Fajrin, A., Hari Ginardi, R.V.: Sentiment analysis and topic modelling for identification of government service satisfaction. In: Proceedings of the 2018 5th International Conference on Information Technology, Computer and Electrical Engineering. ICITACEE 2018, pp. 125–130 (2018). https://doi.org/10.1109/ICITACEE.2018.8576974
2. Morgan: The role of website interactivity and Web 2.0 applications in building relationships between the city of Longmont and its citizens Capstone. J. Chem. Inf. Model. **53**(9), 1689–1699 (2019). https://doi.org/10.1017/CBO9781107415324.004
3. Song, C., Lee, J.: Citizens use of social media in government, perceived transparency, and trust in government. Public Perform. Manag. Rev. **39**(2), 430–453 (2016). https://doi.org/10.1080/15309576.2015.1108798
4. Haro-de-Rosario, A., Sáez-Martín, A., del Carmen Caba-Pérez, M.: Using social media to enhance citizen engagement with local government: Twitter or Facebook? New Media Soc. **20**(1), 29–49 (2018). https://doi.org/10.1177/1461444816645652
5. https://www.google.com/search?client=firefox-b-d&q=Gilang+Rahmadan%2C+%E2% 80%9COmnibus+Law+Sebagai+Sarana+Utama+Penataan+Regulasi%2C%E2%80%9D+ vol.+01%2C+no.+1%2C+pp.+1%E2%80%937%2C+2017%2C+%5BOnline%5D.+Availa ble%3A+http%3A%2F%2Fwww.albayan.ae
6. Suriadinata, V.: Penyusunan Undang-Undang Di Bidang Investasi: Kajian Pembentukan Omnibus Law Di Indonesia. Refleks. Huk. J. Ilmu Huk. **4**(1), 115–132 (2019). https://doi. org/10.24246/jrh.2019.v4.i1.p115-132
7. Michael, T.: Bentuk Pemerintahan Perspektif Omnibus Law. J. Ius Const. **5**(1), 159 (2020). https://doi.org/10.26623/jic.v5i1.2222
8. Law, O., Teknik, S., Anggono, B.D.: Peluang Adopsi Dan Tantangannya Dalam Sistem Perundang-Undangan Indonesia, vol. 9, no. April, pp. 17–37 (2020)
9. https://ejournal2.undip.ac.id/index.php/gk/article/view/6751
10. Feeney, M.K., Welch, E.W.: Technology–task coupling: exploring social media use and managerial perceptions of E-government. Am. Rev. Public Adm. **46**(2), 162–179 (2016). https:// doi.org/10.1177/0275074014547413
11. Guillamón, M.D., Ríos, A.M., Gesuele, B., Metallo, C.: Factors influencing social media use in local governments: the case of Italy and Spain. Gov. Inf. Q. **33**(3), 460–471 (2016). https:// doi.org/10.1016/j.giq.2016.06.005
12. Dini, A.A., Sæbo, Ø., Wahid, F.: Affordances and effects of introducing social media within eParticipation—findings from government-initiated Indonesian project. Electron. J. Inf. Syst. Dev. Ctries. **84**(4), 1–4 (2018). https://doi.org/10.1002/isd2.12035
13. Malawani, A.D., Nurmandi, A., Purnomo, E.P., Rahman, T.: Social media in aid of postdisaster management. Transform. Gov. People Process Policy **14**(2), 237–260 (2020). https:// doi.org/10.1108/TG-09-2019-0088
14. Heravi, B.R., Harrower, N.: Twitter journalism in Ireland: sourcing and trust in the age of social media*. Inf. Commun. Soc. **19**(9), 1194–1213 (2016). https://doi.org/10.1080/1369118X. 2016.1187649

15. Kosasih, I.: Peran Media Sosial Facebook dan Twitter dalam Membangun Komunikasi (Persepsi dan Motifasi Masyarakat jejaring Sosial Dalam Pergaulan). Lembaran Masy. J. Pengemb. Masy. Islam 2(1), 29–42 (2016)
16. https://ejournal.unsub.ac.id/index.php/FIKOM/article/view/531
17. Sosiawan, E.A.: Tantangan Dan Hambatan Dalam Implementasi E-Government Di Indonesia. In: Seminar Nasional Informatika, vol. 2008, no. semnasIF, pp. 99–108 (2018)
18. Putri, K.D., Komunikasi, D.I., Pusat, J., Komunikasi, D.I., Pusat, J.: Twitter, Media Sosial, Humas, Komunikasi, vol. 77 (2018)
19. https://library.universitaspertamina.ac.id/xmlui/handle/123456789/815
20. Nurmandi, A., et al.: To what extent is social media used in city government policymaking? Case studies in three asean cities. Public Policy Adm. 17(4), 600–618 (2018). https://doi.org/10.13165/VPA-18-17-4-08
21. Gusri, L., Lestari, Y., Amenike, D., Hanana, A.: "Penggunaan Media Baru Untuk Komunikasi Politik Pemerintah di Sumatera Barat, vol. 01, no. 01 (2017)
22. https://ejournal.bsi.ac.id/ejurnal/index.php/cakrawala/article/view/1283
23. Tagar: Beda Cara Pembuatan Omnibus Law di Indonesia dan Negara Lain (2020)
24. Zahra, A.A., Purnomo, E.P., Kasiwi, A.N.: New democracy in digital era through social media and news online. Humaniora 11(1), 13 (2020). https://doi.org/10.21512/humaniora.v11i1.6182
25. https://mediakernels.com/author/andist3/page/4/
26. Nugroho, R.A.: Analisis Implikatur Percakapan dalam Tindak Komunikasi di Kelompok Teater Peron FKIP UNS, pp. 1–8 (2017)
27. Dyah Putri Makhmudi, M.M.: Prasarana Lingkungan Pada Program Penataan Lingkungan Permukiman Berbasis Komunitas (Plpbk) Di Kelurahan Tambakrejo, Kota Semarang. J. Pengemb. Kota 6(2), 108–117 (2018). https://doi.org/10.14710/jpk.6.2.108
28. Rahman, A., Sjoraida, D.F.: Strategi Komunikasi Pemerintah Kabupaten Subang Menyosial-isasikan Gerakan Pembangunan Untuk Rakyat Infrastruktur Berkelanjutan. J. Kaji. Komun. 5(2), 136 (2017). https://doi.org/10.24198/jkk.v5i2.8443
29. Adiputra, M.P., Aryani, N.P.E., Werastuti, D.N.S.: Pengaruh Kompensasi Aparatur Pemerin-tah Desa, Efektivitas Kinerja Pendamping Lokal Desa Dan Keterlibatan Masyarakat Terhadap Pengoptimalan Pengelolaan Dana Desa. J. Ilm. Akunt. 11(1), 16 (2020). https://doi.org/10.30591/monex.v8i1.1064

How Does Social Media Affect Money Politics Campaign Rejection in the 2020 Regional Head General Election Social Media? A Case Study of Indonesia

Misran[1][(✉)] [iD], Achmad Nurmandi[1] [iD], Dyah Mutiarin[1] [iD], Suswanta[1] [iD], and Salahudin[2] [iD]

[1] Department of Government Affairs and Administration,
Universitas Muhammadiyah Yogyakarta, Yogyakarta, Indonesia
misranalfarabi@gmail.com
[2] Department of Government Studies, Social and Political Science Faculty,
Universitas Muhammadiyah Malang, Malang, Indonesia

Abstract. The integrity of the regional head elections process in Indonesia has become a concern for the government and the Indonesian people. Thus, to realize regional head elections with integrity, the implementation must be free from various types of violations such as election malpractice. One of the election violations that often appear in every regional head election is the practice of money politics. This paper intends to explain the election supervisory body in preventing money politics in regional head election contestation. This research's data sources included: website, Twitter, and previous research related to money politics. The data were obtained using the Ncapture feature in the Nvivo 12 plus where the Ncapture feature is a web browser extension developed to capture web content in the form of the website content, social media, and other document content such as scientific articles and a collection of opinions from observers about money politics in Indonesia. This paper reveals that social media is a new transformation for election supervisory bodies to disseminate and prevent money politics. The authors also need to emphasize that this time is money politics itself because democracy/general election parties in any part of the world require money/capital both by the organizers and election participants themselves, so a common understanding is needed. What and how money politics can be categorized as an action. Election violations and even fall into the category of election crimes. This paper is reviewed based on an analysis of scientific articles and each election organizer's official Twitter accounts.

Keywords: Social media · Money politics · Local elections · Democracy

1 Introduction

The development of democratic processes in various parts of the world from time to time shows that free, transparent, and fair regional head elections are the foundation

T. Antipova (Ed.): ICADS 2021, AISC 1352, pp. 511–522, 2021.
https://doi.org/10.1007/978-3-030-71782-7_45

for strengthening healthy democracy. According to one publication of the International Institute for Democracy and Electoral Assistance (IDEA) in a book entitled Improving Electoral Practices: Case Studies and Practical Approaches, it fulfills this need for strengthening democracy. The election process must be based on two basic standards: credibility and integrity. Both the regional head elections organizers' credibility and integrity are organizing the elections [1].

In realizing regional head elections with integrity, the process of organizing Regional head elections must be free from various kinds of violations in regional head Elections malpractice or what is also known as electoral malpractice. One form of election violations that often arises is the practice of money politics. This money politics affects the outcome of the regional head elections, determining the election results' quality and integrity. If money politics continues, it is sure that people will feel the harmful effects. Democracy will only be a land for mediocre people, namely those who do not have sufficient achievements to gain power.

Another issue that the authors need to emphasize in his writing this time is money politics itself because democracy/general election parties in any part of the world require money/capital both by the organizers and election participants themselves, so a common understanding is needed. What and how can money politics be categorized as an action? Election violations fall into the category of election crimes. For example, during the campaign period, regional elections participants are given time and space to carry out open campaigns and large numbers of people. On the other hand, when the people come to the campaign venue, the election participants are given a sum of money. This condition is then called money politics activities.

Excellent and correct confirmation is needed from the above phenomena through the election organizer, namely the election supervisory board. The election supervisory board acts as a partner providing political education to the public by involving political parties to educate the public about preventing money politics from implementing clean and honest regional elections. This paper is prepared to explain the election oversight body in preventing money politics in elections through social media on Twitter accounts. Furthermore, this research has not studied much about social media use in preventing fraudulent practices, especially money politics.

2   Literature Review

2.1   Social Media

Social media is a modern technology application in which many features facilitate the interaction and communication between individuals (personal communication) and groups with groups (group communication) that take place across time and space [2]. Social Media's emergence has further revolutionized digital platforms' capacity to allow for constant interaction and cooperation, bringing their voices to the broader public. According to Cangara, "media is a tool or means used to convey messages from communicators to the public [3]. In Shirky's view, social media provides ample space for everyone to convey their ideas and opinions to show their voice through social media [4]. It is generally assumed that social media plays a crucial role in disseminating political groups' information and claims [5].

Through social media, communication and human interaction can occur at any time and anywhere, where everyone has access to build communication and interaction with all parties in various kinds of matters, including business affairs and socio-political affairs [4]. Social media is no longer only a virtual world but turns things of protest, criticism, and disappointment into the real world [6].

Even though social media is entirely given full freedom of opinion, the media also presents alternatives in shaping public opinion and a medium for interaction between parties and politicians and their constituents [7]. Social media has five essential aspects: public participation, openness, communication, community development, and connectivity between social media users [8]. These five aspects are an essential part of social media directly related to human life in various forms of affairs in this modern era.

Users can optimally manipulate themselves through social media through text, graphics, or audio-visual [6]. Social media platforms are used effectively by the community in several ways, such as through blogs and sharing educational videos, updates, and academic documents [9]. According to the latest social media statistics, more than 2 billion Facebook users, more than 300 million Twitter users, more than 500 million Google users, and more than 400 million [9]. The emergence of the internet and social media has allowed the enforcement of community norms to have a central role in upholding justice by expanding the enforcement tools available to the community and increasing these tools' effectiveness [10]. Therefore, social media is an agent of political socialization in today's society [11].

Research on the use of new media is growing. Likewise, the debate about new media's effectiveness to support social change has yet to conclude at the academic level. However, as Radloff stated in the journal *Feminist Africa Women Mobilized*, the internet and other related tools can be useful for resistance, social mobilization, and development in the hands of people or organizations working for freedom and justice [12].

2.2 Local Elections

As an archipelagic country, Indonesia consists of 13,466 islands with 34 provinces divided into 416 districts and 96 cities with their own regional/autonomous government system. As a country with a democratic autonomous government system, holding the regional head/regional head elections manifests its democratic character [13]. 2015 is the first time Indonesia has held a new form of regional election simultaneously. The 2015 Regional Head Elections is indeed different from the previous regional head elections [10]. The simultaneous regional elections are certainly far different from the previous ones, where the elections were held during the COVID-19 pandemic. The regional election system was implemented simultaneously to condense electoral fraud [14]. This condition will lead to various speculations related to Regional Head Elections' safety and prevention [15]. Mapping of shifts after political marketing was used in the 2015 regional head elections, based on information from the Election Supervisory Board (Banwaslu) [16]. The information states, election fraud cases have increased compared to last year's election. On the other hand, there is increasing evidence that many regional elections are not under the second-order election approach [17].

Like general elections held based on the manifestation of the principle of equality before the law and equal opportunity in government (equal opportunity principle), regional head elections is also a process of succession to the transition of a regional leader that involves the real role of the public or the people in sovereignty.

Regional Head Elections is an activity of a democratic process that cannot be separated from the implementation of elections because Regional Head Elections has output, namely political officials (elected official), not electing administrative officials (appointed official) [18]. In the context of democracy, the dynamics of politics and local governance like this will determine the face of Indonesian democracy nationally. Therefore, the simultaneous regional elections are also accessed for the public to participate more widely, gathering much trust in prospective candidates.

2.3 Money Politics

In general, money politics is a term that describes the use of money or other rewards to influence a person, group, or institution in making political decisions [19]. In general, money politics can be understood as electoral mobilization by giving money, gifts, or goods to voters to throw in the election. Money politics in elections requires many ways and stages. Among them is political dowry at the candidacy stage, vote-buying and selling at the campaign stage, and voting. This condition includes bribes to election administrators during the counting and recapitulation stages of votes to improvements in dispute resolution over election results.

Several studies refer to money politics in political distribution theory, divided into two forms [20]. First, specific money politics refers to a retail strategy for buying and selling votes (vote-buying). In terms of time, it is usually carried out before the election or what we know as "dawn attacks." Sometimes prepay before election day. Sometimes postpaid after support is provided. Second, a more long-term, collective, and wholesale money policy strategy abuses programmatic policies such as social assistance or grants or electoral interests. The public does not realize the dangers of money politics in political life in the future. It is even considered mediocre, so there is no effort to avoid money politics [21]. Money politics is an instant medium where constituent votes can be bought [22].

In the administration of democracy, it is always dirty in harmful ways. Money politics is not only at the central government level but also reaches areas far from the center. It has been expected and done openly, such as infrastructure donations, road repairs, renovation of social facilities, or everyone getting money, so they have to participate in elections and vote [22]. The practice of money politics is common in Indonesian society.

Money is the source of society's most needed power, which refers to every transaction or individual [23]. During elections throughout the country, it is complicated to avoid the practice of money politics. According to Hamid, who sees it from an economic perspective, money politics arises because of the reciprocal relationship between actors (parties, politicians, or intermediaries) and victims [24].

3   Method

The approach used in this research was a descriptive qualitative approach that described the phenomenon of the election supervisory body in preventing money politics in the regional head general election contest. Descriptive analysis in this study used NVivo 12 plus software. This research's data sources included: website, Twitter, and previous research related to money politics. The data were obtained using the Ncapture feature in the Nvivo 12 plus where the N capture feature is a web browser extension developed to capture web content in the form of the website content, social media, and other document content such as scientific articles and a collection of opinions from observers about money politics. The Nvivo 12 plus analysis used the Chart analysis feature, Word cloud, Cluster analysis, and Chart analysis. Cluster analysis was used to visualize and collect data/words with similarities and differences. This study looked at the similarities and differences in the tweets of organizational groups. Chart analysis was used for charts to study, process data in research, and answer analysis in the form of N capture; meanwhile, Word cloud analysis listed words or concepts that often appear in research files to visualize and collect data/words that have similarities and differences.

Table 1. Twitter accounts, websites, and media online

Institution	account Twitters	Websites	Media online
Bawaslu RI	@Bawaslu RI	https://www.bawaslu.go.id/	https://www.kompas.com/tag/website
			www.detik.com, www.republikaonline.com, cnn.com

This study aims to see the prevention of money politics in a regional head election. This study will answer two questions: RQ1: How does the Election Supervisory Agency use social media? RQ2: How are the Election Supervisory Agency and Regional Election Supervisory Agency related?

4 Result and Discussion

In the 21st century, social media is growing every day, and people use social media for personal and professional use. In Shirky's view, social media provides ample space for everyone to convey their ideas and opinions to show their voice through social media [25]. According to Cangara, "media is a tool or means used to convey messages from communicators to the public [3]. Social media is no longer a virtual world, but it shifts protests, criticism, and disappointment into the real-world [6]. The use of Twitter in Indonesia has enormous potential. This situation shows a significant number, even the CEO of Twitter himself said that Indonesia plays a central role in business travel on Twitter [26]. Therefore, Twitter accounts have a part in the distribution era to accelerate information delivery to the public. Seeing this potential, the Election Supervisory Agency uses the Twitter account as a tool to socialize prevention of money politics.

Figure 1 shows the data from the Twitter account of @Bawaslu RI. The first tweet was done in January 2017. The most vulnerable time to tweet was April-March 2020. The @Bawaslu RI account's tendency to post content was more frequently on the vulnerable

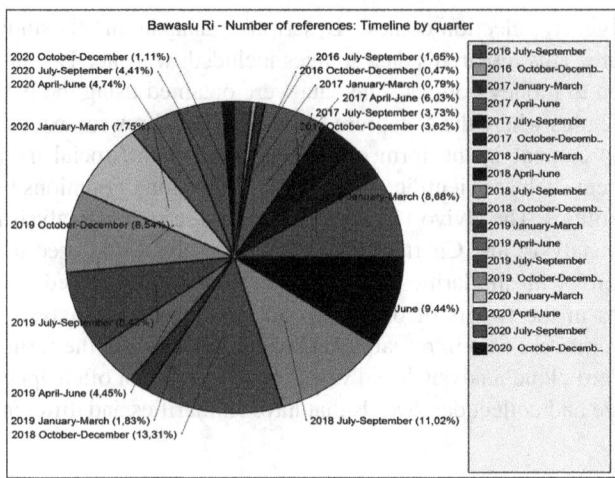

Fig.1. Tweet intensity of @ Bawaslu RI.

time between January-March and July-September 2020. If traced more deeply, June and December coincided with the Regional Heads Election. This situation affected the density of information provided by the election supervisory body through social media and one way for the Election Supervisory Agency to disseminate supervision to the public, also avoid money politics attacks and protect their voting rights. By socializing with the community, they were helped by updated information through social media on Twitter.

Furthermore, the foreign minister will see the extent to which the @Bawaslu RI accounts use social media based on frequently discussed topics. The authors used the word cloud feature found in NVivo 12 Plus software. Word cloud is included in the tag word application category, which is used to learn about the number and types of topics in the text body [27]. Besides, it used a tag cloud to describe text content that develops over time. Typically, this statistical overview is achieved by positively correlating the font size of the described tags with word frequency [27] (Fig. 2).

Based on the visualization of the word cloud results from the social media account @Bawaslu RI (see Fig. 3), placing the phrase Bawaslu, #Sahabatbawaslu, #Bawasluja-gahakpilih, and #bawaslumengawasi as words that are often used in every tweet. At a glance, there are taglines #Sahabatbawaslu and #Bawaslumengawasi as words that have great visualization. Other words with a large enough visualization include #bawaslujaga-hakpilih, #bawasluterbuka, #Pilkada today, #Pemiluterpercaya, #cegahawasitindaki, and other terms. However, statistically, each story that has been visualized has a value accord-ing to how often the sentence appears in the tweet. The data for Word Frequency shows the words that dominate the delivery of information to @Bawaslu RI (see Fig. 3). According to the content and purpose of delivering the information, the classification becomes the movement's variable to reject money politics. Content that dominates #Sahabatbawaslu, #Bawaslujagahakpilih, and #bawaslumengawasi (Table 2).

Table 1 displays the value of each visualized word from the @Bawaslu Ri account. Of the 100 words that were envisioned, the authors took the largest 20 words, which

Fig. 2. The topic of conversation on the Twitter account @Bawaslu RI

became the topic of conversation. The authors believe that the @Bawaslu RI account focuses more on reporting related to actual conditions, seen from several suitable words to describe this, such as #Sahabatbawaslu, #Bawaslumengawasi. Said, prevent, watch action confirming that the @Bawaslu RI account will often provide information via Twitter during the conditions leading up to the regional elections to avoid money politics practices.

Furthermore, the authors conducted a cluster analysis related to the @Bawaslu RI account using the word similarity feature. It is done to visualize the patterns in this study by classifying the sources of sharing similar words. It will be easier to determine an account's information distribution pattern and simultaneously visualize the report's interactions.

Based on the cluster results above, the @Bawaslu RI account's information dissemination has words with other accounts. Where these accounts have the same duties and functions, namely preventing money politics. The tendency of the @Bawaslu RI account to be accountable with other accounts that are also part of the KPU, Komendagri, Electoral House, Dkpp Ri, Bawaslu Riau, Bawaslu Pangkep, Bawaslu Makassar, Bawaslu Central Java, Bawaslu East Jakarta, Bawaslu Banten, Bawaslu Tuban, Bawaslu South Jakarta, Bawaslu West Java. Besides, information about the @Bawaslu RI account spread to local media, namely News CNN Daily, Radielshinta.

The analysis of tweet intensity @Bawaslu RI provides educational information about harassment and monitoring related to election violations. In offering information, @Bawaslu RI links tweets with other accounts. The information provided by @Bawaslu RI is only brief information. The public must first access the report and website to get a complete feature. The authors believe that this method can maximize the information channels by @Bawaslu RI. However, on the one hand, connecting to other information channels to local media can speed up the distribution of the information provided.

Table 2. The frequency of the word @Bawaslu RI account

Word	Length	Count
Bawaslu	7	2255
#sahabatbawaslu	15	846
#bawaslujagahakpilih	20	229
#bawaslumengawasi	17	585
# bawasluterbuka	15	148
Pilkada	7	438
#Pemiluterpercaya	16	112
#pemilutepercaya	16	112
#cegahawasitindaki	17	106
#salamawas	10	56
pemungutan	10	51
awasi	5	54
mengawasi	9	72
laporan	7	90
demokrasi	9	114
dugaan	6	53
penyelesaian	12	56
@kpu	4	113
kampanye	8	113
politik	7	165

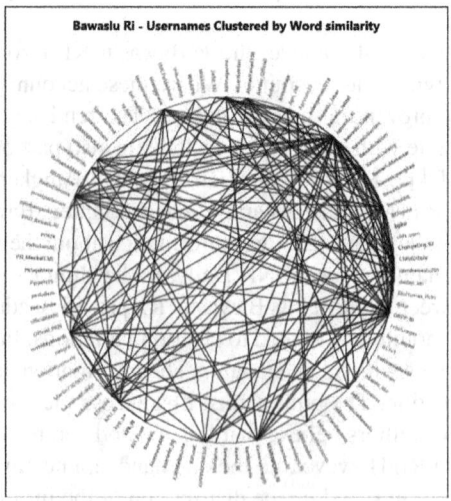

Fig. 3. Account Connectivity and Interaction and Resist money politics

Social media characteristics, which tend to be instantaneous in obtaining information, are why people tend to be more interested in accessing information through social media. Furthermore, when people access information requiring entry on the website, it will increase community participation and add much information. Community involvement on social media platforms makes it possible for the government to use it as a policy basis [28].

In the analysis using the word cloud, the @Bawaslu RI account topic related to the condition of the Election Supervisory Agency. The topic stating that the Election Supervisory Agency oversees people's voting rights to be purchased, and Regional Head Elections prevents and acts on social media, allowing the Election Supervisory Agency to socialize and educate and the effectiveness of their preparedness communication to follow up actions money politic. The authors argue that the Election Supervisory Agency has had the advantage and transformation in preventing and campaigning against money politics using social media. According to Brandt, Bendler, & Neumann, social media analysis can capture campaign patterns related to user elections and environmental involvement based on the discussion. In the cluster analysis, the @Bawaslu RI account has healthy patterns of interaction and communication with institutions. Like [29], social media allows for real-time interaction between government and society. Clarke also emphasizes that social media can also increase the effectiveness in overcoming the problem of practices that violate the rules in the elections [30]. The authors argue that ideally, communication carried out in social media is two-way, making it possible for the actors to exchange information, either the government to the community, or vice versa. This statement is in line with what Nabatchi stated regarding public communication [31]. He noted that two-way communication facilitates the two actors to be both givers and to receive information. Lee & Hoon also stated that two-way communication is the ideal model of public communication. Ideally, the distribution of information is not limited to other government accounts. It will be better if the information distribution also touches the community level to the bottom [32].

5   Conclusion

The efforts made by the Election Supervisory Agency (Bawaslu Ri) are one of the right ways in the current era of disruption and transformation. By leveraging on-going information and communication technology, election supervisors are needed to succeed in a brave world—especially social media, whose users are increasing all the time. The election supervisory agency has made a new transformation or breakthrough because it has used social media as a forum for socialization.

This paper contributes to the finding that social media is the easiest and most effective way to socialize money politics. Several findings were presented to prove that election supervisors succeeded in disseminating money politics through social media. Social media's strength lies in the number of users and the short time to share. Accessers on social media, at one point, act as spectators, but at other times can convey information. The sharing method (share) allows everyone to expand the reach of information circulating in new media. Until finally, a large number can be achieved in a relatively short time. The invasion of hashtags, creative general election supervisory agency, and many followers are several strategies to gather social media strength. This power has the potential to prevent money politics. This paper also contributes to the general election supervisory mapping network in the 2020 regional head election contest. The findings provide a new picture of general election supervisors' development in disseminating the prevention of money politics. This paper describes the map of the Bawaslu Ri movement towards other institutions in preventing money politics.

This paper has limitations on the applied data analysis approach. Namely, data analysis has not described the triangulation analysis between qualitative and quantitative data. For this reason, the issue of money politics needs to be studied more deeply through mixed-method research that focuses on the study of the behavior of the political elite in each contestation and the implications of money politics for the future in Indonesia.

References

1. International Institute for Democracy and Electoral Assistance, Improving Electoral Practices: Case Studies and Practical Approaches (2014)
2. Syahputra, I.: Expressions of hatred and the formation of spiral of anxiety on social media in Indonesia. Search (Malays.) **11**(1), 95–112 (2019)
3. Ike, A.R., dan Maksudi, B.I.: The role of social media in the improvement of selected participation of students based on students in bogor regency, vol. 20, no. 2, pp. 154–161 (2018)
4. Rahmawati, D.: Media Sosial Dan Demokrasi Di Era Informasi. J. Vokasi Indones. **2**(2), 3–4 (2016). https://doi.org/10.7454/jvi.v2i2.40
5. Outhwaite, W., Turner, S., Calderaro, A.: Social media and politics. In: The SAGE Handbook of Political Sociology: Two Volume Set, no. May, pp. 781–795 (2018). https://doi.org/10.4135/9781526416513.n46
6. Jauhari, M.: Media Sosial: Hiperrealitas dan Simulacra Perkembangan Masyarakat Zaman Now dalam Pemikiran Jean Baudrillard. J. AL-'Adalah **20**(1), 117–136 (2017)
7. Hasan, K.: Bahan ajar handout Komunikasi Politik NEW MEDIA DAN KOMUNIKASI POLITIK KONTEMPORER 1, p. 13 (2015)
8. Rossi, A., Lestari, T., Perdana, R.S., Fauzi, M.A.: Analisis Sentimen Tentang Opini Pilkada Dki 2017 Pada Dokumen Twitter Berbahasa Indonesia Menggunakan Naïve Bayes dan Pembobotan Emoji, vol. 1, no. 12 (2017)
9. Arshad, M., Akram, M.S.: International Review of Research in Open and Distributed Learning Social Media Adoption by the Academic Community: Theoretical Insights and Empirical Evidence From Developing Countries Social Media Adoption by the Academic Community: Theoretical Insights, vol. 19, no. 3 (2018)
10. Fish, G.: Social justice and social media: responding to the enforcement of justice by the internet community. Mich. Law Rev. (2017)
11. Octafitria, Y.: Media Sosial Sebagai Agen Sosialisasi Politik Pada Kaum Muda Yovita Octafitria. Indones. J. Sociol. Educ. Policy (2014)
12. Barus, R.K.: Pemberdayaan Perempuan melalui Media Sosial. Simbolika **I**(2), 113–123 (2015). No. September
13. Sadiqi, F.: Morocco's Emerging Democracy : The 2015 Local and Regional Elections, no. 83, pp. 0–4 (2015)
14. Amrizal, D.: Model Countermeasures of Golput (Election Abstainers) Based on Change of Voters' Behavior in Simultaneous Regional Elections in Sumatera Utara, no. April 2018

15. Hilman, Y.A., Khoirurrosyidin, K., Lestarini, N.: Peta Politik Pemilukada Kabupaten Ponorogo 2020 di Tengah Pandemi Covid-19. Polit. J. Ilmu Polit. **2**(2), 129–148 (2020). https://doi.org/10.15575/politicon.v2i2.8983

16. Cocco, R.G., Guasch, C.M.: Demonstrations about the public transportation in Brazil: A different reading from the great Florianópolis case, in Santa Catarina State, Scr. November 2016. https://www.scopus.com/inward/record.uri?eid=2-s2.0-84955569341&partnerID=40&md5=97ea31c66ce6129e273c46424889cfa7

17. Cabeza, L.: 'First-order thinking' in second-order contests: a comparison of local, regional and European elections in Spain. Elect. Stud. **53**, 29–38 (2018). https://doi.org/10.1016/j.electstud.2018.03.004

18. Chaniago, P.S.: Evaluasi Pilkada Pelaksanaan Pilkada Serentak Tahun 2015. Polit. Indones. Indones. Polit. Sci. Rev. **1**(2), 196 (2016). https://doi.org/10.15294/jpi.v1i2.6585

19. Kusmanto, H.K., Prayitno, H.J., Ngali, A., Rahmawati, L.E.: Realisasi Kesantunan Berkomunikasi Pada Media Sosial Instagram @Jokowi: Studi Politikopragmatik. PARAFRASE J. Kaji. Kebahasaan Kesastraan **19**(2), 119–130 (2019). https://doi.org/10.30996/parafrase.v19i2.2648

20. Muhtadi, B.: Politik Uang dan New Normal dalam Pemilu Paska-Orde Baru. Integritas J. Antikorupsi **5**(1), 55–74 (2019). https://jurnal.kpk.go.id/index.php/integritas/article/view/413

21. Harianto, S., Rahardjo, M., Baru, B.M.: Politik Uang (Money Politic), Dan Pengaruhnya Terhadap Demokrasi Dalam Pemilihan Kepala Desa. J. Sos. **19** (2018). https://unmermadiun.ac.id/ejurnal/index.php/sosial/article/view/362/565

22. Solihah, R.: Politik transaksional dalam pilkada serentak dan implikasinya bagi pemerintahan daerah di indonesia, vol. 2, no. 1, pp. 97–109 (2016)

23. Mokodompis, S., Bukido, R., Salim, D.P., Makka, M.M.: Money politic in elections: islamic law perspective. J. Ilm. Al-Syir'ah **16**(2), 126 (2018). https://doi.org/10.30984/jis.v16i2.708

24. Ppkn, P.S., Adi, A.S.: HUBUNGAN PENERIMAAN MONEY POLITIC DENGAN TINGKAT PARTISIPASI JATIKALEN KABUPATEN NGANJUK Keyword: Keywords: acceptance of money politics , the level of political participation (2016)

25. Wulan, I.K., Arief, Y.S., Kristiawati, K.: The relationship of mother behavior in breastfeeding with baby's nutrition status aged 0–6 months. Eurasian J. Biosci. **14**(1), 2507–2512 (2020). https://www.scopus.com/inward/record.uri?eid=2-s2.0-85093902030&partnerID=40&md5=16b9a99dc71f504d6725af8dd8655117

26. Pratiwi, F.I.: The Role of Identity in Politics and Policy (Issue November). no. November (2019)

27. Heimerl, F., Lohmann, S., Lange, S., Ertl, T.: Word cloud explorer: text analytics based on word clouds. In: Proceedings of the Annual Hawaii International Conference on System Sciences, pp. 1833–1842 (2014). https://doi.org/10.1109/HICSS.2014.231

28. Nurmandi, A., et al.: To what extent is social media used in city government policymaking? Case studies in three asean cities. Public Policy Adm. **17**(4), 600–618 (2018). https://doi.org/10.13165/VPA-18-17-4-08

29. Clarke, A., Margetts, H.: Governments and citizens getting to know each other? Open, closed, and big data in public management reform. Policy Internet **6**(4), 393–417 (2014). https://doi.org/10.1002/1944-2866.POI377

30. Allen, R., Nakonechnyi, A., Benedetti, M.S.: Anna's story: how a ukrainian orphan's acquisition of english as a second language transformed her life. J. Cogn. Educ. Psychol. **19**(2), 107–119 (2020). https://doi.org/10.1891/JCEP-D-19-00044

31. Schmidthuber, L., Ingrams, A., Hilgers, D.: Government openness and public trust: the mediating role of democratic capacity. Public Adm. Rev. (2020). https://doi.org/10.1111/puar.13298
32. Kingston, K.L., Furneaux, C., de Zwaan, L., Alderman, L.: Avoiding the accountability 'sham-ritual': an agonistic approach to beneficiaries' participation in evaluation within nonprofit organizations. Crit. Perspect. Account. (2020). https://doi.org/10.1016/j.cpa.2020.102261

Public Policies for Creating Sustainable and Integrated Transport in Jakarta

Bhimo Widyo Andoko[1], Deni Setiawan[2], Eko Priyo Purnomo[2(✉)], Lubna Salsabila[2], and Kalsum Fais[3]

[1] Universitas Muhammadiyah Yogyakarta, Yogyakarta, Indonesia
ses.11dikti851@gmail.com
[2] Department of Government Affairs and Administration, Jusuf Kalla School of Government, Universitas Muhammadiyah Yogyakarta, Yogyakarta, Indonesia
deni.setiawan.psc19@mail.umy.ac.id, eko@umy.ac.id, lubna.salsa@gmail.com
[3] Magister of Law, Universitas Muhammadiyah Yogyakarta, Yogyakarta, Indonesia
kalsum.fais.pasca17@mail.umy.ac.id

Abstract. This study discusses how the Jakarta government improves public transportation policies to create sustainable and integrated public transportation in Jakarta with a sustainable development approach (SDGs) at point 11 (eleven). The high demand for public transport and unintegrated management has led the DKI Jakarta transport system to become a complex issue and seems to have failed to meet its objectives. The lack of a sustainable transport policy has made congestion that can never be resolved. This study will identify the root problem that causes congestion in DKI Jakarta and find the best way to address the issues by considering the program's sustainability using an explorative qualitative method. Sustainable and integrated public transportation is a refinement of existing public transportation that involves the government, private sector, and society by including other important actors. The concept of sustainable and integrated public transportation development is also a form of innovation towards changes in the transportation sector by involving various actors in building public transportation.

Keywords: Public policy · Transportation · Sustainable · Integrated · Jakarta

1 Introduction

There are a few ways to stay away from a transportation issue, which makes the urgency to achieve the 17 Sustainable Development Goals (SDGs) greater [1]. As the transportation aspect, it is directly or indirectly linked to the 8 goals of SDGs. Walkways, road, and metros are responsible for forming a sustainable city [2]. Simultaneously, transportation is responsible for the greenhouse effect as it is responsible for the air pollution in most Asian cities [3]. The uncontrolled flow of urbanization [4], followed by the high demand for the public transportation industry in Jakarta, Indonesia, has led the city into complex and problematic conditions due to the congestion [5]. Due to this situation, Jakarta is

T. Antipova (Ed.): ICADS 2021, AISC 1352, pp. 523–530, 2021.
https://doi.org/10.1007/978-3-030-71782-7_46

now expanding more road infrastructure in various areas. Meanwhile, the urban transport policies, particularly in Jakarta, tend to classify certain parties. It leads to private vehicle users neglecting the interest of those who use most public transport [6].

Currently, the standard policies are shaped by traffic demand restrictions, including but not limited to unusual policies [7], passenger restriction policies [8], car-free day policies and parking restrictions [9, 10]. Jakarta is estimated to lose US$3 billion a year due to traffic congestion that cannot be separated from high vehicle ownership growth [11]. According to the number of vehicles in Jakarta, over 200 new cars and over 1,000 new motorcycle licenses were approved in 2017 [12]. Additionally, as projected by the Department of Population and Civil Registration, the DKI Jakarta population increased by 0.73% in 2019 from 2017 with a density of 25.270/sq mi [13]. It was then projected to increase to 0.7–0.8% in the coming years [14]. Given the population, density and land availability, the DKI Jakarta Government needs to rethink and take immediate action to minimize the loss.

As projected by the Department of Population and Civil Registration, in 2019, the DKI Jakarta population has increased by 0.73% from 2017 with 25,270/sq mi [13], which then is projected to increase to 0.7–0.8% in upcoming years [14]. Due to the population, density, and land availability, the DKI Jakarta government needs to rethink and take immediate action to minimize the loss.

The motorcycle growth trend will go on to achieve a more sustainable transport policy [11]. Transport is an essential tool in supporting the success of a country's development, particularly in supporting society's economic activities [15]. Transportation problems are crucial for government actors because of the community's high diversity and characteristics [15, 16]. The first step in addressing the transport problem is structured and integrated planning in policymaking. Adequate, safe and convenient public transportation in Jakarta has not yet equated with high mobility of persons and goods [17].

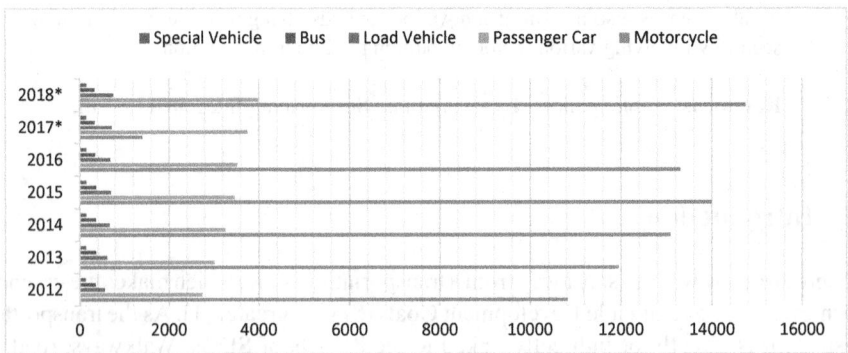

Source: Databoks [18]

Fig. 1. The number of vehicle in Jakarta 2012–2018

Figure 1 shows that the urge to create a new sustainable transportation policy has become more serious. Many solutions are available to resolve transport problems associated with creating sustainable public transport. Sustainable transport covers a wide range of industries. As Sebhatu (2011) explains, public transport must be environmentally, eco-efficiently and socially sustainable to provide better service and efficiency by involving all stakeholders [19]. Jakarta has not made the development of sustainable public transport, the primary focus of solving transportation challenges. Understanding the main problems in the transport system in Jakarta, new Jakarta instruments relating to decision making in sustainable transport policy will be more easily formulated by the government [20]. The use of three essential policy keys to creating sustainable transport reduces adverse environmental effects [21].

Following that, Redman et al. (2013) further explain that the supply of transportation in the form of road networks and service levels offered by private or public transport operators is subject to government control and potentially an ideal policy instrument for influencing the demanding transport. Supply and demand is an inseparable unit in the transport sector [21]. Therefore the supply of transport always follows the demand for transport. For example, as the central public transportation controller, the government will follow passenger transport demand by supplying appropriate buses. Adding busses is never an efficient way to satisfy customers [22]. It must be adapted to the balance of what passengers want (subjective measurement) [19, 21, 22].

2   Discussion

The high demand for public transportation is inseparable from travel effectiveness and efficiency due to congestion [23]. The implementation of sustainable transportation must also pay attention to the primary keys of sustainable transport by considering: (1) Environmental aspects because the sustainable transportation system is how effective; (2) the impact of a sustainable transportation system on economic development; (3) the impact of a sustainable transport system on the quality of social life. If this strategy is implemented, it will positively affect a sustainable and integrated transportation system. The effect of sustainable and integrated public transportation can be seen in three aspects [24].

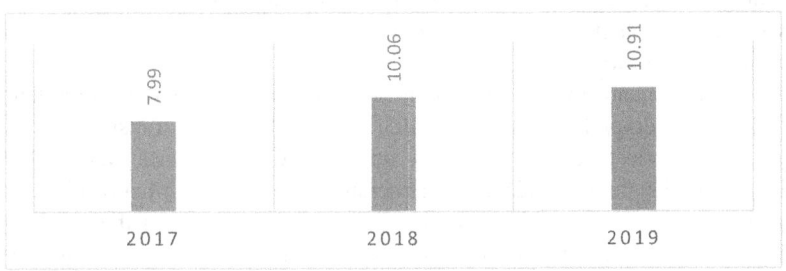

Source: Department of Population and Civil Registration

Fig. 2. DKI Jakarta population 2017–2019 (in a million)

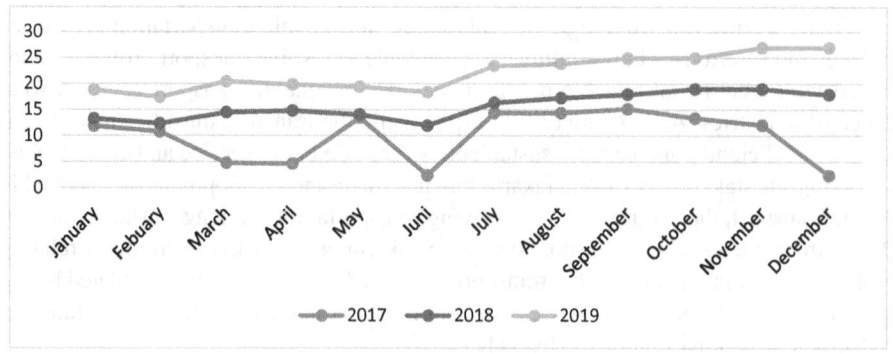

Source: Department of Transportation

Fig. 3. TransJakarta passenger 2017–2018 (in a million)

A significant increase in the population of DKI Jakarta has also affected the number of trans-Jakarta passengers in the last three years (from 2017–2019) (Fig. 2 and 3) [13, 25, 26] which also shows the reason why TransJakarta is needed. The minimum service standard (MSS) is an essential requirement for sustainable public transport to satisfy passengers with their services. MSS is defined as the minimum level of service to be delivered to passengers. TransJakarta did not have a minimum service standard since it was first established in 2004 to provide better public transport services [27]. The TransJakarta system offers a summary of minimum service standards, such as service reliability (timeliness, appropriate speed, minimum bus speed and time limits), safety, accessibility, affordability and convenience [27].

Therefore, only three types of implementation technology are directly affected by Jakarta based on the transport pattern strategy in Regulation No 12/2002 and Governor Regulation No. 103/2007, which are: (1) Electronic Road Pricing (ERP), (2) ITS implementation, and (3) public transport development in MRT, LRT and BRT, each with its technological evolutions. In developing technology for the Transjakarta System, the authors indicated that the transport policy on the use of technology concerns the Bus Tracking System (BTS) use, enabling passengers to properly know the bus departing update or arriving times [28]. In technological development, there is no integrated transport strategy relating to the provision of sustainable transport. The government has no policy restrictions on creating technology as an instrument to protect the environment from transport pollution [29].

On the other hand, the identification, in the context of Regional Regulation No 12/2002 and Governor Regulation No. 103/2007 of the macro-transportation pattern strategy and the latest transport policy to develop the Transjakarta system almost all policies are on development [28]. Infrastructure transportation. Meanwhile, customer satisfaction can be determined by providing the dominant logic of the goods and the dominant service logic [30]. Transjakarta can also integrate all public transportation resources to provide better services. This situation enables common values and systems to respond to the challenges posed by the climate and environmental crises [30, 31].

As a transport authority, the government can create good travel demand management and involve all stakeholders to provide sustainable public transport [32]. Contracts are mainly related to securing resources and therefore achieving a focus on capacity [33]. The role of public transport contracts includes governance to integrate all stakeholders to deliver better public services. The idea of stakeholder management is an efficient way of achieving sustainability. The stakeholder management concept uses the stakeholder view, according to Klein et al. (2012), claiming that the implementation of stakeholder management should be seen as an interactive learning process. In TransJakarta, the provincial government of Jakarta has reached agreements with various operators in various corridors. Therefore, contract governance can no longer be postponed to create efficient and effective management between each party (operator) [34]. The policy is still aimed at reducing travel requirements through better management of demand.

Yet, the government still has no effort to reduce the externalities created by public transportation itself. The number of TransJakarta vehicles, which keep increasing, has led to another congestion problem. During the past three years, TransJakarta added approximately 1.500 new vehicles to overcome the demanded number [35]. Hence, the number of congestion in Jakarta seems to not flat yet. The data from Databoks.com show that the number of congestions makes no changes (Fig. 4) [36].

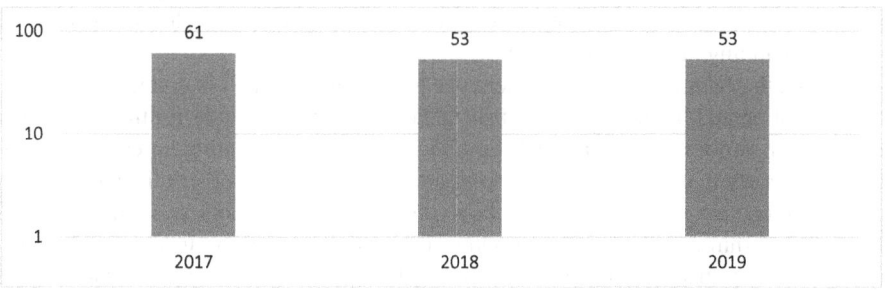

Source: Databoks.com

Fig. 4. DKI Jakarta congestion level 2017–2019

The data presented shows that the Government of DKI Jakarta has not yet been able to overcome congestion problems by adding new vehicles or adding other public transport options. Adding more vehicles without proper management and policy will only increase the population number produced daily. Considering that vehicle emissions cause almost 50% of Jakarta's air pollution, fuel control is urgently needed [37]. Unfortunately, air pollution has not been taken seriously by the DKI Jakarta Environment Agency (DLH). In line with that, the Air Quality Index presented by iqair.com shows that as far as Jakarta's 2019 reading of 49.4 $\mu g/m^3$ is concerned. This number will be included in the 'unhealthy for sensitive groups' bracket, which requires a PM2.5 reading of between 35.5 and 55.4 $\mu g/m^3$ to be classified as such. As the name suggests, such a reading will examine adverse health effects on some demographics of the population, with the most vulnerable young children, the elderly and those with poor health or pre-existing conditions. Although these demographics are particularly at risk, average PM2.5 reading this high will be detrimental to even those with good health when exposed over a long period [38].

3 Conclusion

Jakarta faces a significant challenge in integrating sustainable transport policies. Jakarta understands what Jakarta needs to develop a sustainable transport policy. In the context of this analysis and research, it seeks to conclude the main problems in the development of sustainable transport policies in Jakarta. The new transport policy has a crucial role to play in creating sustainability in the transport sector. First of all, transport policy acts to reduce the environmental impact by applying transport technology correctly. Second, transport policy acts as an instrument to increase service levels. It is not only building new transport infrastructure. The third role of transport policy is to provide an instrument for sustainable transport management. Balancing the supply and demand for transport is essential to make traffic stable. Planning land use, organizational leadership is one of the keys to reducing mobility by creating sustainable transport management.

The main problem with sustainable transport policy in Jakarta is the lack of adequate policy implementation, mainly due to the intense political appeal for ambitious policy goals; problematic land use and intermodal negotiation processes; lack of policies to restrict the use of cars and motorbikes; lack of improved service levels in the BRT system, including lack of bus service.

Lastly, the Jakarta City Government needs to focus on: (1) it is crucial to be aware of the creation and implementation of sustainable transport not only for the community but also for senior officials in the Jakarta Government; (2) creating an easily accessible bureaucracy that facilitates the process and implementation of sustainable transport policies; (corruption). Based on the above discussion, the authors argue that the DKI Jakarta Government, as the leading authority, must pay attention to the development of sustainable transport policies. It is essential to reduce high traffic jams, but it makes sustainability even more important.

References

1. Abeyratne, R.: Air transport and sustainable development. Environ. Policy Law **33**(3–4), 138–143 (2003)
2. UN-Habitat, UNEP, and SLoCaT: Analysis of the Transport Relevance of Each of the 17 SDGs, no. September, pp. 1–20 (2015)
3. GSMA: 2020 Mobile Industry Impact Report: Sustainable Development Goals, no. September (2020)
4. Muluk, S.: Jakarta Menuju Kota Yang Berkelanjutan. In: Analisis Pembangunan, pp. 0–8 (2017)
5. Joewono, T.B., Tarigan, A.K.M., Susilo, Y.O.: Road-based public transportation in urban areas of Indonesia: what policies do users expect to improve the service quality? Transp. Policy **49**, 114–124 (2016)
6. Sitanggang, R., Saribanon, E.: Transportation policies for Jakarta's congestion. In: Proceedings of the Conference on Global Research on Sustainable Transport (GROST 2017) (2018)

7. Li, R., Guo, M.: Effects of odd–even traffic restriction on travel speed and traffic volume: evidence from Beijing Olympic Games. J. Traffic Transp. Eng. (Engl. Ed.) **3**(1), 71–81 (2016)
8. Daganzo, C.F., Cassidy, M.J.: Effects of high occupancy vehicle lanes on freeway congestion. Transp. Res. Part B Methodol. **42**(10), 861–872 (2008)
9. Santos, G., Behrendt, H., Maconi, L., Shirvani, T., Teytelboym, A.: Part I: externalities and economic policies in road transport. Res. Transp. Econ. **28**(1), 2–45 (2010)
10. Yudhistira, M.H., Kusumaatmadja, R., Hidayat, M.F.: Does Traffic Management Matter ? Evaluating Congestion Effect Of Odd-Even Policy in Jakarta, no. January (2019)
11. Rukmana, D.: Rapid urbanization and the need for sustainable transportation policies in Jakarta. In: IOP Conference Series: Earth and Environmental Science, vol. 124, no. 1 (2018)
12. Central Bureau of Statistics (BPS): Jenis Transportation
13. Dinas Kependudukan dan Pencatatan Sipil: Data Jumlah Penduduk Provinsi DKI Jakarta Berdasarkan Kelompok Usia Per Kelurahan, DKI Jakarta (2020)
14. Badan Pusat Statistik: Jumlah Penduduk dan Kepadatan Penduduk DKI Jakarta 2019 (2019)
15. Sinaga, S.M., Hamdi, M., Wasistiono, S., Lukman, S.: Model of implementing Bus Rapid Transit (BRT) mass public transport policy in DKI Jakarta Province, Indonesia. Int. J. Sci. Soc. **1**(3), 261–271 (2019)
16. Hanna, R., Kreindler, G., Olken, B.A.: Citywide effects of high-occupancy vehicle restrictions: evidence from 'three-in-one' in Jakarta. Science (80-.). **357**(6346), 89–93 (2017)
17. Pinagara, F.A., Khamtanet, S.: Will Jakarta still have traffic congestion after MRT?, vol. 36, no. Icbmr, pp. 311–321 (2017)
18. Databoks: Berapa Jumlah Kendaraan di DKI Jakarta (2019)
19. Sebhatu, S.P., Enquist, B., Johnson, M., Gebauer, H.: Sustainability and Innovating Value-Configuration Spaces for Innovative Public Transit Services, p. 13 (2011)
20. Dewi, A., et al.: Global policy responses to the COVID-19 pandemic: proportionate adaptation and policy experimentation: a study of country policy response variation to the COVID-19 pandemic. Health Promot. Perspect. **10**(4), 359–365 (2020)
21. Redman, L., Friman, M., Gärling, T., Hartig, T.: Quality attributes of public transport that attract car users: a research review. Transp. Policy **25**, 119–127 (2013)
22. Nidumolu, R., Prahalad, C.K., Rangaswami, M.R.: Why sustainability is now the key driver of innovation. Harvard Bus. Rev. **87**(9), 85–91 (2010)
23. Yudhistira, M.H., Koesrindartono, D.P., Harmadi, S.H.B., Pratama, A.P.: How congested Jakarta is? Perception of Jakarta's citizen on traffic congestion. Econ. Finance Indones. **62**(3), 141 (2017)
24. Jeon, C.M., Amekudzi, A.: Addressing sustainability in transportation systems: denitions, indicators, and metrics. J. Infrastruct. Syst. **11**, 31–50 (2005)
25. Dinas Perhubungan, Jumlah Penumpang TransJakarta Tahun 2018–2020, DKI Jakarta (2020)
26. Dinas Perhubungan, Data Jumlah Penumpang TransJakarta Tahun 2017, DKI Jakarta (2018)
27. Wibowo, R.S.S., Weningtyas, W., Rahma, S., Magister, P., Sipil, T.: Kualitas pelayanan sistem informasi pada angkutan umum transjakarta. J. Transp. **18**(1), 67–76 (2018)
28. Dishub, No, (2020)
29. Albalate, D., Bel, G., Calzada, J.: Governance and regulation of urban bus transportation: using partial privatization to achieve the better of two worlds. Regul. Gov. **6**(1), 83–100 (2011)
30. Vargo, S.L., Lusch, R.F.: Institutions and axioms: an extension and update of service-dominant logic. J. Acad. Mark. Sci. **44**(1), 5–23 (2016)
31. CIFOR, Hutan dan mitigasi perubahan iklim: apa yang perlu diketahui oleh para pembuat kebijakan, (2013)
32. Purnomo, E.P., Anand, P.B., Choi, J.W.: The complexity and consequences of the policy implementation dealing with sustainable ideas. J. Sustain. Forest. **37**(3), 270–285 (2018)

33. Ramírez, R., Roodhart, L., Manders, W.: How Shell's domains link innovation and strategy. Long Range Plan. **44**(4), 250–270 (2011)
34. Klein, P.G., Mahoney, J.T., McGahan, A.M., Pitelis, C.N.: Who is in charge? A property rights perspective on stakeholder governance. Strategic Organ. **10**(3), 304–315 (2012)
35. Jak Lingko, Transjakarta dalam Angka Tahun 2019, DKI Jakarta (2019)
36. databoks.com, Tingkat Kemacetan DKI Jakarta Stagnan, tapi Peringkatnya Turun (2020). https://databoks.katadata.co.id/datapublish/2020/02/20/tingkat-kemacetan-dki-jakarta-sta gnan-tapi-peringkatnya-turun#. Accessed 11 Dec 2020
37. Media Indonesia, Polusi Udara, DKI Rugi Rp 40 T (2017). https://mediaindonesia.com/meg apolitan/99668/polusi-udara-dki-rugi-rp40-t. Accessed 11 Dec 2020
38. IQAir, Air quality index (AQI) and PM2.5 air pollution in Jakarta (2019). https://www.iqair. com/us/indonesia/jakarta. Accessed 11 Dec 2020

Author Index